The Component Identifier & Source Book

The Component Identifier & Source Book

The Ultimate Cross Reference for the Electronics Industry

By Victor Meeldijk

A Division of
Howard W. Sams & Company
Indianapolis, IN
A Bell Atlantic Company

©1996 by Victor Meeldijk

PROMPT® Publications is an imprint of Howard W. Sams & Company, A Bell Atlantic Company, 2647 Waterfront Parkway, E. Dr., Indianapolis, IN 46214-2041.

All rights reserved. No part of this book shall be reproduced, stored in a retrieval system, or transmitted by any means, electronic, mechanical, photocopying, recording, or otherwise, without written permission from the publisher. No patent liability is assumed with respect to the use of the information contained herein. While every precaution has been taken in the preparation of this book, the author, the publisher or seller assumes no responsibility for errors or omissions. Neither is any liability assumed for damages resulting from the use of information contained herein.

International Standard Book Number: 0-7906-1088-4

Editor: Candace M. Hall
Assistant Editors: Pat Brady, Natalie F. Harris
Typesetting: Natalie Harris
Pasteup: Suzanne Lincoln, Christy Pierce, Phil Velikan
Cover Design: Christy Pierce

Trademark Acknowledgments: All product names and logos are trademarks of their respective manufacturers. All terms in this book that are known or suspected to be trademarks or services have been appropriately capitalized. PROMPT® Publications, Howard W. Sams & Company, and Bell Atlantic cannot attest to the accuracy of this information. Use of a term or logo in this book should not be regarded as affecting the validity of any trademark of service mark.

Printed in the United States of America

9 8 7 6 5 4 3 2 1

TABLE OF CONTENTS

Preface vii

How To Use This Book 1

Manufacturers, Prefixes,
Part Number Types, Logo Descriptions & Family Types 3

Logos 265

Product Listings 285

Sources of Hard-to-Find or Obsolete Components 313

Sources of Further Information on Parts 337

Appendix 341

PREFACE

The Component Identifier & Source Book

This book will help you identify components and locate manufacturers in the following ways as it:

1. Contains logos for manufacturer identification;
2. Cross references abbreviated manufacturer names to the proper name of the manufacturer;
3. Has component prefixes and part number types to identify parts from schematics or parts listings;
4. Cross references to new manufacturer after mergers and acquisitions;
5. Lists nationwide manufacturer offices;
6. Contains worldwide addresses;
7. Lists manufacturers' products with the manufacturer name;
8. Has a product locator section so users can quickly locate sources of various parts;
9. Lists manufacturers and distributors that can manufacture or locate obsolete parts;
10. Contains a list of other publications and data sources that can help you get component specifications.

If you need specifications on a part you can either contact the manufacturer of the device or reference other publications that include such material (i.e., IC Master, Electronic Engineering Master [EEM], D.A.T.A. handbooks, etc.)

Restrictions on Use

All information contained in the publication is for the exclusive use of the purchaser only and is to be used solely by the purchaser. None of the information in this publication may be used to create, in whole or in part, any mailing list or other data compilation in written or electronic form, that is to be sold or otherwise distributed to a third party. The purchaser may use the information as a listing as part of its marketing materials but may not include the information herein as part of any marketing materials. Failure to comply with these restrictions will constitute violations both of this agreement and the copyright laws of the United States. Purchase of this publication constitutes acceptance of the restrictions set forth herein.

Although the author has made every reasonable effort to assure the accuracy of the information contained herein, the author must rely on others over which the author has no control for certain information. The purchaser acknowledges that neither the publisher nor the author does not guarantee or warrant that the information contained in this publication is complete, correct or current, except as specifically set forth herein. The publisher and author disclaim all warranties, expressed or implied, relating to the information in any manner, including (without limitation) implied warranties of merchantability and fitness for a particular purpose.

Techniques for locating components are presented in this book which both the author and publisher believe to be accurate. However, the information provided herein is intended solely as an aid by the user and independent evaluation of any of the sources listed herein should be made by the user. The author and publisher expressively disclaim the implied warranties of merchantability and of the fitness for any particular purpose, even if the author has been advised of the suitability of any source contained in this book the author and publisher also disclaim all liability for direct, indirect, incidental, or consequential damages from the result of the use of any information in this book.

Preface

Neither the publisher nor the author shall be liable to any purchaser, or third party, for any loss or injury allegedly caused, in whole or in part, by the publisher or the author, or for consequential, exemplary, incidental, or special damages, even if the publisher is advised in advance of such damage. All sources listed herein are for informational purposes only and no claims are made by either the author or publisher on the business practices of any sources listed herein. Furthermore, in any event, the author's or publisher's liability (if any) shall never exceed the amount paid by the original purchaser for this publication.

HOW TO USE THIS BOOK

This book can be used in a variety of ways to help you locate the component or manufacturer you need. Some of these ways are:

1. Identify a manufacturer via the logo section.

2. Identify a manufacturer by component prefix.

3. Identify manufacturer by part number type.

4. Identify a manufacturer by abbreviated name.

5. Find the current manufacturer after a merger or acquisition.

6. If you need specifications you can either call the manufacturer using the headquarters and sales locations contained in this book or refer to one of source books that contain such information. (A listing of these books is provided).

7. Find a manufacturer that may still produce parts obsoleted by the original manufacturer. (A listing is provided).

8. Use a distributor or broker that will search for hard to find or discontinued parts. (A listing is provided).

9. Locate a source of a particular component type (i.e., such as manufacturers of DRAMs).

Appendix A contains listings that were omitted due to insufficient information (possibly as a result of mergers, address or phone number changes, discontinuations, etc.) during the book's preparation. Please refer to this section if you cannot find the reference you are looking for.

Manufacturers, Prefixes, Part Number Types, Logo Descriptions & Family Types

Contains manufacturer names and product summaries, abbreviated manufacturer names matched to proper names, component number prefixes, part number types, cross references to new manufacturers after mergers and acquisitions, and component trade names.

Phone numbers are country code, followed by the area code, then the phone number. For example, 64 9 1234567 is for Australia (64), Auckland (9). What is NOT shown are any numbers needed to dial outside the country you are calling from. For example, to call a company in Royston, England, from the U.S., you would dial 011 44 763 123456: where 011 is an overseas call from the U.S., 44 is the country code for England, and 763 is the area code for Royston.

To look up phone numbers in the United States, you can use the following Internet services:

AT&T 800 number directory, http://att.net/dir800/

NYNEX: http://www.bigyellow.com

"The Switchboard" for U.S. companies:
http://www.switchboard.com

1AXXX — 28C

1AXXX part number
See *ABB Hafo AB*

(LEDs, PINS, duplex devices)

1DXXXX-X part number
See *Anaren Microwave, Inc.*

1H prefix
See *Harris Semiconductor*

(Custom part)

1T prefix
See *Sony Semiconductor*

(Variable capacitance diodes)

10E prefix
See *Motorola Semiconductor, Inc.*

10K prefix
See *Motorola Semiconductor, Inc.*

100E prefix
See *Motorola Semiconductor, Inc.*

100K prefix
See *Motorola Semiconductor, Inc.*

100XXX part number
See *National Semiconductor*
(ECL family)

102AXXX-XX part number
See *Loral Corporation*
(Devices used in their products)

112AXXX-XX part number
See *Loral Corporation*
(Devices used in their products)

121AXXX-XX part number
See *Loral Corporation*
(Devices used in their products)

14XXX part number
See *Motorola, Inc.*

(4000 series CMOS devices)

142SCXXX,
143SCXXX part number,
See *SenSym, Inc.*

1527XXX part number
See *Harris Semiconductor*
(Custom part)

16AM prefix
See *GHz Technology, Inc.*

16V8 series
National Semiconductor
(Discontinued 1995), *Lattice, Atmel*

163SCXXX part number
See *SenSym, Inc.*

18XXXT part number
See *National Semiconductor,*
Quality Semiconductor

1800 series, pressure sensors
See *Foxboro/ICT, Inc.*

18PNP prefix
See *Future Domain*

(Plug and play IC)

1XXRP part number
See *Space Electronics*

(Space qualified ICs)

21XXX part numbers
See *DEC, Digital Equipment Company*

2114 part number
See *Intel Corp.*

(Discontinued memory device)

2148 part number
See *Intel Corp.*

(Memory device)

22V10 type parts
See *Atmel Corp.,*
Advanced Micro Devices (AMD),
Cypress Semiconductor,
American Microsystems, ICT,
National Semiconductor,
Aspen Semiconductor, UTMC,
Dense-Pac, Philips, TriQuint

23XXXX part number
See *Space Research Technology, Inc.*

2302XXX part number
See *Teledyne Electronic Technologies*

(Fiber data link products)

24XXXX part number
See *Space Research Technology, Inc.*

24LC devices
See *Microchip Technology, Inc.,*
Catalyst

26CXX/XXX part number
See *Philips Semiconductor*

26LS prefix
See *Philips Components, Motorola*

(AM26LS by Texas Instruments, DS26LS by National Semiconductor)

27L logic
Formerly Advanced Micro Devices (AMD)
See *Rochester Electronics*

28XX prefix
See *Dymec, Inc.*

28C prefix
See *Turbo IC, Atmel, SEEQ, Xicor, Microchip Technology, NEC, Telcom, Space Electronics*
(Also Excel EX28C, TI TSM28C, OKI MSM28C, Catalyst CAT28C, Samsung KM28C)

Manufacturers, Prefixes, Part Number Types, Logo Descriptions & Family Types

28F prefix
See *Intel Corporation* (Also A28F)
(And AMD AM28F, Catalyst CAT28F, Excel XL28F, Fujitsu MBM28F, Texas Instruments TMS28F, MSIS MSI28F, National Semiconductor NM28F, SGS-Thomson M28F, Hitachi HN28F, Seiko Instruments S28F, UTMC UT28F, Atmel AT28F)

28LD prefix
See *M/A-COM*

28LV prefix
See *Turbo IC*
(See also 28C prefix for other sources)

29XXXX part number
See *Advanced Micro Devices*

29C prefix
See *Intel, Matra MHS, Ideal Semiconductor*
(And American Microsystems MG29C, Atmel AT29C, Catalyst CAT29C, Samsung KM29C, Logic Devices L29C, AMD AM29C Quality Semiconductor QS29C, Performance Semiconductor P29C, Texas Instruments TCM29C)

2917753-XXX part number
See *Harris Semiconductor*
(Custom part)

29CXXX part number
See *Matra MHS* or *Temic*
(See Siliconix listing), *Atmel*
(Samsung KM29C, AMD AM29C, Logic Devices L29C, Texas Instruments TCM29C)

29CPLXXX part number
Formerly Advanced Micro Devices (AMD)
See *Rochester Electronics*

3SK prefix
See *Sony Semiconductor*

(GaAs MESFETs)

300SX
See *3DLabs*

3113 part number
See *Advanced Micro Devices*
(Discontinued)

32D prefix
See *Silicon Systems*

32F prefix
See *Silicon Systems*

32R prefix
See *Silicon Systems*

3200DX part number
See *Actel*

34096956-XXX part number
See *Honeywell, Inc.*

(Radiation hardened part)

342A prefix
See *Harris Semiconductor*
(Custom part)

36C prefix
See *Future Domain Corp.*

(Parts pin for pin compatible with NCR Microelectronics 53C series. NCR now Linfinity Microelectronics).

36111XX part number
See *Allied Signal Aerospace Company*

38 prefix
See *Texas Instruments*

3042PXXXX part number
See *Xlinx*

3090PXXXX part number
See *Xlinx*

32C prefix
See *Silicon Systems*

32H prefix
See *Silicon Systems*

4LP (CMOS 4LP process)
See *IBM Corp.*

4M prefix
See *Irvine Sensors*

(DRAM modules)

4NXX part number
Optoelectronic device, made by a variety of manufacturers (Hewlett-Packard, Micropac Industries, Texas Instruments, etc.)

4000 series
See *Xlinx*

(Programmable gate arrays)

404XXXX-XXX part number
See *Honeywell, Inc.*
(parts used in their systems)

44S
See *Samsung*

49BC
See *Philips*
(Tantalum Capacitor)

5H prefix
See *Harris Semiconductor*
(Custom part)

5L (CMOS 5L process)
See *IBM Corp.*

50XX part number
See *Optical Electronics, Inc.*

515XXXX part number
See *Harris Semiconductor*
(Custom part)

5 - 8   53C — 86C

53C (and 53CF) prefix
See *Symbios Logic*
(NCR Microelectronics)

54 series devices
(i.e., 54BCTXXX)
(Denotes military temperature range of -55°C to +125°C)

56XXX part number
See *Motorola*

(DSP ICs)

565XXX-X part number
See *Harris Semiconductor*
(Custom part)

568X part number
See *Philips Corporation*

587R prefix
See *Harris Semiconductor*

5962 prefix
This denotes a standard military drawing part (SMD) as issued by the Defense Electronic Supply Center (DESC) in Dayton, Ohio.

6NXXXX part number
See *Micropac Industries*, *Hewlett Packard*

(Military qualified components)

602X
See *Philips*

620
(The successor to the PowerPC microprocessor)

64 series devices
(i.e., 64BCTXXX)
(Generally denotes industrial temperature range of -40°C to +85°C)

643XXXX part number
See *LSI Logic*

65-XXX-XXX part number
See *Sciteq Electronics, Inc.*

65XXX part number
See *Chips and Technologies, Inc.*

660XX-XXX part number
See *Micropac Industries, Inc.*

67C prefix
See *Advanced Micro Devices*

67F prefix
See *Silicon Systems, Inc.*

67XX part number
See *Optical Electronics, Inc.*

68XX (and 68XXX) part number
See *Motorola*
(For optoelectronics item see Micropac Industries, Inc., Philips, Signetics)

68HCXXX part number
See *Motorola*

710XXX-XXX part numbers
(Discontinued Fairchild Semiconductor parts)

73D, **K**, or **M** prefix
See *Silicon Systems, Inc.*

733W prefix
(Xerox part number)

74 series devices
(i.e., 74BCTXXX)
(Denotes commercial temperature range of 0°C to +70°C)

75XX part number
See *Optical Electronics, Inc.*

7870-XX part number
See *Lambda Advanced Analog*

(DC/DC converters)

78P prefix
See *Silicon Systems, Inc.*

78Q prefix
See *Silicon Systems, Inc.*

79SR prefix
See *IDT*

8XC prefix
See *Philips Semiconductors*

80XX or **80XXX** part number
See *Intel*, *AMD*, *IIT*, *Cyrix*
(And other alternate source microprocessor IC manufacturers)

800XXXX-XXX part number
See *Honeywell, Inc.*
(Parts used in their systems)

80C prefix
See *SEEQ Technology*, *Philips Semiconductors*, *Matra MHS* (Temic Semiconductor Division), *Siemens*, *AVG Semiconductors*
(Second source)

82 prefix
See *Signetics*, *Opti, Inc.*
(82C PC and sound controllers), *Chips and Technologies* (82C9001 PC Video), *Samsung* (82C chip sets), *Omega Micro* (82CXXXX PCMCIA host adaptor controllers), *AVG Semiconductors*
(Second source some ICs)

82XX (and 82XXX) part number
See *Intel Corp.*

83C prefix
See *Siemens Corp.*

(EEPROM microprocessors)

86XXX part number
See the *ZIA marketing alliance*

86C prefix
See *S3*

87X prefix (X is a letter)
See *Philips Semiconductors*
(See also Intel)

87XX part number
See *Intel Corp.*

88XXX part number
See *Motorola, Inc.*

8890 series
See *United Microelectronics Corp. (UMC)*

(Pentium class chip set)

9H prefix
See *Harris Semiconductor*
(Custom part)

90C prefix
See *OnSpec Electronic, Inc.*

91C prefix
See *Standard Micro Systems Corp.*

9100 series
See *Weitek Corporation*

92C prefix
See *Opti, Inc., NCR Microelectronics*

93 prefix
See *Microchip Technology, Inc.*

93S logic
(Formerly Advanced Micro Devices [AMD])
See *Rochester Electronics*

95XXX-XXX part numbers
(These are USAF Monitored Line office parts, available from vendors such as Siliconix, VLSI and Microsemi Corporation. National Semiconductor Corporation line discontinued)

96XXX-XX part number
See *Motorola, Inc.*

97XX part number
See *Stac Electronics*

(Data compression coprocessors)

971XXX-XXXX part number
See *Harris Semiconductor*
(Custom part)

90009XXXXX-X part number
See *Harris Semiconductor*
(Custom part)

A logo
See *Allegro Microsystems, Inc.*

a prefix
See *Array Microsystems*

A prefix
See *Actel Corp.,
Allegro Microsystems, Inc.,
Harris Semiconductor
(Custom part),
EM Microelectronic*

(Real time clocks, voltage regulators and microprocessor surveillance ICs)

AAI Corporation
York Lane
P.O. Box 126
Cockeysville, MD 20130-0126
410-666-1400
Fax: 410-628-3215

(Hybrids, multichip modules, printed circuits, simulators and testers)

AAK Corporation
747 River Street
Haverhill, MA 01832
800-886-2225
508-373-3769

(Power supply input filters)

ABB Hafo AB
Box 520
S-175 26 Jarfalla
Sweden
46 (0)8 58 02 45 00
Fax: 46 (0)8 58 02 01 10

ABB Hafo, Inc.
11501 Rancho Bernardo Road
San Diego, CA 92127
619-675-3400, 619-485-8200
Fax: 619-675-3450, 619-484-2973
TLX: 69-5626

(Note: ABB Hafo AB was acquired from Asea Brown Boveri by the Mitel Corporation [Kanata, Canada] in early 1996. LED, PINs, photodiodes, duplex devices, custom CMOS ICs including RAD hard [radiation hardened] CMOS/SOS parts including ASIC parts based on Harris Semiconductor Designs.)

Abbott Electronics, Inc.
2727 S. La Cienega Blvd
Los Angeles, CA 90034-2643
310-202-8820
Fax: 310-836-1027
520 Main Street, Suite 301
Fort Lee, NJ 07024-4501
201-461-4411
Fax: 201-461-2936

(Military and commercial power supplies, DC/DC converters)

Aborn Electronics
2108 Bering Drive
Suite D
San Jose, CA 95131
408-436-5444
Fax: 408-436-0969

(Custom IR lamps, LEDs, phototransistors)

ABT — ACON

ABT (Advanced BiCMOS [bipolar CMOS] TTL) prefix (family)
See *Texas Instruments*,
Philips Components,
National Semiconductor,
Motorola Semiconductor

ABTE prefix
See *Texas Instruments*

AC logic family
See *National Semiconductor*,
Texas Instruments,
AVG Semiconductors, VTC. Inc.

AC prefix
See *Cougar Components*

Acapella Ltd.
Epsilon House
Chilworth Research Centre
Southampton, Hampshire SO16 7NP
United Kingdom 44 1703 769008
Fax: 44 1703 768612
http://www.acapella.co.uk

(Fiber optic [single fiber links] and telecommunication ICs)

ACC Electronics Company
1300 West Oak Street
Independence, KS 67301
316-331-1000
Fax: 316-331-0449

(Hybrids, voltage regulators, subcontract electronic assembly work)

ACC Microelectronics Corp.
2500 Augustine Drive
Santa Clara, CA 95054
408-980-0622
Fax: 408-980-0626

(PC Chip sets, memory controllers, system bus controllers)

Acculin, Inc.
214 North Main Street
Suite 204
Natwick, MA 01760
508-650-1012
Fax: 508-650-1457

(Track and hold amplifiers)

Accurel Systems International Corporation
785 Lucerne Drive
Sunnyvale, CA 94086-3848
408-737-3892

(Multichip modules)

Accutek Microcircuit Corp.
Business Center at Newburyport
2 New Pasture Road
Newburyport, MA 01950-4054
508-465-6200
Fax: 508-462-3396

(This company can design and test custom modules, SRAM modules)

ACD prefix
See *Anadigics Corporation*

Acer, Inc.
(ALI Acer Laboratories, Inc.)
5F, 156 Min Sheng East Road, Sec. 3, Min
Taipei, Taiwan R.O.C.
886 02 545 1588
Fax: 886 02 719 8690 or 8691

Acer Group
(ALI Acer Laboratories, Inc.)
6Fl., 156, Section 3
Min Sheng E. Rd.
Taipei, Taiwan R.O.C.
886 2 545 5288
Fax: 886 2 545 5308

Acer America Corporation
(ALI Acer Laboratories, Inc.)
2641 Orchard Parkway
San Jose, CA 95134
1-800-239-ACER
Fax: 408-922-2953

North American/European Sales and marketing Operations for ALi:
Pacific Technology Group
4701 Patrick Henry Drive
Building 2101
Santa Clara, CA 95054
408-764-0644
Fax: 408-496-6142, 408-727-9617

(ASICs and PC chip sets [ALi chip sets, Aladdin and FINALI and Genie ICs. See also *Texas Instruments-Acer Inc.* (a joint venture to manufacture DRAMs). VGA controllers, super I/O ICs, PC on a chip, Video Electronic Standards Association United Memory Architecture chip sets, MPEG-1, CD-ROM sound ICs and VGA controller ICs

ACF prefix
See *MICRA Corp.*

ACH prefix
See *Harris Semiconductor*
(Custom part)

ACK Technology, Inc.
7372 Walnut Avenue, Unit D
Buena Park, CA 90620
714-739-5797
Fax: 714-739-5898

(Thermoelectric coolers and heat sink products)

ACON, Inc.
22 Bristol Drive
South Easton, MA 02375
508-230-8022
Fax: 508-230-2371

(DC/DC Converters)

Manufacturers, Prefixes, Part Number Types, Logo Descriptions & Family Types

ACOPIAN — ADAPTEC

Acopian
P.O. Box 638
Easton, PA 18044
800-523-9478
610-258-5441
Fax: 610-258-2842

(Power supplies, DC/DC converters)

ACQ logic family
See *National Semiconductor*

Acrian Inc.
See *GHz Technology* for direct replacements.

(All parts discontinued per DATA)

ACS logic (and prefix)
See *Harris Semiconductor*

Acsist Associates, Inc.
3965 Meadowbrook Road
Minneapolis, MN 55426
612-931-1380

(Hybrids and multichip modules)

ACT logic family
See *Harris Semiconductor*,
National Semiconductor,
Texas Instruments,
AVG Semiconductors, VTC, Inc.

ACT prefix
See *Texas Instruments*
(If ACT-S-XXXXXXX-XXXXXX,
See Aeroflex Circuit Technology Corporation)

ACT3 prefix
See *Actel Corp.*

(Microprocessors)

Actel Corporation
955 E. Arques Ave.
Sunnyvale, CA 94086-4533
1-800-228-3532
800-262-1060 (Literature)
408-739-1010
Fax: 408-739-1540
Internet: http://www.actel.com
E-mail: tech@actel.com

6 Venture, Suite 100
Irvine, CA 92718
714-727-0470

6525 The Corners Parkway, Suite 400
Norcross, GA 30092
404-409-7888

3800 N. Wilke Road, Suite 300
Arlington Heights, IL 60004
708-259-1501

1740 Mass Ave.
Boxborough, MA 01719
508-635-0010

2350 Lakeside Blvd., Suite 850
Richardson, TX 75082
214-235-8944

Actel Europe Ltd.
Intec 2, Unit 22
Wade Road
Basingstoke
Hants RG24 8NE
England
44 0 256 29 209

Actel GmbH
Bahnhofstrasse 15
85375 Neufahrm Germany
49 0 8165 66101

(A fabless company supplying one time antifuse programmable FPGAs, PLDs, Field Programmable Logic Arrays [using anti-fuse technology, the connecting of logic gates by causing two sides of a circuit element to fuse together], military qualified devices available)

Active Design
(This company, a supplier of 2-D and 3-D graphics accelerator ICs, was acquired by Sigma Designs, Inc. of Fremont CA in early 1996).

ACTQ parts
See *National Semiconductor*

ACTS logic
See *Harris Semiconductor*

ACU prefix
See *Anadigics Corporation*

Acuity
(Trademark of General Microchip, Inc.)

Acumos, Inc.,
See *Cirrus Logic*

(Foster City, CA; graphics controller ICs. Became a subsidiary of Cirrus Logic April 1992)

AD prefix
See *Analog Devices*

A&D Co., Ltd.
23-14, Higashi-Ikebukuro 3-Chome
Toshima-ku
Tokyo 170
Japan
81 3 5391 6132

(Hybrid ICs)

Adaptec, Inc.
691 South Milpitas Blvd.
Milpitas, CA 95035
408-945-8600
Fax: 408-262-2533
http://www.adaptec.com/

Belgium, Brussels: 011-322-675-29-30
Hong Kong: 011-852-722-5441
Singapore: 011-65-278-5213
United Kingdom, Basingstroke:
011-44-256-468-186
Germany, Munich:
011-49-89-46-50-32

(A fabless company supplying Peripheral Component Interconnect

ADAPTEC — ADVANCED DETECTOR CORP.

(PCI) ICs, SCSI interface ICs. See Initio Corporation for second sources to I/O ICs)

Adaptive Logic, Inc.
800 Charcot Ave., Suite 112
San Jose, CA 95131
408-383-7200
Fax: 408-383-7201
http://www.adaptivelogic.com

(Formerly called NeuraLogix, a fabless manufacturer of fuzzy logic ICs, including analog microcontrollers)

Adaptive Networks, Inc.
1505 Commonwealth Ave.
Brighton, MA 02135
Box 1020
Cambridge, MA 02142
617-497-5150
Fax: 617-787-8168

(Power line communications ICs [permits spread-spectrum communications over power lines])

Adaptive Solutions, Inc.
1400 N.W. Compton Drive, Suite 340
Beaverton, OR 97006
503-690-1236
Fax: 503-690-1249
Newsletter: editor@asi.com

Southern CA: 714-432-6577
Northern, CA: 408-749-0423
New England: 617-229-5810
NY, NJ, OH, PA: 609-768-7502
Washington, DC: 703-847-6746
TX: 214-733-6505
Central U.S.: 503-690-1236
France: 33 1 40 80 03 70

(Parallel processing chips [such as their trademarked CNAPS], and chip sets, for digital neural-network implementation platforms [such as for pattern image recognition])

ADC prefix
See *Thaler Corporation*, *Analogic Corporation*, *Datel, Inc.*, *National Semiconductor*, *Burr-Brown Corp.*

ADC-XXXXX (or XXXXXX) part numbers
See *ILC-DDC*

Added Value Electronics Distribution, Inc. (AVED)
1582 Parkway Loop, Unit G
Tustin, CA 92680
800-439-7073
714-259-8258
Fax: 714-259-0828

AVED Memory Products
14192 Chambers Road
Tustin, CA 92710
800-778-7928
714-573-5000
Fax: 714-573-5047
E-mail: sales@avedmemory.com

(Distributor for various companies and custom designer and manufacturer of mixed technology assemblies and Flash Memory SIMMs)

ADG prefix
See *Analog Devices*
(Second source to DG parts)

ADI/3
See *Appian Technology, Inc.*

Adlas GmbH & Co. KG (Advanced Laser Designs)
SeelandstraBe 67
D-2400 Lubeck 14
Germany
49 451 3909 300
Fax: 49 451 3909 399

ADLAS, Inc.
636 Great Road
Stow, MD 01775
508-897-0800
Fax: 508-897-0811

(Diode pumped lasers)

ADM prefix
See *Analog Devices*

ADNT prefix
See *Numa Technologies*

ADS prefix
See *Datel, Inc.*, *Burr-Brown Corp.*

Adstar
The disk drive operation of IBM Corporation, which also sells component parts.

A.D.T.
See *Advance Data Technology, Inc.*

ADV prefix
See *Anadigics, Inc.*, *Analog Devices, Inc.*

Advance Data Technology, Inc. (A.D.T.)
P.O. Box 791
Gloucester, MA 01930
508-281-6590
Fax: 508-283-8005

(DRAMs, SRAMs, EPROMs, EEPROMs, microprocessors, multi-port RAMs, VRAMs, NV-RAMs, ECL RAM, FIFOs and modules. They can also manufacture some of the semiconductors that were supplied by Omni-Wave Semiconductor and can supply material compliant to MIL-STD-883 requirements.)

Advanced Detector Corporation
See *Advanced Optoelectronics*

Manufacturers, Prefixes, Part Number Types, Logo Descriptions & Family Types

ADVANCED ANALOG — ADVANCED MICRO DEVICES

Advanced Analog
See *Lambda Advanced Analog*

Advanced Electronic Packaging
21562 Surveyor Circle
Huntington Beach, CA 92646
714-969-1150

(Memory ICs, DRAMs)

Advanced Hardware Architectures, Inc. (AHA)
2635 N. E. Hopkins Ct.
Pullman, WA 99163-5601
509-334-1000
Fax: 509-334-9000
http://www.aha.com

(A fabless company supplying Reed-Solomon error correction ICs, image processor ICs)

Advanced High Voltage Company
P.O. Box 32408
Tucson, AZ 85751
2701 Dos Mujeres Avenue
Tucson, AZ 86715
800-494-2424
520-721-2424
Fax: 520-721-2322

(High voltage power supplies)

Advanced Linear Devices (ALD)
415 Tasman Drive
Sunnyvale, CA 94089
408-747-1155
Fax: 408-747-1286

(CMOS devices including amplifiers, bipolar DACs, rail to rail analog switches)

Advanced Memory Systems
Was a division of Intersil
See *Harris Corp.*

Advanced Micro Devices, Inc. (AMD)
One AMD Place
P.O. Box 3453
Sunnyvale, CA 94088-3453
1-800-538-8450
1-800-222-9323
(Technical support, press 6 for parts the 2 for a tech engineer)

1-800-222-9263
408-732-2400
408-987-3119 (PLDs, FPGAs)
Fax: 408-987-3102
TWX: 910-339-9280
TLX: 34-6306
http://www.amd.com
http://www.amd.com/html/locations/hq.html
http://www.amd.com/html/locations/locations.html
http://www.amd.com/html/products/EPLD/techdocs/litlist.html

(Embedded processor data sheets)

Applications and Literature:
1-800-222-9323
408-749-5703

Sales Offices:
Al: 205-882-9122
AZ: 602-242-4400

California:
Calabasas: 818-878-9988
Sacramento (Roseville): 916-786-6700
San Diego: 619-560-7030
San Jose: 408-922-0300

CO: 303-741-2900
CT: 203-264-7800

Florida:
Clearwater: 813-530-9971
Ft. Lauderdale: 954-938-9550

Latin America and Florida:
Orlando (Longwood): 407-862-9292
Ft. Lauderdale: 305-484 8600
Fax: 305 485 9736

GA: 404-449-7920
ID: 208-377-0393
IL: Chicago (Itasca): 708-773-4422
KY: 606-224-1353
MD: 301-381-3790
MA: 617-273-3970
MN: 612-938-0001

New Jersey:
Cherry Hill: 609-662-2900
Parsippany: 201-299-0002

New York:
Brewster: 914-279-8323
Rochester: 716-425-8050

North Carolina:
Charlotte: 704-875-3091
Raleigh: 919-878-8111

Ohio:
Columbus (Westerville): 614-891-6455
Dayton: 513-439-0268

OR: 503-245-0080
PA: 215-398-8006

Texas:
Austin: 512-346-7830
Dallas: 214-934-9099
Houston: 713-376-8084

Applications and Literature:
UK and Europe: 44 0 256 811101
France: 0590 8621
Germany: 0130 813875
Italy: 1678 77224
Far East Fax: 852 2956 0599

Australia:
N. Sidney: 61 2 9959 1937,
Fax: 61 2 9959 1037

Canada:
Ontario, Canada: 613 592
0060 Woodbridge: 905 856 3377

ADVANCED MICRO DEVICES — ADVANCED PHOTONIX

China:
Beijing: 8610 501 1566
Fax: 8610 465 1291
Shanghai: 8621 6267 8857, 9883
Fax: 8621 6267 8110

Belgium:
Antwerpen: 03 248 4300
Fax: 03 248 46 42

Finland:
Helesinki: 358 0 881 3117
Fax: 358 0 804 1110

France:
Paris: 1 49 75 10 10
Fax: 1 49 75 10 13

Germany:
Bad Homburg: 6172 92670
Fax: 06172 23195
Munchen: 089 45053 0
Fax: 089 406490

Hong Kong:
Kowloon: 852 2956 0388
Fax: 852 2956 0588

Italy:
Milan: 02 381961
Fax: 02 38103458

Japan:
Osaka: 06 243 3250
Fax: 06 243 3253
Tokyo: Shinjuku NS Building, 5 Fl.
2-4-1, Nishi-shinjuku
Shinjuku-ku, Tokyo 163-08
03 3346 7600, 7550
Fax: 03 3342 5197, 5196;
03 3346 7848

Korea:
Seoul: 82 2 784 0030
Fax: 82 2 784 8014

Scotland:
Stirling: 44 7186 450024
Fax: 44 1786 446188

Singapore: 65 337 7033
Fax: 65 338 1611

Switzerland:
Geneva: 41 22 788 0251
Fax: 41 22 788 0617

Sweden:
Stockholm (Bromma):
08 629 2850
Fax: 08 98 09 06

Taiwan:
Taipei: 886 2 7153536
Fax: 886 2712 2182 or 2183

United Kingdom:
London (Woking):
0483 740440
Fax: 0483 756196
Manchester (Warrington):
0925 8030380
Fax: 0925 830204

(Programmable logic devices [PLDs], LAN ICs, microprocessors, 10base-T Ethernet ICs, flash memory, voltage comparators, telephone circuits [subscriber line interface and audio processors and codecs], etc. Note, military-qualified product no longer available)

Advanced Micro Systems, Inc. (AMS)
2 Townsend West
Nashua, NH 03063-1277
1-800-234-2001
603 882-1447

(Motor controller ICs, microstep positioning systems)

Advanced Microelectronic Products, Inc.
No. 20, R&D Road II
Science-Based Industrial Park
Hsin-Chu City, 300
Taiwan
886 35 770 030

(Transistors)

Advanced Miliwave Laboratories, Inc.
AML Communications
900 Calle Plano, Suite K
Camarillo, CA 93012-9013
805-388-1345

(Wideband amplifiers)

Advanced Optoelectronics
15251 Don Julian Road
P.O. Box 1212
City of Industry, CA 91749
818-369-6886
Fax: 818-336-8694
TWX: 910-584-4890

(Photosensors and detectors)

Advanced Orientation Systems
6 Commerce Drive
Cranford, NJ 07016-3515
908-272-7755
Fax: 908-272-7758

(Polymer based electrolytic tilt sensor, angle conversion ICs, inclination sensors)

Advanced Photonix, Inc.
1240 Avenida Acaso
Camarillo, CA 93012
805-484-2884, 805-987-0146
Fax: 805-484-9935

(PIN and avalanche photodiodes, integrated optical filters, amplifiers. See also *Silicon Detector Corporation*)

ADVANCED POWER SOLUTIONS — AEG

Advanced Power Solutions (APS)
1040 Serpentine Lane
Pleasanton, CA 94566
510-485-1280
Fax: 510-485-1299
E-mail: MSAPS@AOL.COM

(External power supplies [wall mount types])

Advanced Power Technology (APT)
405 S.W. Columbia Street
Bend, OR 97702
1-800-222-8APT,
800-522-0809
503-382-8028
Fax: 503-388-0364

Northern Europe:
44 1635 582358
Fax: 44 1635 582458

Southern Europe:
33 57 92 15 15
Fax: 33 56 47 97 61

(FRED, FREDFET, Power MOSFETs and IGBT devices)

Advanced Power Solutions
1040 Serpentine Lane #201
Pleasanton, CA 94566
510-485-1280
Fax: 510-485-1299

(External desktop and wall mount power supplies)

Advanced RISC Machines, Ltd. (ARM)
Fulbourn Road, Cherry Hinton
Cambridge, CB1, 4JN
England
44 1223 400400
info@armltd.co.uk
http://www.arm.com/
http://www.cs.man.ac.uk/amulet/ARM_description.html

ARM architecture:
http://www.acorn.co.uk/acorn/store/ftp/documents/ARM/
ftp://ftp.acorn.co.uk/
ftp:ftp.acorn.co.uk/pub/documents/ARM/

20261 Beatty Ridge Road
Los Gatos, CA 95030
408-399-5195, 5199
408-399-8854 (faxback)
http://www.systemv.com/armitd.index.html
info@arm.com

(Licensed designs for 32 bit microprocessors, video controllers)

Advanced Semiconductor, Inc. (ASI)
7525 Ethel Avenue
North Hollywood, CA 91605
800-423-2354
818-982-1202

(Diodes, optoelectronics, thyristors, transistors, rectifiers, SCRs, triacs, microwave diodes)

Advani Oerlikon Ltd.
Advani Semiconductors Ltd.
Ahmednager
Mile 4/5
Poona, 411 014
India
0212 6748167522

(Diodes, optoelectronics, thyristors, transistors)

AdvanSys
Advanced System Products
1150 Ringwood Ct.
San Jose, CA 95131-1726
1-800-719-1783
408-383-9400

(SCSI controllers)

Advantech
750 East Arques Ave.
Sunnyvale, CA 94086
1-800-800-6889
408-245-6678
Fax: 408-245-8268

(Multifunction data acquisition circuit cards with ASIC IC's)

ADXL prefix
See *Analog Devices*

(Silicon accelerometers)

AE logo
("a" in a diamond outline above an "e" in a diamond outline)
See *Anderson Electronics, Inc.*

(Crystals and oscillators)

AE prefix
See *Amber*

AE-XXXXX/XXX
See *Harris Semiconductor*
(Custom part)

AEG Corporation - Telefunken
(Also **AEG-EUPEC**)
P.O. Box 3800
3140 Route 22 at Orr Drive
Somerville, N.J. 08876
201-231-8300
Telex: 833409

Algeria: 213-2607024
Belgium, Bruxelles: 02 3700-862
Peoples' Republic of China, Beijing: 1 5127516
Denmark, Albertslund:
45 42 64 85 22
France, Clamart Cedex:
1-45 37 9600

AEG — AIM

Germany:
Berlin: 030-828-2063
Eching: 089-3195002-04
Essen: 0201-244-242
Frankfurt: 069-7507-382
Hamburg: 040-3498-513
Hannover: 0511-6304-644

(Fiber optic components including emitters, detectors, LEDs [light emitting and IR types], optical switches, optocouplers, photo-darlingtons, photodiodes, etc.)

AEL Industries, Inc.
305 Richardson Road
Lansdale, PA 19446
215-822-2929
Fax: 215-822-9165
TWX: 510-661-4976

(Microwave and hybrid ICs for military products)

Aeronics, Inc.
12741 Research Blvd.
Suite 500
Austin, TX 78759
512-258-2303
Fax: 512-258-4392

(SCSI differential terminators)
Aeronutronics [CCD Group]).
See *Lockheed-Martin* (formerly Loral) *Fairchild Imaging Sensors*

Aeroflex Laboratories, Inc.
(Aeroflex Circuit Technology)
An ARX Company
Microelectronics Division
35 South Service Road
Plainview, NY 11803
1-800-THE-1553
516-694-6700, 516-752-2418
Fax: 516-694-6771, 6715
TWX: 510-224-6417

Europe:
44 793 850 720
Fax: 44 793 850 730

(Voltage regulators, MIL-STD-1553 bus interface units, MacAir or "Universal" designs, protocol modules, products for the H009 bus, MCM modules)

Aerotek Company Ltd.
1756 Sol Sukhumvit 52
Sukhumvit Road
Bangjark
Prakanong, Bankok 10250
Thailand
66 2 311 4448, 66 2 311 6783 5, 66 2 322 5035
Fax: 66 2 322 5034

(Isolators, circulators, microwave ferrite devices for telecommunications)

Aerovox Group
740 Beleville Avenue
New Bedford, MA 02745-6194
508-994-9661
Fax: 508-999-1000

(EMI/RFI filters)

AFL prefix
See *Lambda Advanced Analog*

(DC/DC converters)

AFW prefix
See *Lambda Advanced Analog*

(DC/DC converters)

ALO prefix
See *Lambda Advanced Analog*

(DC/DC converters)

AGX prefix
See *Xtechnology, Inc.*
A subsidiary of *Integrated Information Technology (IIT)*

AH prefix
See *National Semiconductor, Optical Electronics Inc., KDI/Triangle Electronics, Inc.* (Not a second source), *Advanced Hardware Architectures, Inc.*

AHA logo and prefix
See *Advanced Hardware Architectures*

AHC logic
A TI logic family debuted in 1996. It is faster than HCMOS logic.

AHD prefix
See *KDI/Triangle Electronics, Inc.*

AHE prefix
See *Lambda Advanced Analog*

(DC/DC converters)

AHF prefix
See *Lambda Advanced Analog*

(DC/DC converters)

AHL prefix
See *KDI/Triangle Electronics, Inc.*

AHV prefix
See *Lambda Advanced Analog*

(DC/DC converters)

AIC
See *Amaoto Industrial Co., Ltd.*

AIC prefix
See *Adaptec, Inc.*

AIM prefix
See *Optotek Ltd.*

AIM (Microcomputer family)
See *Rockwell*

AIRMAR — ALLEN-BRADLEY

Airmar Technology Corporation
69 Meadowbrook Drive
New Hampshire 03055-4618
603-673-9570
Fax: 603-673-4624

(Ultrasonic transducers)

AIT prefix
See *AITech International, Inc.*

AITech International, Inc.
47971 Fremont Blvd Fremont, CA 94538 800-882-8184
510-226-8960
Fax: 510-226-8996
Internet: info@aitech.com
sales@aitech.com
http://www.aitech.com

(Video processor IC to process RGB to NTSC/PAL and multimedia products)

AK prefix
See *Asahi Kasei Microsystems Co., Ltd., Accutek Microcircuit Corp.*

AKI
See *American Micro Systems*

(Original American Korean International, which was part of American Micro Systems)

AKM logo
Asahi Kasei Microsystems Co., Ltd.

AKM Semiconductor, Inc.
2001 Gateway Place, Suite 195 East
San Jose, CA 95110
408-436-8580
Fax: 408-436-7591
http://www.akm.com
E-mail: ICinfo@akm.com

Avenue Louise 326 Bte 056
1050 Brussels, Belgium
32 2 649 3062
Fax: 32 2 640 1089

(Telecommunications, mass storage, consumer audio and semicustom/custom ASICs including ASSPs [Application specific ICs], Base-Band ICs for cordless [wireless] and cellular telephones, audio interface ICs, CODECS, continuous tone controlled squelch system encoder/decoders Fax image scanner ICs, DACs, logic ICs)

ALC prefix
See *C&S Electronics*

ALD prefix
See *Advanced Linear Devices*

ALDC prefix
See *IBM*

Alden Scientific, Inc.
P.O. Box 267
Alden, NY 14004
716-681-3390
Fax: 716-681-1606

(Custom thick film hybrid circuits)

Aleph International Corp.
1026 Griswold Avenue
San Fernando, CA 91340
800-423-5622
818-365-9856
Fax: 818-365-7274

(Photosensors, gap detectors, reflective detectors, relays and solid state relays)

ALG prefix
See *Advance Logic, Inc.*

Algotronix
King's Buildings-TTC
Mayfield, Rd
Edinburgh EH9 3JL
Scotland
31 668-1550
Fax: 31 662-4678

(FPGAs)

ALI logo
See *Acer, Inc.*
(ALI Acer Laboratories, Inc.)

Allegro MicroSystems, Inc.
(Prior to 12/90, was Sprague Semiconductor, now owned by Sanken Electric Co. Ltd.)
Suffolk Building
53 Regional Drive
Concord, NH 03301
603-228-5533
http://www.allegrosys.com

115 Northeast Cutoff
Worcester, MA 01615
1-508-ALLEGRO

Swada Bldg
15-9, Shinjuku 1-chome
Shinjuku-ku
Tokyo 160
Japan
81 3 3226 1761

(Linear devices, voltage comparators, voltage regulators, switches and latches with temperature ranges up to +170°C for automotive applications. Screening to MIL-STD-883 is also available.)

Allen Avionics, Inc.
224 E. Second St.
Mineola, NY 11501
516-248-8080
Fax: 516-747-6724

(Inductors, delay lines [video, general purpose and digital types], passive [video] filters, video hum eliminators, isolation transformers)

Allen-Bradley Company
A Rockwell International Company
1201 S. 2nd St.
P.O. Box 2086
Milwaukee, WI 53201

ALLEN-BRADLEY — ALPHA SEMICONDUCTOR

414-382-2000
Fax: 414-382-4444

747 Alpha Drive
Highland Heights, OH 44143
216-646-5000
Fax: 216-646-3075

109 Etna Road
Lebanon, NH 03766
603-448-6300
Fax: 603-448-5659

460 Elm St.
Manchester, NH 03101
603-625-8299
Fax: 603-668-6028

8440 Darrow Road
Twinsburg, OH 44087
800-932-7322
Fax: 216-487-6117

1500 Peebles Drive
Richland Center, WI 53581
608-647-6376
Fax: 608-647-7186

(Resistors, resistor networks, etc. Note: Allen-Bradley also has a separate Industrial Controls Division)

Alliance Semiconductor Corporation
3909 N. 1st Street
San Jose, CA 95134
408-383-4900
Fax: 408-383-4999

Boston, MA:
617-239-8127

Taiwan (Technical Center):
886 2 723 9944

(A fabless company supplying multimedia ICs, graphics/video controllers, high speed SRAMs, flash memory, DRAMs, memory modules)

Allied Electronic & Semiconductor Technology
16F, No 629, Ming Shen E. Rd.
Taipei, 100
Taiwan
886 2 718 1918

(Diodes)

Allied Electronics GmbH
Maariveg 231-233
Koln 50825, Germany
0221 499 3084
Fax: 0221 497 3080

(Encoder/pulse converter LSI)

AlliedSignal Aerospace Company
Microelectronics Center
9140 Old Annapolis Road/MD 108
Columbia, MD 21045-1998
410-964-4000
Fax: 410-992-5813

(SRAMs, microprocessors, CPUs)

AlliedSignal Instrument Systems/Commercial Aviation
15001 NE 36th St.
P.O. Box 97001
Redmond, WA 98052
206-885-3711
Fax: 206-885-2061

(Thermal switches)

Alpha IC
As in *DEC Alpha*
See *Digital Equipment Corp.*

Alpha Components, Inc.
1106 E. Simpson Road
Mechanicsburg, PA 17055
717-697-8985
Fax: 717-697-8608

(Quartz crystal and LC filters for military, consumer and medical products)

Alpha Industries, Inc.
20 Sylvan Road
Woburn, MA 01801
617-935-5150
617-933-0159, 617-935-2359

19/21 Chapel Street
Marlow, Bucks SL73HN
England
44 6824 75562
Fax: 44 6824 74078
TLX: 846331

Stefan-George-Ring 19
D-8000 Munchen 81
Germany
089 93 20 12
Fax: 089 931123
TLX: 5 213 581

(Semiconductors such as diodes [including Schottky and varactor types], thyristors, transistors, analog switches, MMIC amplifiers, voltage variable attenuators, GaAs FETs, PIN diodes, chip capacitors and military qualified devices)

Alpha Products, Inc.
Alpha Center, 351 Irving Drive
Oxnard, CA 93030
805-981-8666
Fax: 805-988-6555

(Optoelectronics, cables, connectors, sockets)

Alpha Semiconductor, Inc.
1031 Serpentine Lane
Pleasanton, CA 94566
510-417-1391
Fax: 510-417-1390

(Voltage references, and low dropout regulators, including second sources for some National Semiconductor and Linear Technology devices)

ALPHA TERMISTOR & ASSEMBLY — AMELCO SEMICONDUCTOR

Alpha Thermistor & Assembly, Inc.
7181 Constitution Court
San Diego, CA 92121
800-235-5445
619-549-4660
Fax: 619-549-4791

(Thermistors)

Alpine Image Systems, Inc.
821 Riverside Drive
Los Altos, CA 94024-4824
415-941-3247

(Boundary scan ICs)

ALPS Electric, Inc.
3553 N. First Street
San Jose, CA 95135
408-432-6000

(Various consumer product subassemblies [TV tuners, etc.] and ICs within them)

ALR
Computer manufacturer with custom ICs

ALS
See *Applied Laser Systems*

ALS logic family
See *Hitachi*, *Motorola, Inc.*, *National Semiconductor*, *SGS-Thomson*, *Lansdale*, *LG Semicon* (Formerly Goldstar Technology), *Inc.*, *Rochester Electronics*, *AVG Semiconductors*

ALS prefix
See *Avance Logic, Inc.*

Altair Corporation
24922 Anza Drive
Unit F
Valencia, VA 91355
800-824-5620
Fax: 805-257-6490

(Subminiature lamps, Note: LEDs now available from the manufacturer, Ledtech)

Altech Corp.
35 Royal Road
Flemington, NJ 08822-6000
908-806-9400
Fax: 908-806-9490

(European fuses, terminal blocks, DIN [miniature] circuit breakers, proximity switches, DIN motor protectors, interface modules, enclosures including DIN rail mount types and metallic, nonmetallic types; foot switches, liquid tight strain reliefs, safety relays and modules)

Altera Corporation
2610 Orchard Parkway
San Jose, CA 95134-2020
800 9 Altera (800-925-8372)
(This off site customer service)
1-800-800-EPLD (Applications)
800-800-7256
(PLDs and FPGAs)
408-984-2800, 894-7000
Fax: 408-435-1394, 408-428-9220

BSDL (Boundary scan) files on BBS:
408-954-0104
http://www.altera.com/html/products/

Solar House
Globe Park
Fieldhouse Lane
Marlow
Bucks SL7 1TB,
England
044 0628 488800
Fax: 044 0628 890078

Dai Tokyo Kasai Shinjuku Bldg.
4 fl.
3-25-3 Yoyogi, Shibuya-ku
Tokyo 151, Japan
03 3375 2281
Fax: 03 3375 2287

(A fabless company supplying programmable memory and logic devices such as PLDs, EPLDs [Erasable Programmable Logic Devices] and FPGAs. Note: Military qualified PLD devices were discontinued by Altera in Nov. 1995)

ALVC logic
See *Texas Instruments*

AM prefix
See *AMD* (Advanced Micro Devices), *Datel Inc.*, *Harris Semiconductor* (Custom part), *M/A-COM Inc.*, *Anzac*, *Ascom Microelectronics*

AMA prefix
See *GEC Plessy*
(Marconi Circuit Technology Inc.)

AMBE-XXXX part number
See *Digital Voice Systems, Inc.*
(DVSI)

AMCC
See *Applied Micro Circuits Corporation*

AMD
See *Advanced Micro Devices*

Amega Technology
London Business Center
Roentgen Rd.
Daneshill East
Basingstroke, Hants RG24 8NG
01256-330301
Fax: 01256-330302

(Data compression IC for modems)

Amelco Semiconductor Co.
Became Teledyne Components and is now known as TelCom Semiconductor, Inc.

AMERICAN ELECTRONIC COMP. — AMERICAN MICROSYSTEMS

American Electronic Components, Inc.
P.O. Box 280
1010 Main Street
Elkhart, IN 46515
219-264-1116
Fax: 219-264-4681

(Hall effect sensors, variable reluctance sensors, inertia sensors)

American High Voltage (AHV)
1957A Friendship Drive
El Cajon, CA 92020
619-258-5804

AHV Commercial
Div. of Hanington International, Inc.
4411 Willows Road
Alpine, CA 91901
619-445-0880

(Military and commercial power supplies)

American LED-gible, Inc.
1776 Lone Eagle St.
Columbus, OH 43228
614-851-1100
Fax: 614-851-1121

(Displays, ETIs [elapsed time indicators], graphic boards, etc.)

American Megatrends International (AMI)
6145 Northbelt Parkway
Norcross, GA 30071-2972
770-263-8181
LaHabra, CA 90631
310-690-8997
http://www.megatrends.com/

(AMI BIOS ICs)

American Microsemiconductor, Inc.
133 Kings Road
Madison, NJ 07940
201-377-9566
Fax: 201-377-3078

(Diodes, and transistors [of their own manufacture], high voltage rectifiers, optoelectronics, thyristors, transistors)

American Microsystems, Inc. (AMI)
2300 Buckskin Road
Pocatello, ID 83201
1-800-639-7264
(1-800-NEWS AMI)
Fax: 208-234-6795

Eastern Area:
237 Whooping Loop
Altamonte Springs, FL 32701
305-830-8889
TWX: 810-853-0269

Axe Wood East
Butler & Skippack Pikes, Suite 230
Ambler, PA 190002
215-643-0217
TWX: 510-661-3878

Central Area:
500 Higgins Road, Suite 210
Elk Grove Village, IL 60007
312-437-6496
TWX: 910-222-2853

Suite No. 204
408 South 9th St.
Noblesville, IN 46060
317-773-6330
TWX: 810-260-1753

29200 Vassar Ave., Suite 303
Livonia, MI 48152
313-478-9339
810-242-2903

725 So. Central Expressway,
Suite A-9
Richardson, TX 75080
214-231-5721, 5285
TWX: 910-867-4766

Western Area:
100 East Wardlow Rd., Suite 203
Long Beach, CA 90807
213-595-4768
TWX: 910-341-7668

3800 Homestead Road
Santa Clara, CA 95051
408-249-4550
TWX: 910-338-0018

20709 N.E. 232nd Ave.
Battleground, WA 98604
206-687-3101

England:
AMI Microsystems, Ltd.
108A Commercial Road
Swindon, Wiltshire
793 31345 or 25445
TLX: 851-449349

France:
AMI Microsystems, S.A.R.L.
124 Avenue de Paris
94300 Vincennes France
01 374 00 90
TLX: 842-670500

Holland:
AMI Microsystems, Ltd.
Calandstraat 62
Rotterdam, Holland
010-36 14 83
TLX: 844-27402

Italy:
AMI Microsystems, S.p.A.
Via Pascoli 60
20133 Milano
29 37 45 or 2360154
TLX: 843 32644

AMERICAN MICROSYSTEMS — AMKOR

Japan:
AMI Japan Ltd.
502 Nikko Sanno Building
2-5-3, Akasaka
Minato-ku, Tokyo 107
Tokyo 586-8131
TLX: 781-242-2180 AMI J

Germany:
AMI Microsystems, GmbH
Rosenheimer Strasse 30/32, Suite 237
8000 Munich 80
Germany
89 483081
TLX: 841 522743
Fax: 89 486591

(Gate arrays, ASICs, wireless communication ICs, various MOS semiconductors such as silicon unilateral switches, transceivers, thyristor triggers, 1 meg ROMs for disk drive and modem applications, etc. Gould Electronics was acquired by Nippon Mining [now Japan Energy] in 1988 and renamed American Microsystems, Inc.)

American Neuralogix, Inc.
See *Adaptive Logic, Inc.*

American Power Devices, Inc.
69 Bennett Street
Lynn, MA 01905-3067
617-592-6090
Fax: 617-592-0677

(Diodes, thyristors, including military qualified devices)

American Zettler, Inc.
75 Columbia
Aliso Viejo, CA 92656
714-831-5000
Fax: 714-831-8642

(LCDs, LEDs, transformers [including horizontal power types, relays])

Ametek Rodan Division
See *Ketema Rodan Division*

AMG
See *AMG Systems, Ltd.*

AMG Systems, Ltd.
3 The Omega Centre
Stratton Business Park
London Road Biggleswade
Bedfordshire SG18 8QB
44 0 1767 600777
Fax: 44 0 1767 600077

(Optical gas detection sensors, using LED based infrared sensor cells)

AMI
See *American Microsystems, Inc.*

Amaoto Industrial Co., Ltd.
12F-1 No. 235
Chung Cheng 4th Road
Kaohsiung, Taiwan, R.O.C.

P.O. Box 1822
Kaohsiung, Taiwan, R.O.C.
886-7-251-0691
Fax: 886-7-272-8143

(Speakers)

American Bright Optoelectronics Corporation (LED BRT)
460 W. Lambert Rd.
Suite H
Brea, CA 92621
714-257-0800
Fax: 714-257-1310

(LEDs, LED displays, IR, photodiode LEDs, SMT, backlight and axial lead LEDs)

American Precision Industries
Delevan/SMD Divisions
270 Quaker Road
East Aurora, NY 14502
716-652-3600
Fax: 716-652-4814
http://www.delevan.com
E-mail: apisales@delevan.com

(Magnetics including chokes, inductors and transformers)

Amber
A Raytheon Company
5756 Thronwood Drive
Goleta, CA 93117-3802
805-692-1348

(FPAs, infrared focal plane arrays)

Ambit Microsystems Corporation
5F-1, No. 5, Hsin An Rd.
Science-based Industrial park
Hsinchu, Taiwan R.O.C.
886 35 784 975
Fax: 886 35 775 100

(Hybrid ICs and multichip modules)

Amkor Electronics
(Anam)
1347 N. Alma School Rd., Bldg 100
Chandler, AZ 85224
602-821-5000
Fax: 602-821-2389, 602-821-2044

Western US:
408-496-0303,
Fax: 408-496-0392

Eastern/Mid-Western US:
214-580-1879,
Fax: 214-518-0293

Asia:
632 815 4777,
Fax: 632 842 1841

Europe:
33 50 40 9797,
Fax: 33 50 40 9888

Korea:
822 460 5280,
Fax: 822 469 8739

AMKOR — ANAHEIM AUTOMATION

Japan:
813 3 3446 1881,
Fax: 813 3 3446 1869

Singapore:
65 324 0722,
Fax: 65 324 1183

Taiwan:
886 35 733 717,
Fax: 886 35 717 576

(IC packaging, such as BGA packages)

AMO-HC prefix
See *AVX Corp.*

(Clock oscillators)

AMP-XX part number
See *Analog Devices, Inc.*

AMP, Inc.
Harrisburg, PA 17105-3608
717-564-0100
(Product information center)

AMP FAX (press 1)
and product information (press 2) 24 hrs/day 1-800-522-6752,
717-986-7575
800-722-1111

Application/Tooling
Assistance Center
http://connect.amp.com
http://www.amp.com/

Canada: 905-470-4425

(Serial interface modules [ARINC 629 SIM], connectors, etc.)

Ampere
See *Dimolex Corp.*

Amperex Electronic Corp.
See *Philips Components*

American KSS, Inc.
See *Kinseki, Ltd. Amplica, Inc.*
950 Lawrence Drive
Newbury Park, CA 91320
805-498-9671
Fax: 805-498-4925

(Solid-state low-noise amplifiers and sub systems for microwave applications, they also have a 2-18GHz surface mount amplifier)

Amplifonix
2707 Black Lake Place
Philadelphia, PA 19154
215-464-4000
Fax: 215-464-4001

(RF components including Bipolar amplifiers, GaAs and MMIC wideband amplifiers, GaAs MMIC and PIn diode switches, digital attenuators, VCOs voltage controlled oscillators, voltage controlled attenuators, limiters and linearizers)

AMPO Technology, Inc.
40100 San Carlos Pl
Fremont, CA 94539-3612
510-651-5587

(Laser printer and typesetter controller IC)

AMPSS (R)
See *Astec America, Inc.*

Amptek, Inc.
6 DeAngelo Drive
Bedford, MA 01730
617-275-2242
Fax: 617-275-3470

(Charge sensitive preamplifiers, voltage amplifiers, peak detectors, preamplifiers, etc.)

Amps Abundant
1891 N. Gaffey St.
Units K&L
San Pedro, CA 90731

1-800-233-0559
Fax: 310-833-9154

(This company stocks replacement power semiconductors, including obsolete devices)

AMS
See *Austria Mikro Systems International GmbH*,
Advanced Micro Systems (Motor ICs),
Harris Corp. (Formerly Advanced Memory Systems division of Intersil),
American Microsemiconductor

AMS prefix
See *Lambda Advanced Analog* (DC/DC converters),
Aptek Williams, Inc.

AN logo
See *Adaptive Networks*

AN prefix
See *Matsushita* (Panasonic);
The AN is for linear (analog) ICs.,
AuraVision Corp.

Anadigics, Inc.
35 Technology Drive
Warren, NJ 07060
201-668-5000
Fax: 201-668-5068
TLX: 510-600-5714
email: mktg@anadigics.com
http://www.anadigics.com

(GaAs ICs including amplifiers and microwave devices such as frequency dividers and up/down conversion of digital CATV, DC/DC converters [in SOT-25 packages])

Anaheim Automation
910 E. Orangefair Lane
Anaheim, CA 92801
800-345-9401
Fax: 714-992-6990

(Stepper motor ICs)

Manufacturers, Prefixes, Part Number Types, Logo Descriptions & Family Types

ANALOG DEVICES

Analog Devices
(Also one division Computer Labs; see address below)
One Technology Way
P.O. Box 9106
Norwood, MA 02062-9106
1-800-262-5643, 5645
(1-800-ANALOGD)
617-329-4700
617-461-3771, 617-461-3881
617-461-3672
DSP Applications Assistance Line

Other Applications lines:
404-263-3722 (Atlanta, GA)
408-879-3037 (Campbell, CA)
(DSPs)

314-921-4429 (St. Louis, MO)
617-937-1428
(Central Applications Technical Support Line)

617-461-4258
(8 data bits, no parity, 1 stop bit, 300-14400 baud), DSP Bulletin Board

Fax: 617-326-8703
AnalogFax (tm):
1-800-446-6212 (For data sheets)
Telex: 924491

Internet: ftp ftp.analog.com or ftp 137.71.23.11, log in as anonymous then use your Email address for your password.
WWW: http://www.analog.com

Three Technology Way
Norwood, MA 02062-9106
617-461-3789
Fax: 617-329-1241

181 Ballardvale St.
Wilmington, MA 01887
617-937-1428
(Codecs)

Ray Stata Technology Center 804 Woburn St. Wilmington, MA 01887
(DC/DC converters)

Documentation Control Department
804 Woburn Street
Wilmington, MA 01887-3462
617-935-5565
Fax: 617-937-2012
Computer Labs
(Analog Devices Assembled Products Division)

7910 Triad Center Drive
Greensboro, NC 27409-9605 (or 9681)

Customer Service Center
1500 Space Park Dr.
Santa Clara, CA 95054-3499
408-562-7592
Fax: 408-562-2654

Sales Offices:
Alabama: 404-263-3719
Alaska: 206-575-6344, 714-641-9391
Arkansas: 214-231-5094
California: 714-641-9391, 408-559-2037
Colorado: 714-641-9391
Connecticut: 516-366-0765, 617-329-4700
Delaware: 215-643-7790
Florida: 407-660-8444
Georgia: 404-263-3719
Hawaii: 714-641-9391
Idaho: 206-575-6344
Illinois: 708-519-1777
Indiana: 219-489-3083
Iowa: 708-519-1777
Louisiana: 214-231-5094
Maine: 617-461-3000
Maryland: 410-992-1995
Massachusetts: 617-461-3000
Michigan: 313-348-5795
Mississippi: 404-263-3719
Montana: 714-641-9391
Nevada: 408-559-2037, 714-641-9391
New Hampshire: 617-461-3000
New Jersey: 516-366-0765, 215-643-7790
New York: 516-366-0765, 315-682-5571, 716-425-9100

North Carolina: 404-263-3719
Oklahoma: 214-231-5094
Oregon: 206-575-6344
Pennsylvania: 215-643-7790
Rhode Island: 617-461-3000
Tennessee: 404-263-3719
Texas: 214-231-5094
Vermont: 617-461-3000
Virginia: 410-992-1995
Washington State: 206-575-6344
Wisconsin: 708-519-1777
Wyoming: 714-641-9391
Puerto Rico: 407-660-84444
Mexico: 617-461-3000

European Headquarters:
Edelsbergstrabe 8-10
8000 Munchen 21
Germany
(89) 57 005-0
Fax: (89)57 005-257
TLX: 841 523712

India: 812-567201, 80 26 72 72

Japan:
Daini Jibiki Building
4-7-8 Koji-Machi
Chiyoda-ku, Tokyo 102
(3) 3263 6826
Fax: 3 3234 2524
TLX: 781 28440

Osaka: (6) 3721814
Korea: (2) 5543301
Austria: (222) 88 55 04-0
Belgium: (3) 237 16 72
Denmark: (42) 84 58 00

France:
Antony Cedex: (1) 64 66 25 25
Meylan: (76) 41 91 43

Germany:
Munich: (89) 57 005-0
Berlin: (30) 391 90 35
Hamburg: (4181) 8051

21

ANALOG DEVICES — ANDERSEN LABORATORIES

Cologne: (221) 68 60 06
Stuttgart: (711) 88 11 31

Hong Kong:
2102 Nat West Tower
Times Square
One Matheson Street
Causeway Bay
Hong Kong: 852 506 9336
Fax: 852 506-4755

Israel: (52) 911 415

Italy:
Milan: (2) 665 00 120
Turin: (11) 287 789
Rome: (6) 86 200 306

Netherlands: (1620) 815 00
Sweden: (8) 282 740

Switzerland:
Zurich: (1) 820 01 02
Morges: (21) 803 25 50
Baar: (42) 330 710

Taiwan: 02 740 0099

United Kingdom:
Head Office, Walton:
(0932) 247 401
Birmingham: (021) 501 1166
Scotland: (0506) 303 06
Southern Area, Walton:
(0932) 246 200
Sales, Walton: (0932) 253 320
Eastern Sales, Harlow:
(0279) 418 611
Newbury (0635) 353 35

(Op amps, difference amplifiers, comparators, references, voltage regulators, ADCs, cross point switches, analog switches, solid state thermostatic switches, audio compression ICs and PC Audio ICs, Hall Effect sensors for automotive applications, digital signal processors [DSPs], MODEM ICs, GSM [Global System for Mobile Communications digital cellular radio] ICs, data telecommunications ICs, IC temperature transducers; RS-232/422/423, RS485, interface ICs; microprocessor generator/support circuits, accelerometers, RGB to NTSC or PAL analog encoders, video compression ICs [including Wavelet compression ICs], military qualified devices available.)

Analog Modules, Inc.
126 Baywood Avenue
Longwood, FL 32750
407-339-4355
Fax: 407-834-3806

(Laser electronics, optical receivers, DC/DC converters [military units available])

Analog Systems
P.O. Box 17389
8075E Research Ct., #104
Tucson, AZ 85731-7389
602-290-1818
(520 area code after 12/31/96)
Fax: 602-885-1189
(520 area code after 12/31/96)

(D/A converters, drivers, power drivers, amplifiers, controllers, power supply ICs, telecommunications ICs, voltage references and comparators)

Analogic Corp.
360 Audubon Road
Wakefield, MA 01880
1-800-446-8936
508-977-3000
Fax: 617-245-1274

8 Centennial Drive B-12
Peabody, MA 01960
508-977-3000
Fax: 508-531-7356

(A/D converters, digital signal processing products, test instrumentation, data acquisition systems and custom system development programs)

Analytic Instruments Corp.
9995 Monroe Drive, Suite 205
P.O. Box 542845
Dallas, TX 75220
214-357-3882

(Frequency synthesizer ICs)

Anaren Microwave, Inc.
6635 Kirkville Road
East Syracuse, NY 13057
800-544-2414
315-432-8909
Fax: 315-432-9121

(Surface mount hybrid couplers and in-phase power dividers)

AND
Purdy Electronics
(formerly W. J. Purdy Co.)
770 Airport Boulevard
Burlingame, CA 94010
415-347-9916
Fax: 415-340-1670
TWX: 910-374-2353

(LED and LCD products)

Anders Electronics, Ltd.
48-56 Bayham Place
London, NW1 OEU
United Kingdom
44 71 387 9094 or 388-7171

(LED displays)

Andersen Laboratories
1280 Blue Hills Ave.
Bloomfield, CT 06002
203-242-0761
45 Old Iron Ore Road
Bloomfield, CT 06002
203-286-9090
Fax: 203-242-4472

(Delay Lines, VCOs, SAW filters and oscillators, IFF transmitters and receivers, etc.)

Anderson Electronics, Inc.
P.O. box 89
Scotch Valley Road
Hollidaysburg, PA 16648
814-695-4428
Fax: 814-696-0403

(Quartz crystals and crystal oscillators)

Anoma Electric Company Ltd.
See *Operating Technical Electronics, Inc.*
(As a source for these power supplies)

Anritsu America, Inc.
365 West Passaic St.
Rochelle Park, NJ 07662-3014
201-843-2690
Fax: 201-843-2665

(PCB electromechanical relays)

Antel Optronics, Inc.
3310 S. Service Road
Burlington, ON L7N 3M6
Canada
416-637-9990

(Photo ICs, photodiodes, laser photodiodes, PIN photodiodes)

Anzac
Div. M/A-COM
1011 Pawtucket Blvd
P.O. Box 3295
Lowell, MA 01853-3295
508-442-5000
Fax: 508-442-4800

(Isolation amplifiers, electronic switches for microwave frequencies)

AOP prefix
See *Anadigics, Inc.*

AOSI
See *Advanced Orientation Systems, Inc.*

AP prefix
See *KDI/Triangle Electronics, Inc.*, *Aptos Semiconductor*

APA Optics, Inc.
2950 N.E. 84th Lane
Blaine, MN 55434
612-784-4995
Fax: 612-784-2038

(GaN UV sensors, a UV radiation detector)

APC Ltd.
47 Riverside, Medway City Estate
Strood, Rochester, Kent ME2 4DP
United Kingdom
44 1634 290588
01634 290588 (within the U.K.)
Fax: 44 1634 290591

In the U.S.:
TMI
18 W. Pamprapo Ct.
Glen Rock, NJ 07452
201-447-8821
Fax: 201-670-4818

(ISDN transformers and common mode chokes)

APD-XXXXXXXX part number
See *Dale Electronics*

(Plasma Displays)

Apex Microtechnology Corp.
(Apex u tech)
5980 N. Shannon Road
Tucson, AZ 85741
1-800-862-1015
1-800-421-1865
(Applications/product selection assistance)

ANDERSEN LABORATORIES — APEX

1-800-862-1021
(Product information/orders)
1-800-546-APEX
(Applications hotline)
520-690-8600, 8603
520-742-8601 (Orders)
520-742-8609 (Sales)
520-742-8619 (Quality assurance)
520-742-8606
(APEX F.S.C., Inc., Export)
520-742-8662, 8663
(Applications engineering)
520-742-8601 (Customer service)
1-800-448-1025 (Literature hotline)
Fax: 520-888-3329
520-888-7003 (Applications Fax)
TLX: 170631

Australia/New Zealand:
08 277-3288
Austria: 222 505 15 220
Belgium/Luxemburg: 03 458 3033
Canada: 416 821-7800
Daehan Minkuk: 02 745-2761
Denmark: 42 24 48 88
Germany: 6172 488510,
0130 81 3599 (Deutschsprachig)
Spain: 1 530 4121
France: 1 69 86 92 89
Hong Kong: 8339013
Italy: 2 6640 0153
Israel: 3 934517
India: 212 339836
Japan: 3 3244-3787
The Netherlands: 10 451 9533
Norway: 2 50 06 50
Peoples Republic of China:
86 500 7788
Republic of South Africa:
021 23 4943
Singapore: 284-8537
Switzerland: 56 26 54 86
Schweiz: 0049 6172 488510
Sweden: 8 795 9650
Taiwan-Republic of China:
02 722 3570
Turkey: 1 337 2245
United Kingdom: 0844 278781

APEX — APTEK WILLIAMS

(Power, pulse width modulator [PWM], and high voltage amplifiers, DC/DC converters, etc.).

API-XXXXX part number
See *ILC-DDC*

A.P.I. Electronics, Inc.
375 Rabro Drive
Hauppauge, N.Y. 11788
516-582-6767
Fax: 516-582-6771

(Semiconductors such as optoelectronics, thyristors and transistors and diodes including military qualified devices)

Apollo Chip Set
See *Via Technology*

Appian Technology, Inc.
477 N. Mathilda Avenue
Sunnyvale, CA 94088-3410
408-730-5400, 5477, 8800

(Local bus IDE interface IC, graphics accelerator cards, single bus 486/386DX controller with Local Bus design, workstations, Peripheral Component Interconnect [PCI] ICs, etc.)

Applications Engineering, Inc.
18 Great Plain Sq., Suite 169
Danbury, CT 06810
203-790-1861

(High voltage diodes with fast recovery)

Applied Laser Systems (ALS)
2160 N.W. Vine Street
Grants Pass, OR 97526
503-474-6560
Fax: 503-476-5105
QWIK-Fax Information:
503-479-1526
Email: ALS@cdsnet.net

(Visible laser line generators, visible laser modules, laser pen pointers. Note: This company, also a manufacturer of computer-aided vision products, sold its laser diode operation [except for the laser aimer line] to Coherent, Inc. [Auburn, CA], in 1995).

Applied Micro Circuits Corporation (AMCC)
6195 Lusk Blvd
San Diego, CA 92121-1792
800-PLL-AMCC (800-755-2622)
(Literature line)
619-450-9333
Fax: 619-450-9885
http://www.amcc.com
E-mail: pciinfo@amcc.com
compinfo@amcc.com/

Northeast:
25 Burlington Mall Rd., Ste. 300
Burlington, MA 01803
617-270-0674
Fax: 617-221-5853

California:
960 S. Bascom Ave., Ste. 1113
San Jose, CA 95128
408-289-1190
Fax: 408-289-1527

Northwest:
Five Centerpointe Drive
Suite 100
Lake Oswego, OR 97035
503-624-4918
Fax: 503-624-9103

Mid-America, Arizona, New Mexico:
840 E. Central parkway, Suite 120
Plano, TX 75074
214-423-7989
Fax: 214-424-6617

South Atlantic:
Atrium Executive Center
80 Orville Dr.
Bohemia, NY 11716
516-563-0401
Fax: 516-563-0406

International:
Applied Micro Circuits Corp.
Saville Court, Saville Place
Clifton, BS8 4EJ
England
011 44 272 239 567
Fax: 011 44 272 237 598

(Clock distribution ICs [drivers, generators and synthesizers], ECL terminators, ASICs, ASSPs [application specific ICs], logic [gate] arrays, data and telecommunication products [HIPPI (High Performance Parallel Interface) ICs, SONET transmitters and receivers, optical fiber channel chip sets, odd/even parity generator/checker IC], crosspoint switches, clock drivers, PCI controller ICs, etc., Military products are available.)

Applied Reasoning Corp.
86A Sherman St.
Cambridge, MA 02140
617-492-0700

(Transistors)

Aptek Microsystems
See *Aptek Williams, Inc.*

Aptek Technologies
See *Aptek Williams, Inc.*

Aptek Williams, Inc.
Div. Williams Controls, Inc.
700 NW 12th Avenue
Deerfield Beach, FL 33442
954-421-8450
Fax: 954-421-8044
TLX: 441020

(Thick film hybrid circuits including telecommunications ICs such as DTMF, subscriber loop and receiver ICs, pulse and tone detectors; tilt sensors, custom smart cables, etc.)

APTIX — ASA

Aptix Corp.
2880 North 1st Street
San Jose, CA 95134
408-428-6200
408-944-0646
sales@aptix.com

(Field programmable interconnect components, combined with field programmable circuit boards and FPGAs, real time emulator [simulation] hardware)

Aptos Semiconductor Corporation
3060 Tasman Drive
Santa Clara, CA 95054
408-986-0188
Fax: 408-986-0331

(A fabless company supplying a combination design house and distributor, founded in June 1994 and concentrates on memory devices including SRAMs, PROMs, FIFOs and custom IC work)

AQZ prefix
See *Aromat Corp.*

(Photo-MOS relay)

AQW prefix
See *Aromat Corp.*

(Photo-MOS relay)

Arcobel Graphics B.V.
P.O. Box 3523
DMs Hertogenbosch 5203
The Netherlands
0031 73 444 144
Fax: 0031 73 444 150

Arizona Microtek, Inc.
225 E. First St.
Mesa, AZ 85201
602-962-5881
Fax: 602-890-2541

(ICs to translate from PECL to CMOS and CMOS to PECL)

ARK prefix
See *ARK Logic, Inc.*

Aristo-Craft
346 Bergen Avenue
Jersey City, NJ 07304-2204
201-332-8100
Fax: 201-332-0521

(Diodes, thyristors and transistors)

ARK Logic, Inc.
1737 North First Street, Suite 680
San Jose, CA 95112
408-467-1988

(Multimedia chip sets including graphics and video accelerators. Note: 51% of this company is owned by ICS, Integrated Circuit Systems.)

ARM prefix
See *Advanced RISC Machines*, *GEC Plessy*

(Microprocessors)

Aromat Corporation
629 Central Avenue
New Providence, NJ 07974
1-800-228-2350
1-800-AROMAT-9
908-464-3550
Fax: 908-464-8714

Marlboro, MA: 508-481-1995
New Providence, NJ:
908-464-3550
Orlando, FL: 407-855-1075
Richardson, TX: 214-235-0415
Elk Grove Village, IL:
708-593-8535
San Jose, CA: 408-433-0466
Garden Grove, CA: 714-895-7707
Mississuga, Ontario, Canada:
416-624-3777

(Relays, photoelectric sensors [NAIS (tm)])

Array Microsystems
1420 Quail Lake Loop
Colorado Springs, CO 80906
719-540-7900
719-540-7999
Fax: 719-540-7950
408-399-1505
Fax: 408-399-1506
http://www.array.com

(A fabless company supplying circuit boards, and video compression ICs including digital video chip sets, image compression coprocessor, motion-estimation coprocessor, and a digital array signal processor [DaSP] chip set. They also have board and application level products to perform FFT [fast fourier transforms], FIR filters and correlation in the frequency domain)

ART prefix
See *Lambda Advanced Analog*

(DC/DC converters)

ARX prefix
See *Aeroflex Laboratories*

AS prefix
See *Austin Semiconductor, Inc.*,
Austria Mikro Systems International GmbH,
Astec Semiconductor,
Alpha Semiconductor
(Voltage references), *Intel*

ASA prefix
See *Lambda Advanced Analog*

(DC/DC converters)

ASAHI CHEMICAL INDUSTRY — ASPEC TECHNOLOGY

Asahi Chemical Industry Co., Ltd.
1-2, Yurakucho 1-Chome,
Chiyoda-ku
Tokyo 100
Japan
81 3 3507 2730

(Converters, car ICs, speech synthesis and recognition ICs, gate arrays and digital signal processor ICs)

Asahi Glass Co. Ltd.
2-1-2, Marunouchi
Chiyoda-ku
Tokyo 100
Japan
81 3 3218 5555
Fax: 03 3201 5390

(Linear ICs, audio, television and VCR ICs, hybrid and logic ICs)

Asahi Kasei Microsystems Co. Ltd.
TS Building, 24-10 Yoyogi 1-chome
Shibuya-ku
Tokyo 151, Japan
03 3320 2062
Fax: 03 3320 2072

Asat Inc.
1010 Corporation Way
Palo Alto, CA 94303-4304
415-969-1141
Fax: 415-969-6580

(A manufacturer of IC packages, such as their trademarked EDQUAD)

ASC prefix
See *AdvanSys*

Ascent
(Dual rate 1773 protocol device)
See *United Technologies Microelectronics Center*

Ascom Microelectronics
Blankenweg 22
Bwarnhem 6827
The Netherlands
31 85 622 122
Fax: 31 85 622 303

CH Chapons des Pres 11
Bevaix 2022
Switzerland
41 38 460 111
Fax: 41 38 461 930

26, Rue du Jura
Ambilly 74100
France
33 5087 0113
Fax: 33 5038 8998

5 Rue Jacques Ruelf
Antony Cedex 92182
France
33 1 4674 2200
Fax: 33 1 4674 2287

Steinstrasse 13
Bad Klosterlainsnitz 07639
Germany
3 6601 32 10
Fax: 3 6601 32 1

Esserheimer Strasse 157
Mainz 55128
Germany
6131 93444 0
Fax: 61341 93444 30

Behringstrasse 2
Halle Westfalen 33790
Germany
5201 16 825
Fax: 5201 10 655

Leharstrasse 19
Bammental 69245
Germany
6223 40 010
Fax: 6223 48 388
Switzerland
(A foundry service?)

(Signal conditioning ICs, servo drivers, proximity detectors, motor drivers)

Asea Hafo, Inc.
See *ABB Hafo*

ASi logo
See *Alden Scientific*, Inc.

ASI
See *Advanced Semiconductors, Inc.*

ASIC Technical Solutions, Inc. (ATS)
See *Microchip Technology*

ASK LCD, Inc.
Unit of Tandburg (NJ and Norway)
1099 Wall Street
WestSuite 396
Lyndhurst, NJ 07071
201-896-8888

(LCD projectors)

ASMC (Asia Semiconductor Manufacturing Company, Ltd.)
B1, #44, Park Avenue, II
Science based Industrial Park
Hsinchu, Taiwan R.O.C.
035 790 358
Fax: 035 779 475

(A new contract fab opened in 1996.)

Aspec Technology, Inc.
850 E. Arques Avenue
Sunnyvale, CA 94086
408-774-2199
Fax: 408-522-9450

(A fabless company supplying HDA, High Density Array Sea of Gates ASICs, licensed to various companies including Samsung Electronics Co. Ltd. and Sanyo Electric Co. Ltd. and engineering design automation tools)

Aspen Semiconductor
4001 North First Street
San Jose, CA 95134
408-456-1800

(A subsidiary of Cypress Semiconductor, products marketed under the Cypress Semiconductor name).

ASPI
See *Atlanta Signal Processors, Inc.*

Associated Components Technology
11575 Trask Avenue
Garden Grove, CA 92643
714-636-2645
Fax: 714-636-8276

(Surface mount EMI suppressors and miniature power line chokes)

Astec America, Inc.
6339 Paseo de Lago
Carlsbad, CA 92009-9717
1-800-451-AMPS
619-757-1880 Fax: 619-930-0698
http://www.astec.com/

401 Jones Road
Oceanside, CA 92054
800-451-2677
Fax: 619-930-0881

High Street Wollaston
Stourbridge
West Midlands
DY8 4PG
United Kingdom
44 01384 440044
Fax: 44 01384 393139

(Power supplies, DC/DC converters, The 401 Jones Road address was formerly AC/DC Power Supplies before the company was acquired.)

Astec Semiconductor
255 Sinclair Frontage Road
Milpitas, CA 95035
408-263-8300
Fax: 408-263-8340

(PWM controllers, current-mode controllers, shunt references, voltage references, secondary control circuits, over-temperature detectors)

Asti Pacific Corp.
1318, Nishijima-Cho, Hamamatsushi
Shizuoka 430
Japan
0534 25 1311

(Hybrid, watch and clock ICs)

Astra Met, Inc.
222 Sherwood Avenue
Farmingdale, NY 11735-1718
516-694-9000
Fax: 516-694-9177

(Thermoelectric coolers)

Astralux Dynamics Ltd.
Red barn Road
Brighlingsea, Colchester, Essex, CO7 0SW
United Kingdom
44 1206 302571
Fax: 44 1206 303450

In the U.S.:
Power Components of Midwest, Inc.
P.O. Box 1348
56641 Twain Branch Dr.
Mishawaka, IN 46545
800-221-9257
Fax: 219-256-6643

(DIP relays)

Astrodyne
300 Myles Standish Blvd
Taunton, MA 02780
1-800-823-8082
508-823-8080
Fax: 508-823-8181

ASPEN SEMICONDUCTOR — AT-XXXX

(Switching power supplies)

AT prefix
See *Atmel, M/A-COM*
(If AT-XXX part number),
Atmos Technology, Inc.

AT-XXXXX part number
See Hewlett-Packard Company, Silicon bipolar transistors for RF and microwave applications.

ATF prefix
See *Atmel*

Atlanta Signal Processors, Inc. (ASPI)
1375 Peachtree Street, NE, Suite 690
Atlanta, GA 30309-3115
404-892-7265
Fax: 404-892-2512
BBS: 404-892-3200 (8,N,1)
Internet: info@aspi.com

(MPEG audio encoder module)

ATMizer
See *LSI Logic*

(Chip sets)

ATR prefix
See *Lambda Advanced Analog*

(DC/DC converters)

ATW prefix
See *Lambda Advanced Analog*

(DC/DC converters)

AT-X part number
See *Mini-Circuits*

AT-XXXX prefix
See *Hewlett-Packard*

ATC — ATMEL KOREA PTE.

ATC prefix
See *Shoreline Electronics, Inc.*

ATC Power Systems, Inc.
45 Depot Street
Merrimack, NH 03054
603-429-0391
Fax: 603-429-0795

(Military and industrial power supplies)

ATF prefix
See *Atmel*

ATI Technologies, Inc.
33 Commerce Valley Drive
Thornhill, Ontario L3T 7N6
Scarborough, Ontario, Canada
905-882-2600
Fax: 905-882-2620
http://www.atitech/ca/
ftp://atitech.ca/
http://www.ati.com/

(Video graphics ICs [including 3D graphics ICs], MPEG ICs, graphics accelerator and controller cards)

ATL prefix
See *Atmel*

(Gate arrays)

ATLV prefix
See *Atmel*

(Gate arrays)

Atmel Corp.
2325 Orchard Parkway San Jose, CA 95131
1-800-292-8635 (USA only)
800-365-3375 (Literature hotline)
408-441-0311 (Headquarters)
408-441-4270
(Northwest region domestic calls)

408-436-4333
(PLD Technical Support Hotline)
Fax: 408-487-2600,
408-436-4200,
4300 (Headquarters),
4314 (Northwest domestic calls)
Fax on Demand:
800-292-8635
408-441-0732
http://www.atmel.com
E-mail: literature@atmel.com

100 Pacifica. Suite 340
Irvine, CA 92718
714-727-3762
Fax: 714-727-3763

1150 E. Cheyenne Mt. Blvd.
Colorado Springs, CO 80906
719-540-1835
719-576-3300
Fax: 719-540-1759

17304 Preston Road
Suite 720, Lock Box 33
Dallas, TX 75252
214-733-3366
Fax: 214-733-3163

1721 Moon Lake Blvd.
Suite 103
Hoffman Estates, IL 60194
708-310-1200
Fax: 708-310-1650

300 Granite Street, Suite 106
Braintree, MA 02184
617-849-0220
Fax: 617-848-0012

101 Carnegie Center
Suite 219
Princeton, NJ 08540
609-520-0606
Fax: 609-520-9175

807 Spring Forest Road
Suite 700
Raleigh, NC 27609
919-850-9889
Fax: 919-850-9894

Atmel U.K. LTD.
Coliseum Business Centre
Riverside Way
Camberley, Surrey
England, GU15 3YL
44-276-686677
Fax: 44-276-686697

55 Avenue Diderot
94100 St. Maur Des Fosses
Paris, France
33 1 48855522
Fax: 33 1 48855595

Atmel GmbH
Ginnheimer Str. 45
6000 Frankfurt 90
Germany
49 69 7075910
Fax: 49 69 7075912

Atmel Asia LTD.
Room 1219
Chinachem Golden Plaza
77 Mody Road
Tsimshatsui East
Kowloon, Hong Kong
852-7219778
Fax: 852-7221369 or 7230651

Atmel Japan K.K.
NT Building, 1-9-12
Uchikanda, Chiyoda-Ku
Tokyo 1, Japan 101
81-3-5259-0211
Fax: 81-3-5259-0217

Atmel Singapore Pte. LTD.
6001 Beach Road
11-03 Golden Mile Road
Singapore 0719
65-2999212
Fax: 65-2992891, 65-2910955

Atmel Korea Pte. LTD.
Room 203, Sungrak Building
682-2, Daelim-Dong
Youngdeungpo-Gu
Seoul, Korea
82-2-8417130
Fax: 82-2-8417131

ATMEL KOREA PTE. — AUSTRIA MIKRO SYSTEME

(Memory and microcontroller chips, flash memory, programmable logic devices [FPGAs], EEPROMs, EPLDs, user configurable and full custom gate arrays, ASIC's, SRAMs, etc.)

Atmos Technology, Inc.
1060 Lincoln Avenue
San Jose, CA 95125
408-292-8066
Fax: 408-292-8241

(Single chip sensor interface circuits)

ATN prefix
See *KDI/Triangle Electronics, Inc.*

ATS
See *ASIC Technical Solutions*

ATT prefix
See *Lucent Technologies*
(Formerly AT&T Microelectronics)

ATTL prefix
See *Lucent Technologies*
(Formerly AT&T Microelectronics)

AT&T Frequency Control Products
See *Vectron Technologies, Inc.*

AT&T Microelectronics
See *Lucent Technologies*

AT&T Power Systems
3000 Skyline Drive
Mesquite, TX 75149
800-526-7819
214-284-2000
Fax: 214-284-8120

(DC/DC converters)

ATR prefix
See *Lambda Advanced Analog*

ATV prefix
See *Atmel Corp.*

(PLDs)

ATW prefix
See *Lambda Advanced Analog*

Audio Digital Imaging, Inc.
511 West Golf Road
Arlington Heights, IL 60005
708-439-1335
Fax: 708-439-1533

(Codecs, image processing, video processing and timing ICs)

Ault, Inc.
7300 N. Boone Avenue
Minneapolis, MN 55485
612-493-1900
Fax: 612-493-1911

(Power supplies including wall mounted units)

AuraVision Corporation
47865 Fremont Blvd.
Fremont, CA 94538
510-440-7180, 510-252-6800
Fax: 510-438-9350

(A fabless company supplying multimedia video record and playback processors for MPEG, JPEG, Intel Indeo, Microsoft Video-1, and Cinepak.)

Aureal Semiconductor
4245 Technology Drive
Fremont, CA 94538
510-252-4245
Fax: 510-252-4400 (Sales)

Media Vision
3185 Laurel View Ct.
Fremont, CA 94538
1-800-348-7116, 510-770-8600
Fax: 510-770-9592

(A fabless company supplying video capture and compression ICs and audio chip sets, prior to late 1995 the company was called Media Vision Technology, Inc. and it also sold multimedia kits and audio boards.)

Aurora Technologies Corporation
7408 Trade Street
San Diego, CA 92121-2410
619-549-4645
Fax: 619-549-7714

(CdZnTe detectors and detector arrays)

Austin Semiconductor, Inc.
8701 Cross Park Drive
Austin, TX 78754-4566
512-339-1188
Fax: 512-8395-8358,
512-339-6641

(Motor control IC's, memory devices [They market military screened memory devices that were sold by Micron Technology. Micron still makes the die but Austin handles all the military screening requirements.])

Austria Mikro Systeme International GmbH
Schloss Premstatten
A-8141 Unterpremstatten, Austria
43 3136 3666 0
Fax: 43 3136 2501/3650
TLX: 312547 ams a
http://www.ams.co.at/

Suite 340
20863 Stevens Creek blvd.
Cupertino, CA 95014
408-865-1217
Fax: 408-865-1219

Direct Mail Department
PO Box 4
Westbury-on-Trym
Bristol BS9 3DS
United Kingdom

(Originally a joint venture of AMI, American Microsystems, Inc. and Voest Alpine an Austrian State owned steel company. ASICs, including those for

AUSTRIA MIKRO SYSTEME — AVX

telecommunications applications [such as phone dialer and hands free monitor/amplifier circuit with tone ringer ICs] and ASSPs, specializing in analog technology)

Austron, Inc.
(See also Datum, Inc.)
P.O. Box 14766-4766
Austin, TX 78761-4766
512-251-2313
Fax: 512-251-9685

(Crystal oscillators including ovenized types)

Autek Power Systems
69 Moreland Rd.
Simi Valley, CA 93065
805-522-0888
Fax: 805-522-8777

(Switching power supplies)

AV prefix
See *Avasem Corporation*,
Lucent Technologies
(Formerly AT&T Microelectronics)

(Audio/video processor)

Avance Logic, Inc.
4670 Fremont Blvd.
Suite 105
Freemont, CA 94538
Taiwan
510-226-9555
Fax: 510-226-8039

(Graphics controllers and GUI [Windows] Accelerator, and audio IC chipsets)

Avantek
(Division of Hewlett Packard)
3175 Bowers Avenue
Santa Clara, CA 95054
(Microwave components)

1-800-227-1817
Canada: 1-800-387-3867
Europe: Fax: 31 2979-8 41 83

(RF and microwave components including GaAs products. Note: in 1995 the UTX and GPX lines of thin film RF amplifiers were sold to Penstock, Inc. [which is part of Avnet's Electronic Marketing Group])

Avasem Corporation
(Acquired by ICS, Integrated Circuit Systems, Inc., December 1992)
1271 Parkmoor Ave.
San Jose, CA 95126
408-297-1201
Fax: 408-925-9460

(Power switch ICs [used to conserve battery life in portable applications], IC system clocks for PCs)

AVC prefix
See *Anadigics, Inc.*

AVED Inc.
See *Added Value Electronics Distribution, Inc.*

Avens Signal Equipment Corp.
256 Lincoln Avenue
Brooklyn, NY 11208
800-452-1966
Fax: 718-235-9805

(Telecommunication active filters)

AVG Semiconductors
363 St. Paul Blvd.
Exec Suite 4004-A
Carol Stream, IL 60188
1-800-AVG-SEMI
708-668-3355
Fax: 1-708-510-7252

(This is the marketing agent/warehouse for semiconductors manufactured in the Republic of Belarus, 4000 CMOS, 7400, 74LS, 74ALS, 74HC/HCT, 74AC/ACT, 74BCT, DRAMs, SRAMs, 4 bit microprocessors, transistors, diodes, regulators, custom ASICs. Processing to MIL-STD-883 also available)

Avionic Instruments
P.O. Box 498
1414 Randolph Avenue
Avenel, NJ 07001
908-388-3500
Fax: 908-382-4996

(DC/DC converters [military units available])

AVP prefix
See *Lucent Technologies*
(Formerly AT&T Microelectronics)

AVX Corporation
A Kyocera Group Company
801 17th Avenue. S.
Myrtle Beach, S.C. 29577
P.O. Box 867
Myrtle Beach, SC 29578
803-448-9411, 803-946-0414, 803-946-0263
Fax: 803-448-1943
http://www.avx.com

Vancouver, WA:
206-696-284o
Fax: 206-695-5836

Olean, NY:
716-372-6611,
Fax: 716-372-6316

Raleigh, NC:
919-878-6200,
Fax: 919-878-6462

Biddleford, ME:
207-282-5111,
Fax: 207-283-1941

AVX Ltd, Fleet, Hants, England:
01252 770000
Fax: 01252 770001

Manufacturers, Prefixes, Part Number Types, Logo Descriptions & Family Types

AVX S.A., France:
1 6918 4600,
Fax: 1 6928 7387

AVX GmbH, Germany:
08131 9004 0,
Fax: 08131 9004 44

AVX s.r.l., Milano, Italy:
02 665 00116,
Fax: 02 614 2576

AVX, Kyocera Singapore Pte, Ltd.:
65 258 2833,
Fax: 65 258 8221

AVX Israel, Ltd.:
972 957 3873,
Fax: 972 957 3853

AVX Kyocera Corp.:
75 593 4518,
Fax: 75 501 4936

(Tantalum, ceramic MLC, microwave, glass, thin film capacitors; clock oscillators; resonators; quartz crystals; resistor chips, arrays and networks; thin film inductors, thin film fuses, transient voltage suppressors, EMI filters, piezo devices, EMI filters, SAW filters and resonators, dielectric filters, hybrid devices)

AWA MicroElectronics Pty. Ltd. (AWAM)
Australia
61 2 763 4105
Fax: 61 2 746 1501

(A wafer foundry and product design center, formerly a subsidiary of AWA Limited [Sidney, Australia] acquired by Quality Semiconductor, Inc. in early 1996 and is now Quality Semiconductor Australia [QSA]. The manufacture hearing aid and implantable medical electronics ICs and mixed signal and analog process ICs.)

Award Software, Inc.
777 East Middlefield Road
Mountain View, CA 94043
415-968-4433
Fax: 415-968-0274
BBS: 415-968-0249

(BIOS ROMs)

AWR prefix
See *Anadigics, Inc.*

AWT prefix
See *Anadigics, Inc.*

Ax logo
See *Astralux Dynamics. Ltd.*

Axis Communications
4 Constitution Way
Woburn, MA 01801
800-444-AXIS
617-938-1188
Fax: 617-938-6161

(LAN controller ICs, CD ROM server, printer server provider)

AXP prefix
See *Digital Equipment Corp.*

AXT prefix
See *Axis Technologies*

AY prefix
See *Microchip Technology*

Aydin Vector
P.O. Box 328
Newton, PA 18940-0328

47 Friends Lane
Newtown Industrial Commons
Newtown, PA 18940
215-968-4271
Fax: 215-968-3214/1015
TWX: 510-667-2320

Aydin International, U.K.
1-3 Hunting Gate
Hitchin, Herts SG4 OTJ
England
44 0462 43455
Fax: 44 0462 420727

(Signal conditioner hybrid circuit and components for data acquisition, signal conditioning and RF telemetry)

Aztech Systems, Ltd.
31 Ubi Road 1
Aztech Building
Singapore 408694
65 7417211

(Multimedia ICs)

B logo
See *Bourns, Inc.*

B prefix,
See *NEC*

(ECL SRAMs)

BA prefix
See *Bus Logic*,
National Semiconductor
(ECL gate array)
American Bright Optoelectronics Corporation
(LED BRT)

(Bar graph and array displays)

BA prefixed Japanese
(Generic linear devices such as operational amplifiers and comparators)

Devices are made by various manufacturers including Panasonic and Rohm.

BABCOCK DISPLAY PRODUCTS — BEAM

Babcock Display Products
14930 East Alondra Blvd.
LaMiranda, CA 90638-5752
714-994-6500
Fax: 714-994-3013

(Plasma displays, display products, relays, power supplies, magnetics)

BAL prefix
See *Zetex*

Balluff, Inc.
8125 Holton Dr.
Florence, KY 41042
800-543-8390
606-727-2200
Fax: 606-727-4823

(Optical sensors on flexible cable)

BAR prefix
See *Siemens*

Barry
See *Barry Industries, Inc.*

Barry Industries, Inc.
67 Mechanic Street
Attleboro, MA 02703
508-226-3350
Fax: 508-226-3317

(Terminators, resistors and attenuators for high power, RF and microwave applications)

BAS prefix
See *Philips Semiconductors*, *AVG Semiconductors*
(Second source some devices)

BAT prefix
See *Zetex*

Baumer Electric, Ltd.
122 Spring St., C-6
Southington, CT 06489
800-937-9336
203-621-2121
Fax: 203-628-6280

(Inductive, capacitive proximity switches, photoelectrics, encoders, mechanical switches, and ultrasonic transducers)

BAY prefix
See *AVG Semiconductors*
(Second source some devices)

BAV prefix
See *Zetex, AVG Semiconductors*
(Second source some devices)

Base2 Systems, Inc.
See *Brooktree Corp.*

BB prefix
See *AVG Semiconductors*
(Second source some variable capacitance diodes),

American Bright Optoelectronics Corporation (LED BRT)

(Blinking LEDs)

BBY prefix
See *Siemens, Zetex*

BC prefix
(Bipolar transistors or ECL gate array)

See *Motorola*,
National Semiconductor
(Gate array),
Zetex (Transistors),
AVG Semiconductor
(Second source transistors),
American Bright Optoelectronics Corporation (LED BRT)

(Clock displays)

BCM prefix
See *Datel, Inc.*,
Broadcom Corporation

BCT
(BiCMOS [bipolar CMOS] Technology family)

See *Texas Instruments, National Semiconductor, AVG Semiconductors*

BCV prefix
See *Zetex*

BCW prefix
See *Motorola, Zetex*

BCX prefix
See *Motorola, Zetex*

BCY prefix
See *Motorola, Zetex*

BD prefix
See *Zetex, Harris Semiconductor*
(PNP transistors)
AVG Semiconductor
(Second source)

BDB prefix,
See *Motorola*

BDC prefix
See *Motorola*

BDS prefix
See *Butterfly DSP, Inc.*

Beam Electronics Industrial Co., Ltd.
3-11-5, Nagisa
Matsumoto-Shi, Nagano 390
Japan
0263 25 7150
Fax: 0263 25 7152

(Hybrid ICs)

BECKER & PARTNER — BETA THERM

Becker & Partner GmbH
Neuenhofstrasse 110
Aachen 52078
Germany
241 928 24 40
Fax: 241 928 24 99

(SRAM modules)

Beckman Industrial Corp.
See *BI Technologies Corporation*

Bedford Opto Technology, Ltd.
The Stables
Lindsayyards, Biggar
Lanarkshire, ML12 6NR
United Kingdom
44 899 212 21
Fax: 44 889 210 09

(Optoelectronics)

Beckman Industrial Ltd.
Queensway, Glenothes
Fila, Scotland KY7 5PU
592 753811
Fax: 44 592 756449

Beckman Industrial Japan, Ltd.
Kakumaru Bldg.
1-10, Toyo 7-Chome
Koto-Ku, Tokyo, Japan
03 3615-1811
Fax: 03 3847-2443

(Hybrid microcircuit D/A converters, trimmers, potentiometers, active and passive networks, and magnetics)

BEI Sensors and Motion Systems Company
Computer Products Division
1755-B La Costa Meadows Drive
San Marcos, CA 92069
619-471-2600
Fax: 619-471-2675

(Optical encoder modules)

Bel
See *Bel Fuse, Inc.*

(Delay lines)

BEL
See *Bharat Electronics, Ltd.*

Bel Fuse, Inc.
198 Van Vorst Street
Jersey City, NJ 07302
201-432-0463
Fax: 201-432-9542

Bel Fuse House
8/F Luk Hop Ind. Bldg.
8 Luk Hop Street
San Po Kong
Kowloon, Hong Kong
852-328-5515
Fax: 852-352-3706

Parc Club Orsay
Universite
4, Rue Jacques Monod
91893 Orsay, France
33 1 69 41 0402
Fax: 33 1 69 41 3320

(Delay lines)

Benchmarq Microelectronics, Inc.
17919 Waterview Parkway
Dallas, TX 75252
800-966-0011
214-437-9195
Fax: 214-437-9198

http://www.benchmarq.com/
http://synapse-group.com/Benchmarq/
http://synapse.onramp.net/benchmarq

Eastern Area:
One Roe Lane
Port Jefferson, NY 11777-1426
516-331-3999
Fax: 516-331-8506

Western Area:
31 Sandpiper
Irvine, CA 92714
714-551-8402
Fax: 714-551-8473

Asia:
Level 36, Hong Leong Building
16 Raffles Quay
Singapore 0104
65 322 8521
Fax: 65 322 8558

(Microprocessor peripheral components, real time clocks [RTCs], NVSRAM modules, SRAM nonvolatile controllers, SRAM modules, battery [energy management] ICs, gas gauge ICs, etc.)

Bertan High Voltage Corporation
121 New South Road
Hicksville, NY 11801
1-800-966-2776
516-433-3110
Fax: 516-935-1799
http://www.li.net/~bertan/
E-mail: info@bertan.com

(High voltage power supplies)

Beta Transformer Technology
40 Orville Drive
Bohemia, NY 11716
516-244-7393, Fax: 516-567-7358
Subsidiary of ILC Data Device Corporation

(Various magnetics including pulse transformers)

BetaTherm Corporation
910 Turnpike Road
Shrewsbury, MA 01545
508-842-0516
Fax: 508-842-0748

(Thermistors)

BETHELTRONIX — BL

BethelTronix, Inc. (BTI)
13825 Cerritos Corporate Drive
Cerritos, CA 90703
310-407-0500
Fax: 310-407-0510
E-mail: btinet@cerfnet.com

(Custom ASICs for wireless/CATV/ satellite optical communications, telecommunications, multimedia and data acquisition/process control. Devices include CDMA ICs, Cellular RF transceivers [700MHz-1GHz], spread spectrum/frequency hopping chip set for cordless phones, GSM RF transceiver, PCS RF transmitter and receiver, PCS VCO/prescaler, CATV descramblers, DBS digital RF receiver/tuner, ATM/SONET/SDH network ICs, burst mode transceivers, Delta-Sigma and Flash A/D converters, etc.)

BF prefix
(ECL gate array and transistors)

See *Motorola*,
National Semiconductor
(Gate array),
Zetex (Transistors),
AVG Semiconductor
(Second source transistors)

BFG prefix, RF transistors
See *Philips Components*

BFP prefix
See *Siemens*

(RF transistors)

BFR prefix
See *Motorola*

BFQ prefix
See *Zetex*

BFR prefix
See *Siemens*

BFS prefix
See *Siemens*, *Zetex*

BFT prefix
See *Siemens*

BFW prefix
See *Motorola*

BFX prefix
See *Zetex*

BFY prefix
See *Zetex*

BG prefix
(ECL gate array)
See *National Semiconductor*

BH prefix
See *Rohm Corporation*

(SCSI terminators)

BI prefix
See *BI Technologies*

BI Technologies Corporation
(Formerly Beckman Industrial Corporation)
Affiliate of Emerson Electric
Electronic Technologies Division
4200 Bonita Place
Fullerton, CA 92635
714-447-2345
Fax: 714-447-2500

BICM
See *Micronetics Wireless*

Big-Sun Electronics Co. Ltd.
F.5, -1, No. 5, Lane 18
Pin Ho Road, Chung Ho City
Taipei Hsien
Taiwan
886 2 233 3090, 5518
Fax: 886 2 223 9215

(Optoelectronics)

Bipolar Integrated Technology (BIT)
1050 NW Compton Dr.
Beaverton, OR 97006
503-629-5490
Fax: 503-690-1498
503-629-6119

(Asynchronous transfer mode ICs)

BIR prefix
See *American Bright Optoelectronics Corporation*
(LED BRT)

(Infrared light-emitting diodes)

BJ prefix,
(ECL gate array with RAM)
See *National Semiconductor*

BKC International Electronics
BKC Semiconductors,
BKC Photo Dector Division
6 Lake Street, P.O. Box 1436
Lawrence, MA 01841
508-681-0392
(Manufacturing facility)

310-431-4220
(Sales/marketing offices)

(Semiconductors, including military qualified devices)

BKL, Inc.
421 Feheley Drive
King of Prussia, PA 19406
800-220-1038, 610-277-2910
Fax: 610-277-2956

(Electroluminescent lamps [for LCD backlighting])

BL prefix
See *Lucent Technologies* (Formerly AT&T Microelectronics),
American Bright Optoelectronics Corporation (LED BRT)

(LEDs)

BLILEY ELECTRIC — BOURNS

Bliley Electric Company
2545 West Grandview Blvd.
P.O. Box 3428
Erie, PA 16508
814-838-3571
Fax: 814-833-2712

(Crystal oscillators, OXCOs)

BIP prefix
See *Robert Bosch Corporation*

BLC prefix
See *National Semiconductor*

Blue Lightning
See *IBM Microelectronics*

(Microprocessors)

Blue Sky Research
4030 Moorpark Avenue
Suite 123
San Jose, CA 95117-1707
408-983-0471
Fax: 408-985-0383
BlueSky@ix.netcom.com

(Pointsource laser diodes)

BM prefix
See *American Bright Optoelectronics Corporation* (LED BRT)

(Dot Matrix displays)

BMR Labs, Inc.
1270 Oakmead Parkway
Suite 215
Sunnyvale, CA 94086
408-733-6655
Fax: 408-733-6685

(A fabless company supplying ASICs and custom gate arrays, standard cell and mixed signal ICs in CMOS, bipolar and BiCMOS technologies, foundry and test services)

BNR
35 Davis Drive
Research Triangle Park, NC 27709
919-991-7109

(Laser diodes)

Bomar Crystal Company
201 Blackford Avenue
Middlesex, NJ 08846
800-526-3935, 908-356-7787
Fax: 800-777-2197, 980-356-7362

(Crystals and oscillators)

Bourns, Inc.
1200 Columbia Avenue
Riverside, CA 92507-2114
909-781-5050
Fax: 909-781-5700
Fax Library 818-837-4341
http://www.bourns.com

Integrated Technologies Division
Networks Products
1400 North 1000 West
Logan, UT 84321
801-750-7200
Fax: 801-750-7253

Bourns Sensors/Controls Division
2533 N. 1500 West
Ogden, UT 84404
801-786-6200
Fax: 801-786-6228

(Optical encoders)

Europe:
353 021 357001
Bourns AG, Zugerstrasse 74
6430 Baar, Switzerland
042 33 33 33,
Fax: 042 33 05 10
TLX: 868 722

Benelux:
070 387 44 00
Fax: 070 387 62 30
TLX: 32 023

Germany:
0711 22 93 0
Fax: 0711 29 15 68

France:
01 40 03 36 04
Fax: 01 40 03 36 14
TLX: 230 381

Ireland:
021 35 70 01
Fax: 021' 35 74 43
TLX: 75 904

United Kingdom:
0276 69 23 92
Fax: 0276 69 10 37
TLX: 859 735

Bourns Asia Pacific PTE. Ltd.
400 Orchard Road, #06-20
Orchard Towers, Singapore 0923
65 738 0290
Fax: 65 733 7960

Asia:
81 3 3980 3313

Hong Kong:
852 7360308
Fax: 852 3170836

Korea:
82 2 556 3619
Fax: 82 2 556 9016

Japan:
03 3 221 9726
Fax: 03 3239 2817

(Trimmer resistors, resistor networks, switches, encoders, panel controls, dials, precision potentiometers, transducers [position and pressure types], pressure sensors [including sapphire sensors])

BP — BTL

BP suffix
See *SenSym, Inc.*, *Rohm Corp.*

(DC/DC converter SIP)

BPD prefix
See *American Bright Optoelectronics Corporation* (LED BRT)

(Photodiodes)

BPR prefix
See *American Bright Optoelectronics Corporation* (LED BRT)

(Photoreflectors)

BPT prefix
See *American Bright Optoelectronics Corporation* (LED BRT)

(Phototransistors)

BPWXX type part number
See *AEG-Telefunken*

bq suffix
See *Benchmarq Microelectronics*

BR prefix
See *Rohm* (memory ICs), *Gentron, Corp.* (solid state relays), *American Bright Optoelectronics Corporation* (LED BRT)

(Resistor LED)

Bright LED Electronics Corporation
2/F., No. 19, Hoping Road
Panchiao
Taipei Hsien
Taiwan
886 2 959 1090 4
Fax: 886 2 954 7006
TLX: 34379 BRTLED

(LEDs)

Brooktree Corp.
9950 Barnes Canyon Rd.
San Diego, CA 92121
800-843-3642
(Technical Hotline)

800-843-3642
(Representative/distributor info)

619-452-7580, 843-3642,
535-3464
619-535-3516
(Literature requests)

619-535-3341
(ATE applications)

Fax: 619-452-1249,
452-7294
TLX: 383596
http://www.brooktree.com

San Diego: 619-535-3622
San Jose: 408-732-0923,
408-980-0812
North Carolina: 919-467-7418
Dallas: 214-490-1945
Montreal: 514-685-8822
England: 44 844 261989
Germany: 49 89 5164 10
Hong Kong: 852 813 0699
Japan: 81 3 588 0414

(A fabless company supplying RAMDACs, video RAMs, timing products such as delay lines and timing edge verniers, video DACs, flash A/Ds, comparators, drivers and loads, video decoders, mixed signal ICs, T1/E1/ISDN, ATM and HDSL transmission ICs. Note: in 1996 this company became a wholly owned subsidiary of Rockwell Semiconductor Systems.)

BS prefix
See *Motorola*, *Zetex*

BSX prefix
Where X is S (single), D (dual), T (three), Q (four), V (five), X (six), F (overflow single) or G (overflow dual) = the number of digits in the display

BSA prefix
See *Robert Bosch Corporation*

BSS prefix
See *Motorola*, *Zetex*

BSV prefix
See *Motorola*, *Zetex*

BSX prefix
See *Motorola*, *Zetex*

Bt prefix
See *Brooktree Corporation*,

For triacs, see *Philips Components*, Discrete Products Division

BT prefix
See *Hunter Components* (LCD display), *BethelTronix, Inc.*

BT&D Technologies
Delaware Corporate Center II
2 Righter Parkway, Suite 200
Wilmington, DE 19803
302-479-0300

(A joint venture founded in 1986 between British Telecom [now known as BT] and E.I. DuPont de Nemours & Company. Products include optoelectronic components, PIN detectors, integrated modules, passive splitters, switches and optical amplifiers.)

BTI
See *BethelTronix*

BTL
Backplane Transceiver Logic for Futurebus +

BTS prefix
See *Siemens Corp.*

BTTC logo
See *Beta Transformer Technology Corporation*

BTWXXXX part number
See *Mullard*

BU prefix
See *Rohm, AVG Semiconductor*
(Second source some transistors)

BUF prefix
See *Analog Devices*

Burr-Brown Corp.
Box 11400
Tucson, AZ 85734
1-800-548-6132
602-746-1111
(520 area code after 12/31/96)
Fax: 602-889-1510,
602-741-3895
(520 area code after 12/31/96)
TLX: 066-6491 BURRBROWN ATU
http://www.bur-brown.com/

3450 So. Broadmont Dr.
Suite 128
Tucson, AZ 85713
1-800-548-6132
Fax: 602-628-1602
(520 area code after 12/31/96)

984 N. Broadway
Suite 402
Yonkers, NY 10701
914-964-5252
Fax: 914-964-0655

Agoura Hills, CA: 818-991-8544
Burlington, MA: 617-229-0300
Dallas, TX: 214-783-4555

France:
Le Chesnay: 33 1 39 54 35 58
Lyon: 33 78 34 88 33
Toulouse: 33 61 40 03 06

Germany:
Filderstadt: 49 711 77040
Bremen: 49 421 25 39 31
Dusseldorf: 49 21 54 85 83
Erlangen: 49 91 31 2 40 36
Frankfurt: 49 61 54 8 20 81
Munchen: 49 89 61 77 37

Italy:
Milano: 02 5801 05 04

Japan:
Atsugi: 81 462 48 4695
Osaka: 81 6 305 3287
Nagoya: 81 52 775 6761
Tokyo: Akasaka Seventh Avenue Building
10-20, Akasaka 7-Chome
Minato-Ku, Tokyo 107
81 33 586 8141
Fax: 81 33 3582 2940

Netherlands:
Maarssen: 31 3465 50204

Switzerland:
Rueschlikon: (ZH) 41 1 724 09 28

United Kingdom:
Watford: 44 1923 233837

(Operational and instrumentation amplifiers, filters, references, optical ICs, power buffers, oscillators, D/A, A/D, Voltage/Frequency converters, sample and hold amplifiers, analog multiplexers, DSP [Digital Signal Processing] products, DC/DC converters, hybrid ICs)

BUS-XXXX (or XXXX)
part numbers
See *ILC-DDC*

BusLogic, Inc.
4151 Burton Drive
Santa Clara, CA 95054
408-492-9090
Fax: 408-492-1542
http://www.buslogic.com/

(ASICs used in their bus interface circuit cards, SCSI I/O products used in network file servers. This company was acquired by Mylex Corporation [Fremong, CA] in January 1996. It is now a wholly owned subsidiary of Mylex.)

Bussman
Cooper Industries
P.O. Box 14460
St. Louis, MO 63178-4460
114 Old State Road
Ellisville, MO 63021-5942
1-800-322-1577
314-394-2877
314-527-3877
(Customer service)

314-527-1270
(Application engineering)

314-527-1450
(Information Fax Line)

Fax: 800-544-2570,
314-527-1445
(International Fax)

TLX: 44-841
http://www.bussman.com

Ismaninger Str. 21
8000 Munchen 80
Germany
89 41 30 06 38
Fax: 89 47 07 228
TLX: 52 13 250

Literbuen 5
DK-2740 Skovlunde, Copenhagen
Denmark
45 42 919900
Fax: 45 42 911151

B - C BUSSMAN — CADDOCK

The Plaza
7500 A Beach Road
No. 14-319/320
Singapore 0719
Republic of Singapore
65 296 8311
Fax: 65 296 3807

Bestwick Works
Frome, Somerset BA 11 1PP
United Kingdom
44(0) 373 64311
Fax: 44(0) 373 73175

(Fuses, fuseholders [in-line, panel mount], fuse clips)

Butterfly DSP, Inc.
2401 SE 161st Ct, Suite A
Vancouver, WA 98684
360-892-5597
Fax: 360-892-0402
email: mflemin@pacifier.com
(to Mike Fleming VP Eng)

(FFT chip sets, high performance DSP ICs and multichip modules, DSP application and development software, DSP development boards. This company is a spinoff of the Sharp Microelectronics digital design center in Camas, WA)

BUX prefix
See *Zetex*

BUY prefix
See *Zetex*

BUZ prefix
See *Harris*

BWR prefix
See *Datel, Inc.*

BYl-XX prefix
See *GHz Technology, Inc.*

BZX prefix
See *Zetex*, *AVG Semiconductors*
(Second source some devices)

C prefix
See *Crystal Semiconductor*,
Space Power Electronics, Inc.

C logo
See *Caddock Electronics, Inc.*

C4 part number
TC4 series of parts by Toshiba

(Single gate logic devices electrically equivalent to 4000B [the Toshiba TC4SXX or C4], and 74HC logic devices [the Toshiba TC7SXX or C7])

C.H. Ting
1306 South B. St.
San Mateo, CA 94402
415-638-5669

(Microprocessors)

C&K Components, Inc.
57 Stanley Avenue
Watertown, MA 02172-4802
800-635-5936
Fax: 617-926-6846

(Trimming potentiometers, switches)

C&S Electronics
2 Christy Street
Norwalk, CT 06850

P.O. Box 2142
Norwalk, CT 06852-2142
203-866-3208
Fax: 203-854-5036

(Microminiature automatic audio level controller)

C-Cube Microsystems
1778 McCarthy Blvd
Milpitas, CA 95035
408-944-6300

Fax: 408-944-6314
http://www.c-cube.com/

(A fabless company supplying video compression IC's including MPEG devices)

CA prefix
See *Harris Semiconductor* (Formerly RCA),
National Semiconductor (Custom part), *Motorola*

Cable and Computer Technology, Inc.
155 S. Sinclair Street
Anaheim, CA
714-937-1341
Fax: 714-937-1225

(Futurebus+ Interface ASICs in commercial and military versions)

Cable Labs, Inc.
400 Centennial Parkway
Louisville, CO 80027
303-661-9100
Fax: 303-661-9199

(Demodulator ICs that can handle VSB and QAB, digital signal modulating techniques)

Caddock Electronics, Inc.
17271 North Umpqua Highway
Roseburg, OR 97470
503-496-0700
Fax: 503-496-0408

Sales Office for the U.S.A. and Canada
1717 Chicago Avenue
Riverside, CA 92507
909-788-1700
Fax: 909-369-1151

(Power film resistors, including devices in TO-220 and TO-126 packages)

Cal Crystal Lab., Inc.
Comclok, Inc.
1142 N. Gilbert
Anaheim, CA 92801
1-800-333-9825
714-991-1580
Fax: 714-491-9825

(Crystals [Cal Crystal Lab., Inc.] and clock oscillators [Comclok, Inc.])

Cal-Sensors
P.O. Box 7219
Santa Rosa, CA 95407
707-545-4181
Fax: 707-545-5113

(Lead salt [lead-sulfide and lead selenide] detectors, silicon photodiodes)

Calex Manufacturing Co. Inc.
2401 Stanwell Dr.
Concord, CA 94520
800-542-3355
510-687-4411
Fax: 510-687-3333
(DC/DC Converters)

California Eastern Laboratories
See *NEC*

California Micro Devices (CMD)
Semiconductor Division
2000 W 14th St.
Tempe AZ 85281
602-921-6000
Fax: 602-921-6298
TLX: TRT 187202

CMD Technology, Inc.
215 Topaz Street
Milpitas, CA 95305
800-325-4966
408-263-3214
Fax: 408-263-7846

(Chip resistors and capacitors, resistor arrays, controller ICs for PCI to IDE and VL to IDE, and custom monolithic resistor and capacitor networks, UARTS)

CALMOS
See *Newbridge Microsystems*

Calogic Corporation
237 Whitney Place
Freemont, CA 94539
510-656-2900
Fax: 510-651-1076

(DMOS/CMOS switches, switch arrays and multiplexers, JFETs, MOSFETs, operational amplifiers, custom devices, etc. This company has equivalents to National Semiconductor parts. They manufacture some of the parts discontinued by Topaz, Intersil and Siliconix [small signal discretes including n and p JFETs, single and duals; n and p channel MOSFETs, singles and duals, enhancement and depletion mode; Monolithic Dielectrically Isolated dual transistors; and monolithic Junction Isolated dual transistors])

Cambridge Accusense, Inc.
1000 Mt. Laurel Circle
Shirley, MA 01464
508-425-2090
Fax: 508-425-4062

(Solid state airflow and temperature switches)

Canadian Instrumentation and Research Ltd.
1155 Appleby Line Unit E8
Burlington, Ontario, Canada L7L 5H9
905-332-1353
Fax: 905-332-1808
Email: cir@netaccess.on.ca

(Fiber optic components including sensors, couplers, spliceless fiber assemblies, inline polarizers, spliceless fiber inferometers, vibration and displacement sensor, servo and piezo modulator, etc.)

Canon, Inc. USA
1 Canon Plaza
Lake Success, NY 11042
516-328-4611
Fax: 516-328-4609

(Various consumer product ICs for their products)

Capacitec, Inc.
P.O. Box 819
87 Fitchburg Road
Ayer, MA 01432
508-772-6033
Fax: 508-772-6036

Capacitec Sarl
16, Rue Sejourne
94044 CRETEIL
France
33 1 43 39 48 68
Fax: 49 80 07 49

(Capacitive sensors for displacements, gaps and thickness measurements)

Capacitor in series with a resistor logic symbol logo
See *Connor-Winfield Corporation*

(Crystal oscillators)

Capital Equipment Corporation
76 Blanchard Road
Burlington, MA 01803
800-234-4232
617-273-1818
Fax: 617-273-9057

(Microprocessor support ICs)

CAPTEUR SENSORS & ANALYZERS — CC

Capteur Sensors & Analyzers, Ltd.
66 Milton Park
Abington, Oxon
OX14 4RY
United Kingdom
44 0 1235 821323
Fax: 44 0 1235 820632

(Gas sensors)

Cardinal Components
Wayne Interchange Plaza II
155 Rte 46 West
Wayne, NJ 07470
201-785-1333
Fax: 201-785-0053
Fax-back: 1-800-881-9331
BBS: 201-812-7082
http://cardinalxtal.com/home/cardinal
E-mail: cardinal@cardinalxtal.com

(Crystals, and crystal oscillators)

Cardio (tm)
See *S-MOS Systems*

Cardon Corporation
222 Berkeley Street
Boston, MA 02116
617-267-1760
Fax: 617-267-4621

(Linear ICs)

Cariger, Inc.
6 Londonderry Commons
Manchester, NH 03053
603-645-4531
Fax: 603-452-4372

(LED bar graph arrays, and programmable bar graph displays, including tri-color units)

Carroll Touch, Inc
P.O. Box 1309 (78680)
811 Paloma Drive
Round Rock, TX 78664
512-388-5614

(Product information center Eastern U.S.)
512-388-5547

(Product information center Western U.S./Canada)
Fax: 512-244-3500, 512-244-7040
BBS: 512-388-5668 (300, 1200, 2400 Baud N81)

(Touch input [controller] ASIC ICs, touch systems)

CAS prefix
See *Celeritek*

CAT prefix
See *Catalyst Semiconductor, Inc.*

Catalyst Semiconductor Inc.
Corporate Headquarters/Western U.S. Sales/Far East Sales:
2231 Calle De Luna
Santa Clara, CA 95054
408-748-7700
Fax: 408-980-0313,
408-980-8209
Telex: 5106017631

U.S. Eastern Sales Office:
455 Douglas Avenue
Suite 2155-1

Altamonte Springs, FL 32714
407-682-1995
Fax: 407-682-6643

Central U.S. Sales Office:
3800 No. Wilke Road
Suite 372
Arlington heights, IL 60004
708-342-0274
Fax: 708-342-0276

Northern/Southern Europe:
Grote Winkellaan 95,
Bus 1
1853 Strombeek
Belgium
32 2 267 7025
Fax: 32 2 267 9731

Central Europe:
Blumenau 166
22089 hamburg
Germany
49 40 209 9955
Fax: 49 40 209 9956

Japan:
Nippon Catalyst K.K.
4th Fl., Shin Nakano, FK Bldg
6-16-12 Honcho
Nakano-ku, Tokyo 164
Japan
81 3 5340 3781
Fax: 81 3 5340 3780

Catalyst Semiconductor U.K.
1 Peckover CT
Great Holm, Milton Keynes
Bucks MK8 9HA
United Kingdom
44 0908 260874
Fax: 44 0908 561233

(A fabless company supplying E2PROMS, flash memories, SRAMs, NVRAMs, microcomputers)

CB prefix
See *National Semiconductor*
(Custom part)

CBT, Crossbar Transceiver Parts
See *TI* and *Quality Semiconductor*

CC prefix
See *Philips Semiconductor*
(Custom part),
National Semiconductor
(Custom part),
Cardinal Components
(Ceramic oscillators)

Manufacturers, Prefixes, Part Number Types, Logo Descriptions & Family Types

CCD — CFC

CCD
See *Cologne Chip Designs*

CCS prefix
See *Celeritek*

CAV prefix
See *Celeritek*

CD prefix
Formerly RCA now Harris Semiconductor
(Also see *National Semiconductor*), *Pioneer New Media Technologies, Inc.*, *Clarkspur Design* (DSP ICs), *Cirrus Logic*
(See also LG Semicon [Formerly GoldStar Technology, Inc.] for 4000B CMOS ICs),
Philips Semiconductors (Custom part)

CDC prefix
See *Texas Instruments*

cdi
See *Conversion Devices, Inc.*

CDP prefix
Formerly RCA, now *Harris Semiconductor*

CDIL
See *Continental Device, India Ltd.*

CE prefix
See *Philips Semiconductor* (Custom part),
National Semiconductor (Custom part),
Fujitsu Microelectronics (Gate array series)

CEC logo
See *Conversion Equipment Corporation*

(Power supplies)

CEI
See *Calvert Electronics International, Inc.*

Celeritek
3236 Scott Blvd
Santa Clara, CA 95051
408-986-5060
Fax: 408-986-5095

(Semiconductors for wireless communications including GaAs MMIC amplifiers, amplifier-switches, GaAs FETs)

CEM prefix
See *On Chip Systems*

Cen logo
See *Central Semiconductor Corp.*

Central Semiconductor Corp.
145 Adams Ave.
Hauppauge, N.Y. 11788
516-435-1110
Fax: 516-435-1824, 3388

(Bridge rectifiers, silicon transistors, power transistors, diodes, This company manufactures surface mount Schottky rectifiers that can be used to replace the MBRS120 series that is no longer available from Motorola Semiconductor.)

Centon Electronics, Inc.
20 Morgan
Irvine, CA 92718
1-800-234-9292, 1-800-836-1986
714-855-9111
1-800-945-8908 (For resellers)
Fax: 714-855-6035
http://www.centon.com

(Memory modules)

Centronic, Inc.
2088 Anchor Court
Anchor Business Park
Newbury Park, CA 91320
805-499-5902
Fax: 805-499-7770

Centronic Ltd.
267, King Henrys Drive
New Addington, Crydon CR9 9BG
United Kingdom
44 689 842 121
Fax: 44 689 845 117

(Silicon photo and fiber optic detectors)

Century Microelectronics, Inc.
4800 Great America Parkway,
Suite 308
Santa Clara, CA 95054
1-888-888-SIMM, ext. 888
1-888-888-DRAM, ext. 888
1-888-4MEMORY, ext. 888
408-748-7788
http://www.century-micro.com

(Memory modules)

Cermetek Microelectronics, Inc.
406 Tasman Drive
Sunnyvale, CA 94089
408-752-5000
Fax: 408-752-5004

(Modem modules, FCC registered DAA [Data Access Arrangement] ICs, etc.)

CF prefix
See *Siemens*, *Robert Bosch Corporation*

CFA prefix
See *Celeritek*

CFB prefix
See *Celeritek*

CFC prefix
See *Celeritek*

CFK — CHIPS

CFK prefix
See *Celeritek*

CG prefix
See *Fujitsu Microelectronics* (Gate array series), *Robert Bosch Corporation*

CGI
See *Component General, Inc.*

CGS prefix
See *National Semiconductor*

(Clock generation and support ICs)

CH prefix
See *Cermetek Microelectronics, Inc., Chrontel, Inc.*

Champion Technologies, Inc.
2553 N. Edgington Street
Franklin Park, IL 60131
708-451-1000
Fax: 708-451-7585
EasyLink: 62931824
TLX: 499-0104 CTI UD

(Crystals and clock oscillators)

Chequers Electronic (China) Ltd.
See *United Chequers* (Hong Kong) *Ltd.*

Cherry Semiconductor Corp.
2000 South County Trail
East Greenwich, RI 02818
1-800-272-3601

(Technical resource center)
401-885-3600
Fax: 401-885-5786
TLX: 6817157
Internet: info@cherry-semi.com

Electrical Products
3600 Sunset Avenue
Waukegan, IL 60087
708-662-9200
Fax: 708-662-2990

(Voltage regulators and semiconductors for automotive applications, pulse width modulators, Hall Effect sensors [including digital vane switches, from the acquisition of Concord Sensors, Inc. in 1995])

Chia Hsin Livestock Company
7F-1, No. 9
Prosperity Road 1
Science Based Industrial park
Hsinchu, Taiwan R.O.C.
886 35 779 987
Fax: 886 35 780 767

(A new fab to manufacture chip sets, memory and ASICs. It opened in 1996).

Chicago Miniature Lamp, Inc.
Chevy Chase Business Park
1080 Johnson Drive
Buffalo Grove. IL 60089
708-459-3400
Fax: 708-459-2708

(Lamps and leds [the LEDs are from Industrial Devices, Inc.])

China Semiconductor Corporation
No. 907 Chung Cheng Rd.
Chung Ho. Taipei Hsien
Taiwan
886 2 223 9696
Fax: 886 2 223 9377

(Optoelectronics)

Chinfa/Hotaihsing Powertech
Chinfa Electronics
Taiwan
02 362 8486
Fax: 02 362 0072

Chinfa/Tumbler Technologies, USA
1340 Fulton Pl.
Fremont. CA 94539
510-651-7724
Fax: 510-651-7840

(DC/DC converters, DC/AC inverters, AC/DC power modules, automobile power adaptors)

Chip Express Corporation
2903 Bunker Hill Ln, #105
Santa Clara, CA 95054-9950
1-800-95CHIPX
408-988-2445
Fax: 408-988-2449
email: moreinfo@Chipx.com
http://www.elron.net/chipx/

Sales Address:
2323 Owen Street
Santa Clara. CA 95054
408-235-7300
Fax: 408-988-0513

(ASICs, laser programmable gate arrays for 24 hour prototypes, formerly part of Elron of Israel. Parts processed to MIL-STD-883 are available.)

Chip Supply, Inc.
7725 N. Orange Blossom Trail
Orlando, FL 32810-2696
407-298-7100
Fax: 407-290-0164

(A supplier of tested "known good die" from Micron Semiconductor, Inc., Allegro, AMD, Analog Devices, Atmel, Brooktree, Harris Semiconductor, IDT, Linear Technology, Micro Power Systems, National Semiconductor, Samsung Semiconductor, Temic, Teledyne, Thomson Components, Texas Instruments and Xilinx. They also have die to support discontinued Motorola military products).

Chips
See *Chips and Technologies*

Manufacturers, Prefixes, Part Number Types, Logo Descriptions & Family Types

CHIPS AND TECHNOLOGIES — CIRRUS LOGIC

Chips and Technologies, Inc.
3050 Zanker Road
San Jose, CA 95134-2100
408-434-0600, 408-894-2090
Fax: 408-526-2275, 408-894-2091, 408-434-0412, 432-9226
http;//www.chips.com/

Norcross, GA:
404-662-5098
Fax: 404-662-5287

Schaumburg, IL:
708-397-4300
Fax: 708-397-4405

FAE Support:
Framingham, MA
508-620-4775
Fax: 508-620-4776

Austin, TX:
512-502-0605
Fax: 512-502-0360

Taiwan:
133 Ming Sheng East Road
Section 3
Taipei, Taiwan R.O.C.

United Kingdom:
Wyvols Court, Old Basingstroke Rd.
Swallowfield, Berkshire RG7 1PY
United Kingdom

Europe:
011 441 734 880 237
Fax: 011 441 734 884 874

Asia Pacific:
011 88 62 717 5595
Fax: 011 88 62 717 5646

(A fabless company supplying video graphics ICs [such as video overlay ICs, image resizing ICs, graphics accelerators], microcomputer system logic, graphics controllers [including LCD graphics controllers for notebook PCs], communication and network devices, PC chip sets)

Chrontel, Inc.
2210 O'Toole Avenue
San Jose, CA 95131-1236
408-383-9328
Fax: 408-383-9338
http://www.docwriter.com/chrontel/

(Mixed signal ICs including clock generator ICs and VGA to NTSC ICs, NTSC/PAL Digital Video Encoder and Clock Synthesizer for Video-CD ICs, multimedia audio codecs)

CIC prefix
See *ERSO*

CII
See *Communications Instruments, Inc.*

Cimetrics Technology
120 West State Street
Ithaca, NY 14850
607-273-5715
Fax: 607-273-5712

(Microcontroller network (uLAN) IC)

Cincinnati Electronics Corporation
DMDL Microcircuits
7500 Innovation Way
Mason, Ohio 45040-9699
1-800-852-5015
513-573-6249
Fax: 513-573-6290

(This subsidiary of Canadian Marconi manufactures thick and thin film microwave multichip modules and surface mount microcircuits and IR detectors. They are qualified to MIL-STD-1772 for hybrid assemblies)

CINOX Corporation
4914 Gray Road
Cincinnati, OH 45232
513-542-5555
Fax: 513-542-5146
TWX: 810-461-2749

(Quartz crystals and oscillators)

Circuit Technology Inc.
See *GEC Plessy Semiconductors* (Marconi Circuit Technology Corporation)

Cirrus Logic Inc.
3100 W. Warren Ave.
Freemont, CA 94538
510-623-8300, 510-226-2252
Fax: 510-226-2240
http://www.cirrus.com/
support@cirrus.com
(customer/technical support)
sales@cirrus.com
ftp://ftp.cirrus.com

Database Marketing Services
P.O. Box 391704
Mountain View, CA 94039-9956
800-858-0487
Fax: 415-940-4337

California:
San Jose: 408-436-7110
Fax: 408-437-8960
Tustin: 714-258-8303
Fax: 714-258-8307
Thousand Oaks: 805-371-5381
Fax: 805-371-5382

Colorado, Boulder:
303-939-9739
Fax: 303-440-5712

Texas:
Austin: 512-794-8490
Fax: 512-794-8069
Plano: 214-985-2334
Fax: 214-964-3119

Mass., Andover:
508-474-9300
Fax: 508-474-9149

Florida, Boca Raton:
407-362-5225
Fax: 407-394-0618

CIRRUS LOGIC — CLINTON ELECTRONICS

Japan:
No. 5 Mizushima Bldg 2F,
16-22, Chuo-Rinkan 4-Chome
Yamato-Shi, Kangawa 242
Japan
0462 76 0601

Singapore:
65 3532122
Fax: 65 3532166

Taiwan, Taipei:
886 2 718 4533
Fax: 886 2 718 4526

Germany, Herrsching:
49 8152 2030
Fax: 49 8152 6211

United Kingdom, Hertfordshire:
44 0727 872424
Fax: 44 0727 875919

(Application specific standard products [ASSPs], including video graphics ICs, to enhance PC graphics and communications capabilities, graphics accelerators, LCD graphics controllers; hard disk digital controller ICs, SCSI Controllers [with Reed-Solomon error correction techniques] mixed signal ICs for Winchester disk drives, Cardbus controllers, Video Electronic Standards Association United Memory Architecture chip sets.)

Citizen Electronics Co. Ltd.
1-23-1 Kamikurechi
Fujiyoshida-Shi
Yamanashi-ken, 403, Japan
81 555 23 4121
Fax: 81 555 242 426
TLX: 3385 468

(Optoelectronics)

CJ prefix
See *Robert Bosch Corporation*

CJSE prefix
See *Solitron*

CK prefix
See *Philips Semiconductor*
(Custom part)

CKE logo
See *CKE Inc.*

CKE Inc.
P.O. Box 211
9 Wood Street
Lucernemines, PA 15754
412-479-3533
Fax: 412-479-3537

(MOV - Metal Oxide Varistors)

CKV prefix
See *International Electronic Products Corporation of America*

CL prefix
See *Cirrus Logic and C Cube Microsystems* (Some MPEG devices second sourced by Texas Instruments and Advanced Micro Devices), *AVG Semiconductors* (Second source VGA graphics parts), *Crosslink Semiconductor* (SRAMs)

CLA prefix
See *GEC Plessy*

(Programmable ASICs)

Clairex Electronics
(Mount Vernon, NY)
See *Clarostat Sensors and Controls*

Clarion Co. Ltd.
2-22-3, Shibuya
Shibuya-ku
Tokyo 150
Japan
81 3 3400 1121
Fax: 03 3400 8505

(Hybrid, linear, analog ICs, communication ICs, FAX ICs)

Clarkspur Design, Inc.
12930 Saratoga Avenue
Saratoga, CA 95070-4600
408-253-3196

(DSP core ICs; this company licenses DSP cores and software to other companies.)

Clarostat Sensors and Controls Group
12055 Rojas Drive
Suite K
El Paso, TX 79936
800-872-0042
Fax: 915-858-8450

(Optoelectronics, photoelectric sensors and controls)

CLC prefix
See *Comlinear Corp.*

CLi prefix
See *Concurrent Logic, Inc.*

Clipper Chip
A communication encryption IC which can be attached to a telephone circuit to scramble and decode voice or data communications. (The Federal government also retains the codes needed to unscramble messages for legal wiretaps.) It was developed by the National Security Agency with NIST assistance in 1993.

CL-SH prefix
See *Cirrus Logic*

CL-PX prefix
See *Pixel Semiconductor*

Clinton Electronics Corporation
6701 Clinton Road
Rockford, IL 61111
815-633-1444
Fax: 815-633-8712

(Custom ASICs used in their high resolution monitors)

CLY prefix
See *Siemens*

CM prefix
See *Mitel Semiconductor*,
California Micro Devices,
Comus International
(Motion sensors and tilt switches),
Gentron Corp.
(Power supply circuits),
Corsair Microsystems
(Memory modules)

CMAC logo
See *C-MAC*

C-MAC Microcircuits
C-MAC Industries, Inc.
C-MAC Microcircuits, Inc.
3000 Industrial Blvd. Sherbrook
Quebeck, Canada JIL 1V8
819-821-4524
Fax: 819-563-1167

C-MAC of America
1601 Hill Avenue (East Wing)
West Palm Beach, FL 33407
407-845-8455
Fax: 407-840-7879
South Denes Great Yarmouth
Norfolk NR30 3PX England
44 0 1493 856122
Fax: 44 0 1493 858536

41 rue Pierre-Brossolette
B.P. 1642 27016 Evreux
Cedex-France
33 32 317010
Fax: 33 32 317194

(MIL-STD-1553 digital driver/receiver hybrid circuits, integrated data bus module a RISC bit slice processor, backpanels, interconnect products, video filters, and contract manufacturing [at other facilities])

C-Mac Quartz Crystals
4709 Creekstone Drive, Suite 311
Riverbirch Building, Morrisville, NC 27560
919-941-0430
Fax: 919-941-0530

Edinburgh Way Harlow
Essex CM20 2DE
England
44 0 1279 626626
Fax: 44 0 1279 454825

(Oscillators [including TCXOs], and quartz crystals)

C-MAC Quartz Crystals, Inc.
840 W. Church Road
Mechanicsburg, PA 17055
717-766-0223
Fax: 717-790-9509

(Formerly Greenray Industries, Inc. [before being acquired by C-MAC, Inc. in August 1995] oscillators including crystal, VCOs, frequency standards and temperature compensated oscillators)

cMCU prefix
See *Texas Instruments*

CMD
See *California Micro Devices*

CMD Technology, Inc.
1 Vanderbuilt
Irvine, CA 92718
800-426-3832
714-454-0800
Fax: 714-455-1656

(Bus controller ICs)

CML logo
See *Consumer Microcircuits, Ltd.*

CMM prefix
See *Celeritek*,
Harris Semiconductor
(Custom part)

CMOS logic
See *Motorola Semiconductor*,
National Semiconductor,
Harris Semiconductor

CMOS 4L, 4LP, 5L
See *IBM*

(Programmable ASICs)

CMOS 8L
See *NEC*

(Programmable ASICs)

CMX XXXX part number
See *C-Mac*

(Crystals)

CNR, on a disc body
A MOV by *PACCOM Electronics*

CNW prefix
See *Hewlett-Packard*

COCO Research, Inc.
1st Maruzen Bldg-16-12, 6-Chome
Nishi-Shin, Jyuku Shinjuku-Ku
Tokyo 160, Japan
3 3348 1021
Fax: 3 3348 1030
TLX: 232 5189 COCO J

(Linear ICs)

Codi Semiconductor, Inc.
(Division of Computer Diode Corporation. This N.J. company ceased operation in 1992)

Cognex Corporation
15 Crawford Street
Needham, MA 02194
617-449-6030
Fax: 617-449-4013

(Image processing ICs)

COHERENT — COMMQUEST TECHNOLOGIES

Coherent, Inc.
5100 Patrick Henry Drive
Santa Clara, CA 95056
408-764-4000, 4983
Fax: 1-800-362-1170,
408-988-6838, 408-764-4800
http://www.cohr.com
E-mail: tech_sales@chhr.com

Benelux: 079 362313
France: 01 6985 5145
Germany: 06071 9680
United Kingdom: 01223 424065

(Laser diodes. Note: Applied Laser Systems [ALS]; a maker of computer-aided vision products, sold its laser diode operation [except for the laser aimer line] to Coherent, Inc. [Auburn, CA] in 1995.)

Coilcraft
1102 Silver Lake Road
Cary, IL 60013
800-322-2645
Fax: 708-639-1469
http://www.coilcraft.com/

(Filters and magnetics including surface mount types of inductors, chokes, current sensors, toroids, transformers, delay lines)

COL-SP-XXXXX part number
See *Harris Semiconductor*
(Custom part)

ColdFire
(Variable length RISC architecture microprocessor)
See *Motorola*

Collins Electronics Corp.
12th Floor, No. 72
Sung Chiang Road
Taipei
Taiwan
886 2 561 5858, 7878
Fax: 886 2 563 6898, 561 0752
TLX: 27246 JVLINCO

(Diodes, optoelectronics)

Collmer Semiconductor
See *Fuji Electric*

Cologne Chip Designs
Eintrachstrasse 113
Koln 50668
Germany
49 221 136 735
Fax: 49 221 134 715

(ISDN FIFO controllers)

Colorado Microcircuits, Inc.
1106-A North Boise Avenue
Loveland, CO 80537
970-663-4145
Fax: 970-663-5589

(A manufacturer of custom high-speed hybrid and MMIC analog microcircuits, military screening is available)

Columbia Research Laboratories, Inc.
1925 MacDade Blvd.
Woodlyn, PA 19094
610-872-3900
Fax: 610-872-3882

(Accelerometers, strain gauge amplifiers)

COM prefix
See *Standard Microsystems Corporation*

Comclok, Inc.
See *Cal Crystal Lab., Inc.*

Comlinear Corp.
4800 Wheaton Dr.
Fort Collins, CO 80525-9483
1-800-776-0500
303-226-0500, 970-225-7421
Fax: 303-226-0564, 970-226-0564, 970-226-0564
FaxCOM: 1-800-970-0102
TLX: 450881

http://www.compdist.com/cdi
(See also the National Semiconductor web page)
clc_apps@cc.com
(Applications group)
Internet: clc_apps@cc.com

The Maples, Kembrey Park
Swindon, Wiltshire SN2 6UT
United Kingdom
+44 (0)203 422958
Fax: +44 (0)203 422961

(Wideband variable gain amplifiers, analog multiplexers, direct resolver to digital converters, military devices available. Note: This company was acquired by National Semiconductor in 1995.)

Commodore Semiconductor
950 Rittenhouse Road
Norristown, PA 19403
215-666-7950
Fax: 215-647-0791

(This division of Commodore Business Machines, which made ICs for Commodore, ceased business [liquidated] in late 1994. Commodore still makes ICs for their own products.)

CommQuest Technologies, Inc.
527 Encinitas Blvd.
Encinitas, CA 92024
619-633-1618
Fax: 619-633-1677

(Wireless, digital cellular mobile communications, and satellite communications chip sets [voice/data satellite links])

COMMUNICATIONS INSTRUMENTS — CONCORD SENSORS

Communications Instruments, Inc. (CII)
P.O. Box 520
Fairview, NC 28730
704-628-1711 (Sales/marketing)
704-681-0404 (Factory)
Fax: 704-628-1439

(TO-5, and MIL-R-39016, MIL-R-5757 qualified relays, mini-grid relays, solenoids, military and commercial products)

Communications Techniques, Inc.
9 Whippany Road
Whippany, NJ 07981
201-884-2580
Fax: 201-887-6245
TWX: 710-986-8265

(Frequency synthesizers, RF and microwave signal sources, voltage controlled oscillators, coaxial oscillators)

Compensated Devices, Inc.
166 Tremont St.
Melrose, MA 02176
617-665-1071
Fax: 617-665-7379

(Semiconductors, including current regulator diodes, commercial and military qualified parts)

Component General, Inc. (CGI)
2445 Success Drive
Odessa, FL 33556
813-376-6655
Fax: 813-372-9096

(Terminations, resistors, power attenuators)

Component Technology
23 Sheer Plaza
Plainview, NY 11803
1-800-COMTEC-1, 516-755-1112
Fax: 516-755-1115

(MOSFET drivers)

Computer Conversions Corporation
6 Dunton Ct.
East Northport, NY 11731
516-261-3300
Fax: 516-261-3308

(Synchro/linear and digital converters, shaft encoders)

Computer Labs
See *Analog Devices*

Computer Management & Development Service
1661 Virginia Avenue
Harrisonburg VA 22801
703-434-5499
Fax: 703-434-5275

(Optoelectronics)

Computer Products, Power Conversion
7 Elkins Street
South Boston, MA 02127
1-800-877-9839
(Ext. 718 for literature)
1-800-733-9288
(East coast, tech assistance)
617-464-6668
617-268-1170
1-800-769-7274
(West coast, tech assistance)
510-657-6700
1-800-624-8999
617-464-6600
Fax: 617-464-6612

National Accounts Division:
47173 Benicia St.
Fremont, CA 94538-7331

P.O. Box 5102
Fremont, CA 94537-5102
510-657-6700
Fax: 510-683-6400

Computer Products
Power Conversion Europe
Youghal,
County Cork
Republic of Ireland
353-24-93130
Fax: 353-24-93257

Computer Products
Power Conversion Asia-Pacific
13-15 Shing Wan Road
Tai Wai Shatin, N.T.
Hong Kong
852 69 92868
Fax: 852-69-91770

(Power supplies, DC/DC converters)

Comset Semiconductors SprL
Quai Aux Pierres
De Taille, 37/39
B-1000 Brussels
Belgium
322 219 0542

(Diodes, optoelectronics, thyristors, transistors, interface, linear and memory support ICs)

Comus International
263 Hillside Avenue
Nutley, NJ 07110
201-667-6200
Fax: 201-667-6837

212 East Spruce St.
Rogers, AK 72756
501-636-1639
Fax: 501-636-2014

(Motion sensors and tilt switches (mercury and non-mercury types))

Concord Sensors, Inc. (CSI)
(See *Cherry Corporation* [Semiconductor], which acquired the Hall effect and sensor design

CONCORD SENSORS & ANALYZERS — COP

manufacturing operation in late 1995. The sensor design and development center is now located in Concord, NH.

Concurrent Logic, Inc.
1290 Oakmead Pkwy
Sunnyvale, CA 94086
408-522-8703
Fax: 408-732-2765
(Owned by Atmel Corporation)

(FPGAs and programmable logic)

Condor DC Power Supplies, Inc.
Subsidiary SL Industries
2311 Statham parkway
Oxnard, CA 93033
800-235-5929 (outside CA)
805-486-4565
Fax: 805-487-8911
http://www.condorpower.com/~condordc

(Power supplies including models for medical device use)

Condutus, Inc.
Instruments and Systems Division
10623 Roselle St.
San Diego, CA 92121-1506
619-550-2700

(Hybrid SQUID [Superconducting Interference Device-capable of sensing extremely minute changes in magnetic flux] ICs)

Connor-Winfield Corporation
1865 Selmarten Road
Aurora, IL 60505
708-851-4722
Fax: 208-851-5040

(Crystal oscillators)

Consumer Microcircuits, Ltd.
1 Wheaton Road
Witham, Essex, CM8 3TD
England
44 0 376 513833
Fax: 44 0 376 518247
TLX: 99382

(Telecommunications ICs including SPM decoders, universal and tone call progress tone decoders, call charge metering [SPM] detector, amplifier arrays, audio filters, codecs, modem ICs, temperature sensor IC)

Contaq Microsystems, Inc.
San Jose, CA
This company was purchased by Cypress Semiconductor in 1994.

(486 microprocessor core logic, PCI-based graphics and multimedia ICs)

Contaq Technologies, Corp.
15 Main Street
Bristol, VT 05443
802-453-3332
Fax: 802-453-4250

(Ultrasonic transducers)

Continental Device, India Ltd. (CDIL)
C-120
Naraina Industrial Area
New Delhi, 110028
India
91 11 536150 53
Fax: 91 011 5410592
TLX: 031 76230, 031 76209

(Diodes, transistors, linear ICs)

Control Electronics Co. Inc.
139 Florida Street
Farmingdale, NY 11735
107 Allen Blvd.
Farmingdale, NY 11735
516-694-0125
516-694-0133

(Delay lines, transponder decoder networks, variable maglines)

Control Sciences, Inc.
9509 Vassar Avenue
Chastworth, CA 91311-4199
818-709-5510
Fax: 818-709-8546
TLX: 4970496 CSI

(Interface ICs, digital/synchro and synchro/digital [D/S and S/D] converters)

Conversion Devices, Inc. (CDI)
15 Jonathan Drive
Brockton, MA 02401
508-559-0880
Fax: 508-559-9288

(DC/DC converters)

Conversion Equipment Corporation (CEC)
330 W. Taft Avenue
Orange, CA 92665
714-637-2970
Fax: 714-637-8654

(Power supplies)

Cooper Industries
See *Bussman Division*

Cooper Instruments & Systems
P.O. Box 3048
Warrenton, VA 22186-9714
703-349-4746
Fax: 703-347-4755

(Strain gauge transducers)

Coors Components
See *Microsemi Corp.*

COP prefix
See *National Semiconductor*

Corcom
844 East Rockland Road
Libertyville, IL 60048-3375
708-680-7400
Fax: 708-680-8169
http://www.cor.com/home.html

(RFI filters, filtered modular jacks, power entry modules)

Cornell Dubilier
Headquarters
1700 N. Route 23
Wayne, NJ 07470
201-694-8600
Fax: 201-694-8873

1605 East Rodney French Blvd.
New Bedford, MA 02744
508-996-8564
Fax: 508-996-3830

California:
2314 Martin Luther King Ave
Calexico, CA 92231
619-357-3441
Fax: 619-357-0689

C-D Marketing, Inc.
140 Technology Place
P.O. box 128
Liberty, SC 29671
803-843-2277
Fax: 803-843-3800

(Relays, mica transmitting capacitors)

Cornes & Company, Ltd.
6F Saito Bldg, 6th Floor
6-6-1 Sotokanda, Chiyoda-Ku
Tokyo 101,
Japan
03 3839 3001
Fax: 03 3839 3005

Cornes, USA
9010 Miramar Road, #100
San Diego, CA 92126
619-586-6200
Fax: 619-586-0529

(Video processing and timing ICs)

Corollary Inc.
412 Cedar St.
Santa Cruz, CA 95060-4369
408-427-0802
Irvine, CA

(PC Chip Sets)

Corsair Microsystems
2005 Hamilton Avenue
San Jose, CA 95125-5917
408-559-1777, 408-383-7240
2961 W. Macarthur Blvd.
Santa Ana, CA 92704
714-556-1818

(Memory modules)

Cosel U.S.A. Inc.
3283 Scott Blvd.
Santa Clara, CA 95054
1-800-888-3526
408-980-5144
Fax: 408-980-9754
http://www.coselusa.com/

(Power supplies, DC/DC convertors)

Coto
See *Coto Wabash*

Coto Wabash
55 DuPont Drive
Providence, RI 02907
401-943-2686
Fax: 401-942-0920

(Relays including solid state DIP relays)

Cougar Components
2225-K Martin Avenue
Santa Clara, CA 95050
408-492-1400
Fax: 408-402-1500

(RF amplifiers)

CP prefix
See *Abbott Electronics*, *Microchip Technology*

(Old General Instrument parts)

CP20K part number
See *Crosspoint Solutions (FPGA)*

CPA prefix
See *M/A-Com, Inc.*

CP Clare Corp. (CPC)
Solid State Products Division
8 Corporate Place
107 Audubon Road
Wakefield, MA 01880
617-246-4000
Fax: 617-246-1356

Western Regional Sales:
10061 Talbert Avenue
Suite 210
Fountain Valley, CA 92708
714-378-1212
Fax: 714-378-1210

Mid-Atlantic Regional Sales:
7281 Sunshine Grove Road
Brooksville, FL 34613
904-596-5886
Fax: 904-596-5987

Mid-American Regional Sales and Latin America:
601B Campus Drive
Arlington Heights, IL 60004
708-797-7000
Fax: 708-797-7023
1-800-99-CLARE

Northeast Region and Canada:
CP Clare Canada, Ltd.
3425 Harvester Rd.
Suite 202
Burlington, Ontario L7N 3N1
905-333-9066
Fax: 905-333-1824

CP CLARE — CRY

European Sales Office:
C.P. Clare International N.V.
Overhaamlaan 40
B-3700 Tongeren
Belgium
32 12 39 04 00
Fax: 32 12 23 57 54
TLX: 39020

CP Clare France s.a.r.l.
9/11, Rue Georges Enesco
F-94008 Creteil Cedex
33 1 43991522
Fax: 33 1 43991524

CP Clare Elektronik GmbH
Muhlstrasse 12
D-71640 Ludwigsburg
49 7141 90089, 49 7141 926972
Fax: 49 7141 90080

Clare Sales
C.l.a.r.e.s.a.s.
Via C. Colombo 10/A
20066 Melzo (Milano)
39 2 95737160
Fax: 39 2 95738829

Clare Sales
Comptronic AB
Box 167
S16329 Spanga
Sweden
46 862 10370
Fax: 46 862 10371

Clare UK Sales
Marco Polo House
Cook Way
Bindon Road
Taunton
Somerset
United Kingdom
TA2 6BG
44 1 823 352541
Fax: 44 1 823 352797

Asian Sales Office:
CP Clare Corporation
Room N1016
Chia-Hsin Bldg. II
10/F., No. 96, Sec. 2

Chung Shan North Road
Taipei, Taiwan
886 2 523 6368
Fax: 886 2 523 6369

(Dry reed and solid state relays, surge arrestors [Arlington Heights] linear optocouplers, integrated telecom circuit [for DAA circuits that combines hookswitch, bridge darlington and ring detect functions in a single IC]. Note: the telephone relay and stepping switch lines were sold to Communications Instruments, Inc. [CII] in 1981.)

CPC
See *CP Clare Corp.*

CQ
See *United Chequers* (Hong Kong) *Ltd.* and subsidiary *Chequers Electronic* (China) *Ltd.*

CQ prefix
See *Optical Imaging Systems, Inc.*

(LCD displays)

CQT prefix
See *CommQuest Technologies, Inc.*

CR-XXXXX part number
See *Teledyne* (RF switch)

Cray Research, Inc.
900 Lowater Road
Chippea Falls, WI 54729-4401
715-726-8334
Fax: 715-726-6713

(Custom IC packaging and stacked IC modules)

Cree Research, Inc.
2810 Meridian Pkwy.
Suite 176
Durham, NC 27713
800-533-2583
919-361-5709
Fax: 919-361-4630

(LEDs, silicon carbide microwave devices [in conjunction with Motorola] and Gallium Nitride [GaN] on Silicon Carbide [SiC] wafers in conjunction with Philips Laboratories.)

Cross-in-a-Circle logo
See *Soltec Corp.*

(Shimadzu products)

Crosslink Semiconductor
2322 Walsh Avenue
Santa Clara, CA 95051
408-982-1896
Fax: 408-982-1898

(SRAMs)

Crosspoint Solutions, Inc.
694 Tasman Drive
Milpitas, CA 95035
408-324-0200
Fax: 408-324-0123
http://www.cadence.com/Crosspoint.html
E-mail: info@xpoint.com

(A fabless company supplying FPGA's)

CRX prefix
See *Hyashi Denkoh Co., Ltd.*

(Platinum RTDs)

CRY prefix
See *Lintel NV/SA*,
Hyashi Denkoh Co., Ltd.

(Platinum RTDs)

CRYDOM — CTS CORPORATION

Crydom Company
6015 Obispo Avenue
Long Beach, CA 90805
1-800-8 CRYDOM
310-865-3536
Fax: 310-865-3318
TLX: 910 250 5756

(Solid state relays)

Crystal Semiconductor
2024 E. St. Elmo Road
Austin, TX 78670

4210 S. Industrial Drive
P.O. Box 17847
Austin, TX 78744
1-800-550-5030
1-800-888-5016
512-445-7222
Fax: 512-445-7581
http://www.cirrus.com/prodtech/crystal.html
http://www.memec.con/selector/Crystal/

(Communications ICs, multimedia, workstation ICs, A/D and D/A converters [including Delta-Sigma A/Ds], PC Audio ICs, etc. [This company became a subsidiary of Cirrus Logic in October 1991])

Crystalonics, Inc.
17A Street
Burlington, MA 01803
617-270-5520
Fax: 617-270-3130

(Formerly Teledyne Crystalonics, a supplier of replacement of military qualified Motorola transistors)

Crystaloid
5282 Hudson Drive
Hudson, OH 44236
216-655-2429
Fax: 216-655-2176
E-mail: crystaloid@aol.com

4170 East Bijou Street
Colorado Springs, CO 80909
719-574-9393
Fax: 719-570-6934

41 Nave 1-1-2
28037 Madrid
Spain
10 288 011 34 1 754 30 01
Fax: 01 288 011 34 1 327 1557

(LCD displays, including custom units)

Crystek Crystal Corp.
P.O. Box 60135
Ft. Myers, FL 33906

1000 Crystal Drive
Fort Meyers, FL 33901
1-800-237-3061
813-936-2109
Fax: 813-939-4226

(Frequency control products such as crystals and oscillators)

CS prefix
See *Crystal Semiconductor*, *Cherry Semiconductor* (CS-XXXX part number), and *Concord Sensors, Inc.*, *Teledyne*

(RF switch CS-XXXXX)

CSC prefix
See *Crystal Semiconductor*

CSC
See *Cherry Semiconductor*

CSdc
See *Conditioning Semiconductor Device Corporation*

CSI logo
See *Catalyst Semiconductor, Inc.*

CSP prefix
See *Lucent Technologies*

(Formerly AT&T Microelectronics)

CSR Industries, Inc.

CSS prefix
See *Celeritek*

(Chip set)

CSW prefix
See *Celeritek*

CT prefix
See *Aeroflex Laboratories, Inc.* (Aeroflex Circuit Technology Corp.), *GEC Plessy Semiconductors* (Marconi Circuit Technology Corporation), *Component Technology*, *Optical Imaging Systems, Inc.*
(LCD displays)

CTI logo
(Circuit Technology Corp.)
See *GEC Plessy Semiconductors* (Marconi Circuit Technology Corporation), *Champion Technologies, Inc.* (Oscillators), *Communications Techniques, Inc.*

CTS logo or prefix
See *CTS (Microelectronics) Corporation*

CTS Corporation
Headquarters
905 N. West Blvd.
Elkhart, IN 46514
219-293-7511
Fax: 219-293-6146

406 Parr Road
Berne, IN 46711
219-589-3111
Fax: 219-589-3243

1100 Roosevelt Street
P.O. Box 2420
Brownsville, TX 78523
210-542-6894
Fax: 210-546-3807

CTS CORPORATION — CXB

1142 W. Beardsley Avenue
Elkhard, IN 46514
219-295-3575
Fax: 219-295-3580

9210 Science Center Drive
New hope, MN 55428
612-533-3533
Fax: 612-536-0349

400 Reimann Avenue
Sandwich, IL 60548
815-786-8411
Fax: 815-786-3600

Microelectronics Division
1201 Cumberland Avenue
West Lafayette, Indiana 47906
317-463-2565

CTS Corporation
Halex Division
1202 McGaw Avenue
Irvine, CA 92714
714-261-6381

CTS Resistor Networks
406 Parr Road
Berne, IN 46711
219-589-3111
Fax: 219-589-3243
http://www.fwi.com/cts

(Thick and thin film hybrid microcircuits, resistor networks)

CTS Components Taiwan, Ltd.
7 Central 6th Road
Chung Lui Road
P.O. Box 26-20
Koahsiung, Taiwan R.O.C.
7 821 6146

CTS Knights, Inc.
Subsidiary of CTS Corporation
Also known as CTS Frequency Control Division

Elkhart Indiana
400 Reinmann Avenue
Sandwich, IL 60548
815-786-8411
Fax: 815-786-3600, 815-786-9743
TWX: 910-642-0860

Cable CTS
Marden Electronics Operation
32100 Droster Ave.
P.O. Box 277
Burlington, WI 53105
414-763-6093

CTS U.K., Ltd.
Blantyre Industrial Estate
High Blantyre
Glasgow, G72 0XA
Scotland, United Kingdom
44 698 824 331
Fax: 44 698 821 944

CTS Singapore Pte. Ltd.
14 Ang Mo Kio Industrial Park 2
Singapore, 2056
Republic of Singapore
65 481 4634
Fax: 65 481 6375

(Crystals, ovens, hybrid clock oscillators, crystal filters)

(CTS) Computer Technology System Corp.
7F, No. 46, Sec. 2
Chung Shan N. Rd.
Taipei, Taiwan, R.O.C.
886-2-531-2317
Fax: 886-2-565-1963

(LCD modules)

Cubic Memory, Inc.
27 Janis Way
Scotts Valley, CA 95066
408-438-1887
Fax: 408-438-1890

(Dense memory modules consisting of stacked and interconnected semiconductor dice and wafer segments, DRAM SIMMs, DIMMs, SO DIMMs)

Cui Stack, Inc.
9640 SW Sunshine Ct. #700
Beaverton, OR 97005
503-643-4899
Fax: 503-643-6129

(Switching power supplies, including wall mount types. This company also sells connectors, rotary encoders, etc.)

Curtis Electromusic Specialities
See *On Chip Technologies* for the CEM chip designs.

Custom Components, Inc.
P.O. Box 334
Lebanon, NJ 08833
908-534-6151
Fax: 908-534-5625

(Diodes)

CV prefix
See *Motorola*

CW prefix
See *Comus International*

(Motion sensors and tiot switches)

CX prefix
See *Sony Semiconductor*,
AVG Semiconductors
(Second source),
Chip Express (ASICs)

CXA prefix
See *Sony, AVG Semiconductors*

(Second source, some ICs)

CXB prefix
See *Sony Semiconductor*

Manufacturers, Prefixes, Part Number Types, Logo Descriptions & Family Types

CXD prefix
See *Sony Semiconductor*

CXG prefix
See *Sony Semiconductor*

CXK prefix
See *Sony Semiconductor*

CY prefix
See *Cypress Semiconductor*,
Cybernetic Micro Systems
(Devices not equivalent),
Robert Bosch Corporation
(Devices not equivalent)

CY7B prefix
See *Cypress Semiconductor*

CY7C prefix
See *Cypress Semiconductor*

Cybernetic Micro Systems
P.O. Box 3000
3000 Highway 84
San Gregorio, CA 94074
415-726-3000
Fax: 415-726-3003
TLX: 910-350-5842
171-135

(Waveform synthesizer, stepper motor controllers, network control ICs, Video driver circuit cards, etc.)

Cypress Semiconductor
3901 North First St.
San Jose, CA 95134
408-943-2600
408-943-2741
TLX: 821032 CYPRESS SNJ UD
TWX: 910-997-0753
Hotline: 1-800-858-1810
408-943-2821, 4090
Telex: 821032 CYPRESS SNJ UD
TWX: 910-997-0753
Fax: 408-943-2701, 2843
Fax hotline: 415-943-4090
Faxback: 800-213-5120
408-943-2798
BBS: 408-943-2954

http://www.cypress.com
E-mail: cyapps@cypress.com
cyonline@sv.poppe.com (product literature)
sze@cypress.com (Cypress:online (tm) newsletter)

Quicklogic:
1-800-842-FPGA (3742)

Europe:
49 89 899 143 28,
Fax: 49 89 857 77 16
E-mail: info@qlogic.com

IC Designs Division
12020 - 113th Avenue N.E.
Kirkland, WA 98034
206-821-9202
Fax: 201-820-8959

Alabama:
555 Sparkman Drive, Ste. 1212
Huntsville, AL 35816
205-721-9500
Fax: 205-721-0230

California:
100 Century Center Court
Suite 340
San Jose, CA 95112
408-437-2600
Fax: 408-437-2699

23586 Calabasas Rd., Ste. 201
Calabasas, CA 91302
818-222-3800
Fax: 818-222-3810

2 Venture Plaza, Ste. 460
Irvine, CA 92718
714-753-5800
Fax: 714-753-5808

12526 High Bluff Dr., Ste. 300
San Diego, CA 92130
619-755-1976
Fax: 619-755-1969

CXD — CYPRESS SEMICONDUCTOR

Colorado:
4704 Harlan St., Suite 360
Denver, CO 80212
303-433-4889
Fax: 303-433-0398

Florida:
10014 N. Dale Mabry Hwy., 101
Tampa, FL 33618
813-968-1504
Fax: 813-968-8474

255 South Orange Ave.
Suite 1255
Orlando, FL 32801
407-422-0734
Fax: 407-422-1976

Georgia:
1080 Holcomb Bridge Rd.
Building 100, Ste. 300
Roswell, GA 30076
404-998-0491
Fax: 404-998-2172

Illinois:
1530 E. Dundee Road, Ste. 190
Palatine, IL 60067
708-934-3144
Fax: 708-934-7364

Maryland:
8850 Stanford Blvd., Suite 1600
Columbia, MD 21045
410-312-2911
Fax: 410-290-1808

Minnesota:
14525 Hwy. 7, Ste. 360
Minnetonka, MN 55345
612-935-7747
Fax: 612-935-6982

New Hampshire:
61 Spit Brook Road, Ste. 110
Nashua, NH 03060
603-891-2655
Fax: 603-891-2676

C - D

CYPRESS SEMICONDUCTOR — D

New Jersey:
100 Metro Park South
3rd Floor
Laurence Harbor, NJ 08878
908-583-9008
Fax: 908-583-8810

New York:
244 Hooker Ave., Ste. B
Poughkeepsie, NY 12603
914-485-6375
Fax: 914-485-7103

North Carolina:
7500 Six Forks Road, Suite G
Raleigh, NC 27615
919-870-0880
Fax: 919-870-0881

Oregon:
8196 S.W. Hall Blvd., Suite 100
Beaverton, OR 97005
503-626-6622
Fax: 503-626-6688

Pennsylvania:
Two Neshaminy Interplex, Ste. 206
Trevose, PA 19053
215-639-6663
Fax: 215-639-9024

Texas:
333 West Campbell Road, Ste. 340
Richardson, TX 75080
214-437-0496
Fax: 214-644-4839

Great Hills Plaza
9600 Great Hills Trail, Ste. 150W
Austin, TX 78759
512-502-3023
Fax: 512-338-0865

20405 SH 249, Ste. 216
Houston, TX 77070
713-370-0221
Fax: 713-370-0222

Virginia:
3151C Anchorway Court
Falls Church, VA 22042
703-849-1733
Fax: 703-849-1734

Europe:
32-2-652-0270
Fax: 32-2-652-1504

Asia:
583 Orchard Road, #11-03 Forum
Singapore 0923
65 735 0338
Fax: 65 735 0228

Canada:
701 Evans Avenue
Suite 312
Toronto, Ontario M9C 1A3
416-620-7276
Fax: 416-620-7279

Japan:
Shinjuku-Marune Bldg.
1-23-1 Shinjuku
Shinjuku-Ku, Tokyo 160
Japan
81 3 526 90781
Fax: 81 3 526 90788

Cypress Online
(Quarterly Newsletter)
Cypress: online Editor, M/S2.2
3901 North First St.
San Jose, CA 95134
408-943-2653
Fax: 408-943-6848
E-mail: sze@cypress.com

(CMOS and BiCMOS products including SRAMs, PROMs, EPROMs, PLDs, FIFOs logic devices, microprocessors and peripherals, multichip modules, memory devices, clock ICs, and one time programmable antifuse FPGAs and ASICs [the QuickLogic line, programmable logic ICs (PLDs)])

Cyrix Corp.
2703 N. Central Expressway
Richardson, TX 75080
800-462-9749
(technical support and sales)
214-234-8387
Fax: 214-699-9857
http://www.cyrix.com
Internet: tech_support@cyrix.com
BBS: 214-994-8610 (2400, 9600, 14.4K, 28.8K baud)

Cyrix International ltd.
603 Delta Business Park
Welton Road
Swindon
Wilts, U.K. SN5 7XF
44 793 417777
Fax: 44 793 417770
Faxback: 44 793 417799

Cyrix K.K.
7F Nisso 11 Bldg. 2-3-4 Shin-Yokohama, Kouhoku-ku
Yokohama, Kangawa 222
Japan
81 45 471 1661
Fax: 81 45 471 1666

Cyrix Asia Pacific (Singapore) Pte. Ltd.
Ang Mo Kio Industrial Park 1
Blook 4008, #02-01 to #02-03
Singapore 2056
65 453 2843
Fax: 65 453 8201

(A fabless company supplying microprocessors including "clones" of Intel devices)

CZ prefix
See *Amplifonix, Inc.*

D prefix
See *Intel, NEC, Destiny Technology* (Laser printer chip sets),
Siliconix, Dionics, Inc.,
Point Nine Technologies (RF units),
Frequency Devices, Inc. (Filters);
DSP Communications, Inc. (DSP

based voice control processor), *M/A-Com* (GaAs MMIC mixers), *Harris Semiconductor* (Unijunction transistors), *AVG Semiconductors* (Second source some devices), *Durel Corporation* (EL lamps and driver ICs).

D XXXCHx part number
See *Westcode Semiconductors*

Three Ds, and the number 3 in the center, inside a triangle (logo)
See *Data Display Devices, Inc.*

Two Ds back to back (with the tail of the "D"'s forming a circle around them) logo
See *Datatronics*

D1 International, Inc.
95 E. Main St.
Huntington, N.Y. 11743
516-673-6866

(DC/DC Converters)

D.O. Industries, Inc.
Laser Products Division
200 Commerce Drive
Rochester, NY 14623
716-359-4000
Fax: 716-359-4999

(Diode lasers)

DA logo
See *Daden Associates, Inc.*

DA prefix
See *Apex Microtechnology*, *GEC Plessy*, *SGS-Thomson*

(Diode arrays)

DA-XXXXX part number
See *ILC-DDC*

DAC prefix
See *Analog Devices*, *Datel Inc.*, *Burr-Brown Corp.*, *Motorola*, and *National Semiconductor*

Daden Associates, Inc.
1001 Calle Amanecer
San Clamente, CA 92673
714-366-1522
Fax: 714-366-9600

(GPS filter/limiter and low noise amplifier combinations)

Daewoo Electronic Components Company Ltd.
60-8 Garibong-Dong, Desig
Guro-Gu
Seoul, Korea
82 863 4859
Fax: 82 864 8200

Daewoo Telecom Ltd.
275-6 Yangjae-Dong
Seoul, Socho-Gu 29219
Korea
812 589 2114
Fax: 812 756 1225

82 Northholt Road, Templar
South Harrow, Middlesex HA2 0YL
44 181 423 7200
Fax: 44 181 864 6070
TLX: 932241

79 Avenue Francois Arago
Nanterre Cedex 92017
France
33 1 4695 3000
Fax: 33 1 4721 7375
TLX: 614646

In the U.S. call distributor
Susco Electronics, Inc.
61A Carolyn Blvd.
Farmingdale, NY 11735
1-800-752-0811
516-249-0811

D — DALE ELECTRONICS

(TV, VCR, telephone and FM stereo ICs, operational amplifiers, comparators, microcomputers, ROMs)

Daico Industries, Inc.
2453 E. Del Amo Blvd.
Rancho Dominguez, CA 90220
310-631-1143
Fax: 310-631-8078

(GaAs switches and MMICs, threshold detectors, attenuators, phase shifters, couplers, modulators, high density MICS, antenna selectors for aircraft, etc.)

Daimler-Benz
See *Siliconix*

Dale Electronics, Inc.
A Company of Vishay
Headquarters:
1122 23rd St.
Columbus, NE 68601-3647
402-564-3131
Fax: 402-563-6418

(Resistors, delay lines, crystal oscillators, plasma displays, infrared touch panels)

Dale Electronics, Inc.
P.O. Box 180
Yankton, SD 57078-0180
605-665-9301
Fax: 605-665-1627

(Inductors, transformers, connectors, surface mounted components)

Dale Electronics, Inc.
Techno Division
7803 Lemona Ave.
Van Nuys, CA 91405-1139
818-781-1642
Fax: 818-781-8647

DALE ELECTRONICS — DATA GENERAL

(Potentiometers, cermet high voltage resistors and dividers, resistor networks, ladder networks)

Angstron Precision, Inc.
P.O. Box 1827
Hagerstown, MD 21740
301-739-8722
Fax: 301-797-6852

(Metal film resistors, power rheostats [formerly the ring rheostats sold by Ward Leonard, Inc.])

Dale Electronics, Inc.
1155 W. 23rd Street, Suite 1A
Tempe, AZ 85282-1883
602-967-7874
Fax: 602-829-9314

(Potentiometers, hybrid crystal oscillators, clock oscillators, and surface mount parts)

Dale Electronics, Inc.
P.O. Box 26728
El Paso, TX 79926-6728
915-592-3253
Fax: 915-595-8199

(Thermistors, surface mounted components)

Note:
There are other Dale divisions that sell power capacitors and metal film resistors.

Electronica Dale de Mexico
S.A. de C.V.
Apartado Postal 3101J
CD Juarex, Mexico
18 11 00, 18 16 24
Fax: 18 10 60

Dale Korea, ltd.
244-83 Sosa 3 Dong
Nam-Gu Bucheon City
Kyunggi-Do, Korea
032 665 2211/2
Fax: 032 653 7200

Dale Israel Electronics Industries, LTD.
Industrial Park,
Dimona, Israel 86000
972 57 5595
Fax: 972 57 54814

Dale Electronics, Asia PTE, LTD.
161 Kampong Ampat
#03-01
Goldlion Building 1336
Singapore
65 284 0522
Fax: 2853634

Dallas Semiconductor
4401 S. Beltway Parkway
Dallas, TX 75244-3292
214-450-0448
Fax: 214-450-0470, 3715
BBS: 214-450-8169
http://www.dalsemi.com/
micro.support@dalsemi.com
ftp://ftp.dalsemi.com

(Silicon timed circuits including delay lines, identification ICs, battery charger ICs, multiport memory, nonvolatile memory, intelligent sockets, timekeeping ICs, battery back-up and battery chargers, microcontrollers, telecommunication ICs (including voice messaging and digital answering machine processors, T1 controllers [and receivers and transceivers], T1/CEPT line interfaces, ADPCM processors, plug and play SCSI terminator, etc.)

Dalsa Inc.
605 McMurray Road
Waterloo, ON
Canada N2V 2E9
519-886-6000
Fax: 519-886-8023

Europe: 49 089 80 402 58,
Fax: 49 089 80 902 16
Japan: 81 3 3419 9190, Fax: 81 3 3419 9239

(CCD image sensors and cameras)

DAS prefix
See *MMC Electronics America, Inc.*

(Surge absorbers)

Databook, Inc.
118 prospect Street
Tower Building
Terrace Hill, Ithaca, NY 14850
716-292-5720, 716-889-4204
Fax: 716-292-5737,
716-889-2593
Danvers MA
508-762-9779

(Dual slot PCMCIA controller IC, PC card readers/writers and software)

Data Delay Devices, Inc.
3 Mt. Prospect Avenue
Clifton, NJ 07013-9990
201-773-2299
Fax: 201-773-9672

(Delay lines, silicon delay lines, voltage controlled oscillators)

Data Display Products (DDP)
445 S. Douglas St.
El Segundo, CA 90245-4630
310-640-0442
Fax: 310-640-7639
1-800-421-6515, ext. 508
(Technical support engineering)

(LED lamp assemblies and surface mount LED assemblies, including blue LEDs)

Data General
4400 Computer Drive
Westborough, MA 01581
508-898-5000
Fax: 508-366-1319

(Used to manufacture microprocessors like the mN series MicroNova)

Data Instruments, Inc.
100 Discovery Way
Acton, MA 01720-3600
1-800-333-DATA (3282)
508-264-9550
Fax: 508-263-0630
http://www.industry.net/data.instruments

(Pressure, displacement and linear position transducers)

Data International Co. Ltd.
(See *Display International for USA*)
2nd Fl., No. 566, Sec. 7
Chung Hsaio E. Rd.
Taipei, Taiwan R.o.C.
886 2 785 1922
Fax: 886 2 7851870

(Character and graphic LCD displays)

Data Vision
See *Display International*

Datakey, Inc.
407 West Travelers Trail
Minneapolis, MN 55337
1-800-328-8828
612-890-6850
Fax: 612-890-2726

United Kingdom: 44 730 816502
Germany: 49-69-578856
Belgium: 32 3 325 19 10
Netherlands: 31 23-31 91 84
Australia: 61-9-370-4488
Korea: 585-1114
Japan: 03-3225-8910

(Keys, cards, tokens and other devices that contain memory ICs)

Datatronics
Datatronics Romoland, Inc.
28151 Highway 74
Romoland CA 92585
909-928-7700
Fax: 909-928-7701, 7728

(Various magnetics including inductors, delay lines and line matching transformers, SMD and military and aerospace devices available.)

Datel, Inc.
11 Cabot Boulevard
Mansfield, MA 02048-1194
1-800-233-2765
(In the East U.S.,
8:30 am - 4:30 pm EST))
1-800-452-0719 (West (PST) U.S.)
508-339-3000
Fax: 508-339-6356
E-mail: datelcomp@aol.com

Datel S.A.R.L.
Zine d'Activities
Du Pas du Lac Nord
9, rue Michael Faraday
78180 Montigny Le Bretonneux
France
1 34 60 01 01, Fax: 1 30 58 21 30
TLX: 689 605

Datel GmbH
Postfach 150826
D-80045 Munchen, Germany
89 54434 0, Fax: 89 536337

Datel (UK) Ltd.
Unit 15,
Campbell Court Business Park
Campbell Road, Bramley
Basingstoke, Hampshire, RG26 5EG
England
256 88 04 44, Fax: 256 88 07 06

Datel KK
Meiji Seimei Gotanda Building, 3F
2-27-4 Nishigotanda
Shinagawa-Ku,
Tokyo 141, Japan

DATA GENERAL — DB

3 3779 1031, Fax: 3 3779 1030

Yachiyo Building, Higashikan
2-Kita 1-21, Tenjinbashi
Kita-Ku, Osaka 530, Japan
6 354 2025, Fax: 6 354 2064

(Data acquisition and conversion devices including sample and hold amplifiers, A/D and D/A converters, multiplexers, Op amps, oscillators, instrumentation amplifiers, tunable active filters, DC/DC converters; VME, PC/AT, EISA, Multibus 1 boards; digital panel meters, calibrators)

Datum, Inc.
34 Tozer Road
Beverly, MA 01915-5510
508-922-1523, Fax: 508-927-4099

(Crystal oscillators including ovenized, precision, ultra-stable and spacecraft qualified types, time standards, GPS down converters and time/frequency receivers, etc.)

David Sarnoff Research Center
201 Washington Road
Princeton, NJ 08540-5300
609-734-2000, Fax: 609-734-2870

(Various custom electronic devices and packaging. See also Sensar, Inc.)

Dawn Electronics, Inc.
1004 Mallory Way
Carson City, NV 89701
702-882-7721
Fax: 702-882-7675

(Ultra low bias input [1fA, 1 x 10-15 Amps])

DB prefix
See *Apex Microtechnology, Databook, Inc.*

DBL — DELTA ELECTRONICS

DBL prefix
See *AVG Semiconductors*
(Telephony ICs), *Daewoo*

DBS Microwave, Inc.
4919 Windplay Drive, S-2
El Dorado Hills, CA 95762
916-939-7545
Fax: 916-939-7540

(Microwave amplifiers, multipliers)

DC prefix
See *Digital Equipment Corp*,
GEC Plessy, *AVG Semiconductors*
(Second source)

DCI Incorporated
14812 West 117th Street
Olanthe, KS 66062
913-782-5672
Fax: 913-782-5766

(LCD and ECD modules)

DCP prefix
See *DCP Research Corporation*
and *Language System Design, Inc.*

DCP Research Corporation
11502-77 Avenue
Edmonton, Alberta T6G 0M1
Canada
403-448-1760, 403-435-1686
Fax: 403-963-1165
http://www.ccinet.ab.ca/dcp.html
info@datacompression.com

(An R&D company specializing in data compression products. The lossless data compression IC they developed is sold by Language Systems Design, Inc.)

DD prefix
See *Intel*

DDC
See *ILC-DDC*

DDC-XXXX part number
See *ILC-DDC*

DDi (Republic of Belarus)
See *AVG Semiconductors*

DDI prefix
See *Dr. Design*

DDP
See *Data Display Products*

DDS-1
See *Sciteq Electronics, Inc.*

DE prefix
See *GEC Plessy*

DEC
See *Digital Equipment Company*

DEECO Systems
Lucas Control Systems Products
31047 Genstar Road
Hayward, CA 94544-7831
510-471-4700

(LCD display monitors)

Deico Electronics, Inc.
48006 Great America Parkway
Santa Clara, CA 95054-1221
408-748-7788

(Memory modules)

Deister Electronics USA, Inc.
9303 Grant Avenue
Manassas, VA 22110-5064
703-631-2595, 703-368-2739

(ASIC for a RF proximity reader)

Delco Electronics Corporation
(Part of GM-Hughes)
1 Corporate Center
Kokomo, IN 46904-9005
http://www.delco.com
800-589-8979
800-824-0154

(Forward sensing assembly)

700 E. Firmin Street
Kokomo, IN 46904
317-451-5011

7929 S. Howell Avenue
P.O. Box 471
Milwaukee, WI 53201
414-768-2000
Fax: 414-768-2086

(Automotive products including pressure sensors, hybrid electronic control modules, ignition systems, engine controls, etc. Other manufacturing and technical centers are in Germany, Mexico, Italy, Singapore, Japan, United Kingdom and Korea.)

Delta Electronics, Inc.
9 Fl., No. 144, Min Chuan E. Rd.
Section 3
Taipei, Taiwan, R.O.C.
886 2 716 4822
Fax: 886 2 716 9764

Delta Products Corporation
3225 Lakeview Court
Fremont, CA 94538
510-770-0660
Fax: 510-770-0122

2000 Aerial Center Parkway
Suite #114
Morrisville, NC 27560
919-380-8883
Fax: 919-380-8383

Delta Electronics Inc.
9th Fl., Asia Enterprise Center, No. 144
Min Chaun E. Road, Sec. 3
Taipei, 10464, Taiwan R.O.C.
886-2-716-4822
Fax: 886-2-716-9764

(DC/DC converters, power supplies, fans, pulse transformers, EMI/RFI filters, delay lines, modular uninterruptable switching power supplies)

Manufacturers, Prefixes, Part Number Types, Logo Descriptions & Family Types

DELPHA COMPONENTS — DH

Delphi Components, Inc.
Division of Aura Systems, Inc.
27721-A La Paz Road
Laguna Niguel, CA 92677
714-831-1771
Fax: 714-831-0862

(Voltage controlled oscillators and phase locked oscillators (CROs, DROs), frequency synthesizers, radar receivers, integrated assemblies)

Dense-Pac Microsystems, Inc.
7321 Lincoln Way
Garden Grove, CA 92641-1428
800-642-4477
714-898-0007
Fax: 714-897-1772

Southern Regional Sales Mgr.
316 Brookewood Drive
Peachtree, GA 30269
404-487-2597
Fax: 404-487-6824

Eastern Regional Sales Manager
1185 Dovington Drive
Hoffman Estates, IL 60194
708-882-7572
Fax: 708-882-6698

Added Value Distribution
1582 Parkway Loop
Unit G
Tustin, CA 92680
714-439-7073
Fax: 714-259-0828

Dense-Pac Europe
18, St. Georges Crescent
Monkseaton, Whitley Bay
Tyne & Wear, England NE25 8BJ
44 91 297 1881
Fax: 44 91 297 1725

(Standard and custom monolithic memories and memory/logic/analog memory modules, including SRAMs, EPROMs, Flash EPROMs, VRAMs, DRAMs, PLDs, stacked chip modules, etc.)

Densitron Corporation
Kyowa Nanabankan
5F No. 1-11-5 Omori-Kita
Ota-Ku, Tokyo 143 Japan
03 767-9701-8
Fax: 03 767 9709 (G2/G3)
TLX: J26914 DENSITRON
Cable: "DENTROSE"

Densitron Corporation America
10430-2 Pioneer Blvd.
Santa Fe Springs, CA 90670
310-941-5000

Western U.S.A.:
3425 W. Lomita Blvd.
Torrance, CA 90505
310-530-3530
Fax: 310-534-8419

Eastern USA:
2039 HWY No. 1 S.
South Camden, SC 29020
803-432-5008
Fax: 803-431-1165

Europe:
Densitron Europe Ltd.
Unit 4, Airport Trading Estate, Biggin Hill
Westerham, Kent TN16 3BW
England
0959 76600
Fax: 0959 71017
TLX: 957353

(Electroluminescent, plasma and LCD displays and control cards, interface cards, single board computers, touch overlays, flat panel monitors, etc.)

Denyo Europa GmbH
Otto-Hanh StraBe 41
Huesenstamm, D-60569
Germany
49 6104 633 32
TLX: 416 011 DENYO

(Optoelectronics, linear ICs)

DEP
(B.V. Delft Electronische Producten)
Postbus 60, 9300 AB Roden
The Netherlands
31 0 5908 18808
Fax: 31 0 5908 13510

(Low level light intensified CCDs)

Destiny Technology
3255-1 Scott Blvd, Suite 201
Santa Clara, CA 95054
408-562-1000
Fax: 405-562-1000

(Chip sets for laser printers)

Devar, Inc.
706 Bostwick Avenue
Bridgeport, CT 06605
203-368-6751
Fax: 203-368-3747

(Optoelectronics)

Dexter Research Center, Inc.
7300 Huron River Drive
Dexter, MI 48130
313-426-3921
Fax: 313-426-5090

(Radiation sensing thermopile detectors [a voltage generating device] and low noise amplifiers)

DG prefix
See *Siliconix* (Analog devices [their ADG series]),
Harris Semiconductor (Second source of switches),
Maxim (Second source of devices)

DGX
See *Chips and Technology*

DH prefix
See *National Semiconductor*

DHC — DINAN

DHC prefix
See *Apex Microtechnology*

(DC/DC converters)

Diablo Industries
2245E Meridian Blvd
Minden, NV 89423
702-782-1041
Fax: 702-782-1044

(Chip resistors, chip capacitors [silicon nitride, tantalum pentoxide, alumina], precision etched thin film microcircuits and devices for commercial and military microwave applications)

Dialight Corp.
1913 Atlantic Avenue
Manasquan, NJ 08736
908-223-9400
Fax: 908-223-5271, 223-8788

France:
33 1 64 30 55 55
Fax: 33 1 60 07 64 65

Japan:
Jepico Corp.
Shinjuku Dai-Ichi, Seimi Bldg.
Nishi Shinjuku 2.7.1
Shinjuku-Ku, Tokyo
Japan
03 3348 0611
Fax: 03 3348 0623

United Kingdom:
Exning Road
Newmarket, Suffolk
44 1638 662317, 0638 665161
Fax: 44 1638 560455, 0638 660718

Singapore:
80 Marine Parkway Road
14-04 Parkway Parade
Singapore 1544
65 4473735
Fax: 65 44752482

(LEDs [including blue LEDs] and infrared emitters and detectors)

Dielectric Laboratories, Inc.
2777 Rt. 20
Cazenovia, NY 13035
315-655-8710
Fax: 315-655-8719

(Chip capacitors, including porcelain chip capacitors)

Digital Equipment Company (DEC)
Semiconductor Operations
77 Reed Street
Hudson, MA 01749
800-332-2717 (US and Canada)
1-508-568-6868
(Outside North America)
TTY: 1-800-332-2515
508-568-5149, 5102
Orders: 1-800-344-4825
(1-800-Digital)
Outside the U.S.: 1-508-568-6868

http://www.dec.com or http://ÿww.digital.com (PCs)
http://www.digital.com/info/semiconductor
e-mail: semiconductor@digital.com or moreinfo@digital.com (PCs)

(RISC microprocessor [the DEC Alpha], PCI bridge ICs, video authoring CODECs)

Digital Research in Electronic Acoustics & Music SA (Dream)
Semur-en-Auxois, France

(Speciality memory and logic electronics for music synthesizers and multimedia computers. This company was acquired by Atmel Corporation [San Jose, CA] in 1996).

Digital Voice Systems, Inc. (DVSI)
617-270-1030
Fax: 617-270-0166
info@dvsinc.com

(Voice codecs)

Digitron Electronic Corporation
1991 Route 22W
Bound Brook, NJ 08805
800-526-4298
908-560-1120
Fax: 908-560-8325

(Diodes, thyristors, transistors)

DII
Diode bridges sold by *Mallory, North American Capacitor Company*

Dimolex Corporation
3800 La Crescenta Ave, Suite 201
La Crescenta, CA 91214
818-957-7001
Fax: 818-957-7005
1-800-877-4068
(Technical support)

This is the U.S. representative of Stable Systems, Inc.
484 Ota-Machi, Tomobe
Nishi-Ibaraki, Ibaraki, Japan 309-17
0296 77-9371
Fax: 0296 77-9369

(Intelligent motion control [stepping motor and AC/DC servo motor controller] ICs, intelligent keyboard controllers, etc.)

Dinan
(Performance Engineering)
150 South Whisman
Mountain View, CA 94041-1512
415-962-9417, 415-962-9401
(Car engine controller ICs)

Diodes, Inc.
3050 East Hillcrest Drive
Westlake Village, CA 91362-3154
805-446-4800
Fax: 805-446-4850

(Rectifiers, bridge assemblies; switching, zener; silicon transistors including low power types, transient voltage suppressors, Schottky rectifiers, surface mount parts available. This company is a sister company of Lite-On.)

Dionics, Inc.
65 Rushmore Street
Westbury, NY 11590
516-997-7474
Fax: 516-997-7479
TWX: 510-222-0974

(Display drivers, diodes, transistors, solid state relays, optoelectronics, thyristors, transistors, display ICs, etc.)

Diotec Electronics Corp.
18020 Hobart Blvd
Unit B
Gardenia, CA 90248
310-767-1052
Fax: 310-767-7958

Diotec Electroniche Bauelmemte GmbH
KreuzmattenstraBe 2
Postfach 11 65
Heitersheim, D-7843
Germany
49 76 34 2393
Fax: 49 76 34 4487
TLX: 07721 497

(Diodes, interface ICs)

Directed Energy, Inc.
2301 Research Blvd., Suite 105
Fort Collins, CO 80526
303-493-1901
Fax: 303-493-1903

(Transistors)

Discovery Semiconductors, Inc.
186 Princeton Hightstown Road
Princeton Junction, NJ 08850-1648
609-275-0011
(Founded in 1992)

(Optoelectronic ICs)

Display International
2973 Carlsbad Court
Oviedo, FL 32765
407-366-7399
Fax: 407-366-7403

(LCD-liquid crystal display modules)

Displaytech, Inc.
2200 Central Avenue
Boulder, CO 80301
303-449-8933
Fax: 303-449-8934
sales@displaytech.com

(ChronoColor ferroelectric LCD on silicon displays and fast switching ferroelectric crystals, used as shutters and spatial light modulators, filters, microdisplays [can be used in eyeglasses and in projection displays])

Display Technologies, Inc. (DTI)
Himeji, Japan
310-407-0500
http://www.dtinet.com

(This is a joint venture between Toshiba Corp. and IBM Japan Ltd. and produces thin film transistor LCDs, spread spectrum and wireless products.)

Displays, Inc.
31 Industrial Park Road
Lewistown, PA 17044
717-242-2541
Fax: 717-248-8680

(Diodes, optoelectronics)

D.I.T.
See *Transistor Co. Inc.*

Diversified Technology (DTI)
35 Wiggins Avenue
Bedford, MA 01730
1-800-443-2667
601-856-4121
Fax: 601-856-2888

Outside U.S.:
201-891-8718
Fax: 201-891-9629

(Single board computers that use their custom ICs)

DLI
See *Dielectric Laboratories, Inc.*

DLZ prefix
See *Protek Devices*

DLP prefix
See *Adaptive Networks, Inc.*

DM prefix
See *National Semiconductor*, *Seeq Technology, Inc.*, *Ramtron International Corp.*

DMC prefix
See *Optrex* if LCD display, *Daewoo*

DME prefix
See *GHz Technology, Inc.*

DMR prefix
See *Daewoo*

DMT
Division of Jay-EL Products
23301 S. Wilmington Ave.
Carson, CA 90745
310-513-7200
Fax: 310-513-0741

(RF switches)

**DMV —
DR. NEUHAUS**

DMV prefix
See *SGS-Thomson Microelectronics*

(Dual diodes for monitor horizontal deflection stages)

DN prefix
See *Matsushita* (Panasonic) where the DN stands for digital IC., *Dawn Electronics, Inc.*

DNE prefix
See *Dr. Neuhaus Engineering GmbH*

DO Industries, Inc.
200 Commerce Drive
Rochester, NY 14623
800-828-6778
716-359-4000
Fax: 716-359-4999
TLX: 200862

(Optoelectronics)

Dolch Computer Systems
3178 Laurelview Court
P.O. Box 5003
Fremont, CA 94538
800-538-7506
Fax: 510-490-2360

(Active matrix TFT LCD screens)

Dolphin Integration
8, Chemin des Clos
B.P. 65 ZIRST
38242 Meylan, France
011 33 76 41 10 96
Fax: 011 33 76 90 29 65

Dophin U.S.
BOC Suite 130
3333 Bowers Avenue
Santa Clara, CA 95054
408-727-7619
Fax: 408-748-1826

(Delta sigma converters, mixed mode VLSI circuits and computer simulation software)

Donnelly Corp.
414 East Fortieth St.
Holland, MI 49423-5368
616-786-6022
Fax: 616-786-6034

(An intelligent Vision System that uses a single video IC to control the reflectivity of solid film electrochromatic mirrors)

Dorado International Corporation
270 South Hanford St., Suite 294
Seattle WA 98134
206-583-0000
Fax: 206-583-0345
TLX: 880212 (DORADO CO UD)

(Ferrite devices including isolators, circulators, filters, phase modulators and switches for microwave applications)

Douglas Randall, Inc.
100 Mechanic Street
Bldg. 26
Pawcatuck, CT 06379
800-447-6799
203-599-2075
Fax: 203-599-1754

(Thyristors)

Dow-Key Microwave Corp.
1667 Walter St.
Ventura, CA 93003
805-650-0260
Fax: 805-650-1734

(Microwave switches and relays)

DP prefix
See *National Semiconductor, Frequency Devices, Inc.*

(Filters)

DPA Labs, Inc.
2251 Ward Avenue
Simi Valley, C 93065
805-581-9200
Fax: 805-581-9790

(This lab which does component screening and failure analysis can also provide form, fit and function military obsoleted components.)

DPD prefix
See *Dense-Pac Microsystems, Inc.*

DP-Tek, Inc.
9920 East Harry
Wichita, KS 67207
316-687-3000
Fax: 316-687-0489

(Image processing ICs)

DPS prefix
See *Dense-Pac Microsystems, Inc.*

DPV prefix
See *Dense-Pac Microsystems, Inc.*

DPZ prefix
See *Dense-Pac Microsystems, Inc.*

DQ prefix
See *Seeq Technology, Inc.*

Dr. Design
5415 Oberlin Drive
San Diego, CA 92121
619-457-4545
Fax: 619-457-1168

(Graphic ICs)

Dr. Neuhaus Engineering GmbH
Haldenstieg 3
D-2000
Hamburg 61, Germany
49 40 55 30 45 10
Fax: 49 40 55 30 45 00

(PCMCIA interface ICs)

DRC-XXXXX part number
See *ILC-DDC*

DRFS prefix
See *Proxim, Inc.*

DRM prefix
See *Telbus GmbH*

DS prefix
See *National Semiconductor*,
Dallas Semiconductor,
Lucent Technologies
(Coprocessors. Formerly AT&T Microelectronics),
Daico Industries, Inc.
(Microwave switches),
MMC Electronics America, Inc.
(Surge absorbers)

DSC-XXXXX part number
(and XXXXX-XXX)
See *ILC DDC*

DSP prefix
See *Motorola*,
Lucent Technologies
(Formerly AT&T Microelectronics),
DSP Semiconductors, Inc.

DSP Communications, Inc.
DSP Corporation
20300 Stevens Creek Blvd.
Cupertino, CA 95015
408-777-2700
Fax: 408-777-2770

29-9, 2-Chome Nishi Gotanda
9th Floor,
Shinagawa-ku, Tokyo, 141 Japan
81 3 5496 1611
Fax: 81 3 5496 1516

(DSP-based voice command processors for cellular equipment)

DSP Group, Inc.
4050 Moorpark Avenue
San Jose, CA 95117
408-986-4300

Fax: 408-985-2108
http://www.dspg.com/

(A fabless company supplying PC voice application ICs. This company licenses DSP cores and software to other companies. This company has a design center in Tel Aviv, Israel and regional sales offices in Paris, France and Tokyo, Japan)

DSS prefix
See *Bourns, Inc.*

(Digital sapphire sensor)

DT prefix
See *Densitron Corporation*

(Flat panel monitors)

DTX prefix
Where X is A, B, C, or D (these are digital transistors)
See *ROHM Electronics Division*

DTC-XXXXX part number
See *ILC-DDC*

DTI
See *Diversified Technology*

DTL prefix
See *Optical Communication Products, Inc.*

DuPont Pixel
See *3Dlabs, Inc.*

Durel Corporation
2225 W. Chandler Blvd.
Chandler, AZ 85224-6155
602-917-6000
(A joint venture of 3M Company [St. Paul, MN] and Rogers Corp. (Rogers, CT)

(A manufacturer of EL lamps and EL lamp driver ICs)

DV prefix
See *National Semiconductor*

DVC prefix
See *DSP Communications*

DW prefix
See *GEC Plessy*

DX prefix
see GEC Plessy

DX-XXX-XXX part number
See *Advanced Orientation Systems*

(Tilt sensors)

DY 4 Systems, Ltd.
1475 S. Bascom Avenue
Suite 202
Campbell, CA 95008
408-377-9822
Fax: 408-377-4725
Email: sales@dy4.com
(Sales support)
support@dy4.com
(Product support)

218 Kerrs Corner Road
Blairstown, NJ 07825
908-362-5557
Fax: 908-362-5821

7255 Ridgedale Drive
Warrenton, VA 22186
703-341-2101
Fax: 703-341-2103

Canada and Asia Pacific:
21 Fitzgerald Road
Nepean, Ontario
Canada
K2H 9J4
613-596-9911
Fax: 613-596-0574

D - E DY 4 SYSTEMS — ECHELON

Europe:
1 Cornflower Close
Lisvane, Cardiff
United Kingdom
CF4 5BD
44 0 222 747927
Fax: 44 0 222 762060

(VMEbus circuits jointly developed with Newbridge Microcircuits, single board computers, memory card, serial/parallel cards, frame grabbers, A/D and D/A I/O boards, intelligent SCSI controllers, LAN controllers, graphics controllers, graphics adapter cards, chassis, MIL-STD-1553 interface software, ethernet communications suite software, real time graphics [RTGS] software, etc.)

DYME prefix,
See *Dymec, Inc.*

Dymec, Inc.
8 Lowell Avenue
Winchester, MA 01890
617-729-7870, Fax: 617-729-1639
TLX: 348 6596

(A/D converters, programmable counters and timers, delta sigma converters, voltage to frequency converters)

Dytran Instruments, Inc.
21592 Marilla Street
Chatsworth, CA 91311
818-700-7818
Fax: 818-700-7880

(Force sensors, impulse hammers, accelerometers, pressure sensors)

E prefix
See *AVG Semiconductors*
(Second source calculator ICs),
Elmwood Sensors, Eurosil Electronic GmbH, Comlinear
(Modular and encased products)

E1-32
See *Hyperstone GmbH*

E-San Electronic Co. Ltd.
15 Floor, No. 658, Tun-Hua
S. Road
Taipei, Taiwan
886 2 755 6788
Fax: 882 2 705 2547
TLX: 23868 ESANINTL

(Digital, linear and microprocessor support ICs)

EA logo
See *NEC Microelectronics*
(Formerly Electronic Arrays)

EAC prefix
See *Integrated Telecom Technology, Inc.*

Eagle Picher Electronics Div.
Box 130
Bethel Road
Seneca, MO 64865
417-776-2256
Fax: 417-776-2257
TWX: 62864271

(Lithium batteries and batteries for memory back-up)

Eastman Kodak Co.
Microelectronics Technology Div.
1669 Lake Ave.
Rochester, N.Y. 14650-2010

P.O. Box 92894
Rochester, NY 14692-9931
716-722-4385
Fax: 716-477-4947

(Solid state image sensors including: color linear CCDs, video interline CCDs, full frame CCDs, Infrared CCDs, and support ASICs)

Eastron Corporation
15 Hale Street
Haverhill, MA 01830
508-373-3824
Fax: 508-373-7051

(Diodes)

EC prefix
See *Ecliptek Corp.*

EC2
See *Engineered Components Company*

ECE (Registered trademark)
See *Excell Cell Electronic Co., Ltd.*

ECG Semiconductors
(Sylvania Electronic Components)
(A North American Philips Company)
Distributor and Special Markets Division
1025 Westminster Drive
P.O. Box 3277
Williamsport PA 17701
800-526-9354

ECG Canada Inc.
Electronic Components and
 Systems
8580 Darnley Road
Montreal, Quebec
Canada H4T 1M6

(Generic replacement semiconductors)

Echelon Corporation
4015 Miranda Avenue
Palo Alto, CA 94304
800-258-4566
415-855-7400
Fax: 415-856-6153
http://www.echelon.com

(Neuron ICs for neural network applications)

Echo Speech Corporation
6460 Via Real
Carpinteria, CA 93013
805-684-4593
Fax: 805-684-6628

(CD ROM interface IC and audio boards)

ECI Semiconductor
(Gamma Inc.) Santa Clara, CA
See *Semtech Corporation*

ECL logic
See *Synergy, Motorola Semiconductor, Fujitsu Microelectronics*

ECLinPS
Trademark of *Motorola, Inc.*

Ecliptek Corp.
3545-B Cadillac Avenue
Costa Mesa, CA 92626
1-800-ECLIPTEK
714-433-1200
Fax: 714-433-1234
ecsales@ecliptek.com
WWW: http://www.ecliptek.com/ecliptek/

(Inductors, crystals and oscillators [including TXCO types])

ECS Inc.
230 n. Monroe Street
P.O. Box 273
Olanthe, KS 66061
800-237-1041
913-782-7787
Fax: 913-782-6991

(Crystals and crystal oscillators)

ECS-II part number
See *Enstore R&D GmbH*

EDC
See *Electro Dynamics Crystal Corporation*

Edal Industries, Inc.
4 Short Beach Road
East Haven, CT 06512
203-467-2591
Fax: 203-469-5929

(Diodes)

Edge Technology
40 Salem Street
Lynnfield, MA 01940
617-246-3800
Fax: 617-246-3888

(A/D Converters for data acquisition and CCD imaging systems, CCD analog processors)

Edgetek SN
7, Avenue des Andes
Les Ulis Cedex
France
33 1 64 46 0650
Fax: 33 1 69 28 4396
TLX: 600333

(SRAM modules)

EDI prefix/logo
See *Electronic Designs, Inc.*
(Memory devices),
Electronic Devices, Inc.
(Diodes and diode assemblies)

Edsun Laboratories, Inc.
564 Main Street
Waltham, MA 02154-4482
617-647-9300
Fax: 617-894-6927
TLX: 853664

(Interface and microprocessor support ICs)

EEV, Inc.
4 Westchester Plaza
Elmsford, NY 10523
914-592-6050
Fax: 914-682-8922

67 Westmore Drive
Rexdale, Ontario M9V 3Y6
Canada
416-754-9494
Fax: 416-745-0618

(Image processing, video processing and timing ICs, CCDs)

EF prefix
See *Thompson Components and Tubes Corporation,*
National Semiconductor
(Custom part)

Efar Microsystems, Inc.
800 Charcot Ave. #110
San Jose, CA 95131
408-943-1688
Fax: 408-943-1689
(This core logic and computer chip set company was acquired by Standard Microsystems Corporation (SMC) in March 1996)

EMF Systems, Inc.
120 Science Park Road
State College, PA 16803
814-237-5738
Fax: 814-237-7876

(VCXO/COMB generators, crystal oscillators, C-Band amplifiers, frequency synthesizers, VCOs, parts for military applications available)

EFL prefix
See *Excel Cell Electronic Co., Ltd.*

(Multilayer chip inductors)

EFM prefix
See *Gentron, Corp.*

EFZ prefix
See *Excel Cell Electronic Co., Ltd.*

(Ferrite chip EMI suppressors)

EG&G FREQUENCY PRODUCTS — EL

EG&G Frequency Products
See *CINOX Corporation*

EG&G IC Sensors
1701 McCarthy Blvd.
Milpitas, CA 95035-7416
800-767-1888
408-432-1800

(Accelerometers)

EG&G Power Systems, Inc.
1330 East Cypress Street
Covina, CA 91724
818-967-9521
Fax: 818-967-3151

(Military power supplies and DC/DC converters, formerly Almond Instruments or EG&G Almond Instruments)

EG&G Solid State Products Group
(Includes EG&G Reticon, manufacturer of image sensors, CCDs, solid state cameras; EG&G Vactec, manufacturer of phototransistors, LDRs, silicon sensors, EG&G Judson, manufacturer of infared detectors and EG&G Optoelectronics [formerly RCA Electro Optics] manufacturer of silicon APDs, diode lasers, LEDs and silicon and InGaAs detectors)

Group Sales Offices:
Massachusetts, Salem
508-745-7400
Pennsylvania, Montgomeryville
215-368-4003
Illinois, Elk Grove Village
708-640-7785
California, Sunnyvale
408-245-2060
El Toro 714-583-2250

EG&G Judson
221 Commerce Drive
Montgomeryville, PA 18936
215-368-6901
Fax: 215-368-6927

Sales Offices:
Eastern Region: 508-745-7400
Mid-Atlantic: 215-368-4003
Southeast: 404-928-1910
Central Region: 708-640-7785
Western Region: 408-245-2060
Southwest: 714-583-2250
Europe, Plaisir France:
33-1/30 54 70 21
Japan, Kanagawa-Ken:
04 66-35-0271

EG&G Reticon
345 Potero Avenue
Sunnyvale, CA 94086-4197
408-738-4266

Sales Offices:
35 Congress Street
Salem, MA 01970-6529
508-745-3200
Fax: 508-745-0894

416 Hungerford Drive, Suite 307
Rockville, MD 20850
301-251-0355

2260 Landmeier Road, Suite J
Elk Grove Village, IL 60007
312-640-7713

United Kingdom:
34/35 Market Place
Wokingham
Berkshire RG11 2PP, England
0734 788666
TLX: 847510 EGGUK G

Europe/Germany:
EG&G Instruments GmbH
Hohenlindener Str. 12
D-8000 Muenchen 80
Germany
089 92692-666
TLX: 528257 EGGID

EG&G Optoelectronics Canada
22001 Durnberry Road
Vaudreull
Quebec 7JV 8P7
Canada
(Si, InGaAs detectors)

EG&G Optoelectronics
(In the US, see EG&G Reticon)

(Automotive accelerometers)

EH prefix
See *Elantec, Inc.*

EIC logo
See *Electronics Industry (USA) Co., Ltd.*

Eight-by-Eight (8 x 8), Inc.
2445 Mission College Blvd.
Santa Clara, CA 95054
408-727-1885
Fax: 408-980-0432

Harleyford Estate
Marlow, Bucks SL7 2DX
England
44 1628 890 984
Fax: 44 1628 890 938

(Prior to early 1996 was known as Integrated Information Technology, Inc. [IIT]. Video codec ICs, graphics accelerators, image processing [compression and controller] ICs, MPEG ICs, video [image] processing, programmable video ICs, timing ICs, etc. Note: GUI [Graphics Accelerator-Interface] ICs being sold by subsidiary Xtechnology Inc., [which was created in late 1993]. The 486 microprocessor line was dropped in early 1996 with the company emphasis on video compression technology.)

EL prefix
See *Elantec, Inc.*, *Planar Systems, Inc.* (Flat panel displays),
Elmos GmbH

Manufacturers, Prefixes, Part Number Types, Logo Descriptions & Family Types

ELANTEC — ELECTRONIC DESIGNS, INC.

Elantec, Inc.
1996 Tarob Court
Milpitas, CA 95035
1-800-333-6314
408-945-1323
Fax: 408-945-9305
ElanFax (Faxback):
1-800-DATA113
TWX: 910-997-0649
http://www.elantec.com
Email: sales@elantec.com

Mark 128 Office Park
140 Wood Road, Suite 410
Braintree, MA 02184
617-849-9181
Fax: 617-849-0285

Europe:
Gordon House Business Centre
First Floor
6 Lissenden Gardens
London, NW5 1LX
England
44-71-482-4596
Fax: 44-71-267-1026

(Amplifiers including video amplifiers, operational amplifiers, monolithic DC/DC converters in SOIC packages, etc. [Note: military devices discontinued in February 1993])

Elbex Video Ltd.
Nihon Seimei Trade Center Bldg., 3Fl
7-25-5, Nishgotanda
Shinagawa-ku
Tokyo 141
Japan
03 3779 5222
Fax: 03 3779 5201

(Hybrid ICs, CCDs)

Elcap Electronics, Ltd.
19, Dai Fu Str.
Tao PO, New Territories
Hong Kong
8520 657 8883
Fax: 8520 650 7535
TLX: 33844

(74HC and 74HCT logic, SRAMs)

Elcut
(Thermal cutoffs)
See *Uchihashi Estec Co., Ltd.*

ELDEC Corporation
Power Conversion Division
P.O. Box 100, M/S M3-20
Lynnwood, WA 98046-0100
206-743-8399
Fax: 206-743-8562

(Commercial and military DC/DC converters)

Elec-Trol, Inc.
612 East Lake Street
Lake Mills, WI 53551
414-648-3000
Fax: 414-648-3001

(Optoelectronics)

Electrodynamics, Inc.
Minelco Products
A Talley Industries Company
1200 Hicks Road
Rolling Meadows, IL 60008
708-259-0740
Fax: 708-255-3827

(LED indicators, panel mount and military specification units)

Electrodyne
11200 SE 21 St.
Milwaukie, OR 97222
503-654-0711
Fax: 503-654-1959

(Miniature and small signal relays for communications applications)

Electromatic Controls Corp.
2495 Pembrook Ave.
Hoffman Estates, IL 60195
708-882-5757
Fax: 708-882-7234

(Solid state relays)

Electron Tubes, Inc.
100 Forge Way Unit F
Rockaway, NJ 07866
800-521-8382
201-586-9594
Fax: 201-586-9771
Email: phototubes@aol.com
Factory: Ruslip, England
Additional offices in San Diego, CA.

(Photomultiplier tubes, avalanche photodiodes, UV/Blue photodiodes, low dark current PINs, quadrant photodetectors, high unifirmity photosensors, high speed photodiodes, linear arrays, IR enhanced photodiodes, high voltage power supplies, and silicon photodetectors [the latter manufactured by Silicon Sensors, GmbH, Berlin Germany])

Electronic Designs Inc. (EDI)
42 South Street
Hopkinton, MA 01748
508-435-2341
Fax: 508-435-6302

One Research Drive
Westborough, MA 01581
508-366-5151
(LCDs)

Dallas, TX: 214-219-7288,
Fax: 214-219-7298
Washington: 206-834-8740,
Fax: 206-693-8652

Electronic Designs Europe, Ltd.
Shelly House,
The Avenue, Lightwater
Surrey GU18 5RF
United Kingdom
0276 72637
Fax: 0276 73748
Telex: 851 858325

ELECTRONIC DESIGNS, INC. — ELECTRO-OPTICAL

(SRAM products including qualified military devices. Note: This company was acquired by Crystallume in October 1995, sunlight readable LCD panels.)

Electronic Devices, Inc. (EDI)
An Electronic Components Company
21 Gray Oaks Ave.
Yonkers, NY 10710
800-678-0828
Fax: 914-965-5531

Germany:
EDI Halbleiter GmbH
Motzinger StraBe 43, POB 340
D-7270 Nagold
074 52/6 50 60, Fax: 7452/1470
TLX: 765946
Ansb: ENDRI.D

(Diode and diode assemblies)

Electronic Measurements, Inc.
405 Essex Road
Neptune, NJ 07753
908-922-9300
Fax: 908-922-9334

(Power supplies)

Electronic Precision Components
519 S. Fifth Ave.
Mt. Vernon, NY 10550
914-664-5591, 914-664-2333
Fax: 914-664-4729

(Inductors, delay lines)

Electronic Techniques, Ltd. (ETAL)
(Anglia)
10 Betts Avenue
Martlesham Heath,
Ipswitch IP5 7RH, England
44 0 1473 611422
Fax: 44 0 1473 611919

(Telecommunications transformers)

Electronic Technology Corporation
154 Research Park
Ames, IA 50010
515-296-7000
Fax: 515-296-7001

(Switched capacitor filters, pin for pin compatible parts for Linear Technology Corp. devices)

Electronics Industry Co., Ltd. (EIC)
(USA)
Bangkok Thailand Plant
and Sales Office:
662 326 0540-1/326 0932/326 0102
Fax: 662 326 0933

EIC Semiconductor, Inc.
16011 Foothill Blvd.
Irwindale, CA 91706
818 969-1315
Fax: 818 969-0965

EIC International Co. Ltd.
852 341 6681/344 4980
Fax: 852 343 9959

(Transient voltage suppressors, zener diodes, Schottky diodes and surface mounted diodes)

Electro Corporatior
1845 57th Street
Sarasota, FL 34243
1-800-446-5762
813-355-8411
Fax: 813-355-3120

(Sensors with built in processing circuitry)

Electronic Arrays
See *NEC Microelectronics*

(Consumer ICs including clock ICs and programmable parts)

Electronic Designs Inc.
See *EDI*

Electronic Solutions
210 Goddard Boulevard
King of Prussia, PA 19406
215-992-0882
Fax: 215-992-0734

(Custom ICs)

Electro Dynamics Crystal Corporation (EDC)
9075 Cody
Overland Park, KS 66214
800-EDC-XTAL
913-888-1750
Fax: 913-888-1260

(Custom and standard crystals, crystal filters, and oscillators)

Electrohome Ltd.
809 Wellington St.
North Kitchener, Ontario
N2G 4J6
519-744-7111

(LCDs)

Electromagnetic Technologies, Inc.
871 Mountain Ave.
Springfield, NJ 07081
201-379-1719
Fax: 201-379-1651

(Lumped element allpass networks, powered dividers and combiners, directional couplers [to 65GHz], beamformers)

Electro-Optical Systems, Inc.
1000 Nutt Road
Phoenixville, Pa 19460
215-935-5838
Fax: 215-935-8548

(IR radiation detectors, amplifiers, etc.)

Manufacturers, Prefixes, Part Number Types, Logo Descriptions & Family Types

ELECTROSTATIC DESIGNS — ELOGRAPHICS

ElectroStatic Designs
P.O. Box 30575
Tucson, AZ 85751
520-296-2868
Fax: 520-296-2904

(StaticBug [tm], static event detectors that are packaged as an IC)

Elektrotechnische Apparate (ETA)
Industriestr 2-8
Altdorf 84032
Germany
9187/10-0
Fax: 9187/10-397

(RISC processors)

Eletech Electronics, Inc.
16019 Kaplan Ave.
Industry Park, CA 91744
818-333-6394
Fax: 818-333-6494

Eletech Enterprise Co.
531-3F Chung Cheng Rd.
Hsin Tien
Taipei Shien
Taiwan
2 218 0068
Fax: 2 218 0254

(Digital sound ICs and voice module boards)

ELH
See *Elantec*
(See also *National Semiconductor* for LH parts)

Elisra Electronic Systems
48, Mivtza Kadesh St.
Bnel Braq, Israel 51203
972 3 7545015
Fax: 972 3 7545650

(SRAMs, including military type devices)

Elite Microelectronics, Inc.
4003 North First Street
San Jose, CA 95134-1599
408-943-0500
Fax: 408-943-0561

(Data controllers, cache DRAM controllers)

Elite Semiconductor Products, Inc.
430 W. Merrick Road
Valley Stream, NY 11580
516-825-1010
Fax: 516-825-1036

(Diodes and thyristors)

Elm State Electronics, Inc.
300 Shaw Road
North Bradford, CT 06471
203-484-7111

(Thyristors, transistors)

ELMA Electronic, Inc.
44350 Grimmer Blvd.
Fremont, CA 94538
510-656-3400
Fax: 510-656-3783

(LEDs, including chip and panel mount types)

Elmec Technology of America
1875 S. Grant Street, Suite 560
San Mateo, CA 94402
800-554-7652
415-341-1611
Fax: 415-341-2370
MCI Mail: 539-2234

(Delay lines)

Elmo Semiconductor Corp.
7590 North Glenoaks Boulevard
Burbank, CA 91504-1052
818-768-7400
Fax: 818-767-7038
TWX: 910-321-2943
TLX: 69-8181

(This company packages and screens IC die, they also have die to support discontinued Motorola military products. Kimball International, Inc. [Jasper, IN] acquired Elmo in early 1996.)

Elmos Electronik GmbH
Emil-Figge Strasse 81
Dortmund, Germany
49 0231 7544979

(ASICs and microcontrollers, primarily for automotive applications)

Elmwood Sensors, Inc.
500 Narragansett Park Drive
Pawtucket, RI 02861
800-356-9663
Fax: 401-728-5390

(Thermal cut-off devices, thermal sensing devices)

Elna America, Inc.
10529 Humbolt
Los Alamitos, CA 90720
800-700-ELNA
714-761-8600
Fax: 714-761-9188

(Surface mount chip LEDs, miniature sound generators, multicolor SMT [surface mount] switches)

Elo Touchsystems
(Formerly *Elographics*)
105 Randolph Road
Oak Ridge, TN 37830
615-482-4100
Fax: 615-482-4943

(Touch Screen ASICs, touch-screens)

Elographics
See *Elo Touchsystems*

ELPAC POWER SYSTEMS — EMI FILTER

Elpac Power Systems
1562 Reynolds Avenue
Irvine, CA 92714-5612
714-476-6070
Fax: 714-476-6075

(Power supplies including wall mounted and external units)

Elpaq
Division of Elmo Semiconductor
7590 North Glenoaks Boulevard
Burbank, CA 91504-1052
818-768-7500, Fax: 818-767-7038
TWX: 910-321-2943
TLX: 69-8181

(Division of Elmo semiconductor, MCM modules including SRAM, DRAM and Flash memory modules)

Eltech Electronics, Inc.
1262 E. Katella Avenue
Anaheim, CA 92805
714-385-1707
Fax: 714-385-1708

(Voice processing ICs)

Eltech Instruments, Inc.
350 Fentress Blvd.
P.O. Box 9610
Daytona Beach, FL 32120-9610
800-874-7780
904-253-5328, 252-0411
Fax: 904-258-3791

(Optoelectronics, linear ICs)

ELY logo
See *Elytone Electronics Co., Ltd.*

(Magnetics)

ELY prefix
See *Elytone Electronics Co., Ltd.*

(Magnetics)

Elytone Electronics Co., Ltd.
No. 9, Lane 210 Wen Chang St.
Taipei 10664,
Taiwan R.O.C.
886 2 7092500, Fax: 886 2 7557261

6175 Commodore St.
Columbia, MD 21045
410-740-8231, Fax: 410-997-7037

(Transformers and inducers, including SMD, ISDN, pulse and custom units)

EM Microelectronic-Marin S.A.
Zone Industrielle des Sors
CH-2074 Marin Switzerland
41 038 35 51 11
(After 11/9/96 41 32 755 51 11)
Fax: 41 038 35 54 03
(After 11/9/96 41 32 755 54 03)
TLX: 952 790

1600 Golf Road, Suite 1200
Rolling Meadows, IL
847-806-1497
Fax: 847-981-5006

(ASICs, mixed mode custom arrays, LCDs, display drivers, real time clocks, watchdog timers, watch and clock ICs, smart reset ICs, 4 bit microcontrollers, voltage surveillance IC [to clear the microprocessor after power up], voltage regulators, linear power supply ICs, dual two channel level shifter-read only contactless memory devices for identification applications, etc.)

EM prefix
See *Sigma Designs, Inc.*,
EM Microelectronic

(4-bit microcontrollers)

EM Research Engineering, Inc.
2705 Highway 40
Suite 301
P.O. Box 1247
Verdi, NV 89439
702-345-2411
Fax: 702-345-1030

(RF and microwave products including phase locked oscillators, synthesizers, multipliers, power amplifiers, broadband amplifiers and microwave assemblies)

EMBE Electronik
Postfach 12 04 08
Munchen 12, 8000
Germany
89 502 5826
Fax: 89 502 6111

(Stepper motor controllers)

EMCO High Voltage Company
11126 Ridge Road
Sutter Creek, CA 95685
P.O. Box 1025
Sutter Creek, CA 95685-9989
e-mail: emco@ix.netcom.com
800-546-3680
Fax: 209-223-2779

(High voltage power supplies)

EMD
See *Emulex Micro Devices*

EMF prefix
See *Elpaq*

Emhiser Micro-Tech
2705 Old Highway 40 West
P.O. Box 708
Verdi, NV 89439-0708
702-345-0461
Fax: 702-345-2484

(Miniature VCOs)

EMI-XXXXX part number
See *ILC-DDC*

EMI Filter Company
Division of Nordquist
Dielectrics, Inc.
9075 A 130th Avenue
N. Largo, FL 34643
813-585-7990
Fax: 813-586-5138

(Miniature EMI filters)

EMS prefix
See *Elpaq*

EMSI
See *Enhanced Memory Systems, Inc.*

EMT logo
See *Emhiser Micro-Tech*

E-mu Systems, Inc.
1600 Green Hills Road
Scotts Valley, CA 95067
408-438-1921
Fax: 408-438-8612

(Subsidiary of parent company Creative Laboratories, audio ICs, multimedia audio digital signal processors)

EMU prefix
See *E-mu*

Emulex Corp.
Emulex Micro Devices (EMD) Division
Renamed Q Logic Corp. in May 1993.
3545 harbor Blvd
Costa Mesa, CA 92626
714-662-5600

Anaheim CA: 714-385-1685
San Jose, CA: 408-452-4777
Rosewell, GA: 404-587-3610
Burlington, MA: 617-229-8880
Saddlebrook, NJ: 201-368-9400
Houston, TX: 713-981-6824
Reston, A: 703-264-0670
Schaumburg, IL: 708-605-0888
Chatsworth, Australia:
61 2 417 8585
Ottawa, Ontario Canada:
613 230 3543
Toronto, Canada: 416-673-1211
Berkshire, England: 44 734 772929
Paris, France: 33 134 65 9191
Munich, Germany: 49 89 3608020
Agrate Brianza, Italy:
39 39 639261
Taipei, Taiwan: 886 2 5 62 3230

(SCSI processor ICs. Note: Firefly fiber channel chip set licensed to VLSI Technology in 1995)

Encore (Gate Arrays)
See *Orbit Semiconductor, Inc.*

Endevco
30700 Rancho Viejo Road
San Juan Capistrano, CA 92675-1785
800-982-6732, 800-309-9090
714-493-8181
Fax: 714-661-7231
TLX: 68-5608 ENDEVCO SJUC

AL: 800-309-9090, 904-383-0477
Fax: 904-383-6558
AK: 415-697-9887
Fax: 415-697-1254 Alaska
AZ (except Yuma): 505-292-8990

(Accelerometers, pressure transducers)

Endicott Research Group, Inc. (ERG)
2601 Wayne Street
P.O. Box 269
Endicott, NY 13760
607-754-9187
Fax: 607-754-9255

(DC/AC inverters for LCD backlights)

Engineered Components Co. (EC2)
3580 Sacramento Dr.
San Luis Obispo, CA 93406-8121
P.O. Box 8121
San Luis Obispo, CA 93403-8121
800-235-4144 (outside CA)
805-544-3800
Fax: 805-544-8091

(Delay lines, encoders, RF/low power toroidal inductors, ECL 100K square wave module)

The Engineering Consortium (TEC)
3130B Coronado Drive
Santa Clara, CA 95054
408-748-1984
Fax: 408-748-0216

(A fabless foundry and design center for mixed signal technology ICs for data and telecommunication use. Products include ICs for the hearing device market.)

English Electric Valve Co. Ltd.
Waterhouse Lane
Chelmsford, Essex, CM1 2QU
United Kingdom
0245 493493
Fax: 0245 492492
TLX: 99103G

(Diodes, optoelectronics, transistors)

Enhanced Memory Systems, Inc.
1850 Ramtron Drive
Colorado Springs, CO 80921
800-545-3276 (800-545-DRAM)
719-481-7000
Fax: 719-488-9095, 719-481-9170
e-mail: info@ramtron.com
web server: http//www.csn.net/ramtron

(The enhanced DRAM operations of Ramtron International Corporation, spun off as a separate subsidiary in May 1995).

Ensign-Bickford Company
Laser Diode Operations
660 Hopmeadow Street
P.O. Box 427
Simsbury, CT 06070-0427
203-843-2286
Fax: 203-843-2675

(Semiconductor laser diodes)

ENSTORE R&D — ERG

Enstore R&D GmbH
Niesenbergergasse 39
A-8020 Graz, Austria
43 316 91 77 55
Fax: 43 316 91 51 01

In the U.S. imported by EnChip Inc.
434 Ridgedale Ave.
Suite 11-327
East Hanover, NJ 07936
201-328-2049
Fax: 201-301-0402

(Battery charger IC which controls and monitors charging current flow)

Entran Devices, Inc.
10 Washington Ave.
Fairfield, NJ 07004
1-800-635-0650
201-227-1002
Fax: 201-227-6865

B.P. No. 59, 78340
Les Clayles-Sous-Bois, France
33 1 30 55 49 85
Fax: 33 1 34 81 03 59
TLX: 695539

26, rue des Dames
78340 Les Clayes-Sous-Bois
France

19-19a, Garston Park Parade
Garston, Watford, Herts, WD2 6LQ
England
0923 893999
Fax: 0923 893434

Benzstr. 27
D-67063 Ludwigshafen
Germany
0621 692061
Fax: 0621 631428

(Pressure transducers, load cells, accelerometers, strain gauges, pressure transmitters)

E-O Communications, Inc.
2 Fl., No. 469, Fu Hsing
N. Road, Taipei
Taiwan, R.O.C.
886 2 5455735
Fax: 886 2 5455652

834 West California Avenue
Sunnyvale, CA 94086
408-720-8608
Fax: 408-720-8606

(LEDs, including high brightness, clusters, dot matrixes and modules)

EP prefix
See *Elantec, Altera Corp.,*
PCA Electronics, Inc.,
Eagle Pitcher (If battery)

EPB prefix
See *Altera Corp.*

EPC prefix
See *Altera Corp.,*
Allied Electronics GmbH

(Encoder pulse converter)

EPF prefix
See *Altera Corp.*

Epitaxx, Inc.
7 Graphics Drive
West Trenton, NJ 08628
609-538-1800
Fax: 609-538-1684, 609-452-0824

West Coast:
310-551-6507
Fax: 310-551-6577

(Laser products, such as PIN photodiodes, laser diodes)

Epitek Electronics, Inc.
Commerce Park
Odgensburg, NY 13669
613-592-2240

(Resistor networks)

Epitek International, Inc.
100 Schneider Road
Kanata, Ontario K2Y 1Y2
Canada
613-592-2240
Fax: 613-592-9449
TLX: 053 3544

(Interface and linear ICs)

EPM prefix
See *Altera Corporation,*
Texas Instruments

EPS prefix
See *Altera Corp.*

Epson America, Inc.
(Seiko Epson Corporation)
20770 Madrona Avenue
P.O. Box 2843
Torrance, CA 90503
800-922-8911
Fax: 310-782-5220

OEM Division:
P.O. Box 2842
800-433-3597
Fax: 310-782-5230

3-5 Owa 3-Chome
Suwa-Shi
Nagano-Ken, 392
Japan
81 0266 52 3131
Fax: 81 0266 58 9861
TLX: 3362-435
See also *Seiko*

EPX prefix
See *Altera Corp.*

ER prefix
See *General Instrument*

ERG logo
See *Endicott Research Group*

Ericsson Components AB
Power Products
S-164 81 Kista
Stockholm, Sweden
46 8721 6356, Fax: 46 8721 7001

http://www.ericsson.com/
http://www.ericsson.com/EPI/BK/index.html
http://www.ericsson.nl/
(The Netherlands)

Ericsson Components, Inc.
(PBL prefix)
403 International Parkway
Richardson, TX 75081
214-669-9900 (Network Division)
214-997-6744
214-952-8800 (Radio Division)
Fax: 214-680-1059

701 N. Glenville Dr.
Richardson, TX 75081
214-997-6561

France, Guayancourt:
33-1-30 64 85 00,
Fax: 33-1-30 64 11 46

Germany, Neu-Isenburg:
49-6102-200 50,
Fax: 49-6102-20 05 33

United Kingdom, Swindon:
44-793-48 83 00,
Fax: 44-793-48 83 01

Hong Kong, East Asia, Wanchai,:
852-519 23 88,
Fax: 852-507 46 84

Italy, Milano:
39-2-33 20 06 35,
Fax: 39-2-33 20 06 41

Norway, Oslo:
47-2-84 18 10,
Fax: 47-2 84 19 09

Sweden, Stockholm:
46-8-757 43 84,
Fax: 46-8-757 44 21

(Telecommunications products, DC/DC converters, line protection circuits, subscriber interface circuits [SLICs], power devices, speaker-phone ICs, stepper motor controllers and drivers, part of the Swedish telecommunications firm L.M. Ericsson. Note: In 1995 the relay production business was sold to Anritsu Wiltron, the company that licensed the relay technology to Ericsson.)

ERSO
8th Floor, 315 Song Chiang Road
Taipei 10477
Taiwan
812 2 502 8212
Fax: 812 2 502 8795
TLX: 12974

(Microcontrollers, DRAMs, ROMs, SRAMs, keyboard controllers, error correction ICs, RTCs, computer ICs, telecommunications ICs [phone dialers, tone ringer, DTMF], modem ICs, phase locked loops, telephone line filters. Part of Taiwan Research Service Organization)

ES logo
See *Electronic Solutions, Inc.*

ES prefix
See *ESS Technology, Inc.*

ES2 Limited
See *European Silicon Structures*

ESC Electronics Corporation
534 Bergen Blvd
Palisades Park, NJ 07650
800-631-0853
Fax: 201-947-0406

(Delay lines, filters)

ESE
See *Elm State Electronics, Inc.*

ESG prefix
See *AVG Semiconductors*
(Second source voltage regulators)

ESP prefix
See *Q Logic Corp.*

ESS Technology, Inc.
46107 Landing Parkway
Fremont, CA 94538
510-783-3100, 510-226-1088

(A fabless company offering PC audio ICs, Codecs)

ET prefix
See *Tseng Labs, Inc., Edge Technology, Eteq Microsystems, Inc., Elytone Electronics Co., Ltd.*

(Magnetics)

E-T-A
1551 Bishop Ct.
Mount Prospect, IL 60056
847-827-7600, Fax: 847-827-7655
http://www.etacb.com

236 Hood Road
Markham, ON L3R 3K8, Canada
905-475-5886
Fax: 905-475-5889

(Panel mounted circuit breakers)

ETA
See *Elektrotechnische Apparate*

ETAL
See *Electronic Techniques* (Anglia), *Ltd.*

Etalon, Inc.
1600 W. Main Street
Lebanon, IN 46052
317-483-2550, Fax: 317-483-2560

(Piezo transducers)

ETC — EVERLIGHT

ETC
See *Electronic Transistors Corp.*

E-TEK
See *E-TEK Dynamics, Inc.*

E-TEK Dynamics, Inc.
1885 Lundy Avenue
San Jose, CA 95131
408-432-6300
Fax: 408-432-8550

(Pumped laser modules, includes laser diode, thermoelectric cooler, bandpass wavelength division multiplexer, monitor photodiode and pigtailed fiber)

Eteq Microsystems, Inc.
1900 McCarthy Blvd
Milpitas, CA 95035
408-432-8147, Fax: 408-432-8146

(PC Chip sets)

ETF Technology
800 Paloma Drive
Round Rock, TX 78664-2400
512-218-1048

(Thin film chip resistors, thick film multilayer circuits, custom networks and hybrids)

ETL, Enhanced Transceiver Logic
A relaxed specification form of BTL or Backplane Transceiver Logic
See *TI*

Etmas Elektrik Tesisati VE
(Turkey)
See *Siemens*

ETRI, Inc.
1503 Rocky River Road
North, Monroe, NC 28110
704-289-5423
Fax: 704-283-7170

(DC/DC converters)

Etron Technology, Inc.
1F, No. 1
Prosperity Road I
Science Based Industrial Park
Hsinchu, Taiwan R.O.C.
886 35 782 345
Fax: 886 35 779 001

Etron Integrated Circuits Corporation
Sunnyvale, CA
408-987-2255
Fax: 408-987-2250

(This fabless company offers SRAMs, DRAMs and speciality memories.)

Eupec
Max-Planck StraBe 5
Warstein-Belecke, 4788
Germany
49 2902 764 160
Fax: 49 2902 764 510

(Diodes, SCRs, thyristors, transistors)

EUROM FlashWare Solutions, Ltd.
Atidim Industrial Park Bldg. 1
P.O.B. 58032
Tel-Aviv 61580, Israel
972 3 490920
Fax: 972 3 490922

556 Mowry Avenue
Ste. 103
Fremont, CA 94536
510-505-9083
Fax: 510-505-9084

4655 Old Ironsides Drive
Suite 200
Santa Clara, CA 95054
408-748-9995
Fax: 408-748-8408

(Multichip modules, including a 28 pin device that stores BIOS and a solid state disk, called DiskOnChip, Smart Flash ROM called SFROM [trade mark, patented in 1993], an in-system reprogrammable device through a serial interface.)

European Silicon Structures (ES2)
Aix-Provence
Rousset, France
Bracknell, Berks, UK
011 44 344 52525

(An ASIC foundry specializing in fast turnaround parts, acquired by Atmel Corporation in early 1995)

Eurosil Electronic GmbH
Erfurter Strasse 16
Eching 85386
Germany
089 319 700
Fax: 089 319 4621
TLX: 522432

77 Mody Road
Tsim Sha Tsui
Kowloon, Hong Kong
0083 722 1306
Fax: 0083 721 4843
TLX: 33 448

(Microcontrollers, EPROMs)

Everlight Electronics Co., Ltd.
No. 25, Lane 76
Chung Yang Rd.
Sec. 3, Tucheng, Taipei 236
Taiwan, R.O.C.
886-2-260-2000
Fax: 886-2-260-6189

Sales Offices:
U.S.
34 Foley Drive
Sodus, NY 14551-0061
1-800-253-3576
315-483-4930
Fax: 315-483-9480
TLX: 200806

Hong Kong: 852-385-8862, 388-0602; Fax: 852-388-1127

Malaysia: 60-4-360398, 361796, 362746; Fax: 60-4-362196

Europe: 01-858-16-11; Fax: 01-858-05-25

(LEDs, IR receiver modules, photo-interrupters, LCD backlights, full color RGB LED lamps)

Exar Corporation

Headquarters:
P.O. Box 49007
2222 Qume Drive
San Jose, CA 95161-9007
1-800-782-5167, 408-434-6400
Fax: 408-943-8245, 408-435-1233
Worldwide Web Site:
http://www.exar.com
Email: converters@exar.com

Southwest:
23 Riverun
Irvine, CA 92714
714-559-6179
Fax: 714-559-0697

South Central:
P.O. Box 260527
Plano. TX75026-0527
214-235-2698

North Central:
800 E. Northwest Hwy, Suite 728
Palantine, IL 60067
708-705-3832
Fax: 708-705-3852

Northeast:
33 Boston Post Road, West, Suite 270
Marlborough, MA 01752
508-624-4400
Fax: 508-624-0799

Mid-Atlantic and Southeast:
293 Sentinel Ave.
Newtown, PA 18940
215-579-7542, Fax: 215-579-7543

Northern Europe/United Kingdom:
Orion House, 49 high Street
Addlestone, Surrey KT15 1TU
United Kingdom
44 932 857315
Fax: 44 932 858761

Japan:
3-18-9 Shin-Yokohama
Kohoku-Ku Yokohama-Shi
Kamagawa 222
Shin-Yokohama IC Bldg. 2F
Japan
81 45 472 4349
Fax: 81 45 472 4601

France:
29 Rue du President Kennedy
91440 Bures/Yvette
France
33 1 692 83131
Fax: 33 1 692 86960

(A fabless company supplying modem, fax/data, phase locked loops, signal generator ICs, ASICs, caller ID ICs, PCM [pulse code modulation] transmission ICs, telephone ICs, UARTs, audio filters and equalizers, VTR tracking filter, disk drive ICs, D/A-A/D converters [Micro Power System product line, which was acquired by EXAR in June 1994], etc.)

Excel Cell Electronic Co., Ltd. (ECE)

No. 20, 25RD
Taichung Industrial Park
Taichung, Taiwan, R.O.C.
886-4-3591253 (Rep.)
Fax: 886-4-3593893
Cable: "EXCELL," Taichung
TLX: 57137 EXCELL

6F-9, No. 1
Wu-Chuan 1 Road
Hsin Chung City
Taipei Hsien, Taiwan, R.O.C.
886-2-2996268 (Rep.)
Fax: 886-2-2994795

(Chip inductors and EMI suppressors, machine pin IC sockets for PGA, SIP, DIP [including solder fork pin DIP sockets] and carriers, reed relays, dip switches and terminal blocks).

Excel Technology International Corp.

Unit #5, Bldg. #4, Stryker Lane
Hillsborough Industrial Park
Belle Meade, NJ 08502
908-874-4747
Fax: 908-874-3278
TLX: 9103339588

(Optoelectronics)

Exel Microelectronics, Inc.

(A Division of ROHM Corporation)
2150 Commerce Drive
P.O. Box 49038
San Jose, CA 95161-9038
800-438-3935
408-432-0500
Fax: 408-432-8710
http://www.exel.com
(For International sales representatives in France, Germany, Hong Kong, and the United Kingdom, see *ROHM Electronics*.)

(E2 products such as E2PROMs, and E2PLDs, flash based microcontrollers and an random code generator for security applications [Surelok])

Exergen Corporation

One Bridge Street
Newton, MA 02158
1-800-422-3006
617-527-6660
Fax: 617-527-6590

(IR thermocouples including focused types to measure temperatures of small targets at long distances)

[-]- EXPONENTIAL TECHNOLOGY — FBH

Exponential Technology, Inc.
2075 Zanker Road
San Jose, CA 95131
408-441-6050
Fax: 408-441-6-51
(Formerly Renaissance Microsystems, founded in 1993)

(Microprocessors such as PowerPC clone ICs)

EY logo
See *Elytone Electronics Co., Ltd.*

(Magnetics)

EY prefix
See *Elytone Electronics Co., Ltd.*

(Magnetics)

EZ prefix
See *Rosetta Technologies, Inc.*, *Semtech Corporation*

(Voltage dropping circuit, such as from 5 to 3.3V)

EZ Communications, Inc.
9939 Via Pasar
San Diego, CA 92126
619-621-2700
Fax: 619-621-2722

(VCOs)

EZ Link Chip Set
(A fiber channel chip set)
See *Vitesse Semiconductor Corp.*
F, between a top and bottom bar logo
See *Fujitsu*

F logic family
See *Philips Corporation*, *Motorola*, *National Semiconductor*, *Mitsubishi*, *Texas Instruments*

F prefix
See *National Semiconductor* (Fairchild Semiconductor), *FOX Electronics* (Crystal oscillators), *AVG Semiconductors* (Second source some ICs)

FA prefix
See *Applica, Inc.*

Fact Logic
A trademark of *National* (formerly Fairchild) *Semiconductor*

Fagor Electronic Components, Inc.
2250 Estes Avenue
Elk Grove Village, IL 60007
800-888-9863
708-981-0161
Fax: 708-981-1311
TLX: 285273 FAGOR UR

Rm 1208 & 1211, Sunbeam Centre
27 Shing Yip Street
Kwun Tong
Kowloon, Hong Kong
852 797 08 61 4
Fax: 852 797 85 19
TLX: 32689 FAGAS HX

(Rectifiers, etc.)

Fairchild Semiconductor
This company was formed in 1958 and acquired by Schlumberger Ltd. in 1979. It was sold to National Semiconductor in 1987 [except for the Fairchild Clipper processor division which went to Intergraph Corporation, Palo Alto, CA]. The name was revived in 1996 when National Semiconductor spun off its standard [CMOS] logic, memory.

(EPROMs, EEPROMs and application specific flash memory and discrete transistor products.)

Faraday, Inc.
See *Western Digital*
(They are a division)

Farnell Advance Power, Inc.
32111 Aurora Road
Solon, OH 44139
216-349-0755
Fax: 216-349-0142

(Power supplies, digital photo-multipliers)

Farnell Relay Products
18 Technology Drive
#200
Irvine, CA 92718
714-727-3001
Fax: 714-727-2109

(Relays)

FAS prefix
See *QLogic Corp.*

Fasco Industries, Inc.
1845 57th Street
P.O. Box 3049
Sarasota, FL 34230-3049
813-355-8411
Fax: 813-355-3120

(Optoelectronics)

Fast Logic
A trademark of *National* (formerly Fairchild) *Semiconductor*

FB prefix
See *Philips Semiconductor*, *Smar Research Corp.*

(Field bus controller)

FB logic family (i.e., 54FBXXX)
See *Texas Instruments*
(For Future Bus applications)

FBH prefix
See *Semtech* (Formerly Lambda)

Manufacturers, Prefixes, Part Number Types, Logo Descriptions & Family Types

FC prefix
See *STC Components, Inc.*, *C-MAC Microcircuits Ltd.*, *STC Components, Inc.*

FCI Components
350 Town & Country Vil.
San Jose, CA 95128
408-985-4571
Fax: 408-261-8289

(Magnetics, BNC connectors, transient voltage suppressors, DC/DC converters)

FCT logic parts
See *Integrated Device Technology (IDT)*, *Harris Semiconductor*, *National Semiconductor*, *Performance Semiconductor* (Now sold by Cypress Semiconductor), *Pericom Semiconductor Corp.*, *Quality Semiconductor, VTC Inc.*, *Motorola Semiconductor*

FCT prefix
See *Integrated Device Technology (IDT)*

FD-XXX part number
(Floppy drives)
See *Teac*

FD prefix
See *Fermionics Opto-Technology*

FDK America, Inc.
2270 North First St.
San Jose, CA 95134
408-432-8331
Fax: 408-435-7478

404 Wyman St.
Suite 300
Waltham, MA 02154
617-487-3198
Fax: 617-487-3199

Las Colinas Executive Suites
320 Decker Drive
Suite 140
Irving, TX 75062
214-719-2536
Fax: 214-719-2591

FDK Corporation
5-36-11, Shimbashi, Minato-ku
tokyo 105
03 3434 1271
Fax: 03 3431 9436

(Optcal isolators, optical circulators, optical switches and optical isolator modules)

FE prefix
See *FEI Microwave, PixTech, Inc.*

(Field emission displays)

FDC prefix
See Standard Microsystems Corp.

(Floppy disc controller)

Federal Systems Company
See *Loral Federal Systems Company*

FEE Fil-Mag
See *Fil-Mag*

FEI Communications, Inc.
Subsidiary of Frequency Electronics, Inc., TRW
55 Charles Lindbergh Blvd.
Mitchel Field, NY 11553
516-794-4500
Fax: 516-794-4340

(GaAs MMIC amplifiers)

FEI Microwave, Inc.
Subsidiary of Frequency Electronics, Inc.
825 Stewart Drive
Sunnyvale, CA 94086
408-732-0880
Fax: 408-730-1622

FC — FERMIONICS OPTO-TECHNOLOGY

Mexico:
Transamerican Industries S.A.
34-A del Centro Comercial
Mission del Sol
Tijuana, Mexico

(Semiconductors, including military qualified devices. Note: The Diode product line was sold to M/A-Com Semiconductor Division, Burlington Semiconductor Operations, 43 South Avenue, Burlington, MA 01803-4903, 617-272-3000, Fax: 617-272-8861)

Fema Electronics Corp.
12-6 Edgeboro Road
East Brunswick, NJ 08816
800-292-3362
908-238-1223
Fax: 908-238-7981

(Diodes, LCD display modules)

Fer Rite Electronics Ind. Co., Ltd.
United Electronics Corp. Ltd.
11F, No. 568, Kuang Fu S. Rd.
Taipei, Taiwan, R.O.C.
886 2 706 7220 2
Fax: 886 2 705 6753

Hong Kong:
852 2602 6050
Fax: 852 2608 0234

(Multilayer ferrite chip beads and inductors, ferrite cores)

Fermionics Opto-Technology
4555 Runway Street
Simi Valley, CA 93603
805-582-0155
Fax: 805-582-1623

(InGaAs photodiodes, laser diodes, photodetectors)

FERRANTI SEMICONDUCTORS — FORMOSA

Ferranti Semiconductors
See *GEC Plessy*

FF prefix
See *Zetex*

Figaro USA, Inc.
1000 Skokie Blvd., Ste 575
Wilmette, IL 60091
708-256-3546, Fax: 708-256-3884

(Gas sensors)

Fil-Mag
See *Pulse*

Film Microelectronics, Inc.
530 Turnpike
North Andovewr, MA 01845
508-975-3385
Fax: 508-975-3506

(Custom multichip modules and hybrid microcircuits, pin diode drivers, solid state relays, MMIC microwave amplifiers, resistor chips, resistor networks. This company is certified to MIL-STD-1772.)

Fincitec OY
Kaylakuja 1
Helsinki, 00610, Finland
35890 799 863
Fax: 358 90 757 2952

Lumikontie 2
Kemi 94600, Finland
358 9698 21 490
Fax: 358 9698 21 561

Teknoliantie 2
Oulu 90570, Finland
358 981 5514 519
Fax: 358 981 5514 518

(Sigma delta ICs, codecs, A/D converters, telephone ICs)

Finisar Corporation
620B Clyde Avenue
Mountain View, CA 94043
415-691-4000
Fax: 415-361-4010

(Optical transceiver, transmitter and receiver modules)

Finlux Inc.
1400 NW Compton Drive
Beaverton, OR 97006
503-690-1100, Fax: 503-645-7024

(Optoelectronics)

FIS, Inc.
2-5-26-208, hachizuka
Ikeda, Osaka, 563, Japan
81 727 61 5886, Fax: 81 727 61 5876

(Gas sensors)

Flat Candle Company
4725 Suite B Town Center Drive
Colorado Springs, CO 80916
719-573-1880, Fax: 719-573-2080

(Fluorescent backlight lamps)

FLDM-TTL-XXXX part number
See *Engineering Components Company*

(Delay line)

FLJ prefix
See *Datel, Inc.*

Floyd Bell, Inc.
PO Box 12327, 897 Higgs Avenue
Columbus, OH 43212
614-294-4000, Fax: 614-291-0823

(Piezo alarms)

FLT prefix
See *Datel, Inc.*

FM prefix, digital transistors
See *ROHM Electronics Division*,
Ramtron International Corp.
(Memory devices, FRAMs),
Ascom Microelectronics,
GHz Technology, Inc.
(Microwave transistors)

FMMD prefix
See *Zetex*

FMMT prefix
See *Zetex*

FMMV prefix
See *Zetex*

FMMZ prefix
See *Zetex*

Focam Technologies, Inc.
3050 Cartier Blvd
West laval, Quebeck H7V 1J4
Canada
514-687-4883
Fax: 514-687-4875

317 Renfrew Drive, Suite 302
Markham, Ontario
Canada

(A fabless company supplying engineering design services in mixed signal designs. They provide a custon designed IC in a turn-key operation.)

Ford Microelectronics, Inc.
9965 Federal Drive
Colorado Springs, CO 80921-3698
800-824-0812
Fax: 719-528-7635
E-mail: fmicos!fmi@uunet.uu.net (In ASCII)

(Custom devices for automotive applications)

Formosa Silicon Semiconductor

(Rectifiers, etc., available from Taitron Components, Inc., 25202 Anza Drive, Santa Clarita, CA 91355-3496, 800-

824-8766, 800-247-2232, 805-257-6060; Fax: 800-824-8329, 805-257-6415)

Fortec Electronic AG
Zenithstrasse 132
Neunkirchen-Seelscheid 56479
Germany
2247 66 60
Fax: 2247 71 59

IS Maningerstrasse 7
Ascheim 85609
Germany
89 903 8581
Fax: 89 903 0384
TLX: 521 4916

(SRAM modules)

Forton/Source
2925 Bayview Drive
Fremont, CA 94538
510-440-0188
Fax: 510-440-0928

(ÿabletop power supplies and DC/DC converters)

FOX Electronics
5570 Enterprise Parkway
Ft Myers, FL 33905
813-693-0099
Fax: 813-693-1554
TLX: 510-951-7386

(Crystal oscillators, programmable clock oscillators, real time clocks [RTCs])

Foxoboro/ICT, Inc.
A Siebe Company
199 River Oaks Parkway
San Jose, CA 95134-1996
1-800-898-2224
408-432-1010
Fax: 408-432-1860

(Pressure sensors in TO cans and dip packages)

FPD Corp.
Eindhoven, The Netherlands

(AM LCDs)

FPM prefix
See *Fujikura, Servoflo Corp.*

FR prefix
See *FCI Semiconductor*

(Transient voltage suppressors-avalanche protected button shaped diode)

FR Industries, Inc.
557 Long Road
Pittsburg, PA 15235
412-242-5903
Fax: 412-242-5908
TLX: 866 719

(Thyristors)

FRE prefix
See *Harris Semiconductor*

(Radiation hardened power tran-sistors)

Frederick Components International Ltd.
Warner-West Plaza
22020 Clarendon St.
Suite 207
Woodland Hills, CA 91364
818-347-4571

(Surface mounted transient suppressors)

Frequency and Time Systems, Inc.
See *Datum, Inc.*

Frequency Devices, Inc.
25 Locust Street
Haverhill, MA 01830
508-374-0761
Fax: 508-521-1839

FORMOSA — FRONTIER ELECTRONICS

(Hybrid active filter products including linear, digital, wideband, programmable, noise elimination, bandreject, multichannel filters, custom filters and telecom-munication filters; digitally programmable oscillators, and filter/amplifier systems)

Frequency Electronics, Inc.
55 Charles Lindburgh Blvd.
Mitchell Field, NY 11553
516-794-4500
Fax: 516-794-4340

(Quartz crystals, crystal oscillators, etc.)

Frequency Products Corporation
(Formerly EG&G Frequency Products, Inc.)
See *CINOX Corporation*

Frontech Industries, Inc.
14115 Goldenwest St.
Suite C
Westminster, CA 92683
714-379-7320
Fax: 714-379-7322

(Founded in 1995, this company is a joint venture with manufacturing firms in mainland China to market LCDs in North America. Products include STN [super twist nematic], TN [twisted nematic] in alpha-numeric, dot matrix and graphics formats)

Frontier Electronics Co., Ltd.
4F, No. 48, Min Chuan Road
Hsin-Tien
Taipei Hsien
Taiwan
886 2 914 7685 9
886 2 918 6304

(Diodes)

FRP — FUJITSU MICROELECTRONICS

FRP prefix
See *Elektrotechnische Apparate (ETA)*

FSI/Fork Standards, Inc.
668 E. Western Avenue
Lombard, IL 68148
800-468-6009
Fax: 708-932-0016
http://www.xnet.com/~fsi

(Optical shaft encoders, photo-electric sensors, industrial counters and controls)

FSY Microwave, Inc.
7125 Riverwood Drive
Columbia, MD 21046
410-381-5700, Fax: 410-381-0140

(RF and microwave filters and multiplexers)

FT prefix
See *Fincitec OY*

FTD logo
See *Future Technology Devices International, Ltd.*

FTD prefix
See *Future Technology Devices International, Ltd.*

FTR prefix,
See *Optical Communication Products, Inc.*

FTS
See *Frequency and Time Systems, Inc.*

Fuji Electric Co., Ltd.
Fuji Semiconductor
New Yurakucho Building
1-12-1, Yuraku-Cho, Chiyoda-Ku
Tokyo 100, Japan
03 3211 7111
Fax: 03 3536 8868

Imported and distributed by:
Collmer Semiconductor, Inc.
14368 Proton Road
Dallas, TX 75244
214-233-1589
1-800-527-0251

(Diodes, transistors logic ICs and linear and hybrid ICs, for cameras, cars, communications, VCRs, etc.)

Fuji Electrochemical Co., Ltd.
36-11, Shimbashi 5-chome
Minato-ku
Tokyo 105
Japan
81 3 3434 1271

(Hybrid ICs)

Fujitsu Ten Limited
2-28, Gosho-dori 1-chome
Hyogo-ku
Kobe-shi
Hyogo 652
Japan
078 671 5081

(Hybrid ICs)

Fujikura America, Inc.
1400-100 Galleria Parkway NW
Atlanta, GA 30339
404-956-7200
Fax: 404-984-6965

(Pressure sensors; see also *Servoflo Corp.*)

Fujitsu General Limited
International Business Operations
1116 Suenaga
Takatsu-ku
Kawasaki 213, Japan
81 044 866 1111
Fax: 81 044 888 4901
TLX: 3842511

(Hybrid microcircuits)

Fujitsu Kiden Ltd.
Yanokuchi Inagi
Tokyo, Japan
0423 77 5111
Fax: 0423 78 3013
TLX: 2832 621

(Optoelectronics)

Fujitsu Microelectronics, Inc.
3545 N. First Street
San Jose, CA 95134-1804
1-800-866-8608
(24 hour customer service line)
408-456-1161
408-922-8000, 9000, 9200
Fax: 408-432-9044, 408-943-9293
http://www.fujitsu.com/fml/

Advanced Products Division
77 Rio Robles
San Jose, CA 95134-1807
LAN Products: 1-800-866-8608
SPARC Products: 1-800—523-0034
Fax: 408-943-9293

10600 N. DeAnza Blvd. #225
Cupertino, CA 95054
408-996-1600

Century Center
2603 Main St. #510
Irvine, CA 92714
714-724-8777

5445 DTC Parkway, #300
Englewood, CO 80111
303-740-8880

3500 Parkway Ln., #210
Norcross, GA 30092
404-449-8539

One Pierce Place, #1130
Itasca, IL 60143-2681
708-250-8580

1000 Winter Street, #2500
Waltham, MA 02154
617-487-0029

FUJITSU MICROELECTRONICS — FULLYWELL

3460 Washington Dr., #209
Eagan, MN 55122-1303
612-454-0323

898 Veterans Memorial Highway
Building 2, Suite 310
Hauppauge, N.Y. 11788
516-582-8700
Fax: 516-582-3855

15220 N.W. Greenbrier Pkwy.
#360
Beaverton. OR 97006
503-690-1909

14785 Preston Rd., #274
Dallas, TX 75240
214-233-9394

Asia:
Fujitsu Microelectronics Pacific
 Asia Ltd.
616-617, Power B,
New Mandarin Plaza
14 Science Museum Road
Tsimshatsui East,
Kowloon, Hong Kong
723-0393
Fax: 721-6555

1906, No. 333 Keelung Rd.
Sec. 1
Taipei, 10548, Taiwan, R.O.C.
02 7576548
Fax: 02 7576571

Fujitsu Limited
Semiconductor Marketing
Furukara Sogo Bldg.
1-6-1 Marunouchi, 2-Chome
Chiyoda-ku, Tokyo 100, Japan
03 3216-3211
Fax: 03 3213 7174, 03 3216 9771
http://www.fujitsu.co.jp/ (Japanese)

Fujitsu Microelectronics Asia PTE
Ltd.
51 Bras Basah Rd., Plaza by the Park
#06-04/07 Singapore 0718
336-1600
Fax: 336-1609

Europe:
Fujitsu Mikroektronik GmbH
Immeuble le Trident
3-5, Voie Felix Eboue
94024 Creteil Cedex, France
1 45131212
Fax: 1 45131213

Am Siebenstein 6-10
6072 Dreieich-Buchschlag,
Germany
06103 6900
Fax: 06103 690122

Carl-Zeiss-Ring 11
8045 Ismaning, Germany
089 9609440
Fax: 089 96094422

Am Joachimsberg 10-12
7033 Herrenberg, Germany
07032-4085
Fax: 07032-4088

Fujitsu Microelectronics Italia
S.R.L.
Centro Direzionale Milanofiori
Strada 4-Palazzo A/2
20094 Assago (Milano), Italy
02 8246170/176
Fax: 02 8246189

Fujitsu Mikroelektronik GmbH
Europalaan 26A
5623 LJ Eindhoven, The Netherlands
040 447440
Fax: 040 444158

Fujitsu Microelectronics Ltd.
Torggatan 8
17154 Solna, Sweden
08 7646365
Fax: 08 280345

Hargrave House
Belmont Road, Maidenhead
Berkshire SL 6 6NE,
United Kingdom
0628 76100, Fax: 0628 781484
http://www.fujitsu-ede.com/
(European site)

Singapore: 65-336-1600
Fax: 65-336-1609
Hong Kong: 852-723-0393
Fax: 852-721-6555
Taiwan, Taipei: 886-2-757-6548
Fax: 886-2-757-6571
Japan: 81-3-3216-3211
Fax: 81-3-3216-9771
Germany, Dreieich-Buschlag:
06103-6900
Fax: 06103-690122

(Various devices including ASICs, Gate Arrays, RAMs [NMOS, CMOS, BiCMOS, ECL, Bipolar, etc. Dynamic and Video], SRAMs, SCSI controllers, programmable ROMs, flash memory cards, PLL frequency synthesizers, ECL, linear devices, voltage comparators and regulators, SAW filters, 10base-T Ethernet ICs, BECN-Backward Error Correction Notification ICs, custom hybrid devices, disk drive controllers, and fuzzy logic controllers, camera, watches, clock and VCR ICs, speech synthesis and recognition ICs. Fujitsu Ltd. also manufactures thin film color LCD panels, plasma displays and relays)

Fujitsu Ten Ltd.
2-28, Gosho-Dori, 1-Chome,
Hyogo-Ku, Kobe-Shi, Hyogo 652
078 671 5081

(Car ICs)

Fullywell Semiconductor Co., Ltd.
No. 7, Alley 6, Lane 61
Chin Li St., Tu-Chen
Taipei Hsien, 23605
Taiwan
886 2 260 6978, 9242
Fax: 886 2 260 6979

(Diodes)

F - G FUTABA — GAO

Futaba
1605 Penny Lane
Schaumburg, IL 60173
708-884-1444
Fax: 708-884-1635

Fukushima Futaba Electric Co., Ltd.
142, Miyanomae, Ganpoji
Tamagawa-Mura
Ishikawa-Gun
Fukushima 979-63
Japan
024757 3121

(Vacuum fluorescent displays and modules)

Future Domain Corp.
2801 McGaw Ave.
Irvine, CA 92714
714-253-0400
Fax: 714-253-0913

Royal Albert House
Sheet Street, Windsor
Berkshire, SL4 1BE
United Kingdom
44 753 831 262
Fax: 44 753 620 184

(SCSI chip sets, plug and play ICs,)

Future Technology Devices International. Ltd. (FTD)
St. Georges's Studios
93/97 St. George's Road
Glasgow
Scotland
G3 6Ja
44 0 141 353 2565
Fax: 44 0 141 353 2656
http://www.ant.co.uk/~ftdi/

(PC chip sets [such as the FTD4000 that incorporates an 82C4591, 82C4592 and 82C4593], custom ASICs)

F.W. Bell
Headquarters:
6120 Hanging Moss Road
Orlando, FL 32807
800-775-2550
407-678-6900
Fax: 407-677-5765

(Hall Effect sensors and related measuring equipment)

FX prefix
See *Consumer Microcircuits, Ltd.*

FXT prefix
See *Zetex*

FZT prefix
See *Zetex*

G logo, with an arrow on the inside of the "G"
See *Germanium Power Devices*

G prefix
See *GTE*, *Harris Semiconductor* (Custom part), *Loral Corporation* (Used in their equipment), *National Semiconductor* (if G16V8, 20V8, 22CV10, 22V10, discontinued, see *Atmel*, *Lattice*, *Cypress Semi.*, *TriQuint Semi.*, *AVG Semiconductors* [second source video DAC], *Seiko Instruments* [LCD module])

G-TAXI chip set
(A fiber channel chip set)
See *Vitesse Semiconductor Corp.*, *Advanced Micro Devices*

(Taxi chips)

GA prefix
See *TriQuint Semiconductor, Inc.*, *Semicoa*

GAL prefix
See *Lattice Semiconductor Corp.*

Galil Motion Control, Inc.
203 Ravendale Drive
Mountain View, CA 94043
1-800-377-6329
415-967-1700
Fax: 415-967-1751
galil@netcom.com

(Custom parts used in their motion controller cards)

Galileo Technology, Inc.
1735 N. First St, #308
San Jose, CA 95112
408-451-1400
Fax: 408-451-1404

(A fabless company offering single IC system controller for IDT R3051 family of processors, core logic ICs, bidirectional synchronous FIFO buffers, ethernet controllers)

Gamma, Inc.
(ECI Semiconductor),
See *Semtech Corporation*

Gamma High Voltage Research, Inc.
1096 No. US Highway #1
Ormond Beach, FL 32174
904-677-7070
Fax: 904-677-3039

(DC/DC converters, high voltage power supplies)

GammaLink
1314 Chesapeake Terrace
Sunnyvale, CA 94089
408-744-1400
Fax: 408-744-1900

(Telecom ICs)

GAO prefix
See *GEC Plessy*
(Marconi Circuit Technology, Inc.)

GATEFIELD — GEC PLESSY SEMICONDUCTORS

GateField
GateField Division of Xycad Corp.
Fremont, CA
510-249-5757
Fax: 510-623-4484

(FPGAs)

Gateway Photonics Corporation
P.O. Box 4153
St. Louis, MO 63042-0753
314-731-1184
Fax: 314-731-1323

(High power laser diodes and diode arrays)

Gazelle
See *TriQuint Semiconductor, Inc.*

GB prefix
See *Gennum Corporation*

GBC, Inc.
P.O. Box 2205
Ocean, NJ 07712
908-774-8502
Fax: 908-774-8041

(Indicators, neon lamps, surge protectors, voltage regulators, electronic switches)

GBL
See (GigaBit Logic) *TriQuint Semiconductor*

GC prefix
See *Graychip*

GCXXXK series part numbers
See *American Microsystems, Inc.*

(Gate arrays)

GDXXXK series part numbers
See *American Microsystems, Inc.*

(Gate arrays)

GD prefix
See *Cirrus Logic, LG Semicon*
(Formerly GoldStar Technology, Inc.)

GD Rectifiers Ltd.
Victoria Gardens, Burgess Hill
West Sussex, RH15 9NB
United Kingdom
44 444 243 452, Fax: 44 444 870722

(Diodes)

GDP prefix
See *Soshin Electric Co., Ltd.*

GE (General Electric Semiconductor)
See *Harris Corporation*
(OEM devices),
Thomson Consumer Electronics
(Generic replacement devices)

GEC-Advanced Optical Products
West Hanningfield Road
Great Baddow
Chelmsford, Essex, England CM2 8HN
44 1245 242464, 44 1245 73331
Fax: 44 1245 475244,
44 1245 75244
TLX: 995016 GEC RES G

(2-18GHz photoreceivers, integrated optical amplitude modulators, integrated optical switches, fibre polarization controller, 18 Ghz bandwidth photodiodes.)

GEC-Marconi Materials Technology, Ltd.
Division of GEC-Marconi ltd.
9630 Ridgehaven Court
San Diego, CA 92123
619-571-7715, Fax: 619-278-0905

Caswell Towcester Northants
United Kingdom
01327 350581
Fax: 01327 356775

(High power 1550nm DFB lasers)

GEC Plessy Semiconductors
P.O. Box 660017
1500 Green Hills Road
Scotts Valley, CA 95067-0017
1-800-927-2772
1-800-TOP-ASIC
E-Mail (Internet):
TopAsic@io.com
408-438-2900
Fax: 408-438-5576, 408-438-7023
ITT Telex: 4940840
Technical Assistance:
408-439-6029

Sales Offices:
Two Dedham Place, Suite 1
Allied Drive
Boston, MA 02026
617-320-9790
Fax: 617-320-9383

1735 Technology Drive, Suite 100
San Jose, CA 95110
408-451-4700
Fax: 408-451-4710

4635 South Lakeshore Drive
Tempe, AZ 85282
602-491-0910
Fax: 602-491-1219

9330 LBJ Freeway, Ste. 665
Dallas, TX 75243
214-690-4930
Fax: 214-680-9753

7935 Datura Circle West
Littleton, CO 80120
303—798-0250
Fax: 303-730-2460
668 N. Orlando Avenue
Suite 1015 B
Gaitland, FL 32751
407-539-1700
Fax: 407-539-0055

GEC PLESSY — GENERAL INSTRUMENT

385 Commerce Way
Longwood, FL 32750
407-339-6660
Fax: 407-339-9355

13900 Alton Parkway #123
irvine, CA 92718
714-455-2950
Fax: 714-455-9671

3608 Boul, St. Charles, Suite 9
Kirkland, Quebeck, H9H 3C3
514-697-0095
Fax: 514-694-7006

UK and Scandinavia:
Cheney Manor, Swindon, Wiltsire
United Kingdom SN2 2Qw
0793 518000, 0793 518510
Fax: 0793 518411, 0793 518582

UK, Ireland and Denmark:
Unit 1
Crompton Road
Groundwell Industrial Estate
Swindon, Wilts, U.K., SN2 5AF
0793 518510
Fax: 0793 518582

Power conversion products:
Carholme Road
Lincoln, U.K., LN1 1SG
0522 510500
Fax: 0522 510550

France and Benelux:
Z.A. Courtaboeuf, Miniparc-6
Avenue des Andes, Bat. 2-BP 142,
91944
Les Ulis
Cedex A. France
1 64 46 23 45
Fax: 1 64 46 06 07

UngererstraBe 129
80805 Munchen 40, Germany
089 36 0906-0
Fax: 089 36 0906-55

Via Raffaello Sanzio, 4,
20092 Cinisello Balsamo
Milan, Italy
02 66040867
Fax: 02 66040993

CTS Kojimachi Building (4th floor)
2-12, Kojimachi, Chiyoda-ku,
Tokyo 102, Japan
03 5276-5501, 03 3296 0281
Fax: 03 5276-5510, 03 3296 0228

No. 3 Tai Seng Drive,
GEC Building,
Singapore 1953
65 3827708
Fax: 65 3828872

Kataring Bangata 79
116 42 Stockholm 1953
Sweden
46 8 7029770
Fax: 46 8 6404736

Australia and New Zealand:
GEC Electronics Division
Unit 1, South Street
Rydalmere, NSW 2116 Australia
612 638 1888
Fax: 612 638 1798

(CMOS, bipolar, mixed signal bipolar, hybrids/MCMs, power semiconductors [diodes, thyristors, and GTO types], microwave and SAW devices, gate arrays, mixed signal ASICs, image processing, video processing, timing, motor speed control, digital signal processing ICs, RISC processors, wireless LANs, global positioning satellite [GPS] chip sets, IIR filter ICs, automotive devices, crystals, quartz crystals, etc.)

GEC Plessy Semiconductors
(Hybrid Products)
Marconi Circuit Technology
Corporation
100 Smith Street
Farmingdale, NY 11735

160 Smith Street
Farmingdale, NY 11735
516-293-8686
Fax: 516-293-0061

(Hybrid circuits, MIL-STD-1553/1760 remote terminal ICs)

GEM Asia Enterprise Co. Ltd.
No. 114, Gow Gung Road
Run Run District, Keelung
Taipai, Taiwan
886 32 561 881, 886 32 572 846
Fax: 886 32 572 846

(Diodes)

Gem Electronics, Inc.
160 Fernwood Dr.
E. Greenwich, RI 02818
800-870-5385
401-885-8454
Fax: 401-885-1741

(Microminiature reed relays)

General Eastern Instruments
20 Commerce Way
Woburn, MA 01801-9712
800-225-3208
Fax: 617-938-1071

(Humidity sensors and boards)

General Electric Company (of Britain)
See *GEC Plessy*

General Instrument
(ICs)
See *Microchip Technology*, *Quality Technologies*
(Optoelectronics)

General Instrument
Power Semiconductor Division
10 Melville Park Road
Melville, NY 11747-9023
516-847-3000
Fax: 516-847-3150
http://www.gi.com/

South Eastern Regional Sales Office
6855 Jimmy Carter Boulevard
Suite 2250
Norcross, GA 30071
404-446-1265
TWX: 510-101-3222

Central Regional Sales Office
2355 S. Arlington Heights Road
Arlington Hts, IL 60005
312-364-5880
TWX: 910-222-0431

Taiwan:
Semiconductor Components
Division
75-1 Tao Tou
Cho Road Hsin Tien
Taipei, Taiwan

General Instrument
Special Product Assembly Group
See *Solid State Devices, Inc.*

(Semiconductors, including transient voltage suppressor arrays and military qualified devices)

General Microcircuits Corporation
780 Boston Road
Billerica, MA 01821
508-663-9101
Fax: 508-663-9132

(MIL-STD-1772 custom military hybrid circuits)

General Microwave
5500 New Horizons Road
Amityville, NY 11701
516-226-8900
Fax: 516-226-8966

(Microwave oscillators, phase shifters, attenuators, switches and MICs [microwave ICs])

General Resistance
A Prime Technology Company
P.O. Box 185
Twin Lakes Road
North Branford, CT 06471
203-481-5721
Fax: 203-481-8937
TWX: 710-452-3092

(Precision wirewound resistors)

General Semiconductor Industries, Inc.
See *General Instrument Power Semiconductor Division,*
ProTek Devices

(Diode assemblies and transient voltage suppressors)

General Transistor Corp. (GTC)
216 W. Florence Avenue
Inglewood, CA 90301
310-673-8422
Fax: 310-672-2905

(Diodes, thyristors, transistors)

Genesis Microchip, Inc.
200 Town Centre Blvd, Suite 400
Markham, Ontario
Canada L3R 8G5
416-470-2742, Fax: 416-470-2447

2111 Landings Drive
Mountain View, CA 94043
415-428-4277
Fax: 415-428-4288

(A fabless company supplying graphics DSP ICs including image resizing ICs)

GENNUM Corporation
P.O. Box 489
Stn. A
Burlington, Ontario, Canada
L7R 3Y3
416-632-2996
Fax: 416-632-2055
Telex: 061-8525

GENERAL INSTRUMENT — GERMANIUM POWER DEVICES

B-201, Miyamae Village,
2-10-42 Miyamae, Suginami-ku
Tokyo 168, Japan
03 3247-8838
Fax: 03 3447-8839

(Analog ASICs, resonant mode controllers, op amps, analog signal devices and derial digital devices, video switches and multiplexers)

GENRES
See *General Resistance*

(Precision wirewound resistors)

Gentron Corporation
7345 East Acoma, Suite 101
Scottsdale, AZ 85260
602-443-1288
Fax: 602-443-1408

(AC solid state relays, power supply components including power factor correction circuits, DC-DC converters, bridge MOSFET circuits)

Genwave
See *MTI - Milliren Technologies*

GEP prefix
See *Germanium Power Devices*

Germanium Power Devices Corp.
P.O. Box 65
Shawsheen Village Station
Andover, MA 01810

P.O. Box 3065 SVS
Andover, MA 01810

Shetland Industrial Center
Building #4
York Street
Andover, MA
508-475-5982
Fax: 508-470-1512

GERMANIUM POWER DEVICES — GOLDEN PACIFIC

(Diodes, optoelectronics [including photodetectors], transistors, thyristors)

GES prefix
See *Harris Semiconductor*

(Unijunction transistors)

GET - GET Engineering Corporation
9350 Bond Avenue
El Cajon, CA 92021
619-443-8295
Fax: 619-443-8613

(NTDS interface circuits, parts designed to meet MIL-STD-1397 [Ships].)

GF prefix
See *GateField Division of Xycad Corp.* (FPGAs), *GFS Magnetics*

GFS Manufacturing Company, Inc.
140 Crosby Road
Dover, NH 03820
603-742-4375
Fax: 603-742-9165

(Magnetic components including surface mount devices)

GH prefix
See *KDI/triangle Electronics, Inc.*

(GH series hall effect generators; see F.W. Bell)

GHz Technology, Inc.
3000 Oakmead Village Drive
Santa Clara, CA 95051-0808
408-986-8031
Fax: 408-986-8120

(RF microwave silicon power transistors)

GI
See *General Instrument*

Giddings & Lewis Advanced Circuitry Systems
1741 Circuit Drive
Round Lake Beach, IL 60073-1304
708-546-8251
Fax: 708-546-5825

(Diodes)

Gigabit Logic
See *TriQuint Semiconductor, Inc.*

Gilway Technical Lamp
800 W. Cummings Park
Woburn, MA 01801-6355
617-935-4442
Fax: 617-938-5867
E-mail: sales@gilway.com

(LEDs, IR LEDs, phototransistors and lamps)

GIM prefix
See *Optotek Ltd.*

GJRF prefix
See *Gran-Jansen AS*

GL prefix
See *LG Semicon* (Formerly GoldStar Technology, Inc.)

GLC prefix
See *LG Semicon* (Formerly GoldStar Technology, Inc.)

GLiNT processors
See *3Dlabs, Inc.*

GlobTek, Inc.
186 Veterans Drive
Northvale, NJ 07647
201-784-1000
Fax: 201-784-0111

(This distributor for international manufacturers has external power supplies, cordsets)

GLP prefix
See *Soshin Electric Co., Ltd.*

GLX prefix
See *Vitesse Semiconductor*

(GaAs gate arrays)

gm prefix
See *Genesis Microchip, Inc.*

GM prefix
See *LG Semicon, Inc.*
(Formerly GoldStar Technology, Inc.), *Gennum Corporation*, *Germanium Power Devices*

GM-Hughes Electronics
(ICE rept.)

GM-Automotive Components Group (ACG)
(Semiconductors for GM cars)

GMF prefix
See *II Stanley Company, Inc.*

(LCD display)

GMS prefix
See *LG Semicon*

GOC prefix
See *Soshin Electric Co., Ltd.*

Golden Pacific Electronics, Inc. (GPE)
Member of the Sceptre Group
560 S. Melrose St.
Placenta, CA 92670
714-993-6970
Fax: 714-993-6023

(Power supplies [including wall, desktop models and internal computer models], buzzers, sound generators [including surface mount devices], speakers, cords, jacks, connectors, foldable keyboards)

Manufacturers, Prefixes, Part Number Types, Logo Descriptions & Family Types

GOLDENTECH DISCRETE — GT

Goldentech Discrete Semiconductor, Inc.
4Fl., No. 82. Pao Kao Road
Hsin Tien City
Taipei Hsien, Taiwan
886 2 917 8496
Fax: 886 2 914 9235

(Diodes)

GoldStar Technology, Inc.

Goldstar Electron America, Inc.
See *LG Semicon*
(Also known as Lucky-Goldstar)

GOP prefix
See *Soshin Electric Co., Ltd.*

Gould Electronics
See *AMI*
(American Microsystems, Inc.)

Goyo Electronics Co., Ltd.
64, Naganuma, Tennoh-Machi
Minami-Akita-Gun
Akita 010-01
Japan
0188 73 2011
Fax: 0188 73 5385

(Hybrid ICs)

GP prefix
See *GEC Plessy, Sharp*

(Optosensors)

GPD
See *Germanium Power Devices*

GR prefix
See *Greenwich Instruments* (and General Resistance for precision wirewound resistors)

Gran-Jansen AS
Oslo, Norway
Fax: 47 22 49 59 03
granjan@oslonet.no

(Single IC VHF/UHF transceiver for a spread spectrum system)

Graseby Optronics
(Graseby Infrared)
12151 Research Parkway
Orlando, FL 32826-3207
407-282-7700
Fax: 407-273-9046

(Infrared detectors, blackbody radiation sources, amplifiers, controllers, multiplexers)

Graychip
2185 Park Boulevard
Palo Alto, CA 94306
415-323-2955
Fax: 415-323-0206
http://www.greychip.com
E-mail: sales@greychip.com

(DSP ICs including crossbar switches, narrowband/wideband tuners, transversal filters, asynchronous resamplers and mixer/carrier removal, quad receiver or transmitter and 10 MB QAM Equalizer/Demodulator ICs, and custom ICs and systems)

GrayHill, Inc.
561 Hillgrove Avenue
LaGrange, IL 60525
708-482-2132
Fax: 708-354-2820

(Optical encoders, switches)

Green Logic, Inc.
189 Bernardo Ave.
Mountain View, CA 94043-5203
415-903-1700

(Card bus [PCMCIA] controllers)

Greenray Industries, Inc.
See *C-MAC Quartz Crystals, Inc.*

Greenwich Instruments Ltd.
The Crescent, Main Road
Sidcup, Kent DA14 6NW
England
081 302 4931
Fax: 081 302 4933

Greenwich Instruments USA
3401 Monroe Road
Charlotte, NC 28205
800-476-4070
704-376-1021
Fax: 704-335-8707

(EPROM Emulators, nonvolatile memories that do not have to be erased and data can be rewritten at any time, pin-outs compatible with normal EPROMS)

Group West International
1166 Redfern Court
Concord, CA 94521
510-672-2188
Fax: 510-672-2570

No. 17, Wu Chuan 6 Rd.
Wu Ku Industrial Zone
Taipei Hsien, Taiwan, R.O.C.
886-2-299-1889
Fax: 886-2-299-1324

(Wall plug-in power supplies and transformers)

GS prefix
See *Gennum Corporation*

GST prefix
See *Semicoa*

GSWA prefix
See *Mini-Circuits*

GT prefix
See *Gennum Corporation*

G - H GOLDENTECH DISCRETE — HAMAMATSU

GT-XXXXX part number
See *Galileo Technology, Inc.*

GTC
See *General Transistor Corporation*

GTE Corporation
Electronic Components and Materials
2401 Reach Rd.
Williamsport, PA 17701
717-326-6591

(Capacitors, quartz crystals. Note: The miniature telephone relay line was sold to Communications Instruments, Inc. [CII] in 1985.)

GTE Microelectronics
GTE Government Systems Corp.
77 "A" Street
MS-15
Needham, MA 02194-2892
1-800-544-0052
617-455-5814, 617-455-5152
Fax: 617-455-5885

(ATM switch with HIPPI module, ASICs, FPGAs, thick film modules, to meet various requirements including undersea and spacecraft conditions.)

GTL, Gunning Transistor Logic
A fast speed logic family
See *TI*

Guardian Electric Manufacturing Co.
1425 Lake Avenue
Woodstock, IL 60098
800-762-0369
815-337-0050
Fax: 815-337-0377

(Relays)

GUI prefix
See *Trident Microsystems*

GX prefix
See *Gennum Corporation*

GXX part number
(On SOT transistor)
See *Rohm*

GY prefix
See *Gennum Corporation*

H logo
See *Hi-Sincerity Microelectronics Corp.*

H prefix
See *SGS-Thompson Microelectronics*,
Quality Technologies (Phototransistor optocouplers),
EM Microelectronic (Watch and clock ICs, watchdog timers, smart reset ICs)

H, on top of a V logo
See *High Voltage Semiconductor Specialists, Inc.*

H4C prefix
See *Motorola*

(Programmable ASICs)

H8/ prefix
See *Hitachi America Ltd.*

HA prefix
See *Harris Semiconductor*,
Hitachi Semiconductor
(Devices are not related to each other, i.e., not second sources.)

HAD prefix
See *Signal Processing Technologies, Inc.*

HADC prefix
See *Signal Processing Technologies, Inc.*

(A/D converters)

Hafo
See *ABB Hafo AB*

HAL prefix
See *AVG Semiconductors*

(VGA port decoder and memory address decoder)

Hamai Electric Lamp Co. Ltd.
1-9-26, Kasuga, Bunkyo-ku
Tokyo 112, Japan
81 3 3813 8811
F: 81 3 383 7054

(Op Amps, car, communications audio, television, watch, clock voltage comparators and voltage regulator ICs)

Hamamatsu Corporation
P.O. Box 6910
360 Foothill Road
Bridgewater, N.J. 08807
1-800-524-0504
908-231-0960
E-mail: hamacorp@interamp.com

8 Rue de Saule Trapu
Massy Cedex 91882
France
1 49 75 56 80, 33 1 69 53 71 00
Fax: 33 1 6953 7110

Arzberger Strasse 10
Herrsching 82211
Germany
49 08152/375-0
Fax: 49 8152 2658

Via Monte Grappa 30
Arese Mailand 20020
Italy
39 2 935 81 733/734/740
Fax: 39 29 358 1741

HAMAMATSU — HARRIS SEMICONDUCTOR

Hamamatsu Photonics K.K.
1126-1, Ichino-Cho
Hamamatsu-Shi, Shizuoka 435
053 434 3311
Fax: 053 434 5184

Spain:
34 3 582 44 30, 3 699 65 5

Farogalan 7
Kista 16440
Sweden
46 8 7032950
Fax: 46 8 750 5895

Lough Point
Gladbeck Way
Enfield
Middlesex EN2 7JA
United Kingdom
44 081 367 3560
Fax: 44 081 367 6384

(Photo-optical ICs, photomultiplier tubes, and detectors)

Hamlin
612 East Lake Street
Lake Mills, WI 53551
414-648-3000
Fax: 414-648-3001

(Reed relays, reed switches, position sensors)

Hammond Manufacturing
4700 Genesee Street
Cheektowaga, NY 14225-2466
716-631-5700
Fax: 716-631-1156

(Enclosures, cases, fan filters, transformers)

Hanning Elektro-Werke GmbH
Holter Strasse 90
Oerlinghausen 33813
Germany
5202 707 0
Fax: 5202 707 301

(Tachometer controller)

Hantronix, Inc.
250 Santa Ana Court
Sunnyvale, CA 94086
408-736-3191
Fax: 408-749-0477

(LCDs, see also Orion Electric Co. Ltd.)

Harris Semiconductor
1301 Woody Burke Road
Melbourne, FL 32902
800-442-7747 (1-800-4-Harris)
Technical Hotline
407-724-3000, 3739, 3576
Answer Fax Service: 407-724-7800
Internet: centapp@harris.com
Home Page:
http://www.semi.harris.com
Technical Assistance E-mail:
centapp@harris.com

Sales Offices:
Alabama:
Suite 103, Office Park South
600 Boulevard South
Huntsville, AL 35802
205-883-2791
Fax: 205-883-2861

California:
Suite 320
1503 So. Coast Drive
Costa Mesa, CA 92626
714-433-0600, 0660
Fax: 714-433-0682

Suite 308
5250 W. Century Blvd.
Los Angeles, CA 90045
213-649-4752
Fax: 310-649-4804

3031 Tisch Way
1 Plaza South
San Jose, CA 95128
408-985-7322
Fax: 408-985-7455

Suite 350, 6400 Canoga Ave.
Woodland Hills, CA 91367
818-992-0686
Fax: 818-883-0136

Harris Corp. Semiconductor Sector
Power R&D
Latham, NY
518-782-1182

Harris Microwave
Semiconductor Products
See *Samsung Microwave Semiconductor, Inc.*

Florida:
1301 Woody Burke Rd.
Melbourne, FL 32902
407-724-3576
Fax: 407-724-3130

300 6th Avenue, North
Indian Rocks Beach, FL 34635
813-595-4030
Fax: 813-595-5780

2401 Palm Bay Road
Palm bay, FL 32905
407-729-4984
Fax: 407-729-5321

Illinois:
Suite 600, 1101 Perimeter Dr.
Schaumburg, IL 60173
708-240-3480
Fax: 708-619-1511

Indiana:
Suite 100, 11590 N. Meridian St.
Carmel, IN 46032
317-843-5180
Fax: 317-843-5191

Massachusetts:
Six New England Executive Park
Burlington, MA 01803
617-221-1850
Fax: 617-221-1866

HARRIS SEMICONDUCTOR

Michigan:
Suite 460
27777 Franklin Rd.
Southfield, MI 48034
313-746-0800
Fax: 313-746-0516

New Jersey:
Plaza 1000 at Main Street
Suite 104
Voorhees, NJ 08043
609-751-3425
Fax: 609-685-6140

Suite 210 North
6000 Midlantic Drive
Mt. Laurel, NJ 08054
609-727-1909
Fax: 609-727-9099

724 Route 202
P.O. Box 591 M/S 13
Somerville, NJ 08876-0591
908-685-6150
Fax: 908-685-6140

Plaza 1000 at Main Street
Suite 104
Voorhees, NJ 08043
609-751-3425
Fax: 609-751-5911

Rahway: 908-381-4210

New York:
Hampton Business Center
1611 Rt. 9, Suite U3
Wappingers Falls, NY 12590
914-298-0413
Fax: 914-298-0425

490 Wheeler Road, Suite 165B
Hauppauge, NY 11788-4365
516-342-0219
Fax: 516-342-0295

New York, Great Neck:
516-342-0291

North Carolina:
4020 Stirrup Creek Dr.
Building 2A, MS/2T08
Durham, NC 27703
919-549-3600
(3603 for switching regulators)
919-361-1500, Fax: 919-549-3660

Texas:
Suite 205, 17000 Dallas Parkway
Dallas, TX 75248
214-733-0800
Fax: 214-733-0819

European Headquarters:
Mercure Center
100, Rue de la Fusee
1130 Brussels, Belgium
32 2 46 21.11
Fax: 32 2 246 22 05, 06
Telex: 61566

United Kingdom:
Riverside Way, Camberley
Surrey GU15 3YQ
44 276 686 886
Fax: 353 4273518
Telex: 43679

France:
2-4, Avenue del Europe
78140 Velizy
33 1 34 65 40 80 (dist.),
27 (sales) or 54
Fax: 33 1 39 64 054
TLX: 697060

Germany:
Putzbrunnerstrasse 69
8000 Muenchen 83
49 89 63813 0
Fax: 49 8 963 8130 (Munich) or 76201
TLX: 529051

Kieler Strasse 55-59
2085 Quickborn
49 4106 5002 04
Fax: 49 4106 68850
TLX: 211582

Wegener Strasse, 5/1
7032 Sindelfingen
49 7031 873 469
Fax: 49 7031 873 849
TLX: 7265431

Hong Kong:
Harris Semiconductor H.K. Ltd.
13/F Fourseas Building
208-212 Nathan Road
Tsimshatsui, Kowloon
852-3-723-6339
Fax: 852-3-739-8946
TLX: 78043645

Italy:
Harris SRL
Viale Fulvio Testi, 126
20092 Cinisello Balsamo
39 2 262 07 61
(Dist and OEM ROSE)
39 2 240 95 01
(Dist and OEM Italy)
Fax: 39 2 248 66 20
39 2 262 22 158 (ROSE)
TWX: 324019

Italy, Milano
39 2 262 22141
Japan (North Asia):
Harris K.K.
Shinjuku NS Bldg. Box 6153
2-4-1 Nishi-Shinjuku
Shinjuku-Ku, Tokyo 163 Japan
81 03 3345 8911
Fax: 81 3 3345 8910

Korea:
RM #419-1 4th FL
Korea Air Terminal Bldg
159-1, Sam Sung-Dong
Kang, Nam-ku, Seoul
135-728 Korea
82-2-551-0931/4
Fax: 82-2-551-0930

Malaysia:
RCA Sendirian Berhad
75 Jalan Ulu Klang
Ampang Industrial Estates
Kuala Lumpur, Malaysia

Manufacturers, Prefixes, Part Number Types, Logo Descriptions & Family Types

The Netherlands:
Benelux OEM Sales Office
Kouterstraat 6
NL 5345 LX Oss
31 4120 38561
Fax: 31 4120 34419

North Asia:
Shinjuku NS Building
Box 6153
2-4-1 Nishi-Shinjuku
Shinjuku-ku, Tokyo 163-08
Japan
81-3-3345-8911
Fax: 81-3-3345-8910

Singapore:
Harris Semiconductor Pte. Ltd.
105 Boon Keng Road
#01-01 Singapore 1233
65-291-0203
Fax: 65-293-4301
TLX: RS36460 RCASIN

Taiwan:
Room 1103
600 Ming Chuan East Road
Taipei
886-2-716-9310
Fax: 886-2-715-3029
TLX: 78525174

(Data acquisition [A/D and D/A converters, multiplexers, crystal oscillators, switches, flash converters, real time counters, display drivers, timers/counters], digital signal processing, signal processing [linear], amplifiers, data converters, interface [including RS232 ICs], analog switches, multiplexers, filters, DSP [down converters and demodulators (PSK, FSK, AM and FM), multipliers, filters, signal synthesizers, video processors], telecom [analog and digital interfaces, subscriber line circuit ICs, CVSD], hold devices, speech synthesis and recognition ICs, power products including power MOSFETs, IGBTs, bipolar discretes, transient voltage suppressors, power rectifiers, MOVs, power ICs, power and mixed signal ASICs, hybrid programmable switches, custom high voltage ICs, linear regulators, step-up converters, voltage inverters, doublers, battery back-up switches, cross point switches, ASICs [full custom, analog, mixed signal] digital ICs [CMOS microprocessors, microcontrollers, peripherals and logic] microwave [GaAs FETs, MMICs, foundry services], video amplifiers and buffers, crosspoint switches, amplifiers, comparators, sample and hold amplifiers, differential amplifiers, radio and television ICs, IF devices, prescalers, horizontal/vertical countdown and sync processors, chroma/luma ICs, security ICs, automotive ICs and military/aerospace and Rad-Hard [radiation-hardened] ICs)

Hasco Components International Corp.
247-40 Jericho Turnpike
Bellerose Village, NY 11001
516-328-9292
Fax: 516-326-9125

(Relays, reed switches, reed relays)

HB prefix
See *Hitachi Semiconductor*

HBC prefix
See *Carroll Touch, Inc.*

HBCS prefix
See *Hewlett-Packard*

HC logic family
See *Harris Semiconductor*,
Motorola, Inc., Texas Instruments,
New Japan Radio Company, Ltd.,
NJR Corporation, Hitachi, Panasonic,
National Semiconductor,
SGS-Thomson, Universal, LG Semicon
(formerly Goldstar Technology, Inc.),
Mitsubishi, Philips, Toshiba,
AVG Semiconductors, Elcap
Electronics, Ltd., Hua Ko Electronics

HARRIS SEMICONDUCTOR — HCT

HC Power, Inc.
17032 Armstrong
Irvine, CA 92714-5716
800-486-4427
714-261-2200
Fax: 714-261-6584

(High power, power supplies)

HC prefix
See *Hycomp, Inc.,*
Honeywell Solid State Electronics Center (The C indicates a Honeywell CMOS process)

HC1 prefix
See *Harris Semiconductor*
(Custom part)

HCMP prefix
See *Signal Processing Technologies, Inc.*

(Comparators)

HCPL-XXXX part number
See *Hewlett-Packard,*
Quality Technologies

HCS logic
See *Harris Semiconductor*

HCS prefix
See *Harris Semiconductor,*
Microchip

(Wireless security products)

HCT (HCMOS) logic family
See *Harris Semiconductor,*
Motorola, Inc., Texas Instruments,
LG Semicon (formerly Goldstar Technology, Inc.), Mitsubishi,
Philips, Toshiba,
National Semiconductor,
GEC Plessy, AVG Semiconductors,
Elcap Electronics, Ltd.,
Hua Ko Electronics

HCT — HEWLETT-PACKARD

HCT prefix
See *Harris Corporation, Syvantek Microelectronic Corporation*

HCTS logic
See *Harris Semiconductor*

HCTS prefix
See *Harris Semiconductor*

HD logo (H intertwined with the D),
See *Hayashi Denkoh Co., Ltd.*

HD prefix
See *Hitachi Semiconductor, Siemens, Natal Engineering Co., Inc.* (Digital to Synchro converters), *Zetex*

HDA prefix
See *Watkins-Johnson*

HDAC prefix
See *Signal Processing Technologies, Inc.*

(D/A converters)

HDC prefix
See *Staktek Corp.*

(Memory staked modules)

HDK prefix
See *LSI Logic*

HDX-XXXX part number
See *Harris Semiconductor*

HDAS prefix
See *Datel, Inc.*

HDF prefix
See *Shindengen America*

HDK Logo
See *Hokuriku Electric Industry Co., Ltd.*

HDL Research Lab, Inc.
P.O. Box 2108
Brenham, TX 77834-2108
409-836-2300
Fax: 409-836-1056

(DC/DC converters)

HDM prefix
See *Hyundai Electronics America* (Hyundai Digital Media Division is now Odeum Microsystems, Inc.), *Watkins-Johnson*

HDMP prefix
See *Hewlett-Packard*

HDN prefix
See *Siemens HD-PAK, Powercube Corp.*

(DC/DC converter)

HDSP prefix
See *Siemens*

HE prefix,
See *Hitachi Semiconductor* (Laser diodes/LEDs), *Silicon Sensors, Inc.* (PIN Si photodiodes)

HE-XXXX part number
See *Cherry Electrical Products*

(Hall effect sensors)

Headland Technology, Inc.
46221 Landing Parkway
Fremont, CA 94538
510-623-7857
Fax: 510-656-0397

(PC and logic chip sets, LSI Logis is the parent company of this firm)

HEI, Inc.
1495 Steiger Lake Lane
P.O. Box 5000
Victoria, MN 55386-5000
612-443-2500
Fax: 612-443-2668

(Optoelectronics)

Hermetic Switch, Inc.
P.O. Box 1325
3100 Highway 92 South
Chickasha, OK 73023
405-224-4046
Fax: 405-224-9423

(Reed switches, proximity sensors, power sensors)

Herotek, Inc.
222 N. Wolfe Road
Sunnyvale, CA 94086
408-730-1702
Fax: 408-730-4015

(Comb generators, limiters, switches, GaAs FET amplifiers, subsystems)

Herrmann KG
Fabrik Fur Electrotechnik
Grunberger Str, 43 Postfach 551167
Nurnberg 55, D-8500
Germany
49 911 834 054
Fax: 49 911 834 671
TLX: 622774 HEKGN

(Diodes, thyristors)

Hewlett-Packard Company
19310 Pruneridge Ave.
Cupertino, CA 95014
800-752-0900
800-452-4844
http://www.hp.com/

P.O. Box 58059
Santa Clara, CA 95052-8059
800-227-1817

Manufacturers, Prefixes, Part Number Types, Logo Descriptions & Family Types

HEWLETT-PACKARD — HITACHI

(MCM devices, fiber channel controllers, solid state switches etc.)

Optoelectronic (Optical Communications) Division
370 West Trimble Road
San Jose, CA 95131-1096
800-537-7715, ext. 2000
800-235-0312

(LEDs [including blue LEDs], prescalers, telecommunications transceivers and receivers, etc.)

Hewlett-Packard Company Inquiries
P.O. Box 58059
5301 Stevens Creek Blvd.
Santa Clara, CA 95052-8059
800-537-7715

(Surface mount optical encoders, optocouplers)

Yokogawa-Hewlett Packard, Ltd.
3-29-21, Takaido-Higashi
Suginami-ku
Tokyo 168 Japan
03 3331 6111
Fax: 03 3335 1478

(Telephone ICs)

Hexawave, Inc.
2 Prosperity Road II
Science Based Industrial Park
Hsinchu, Taiwan
886 35 785100, Fax: 886 35 770512

P.O. Box 7205
Fairfax Station, VA 22039-7205
703-764-2922, Fax: 703-764-2581

(Power FETs, microwave semiconductors, GaAs ICs including SPDT switches)

HFA prefix
See *Harris Semiconductor*

HFC prefix
See *Cologne Chip Designs*

HG prefix
See *Hitachi Semiconductor*

HI prefix
See *Harris Semiconductor*, *Burr-Brown Corp.*, *Holt Integrated Circuits, Inc.*

HiBrec
(High Bandwidth Resource Interface Controller)
See *MacroTek*

HI-G Company, Inc.
See *Struthers Dunn/Hi-G*

HIP prefix
See *White Technology, Inc.*, *Motorola, Inc.*, *Harris Semiconductor*

(Switching regulators)

Hi-Sincerity Microelectronics Corp.
10F, No. 61 Chung Shan N. Road, Sec. 2
Taipei, Taiwan, R.O.C.
886-2-521-2056
Fax: 886-2-563-2712
TLX: 20411 SCTCORP

Factory address:
No. 38, Kuang Fu S. Road
Fu Kou, Hsin Chu Industrial Park
Hsin Chu, Taiwan, R.O.C.

(Diodes, rectifiers and transistors. Available from Taitron Components, Inc., 25202 Anza Drive, Santa Clarita, CA 91355- 3496, 800-824-8766, 800-247-2232, 805-257-6060; Fax: 800-824- 8329, 805-257-6415)

High Power Devices, Inc.
100 Jersey Avenue, Bldg. G
New Brunswick, NJ 08901
908-249-2228
Fax: 908-545-0120
Email: hpdi@aol.com

(High power laser diodes)

High Reliability Components Corporation
Shiba Daimon Makita Bldg
5-8, Shiba-Daimon 2-Chome
Minato-Ku, Tokyo 105, Japan
81 3 3435 9591
Fax: 03 3435 9592

(Analog, logic, memory and hybrid ICs)

High Voltage Component Associates (HVCA Marketing)
P.O. Box 2484
Farmingdale, NJ 07727
908-938-4499
Fax: 908-938-4451

(High voltage assemblies, diodes, bridges, custom assemblies, microwave range parts)

High Voltage Semiconductor Specialists, Inc.
2 Winthrop Drive
Woodbury, NY 11797
516-692-7556

(Optoelectronics)

HIN2 prefix
See *Harris Semiconductor*

Hind Rectifiers, Ltd.
Lake Road
Bhandup
Bombay 400 078, India
91 022 564 4114
TLX: 011 72105 HIRT IN

(Diodes, thyristors)

Hitachi America, Ltd.
Semiconductor and I.C. Division
San Francisco Center
2000 Sierra Point Parkway
Brisbane, CA 94005-1819
415-589-8300
415-244-7286

HITACHI — HITACHI

For Literature: 1-800-285-1601
(8 am to 6 pm MST)
Fax: 415-583-4207
TWX: 910-338-2103
http://www.hitachi.com
http://www.hitachi.co.jp/
ftp://ftp.hitachi.co.jp/

Brisbane CA:
800-285-1601 (Gate arrays)

Customer Inquiry Department
P.O. Box 17464
Denver, CO 80217-9365

Design and Development Facility:
Hitachi Micro Systems, Inc.
179 East Tasman Drive
San Jose, CA 95134

Engineering Facility:
6321 East Campus Circle Drive
Irving, TX 75063-2712

Regional Offices:
Northeast Region
Hitachi America Ltd.
77 South Bedford Street
Burlington, MA 01803
617-229-2150
Fax: 617-229-6554

North Central Region:
500 Park Boulevard, Suite 415
Itasca, IL 60143
708-773-4864
Fax: 708-773-9006

Northwest Region:
1740 Technology Drive, Suite 500A
San Jose, CA 95110
408-451-9570, 408-436-3900

Southeast Region:
5511 Capital Center Drive, Suite 204
Raleigh, NC 27606
919-233-0800
Fax: 919-233-0508

South Central Region:
Two Lincoln Center, Suite 865
5420 LBJ Freeway
Dallas, TX 75240
214-991-4510
Fax: 214-991-6151

Southwest Region:
2030 Main Street, Suite 450
Irvine, CA 92714
714-553-8500
Fax: 714-553-8561

Pacific Mountain Region:
Metropoint
4600 S. Ulster Street
Suite 690
Denver, CO 80237
303-740-6644
Fax: 303-740-6609

Automotive Region:
330 Town Center Drive, Suite 311
Dearborn, MI 48126
313-271-4410
Fax: 313-271-5707

Mid-Atlantic and Telecom Region:
325 Columbia Turnpike, Suite 203
Florham Park, N.J. 07932
201-514-2100
Fax: 201-514-2020

21 Old Main Street
Suite 104
Fishkill, NY 12524
914-897-3000
Fax: 914-897-3007

District Offices:
Hitachi America Ltd.
3800 W. 80th Street, Suite 1050
Bloomington, MN 55431
612-896-3443

10777 Westheimer
Suite 1040
Houston, TX 77036
713-974-0534
Fax: 713-974-0587

9600 Great Hills Trail
Suite 150 W
Austin, TX 78759
512-502-3033

4901 N.W. 17th Way, Suite 302
Fort Lauderdale, FL 33309
305-491-6154
Fax: 305-771-7217

3350 Holcomb Bridge Road
Suite 300
Norcross, GA 30092
770-409-3000

Hitachi Canadian Ltd.
320 March Road, Suite 602
Kanata, Ontario, Canada K2K 2E3
613-591-1990
Fax: 613-591-1994
416-826-4100

Manufacturing Facility:
6321 East Campus Circle Drive
Irving, TX 75063-2712

Engineering Facility:
Hitachi Micro Systems, Inc.
180 Rodes Orchard Way
San Jose, CA 95134

Hitachi Electric Company, Ltd.
4-6 Kanda Surugadai
Chiuyoda-ku
Tokyo 101, Japan
81 3 3212 1111
Fax: 81 3 3258 2375,
81 3 3284 0346

Hitachi Denshi, Ltd.
1 Kanda Izmi-cho
Chiyoda-ku
Tokyo 101 Japan
03 5821 5311
Fax: 03 5821 5394

(DRAMs, SRAMs, ASICs, RISC CPUs, flash memories [some parts jointly developed with Mitsubishi], Ferroelectric memory [FRAMs, technology licensed from Ramtron

International Corp.], memory cards, fuzzy logic, gate arrays and microcontrollers, disk drive ICs [including Hall effect sensors], level-shifters, transceivers, audio amplifiers, data converters, interface ICs, regulators, logic, video devices, multimedia ICs, LCD drivers and controllers, AM-LCDs, optoelectronics, communications ICs including laser diodes, wireless communications ICs, speech synthesis and recognition, VCR, television, watches, clocks, hybrid ICs, etc. Hitachi also has an agreement with Ramtron to develop FRAMs. They have a HiMAP, Hitachi market Application Program on disk to help the user select components for an application. Contact Hitachi America, Ltd., Semiconductor and IC Division, 2000 Sierra Point Parkway, MS250, Brisbane, CA 94005-1835.)

Hi-Tron Semiconductor Corp.
85 Engineers Road
Hauppauge, NY 11788
516-231-1500

(Diodes, transistors, thyristors)

Hittite Microwave Corporation
21 Cabot Road
Woburn, MA 01801
617-933-7267, Fax: 617-932-8903

(Doubled balanced mixers, GaAs MMIC variable gain amplifiers, biphase modulators, transfer switches, frequency doublers)

HI-WIT Electronics Co., Ltd.
No. 8, Alley 16, Lane 337
Ta Tung Road, Hsi Chih
Taipei Hsien, Taiwan
886 2 641 4570, Fax: 886 2 642 8644

(Optoelectronics)

HG prefix
See *Hitachi*

HGTG prefix
See *Harris Semiconductor* (IGBTs)

HKC
See *Hong Kong Crystals Company*

HKC prefix
See *Hua Ko Electronics*

H.K. Crystals Co.
Unit H
22/F Shield Ind. Ctr.
34-36 Chai Wan Kok St., T.H.
Hong Kong
852 4120121
Fax: 852 498 5908

(Crystals and resonators)

HL prefix
See *Hitachi Semiconductor* (Laser diodes/LEDs), *Siemens*

HLC prefix
See *Cologne Chip Designs*

HLD prefix
See *Shindengen America*

HLMP prefix
See *Quality Technologies*, *Siemens*

(Light bars - discontinued)

HM prefix
See *Hualon Microelectronics*, *Hitachi*, *Harris Semiconductor* (If HMX-XXXX), *Temic* (Old Marta Harris devices), *AVG Semiconductors* (Second source crosspoint switches), *KOA Speer* (Resistor networks)

HMC prefix
See *Hittite Microwave Corporation*

HMC Philippines Trade Corporation
See *Hualon Microelectronics Corporation*

HMCS prefix
See *Hitachi*

HML prefix
See *Hualon Microelectronics*

HMM prefix
See *Hitachi*

(Chip sets)

HMP part number
See *Harris Semiconductor* (Microwave), *Samsung Microwave Semiconductor, Inc.*

HMPS
See *Hi-Sincerity Microelectronics Corp.*

HMR prefix
See *Honeywell Solid State Electronics Center*
(MR = Magnetoresistive sensor)

HMS
See *Samsung Microwave Semiconductor*

HMSDC-XXXX part number
See *ILC-DDC*

HN prefix
See *Hitachi Semiconductor*, *KOA Speer*

(Resistor networks)

HNFSD prefix
See *KOA Speer*

(Resistor networks)

Hobbit
An AT&T microprocessor
(The line was discontinued in mid - 1995)

HOBBS — HONEYWELL

Hobbs Corporation
P.O. Box 19424
Springfield, IL 62794-9424
217-753-7773
Fax: 217-753-7789

(Pressure and vacuum switches)

Hokuriku Electric Industry Co., Ltd. (HDK)
3158, Shimo-Okubo,
Osawanomachi,
Kami-Nilkawa-gun,
Toyama 939-22
0764 67 1111
Fax: 0764 68 1508

Foreign Trade Dept.
Nikko Gotanda Bldg, 7th Floor
29-5, Nishi Gotanda 2-Chome
Shinagawa-ku, Tokyo 141
Japan
03 495 1811
Fax: 03 495 1800

Hokuriku U.S.A. Ltd.
8145 River Road
Morton Frove, IL 60053
708-470-8440
Fax: 708-967-1188

117 Jetplex Circle, Suite C1
Madison, Albama 35758
205-772-9620
Fax: 205-772-3475

Hokuriku (Singapore) Pte., Ltd.
3A Joo Koon Circle
Jurong, Singapore 2262
861 1677
Fax: 861-4270

Hokuriku Asia Sales Pte., Ltd.
3A Joo Koon Circle
Jurong, Singapore 2262
861-0985
Fax: 861-3664

Taipei Hokuriku Electric Industry Co., ltd.
5F, no. 274 Song Chiang Road
Taipei, Taiwan
R.O.C.
02 522 2252
Fax: 02 537 4111

Hokuriku Hong Kong Co., Ltd.
UNIT 2-3
Fook Hong Industrial Bldg
19 Sheung Yuet Road
Kowloon, Hong Kong
7558073
Fax: 7957199

Korea HDK Corporation
RMA-54 1F Electro Land
Main Bldg
16-9, Hangangro, 3-Ka
Yongsan-Ku
Seoul, Korea
02 714 4158
Fax: 02 717 4159

(Hybrid ICs)

Holt Integrated Circuits, Inc.
(HI prefix)
9351 Jeronimo Road
Irvine, CA 92718
800-222-4658
714-859-8800
Fax: 714 859 9643

(Power IC drivers, ballast control ICs, etc.)

Holtek Microelectronics, Inc.
No. 5 Creation Road II
Science Based Industrial Park
Hsinchu, Taiwan R.O.C.
886 35 784 888
Fax: 886 35 770 879

Holmate Technology, Corp.
17921 Rowland St.
Industry, CA 91748
818-912-9378
Fax: 818-912-9598

(ASICs and consumer electronic ICs including 4-bit controllers, voltage regulators, lamp and led flasher drivers, liquid crystal display drivers, encoder/decoder ICs, watch ICs, and peripheral and piano-sound ICs)

Honeywell Micro-Switch Division
11309 W. Chetlain Lane
Galena, IL 61036
815-777-2780
Fax: 815-777-3603

(Hall effect sensors, optical encoders)

Honeywell Optoelectronics
830 Arapaho Rd.
Richardson, TX 75081
800-367-6786
214-470-4271
Fax: 214-470-4417

(Fiber optic components and optoelectronics)

Honeywell Solid State Electronics Center (SSEC)
12001 State Highway 55
Plymouth, MN 55441
800-323-8295 (Radiation hardened product line, training hotline)
800-323-8293 (Sensors and ASICs)
800-323-7551
(ASIC Customer Support Hotline)
800-238-1502
(Customer Satisfaction Hotline)
612-954-2692
612-954-2956
(QML contact Bret Rinehart)
http://www.ssec.honeywell.com/

4801 University Square, Suite 29A
Huntsville, AL 35816
204-830-0223

300 Continental Blvd., Suite 340
El Segundo, CA 90245
310-335-0542

Manufacturers, Prefixes, Part Number Types, Logo Descriptions & Family Types

HONEYWELL

721 South Parker Suite 200
Orange, CA 92668
714-953-4821

7837 Convoy Court
San Diego, CA 92111
619-573-0157

4145 North First
San Jose, CA 95134
408-433-3000

1250 Academy Park Loop
Colorado Springs, CO 80910
719-573-6024

1190 W. Druid Hills Drive, N.E.
Atlanta, GA 30329
404-248-2300

11842 Borman Drive
St. Louis, MO 63146
314-569-5400

3320 Candelaria N.E.
Albuquerque, NM 87107
505-880-3204

7485 Henry Clay Blvd.
Liverpool, NY 13088
315-457-7959

4540 Honeywell Court
Dayton, OH 45424
513-237-4000

2600 Eisenhower Avenue
Valley Forge, PA 19482
215-666-8200

14643 Dallas Parkway, Suite 800
Dallas, TX 75240
214-701-2800

1104 Country Hills Drive
Ogden, UT 84403
801-627-1821

7900 Westpark Drive
McClean, VA 22101
703-734-7830

600-108th Avenue N.E.
Honeywell Center
Bellevue, WA 98004
206-453-7541

Honeywell, Ltd. Australia
863-871 Bourke St.
Waterloo, New South Wales 2017

P.O. Box M132
Sydney Mail Exchange 2012
61 699 0155

679 Victoria Street
Richmond, Melbourne
Australia VIC 3067
61 3 420 5555

Honeywell Austria Ges.m.b.H.
Edelsinnstrasse 7-11
1120 Vienna, Austria
43 222 811010

Honeywell Europe S.A.
3 Avenue de Bourget
1140 Brussels
Belgium
32 2 728 2775

Honeywell S.A.
Schipollaan, 1
1140 Brussels, Belgium
32 2 2782/2611

Honeywell Ltd.
155 Gordon Baker Road
North York, Ontario
M2H 3N7, Canada
416-499-6111

Honeywell Limited Aerospace
Division
Highway 17
P.O. Box 1300
Rockland, Ontario
KOA 3AO
Canada
613-446-6011

Honeywell A/S
Hejrevel 26
2400 Copenhagen
Denmark
45 38 34 77 22

18 Sawra Street
Apr. 4
Heliopolise, Cario
Arab Republic of Egypt
20 2 2909665

Honeywell A&D Ltd.
12B Alfred Street
Westbury Wilts
BA13 3DY
England
44 373 858 977

Honeywell Control Systems Ltd.
Head Office
Honeywell House
Charles Square
Bracknel, Berkshire
England
44 344 424 555

Honeywell Avionics Systems, ltd.
Edison Road
Basingstoke, Hampshire RG21 2QD
England
44 256 51111

Honeywell OY
Ruukintie 8
02320 Espoo 32
Finland
358 0 801-01

Honeywell S.A.
Aerospace and Defense
4 Avenue Ampere
Montigny le Bretonneaux
78886 Montigny Bretonneaux
Cedex, France
33 1 30 43 20 91

97

HONEYWELL

Honeywell Aerospace
1 Rue Marcel Doret
31700 Blanac
Toulous (Blagnac)
France
33 6212-1559

Honeywell AG
Honeywellstrasse
6457 Maintal
Germany
49 6181-4011

Kaiserfeistrasse 39
Postfach 100865
D-6050 Offenbach/main
Germany
49 69 80640

Honeywell ELAC Nautic GambH
(Manufacturing Facility)
Westring 425-429
Postfach 2520, 2300 Kiel 1
Germany
49 431-8830

Bonn-Laison office
Honeywell AG
Bonn Center HI 401
Bundeskanzlerplatz 2-10
5300 Bonn
Germany
49 228-213-876

6457 Maintal 1, Germany
Soundertechnik Group , (A&D Germ)
Honeywell Regelsystme GmbH
Honeywellstrasse
Germany
49 6181-4011

Honeywell Asia Pacific, Inc.
30/F Suite 3001, Office Tower
Convention Plaza, 1 Harbour Road
Wanchai
Hong Kong
852 829-8298

11th Floor
Guardian House
21 Oi Kwan Road
Wanchai
Hong Kong
852 838-2883

Honeywell S.p.A.
Via Vittor Pisani, 13
20124 Milan, Italy
39 2 677-31

Honeywell International Mgmt.
Via AVolta, 16
20093 Cologno Monzese
Milan, Italy
39 2 25 30 41

Space Controls, Alenia-Honeywell
Via Diocleziano, 328
80125
Napoli, Italy
39 81 8687328

Honeywell Asia Pacific Region
Totate International Bldg.
2-12-19 Shibuya
Shibuya-ku
Tokyo 150, Japan
81 3 3409-1611

Honeywell Haneda Office
1-6-3 Haneda-Kuko
Jal Kiso Bldg., Ota-Ku
Tokyo 144, Japan
81 3 3747 3981

Yamatake-Honeywell Co. Ltd.
Head Office, Totate International Bldg.
12-19 Shibuya
2-Chome Shibuya-ku
Tokyo 150, Japan
81 3 3486-2506

Honeywell A-P Korea
6th Fl. Citicorp Center Bldg.
Chongro-Ku
Seoul 110-062
Korea
82 2 72-511-6

Honeywell A/S
Askerveien 61
1371 Asker
Norway
47 2 90 20 30

Honeywell, Inc.
111, Somerset Road, 11-01
P.U.B. Building
Singapore
65 731 7608/9

Honeywell AB
Storsatragrand 5
127 86 Skaerholmen
Sweden
46 8 775 55 00

Honeywell Centra Buerkle Ag
Erlenstrasse 31
2503 Bienne
Switzerland
41 32 3253 524

Honeywell B.V.
Aerospace and Defense
Laardenhootweg 18
1100 EA Amsterdam
The Netherlands
31 2 565 65 10

7821 AJ Emmen
Honeywell B.V.
Aerospace and Defense
Phineas Foggstratt 7
The Netherlands
31 5910 959 11

(Silicon sensors-high temperature [to 300C] pressure sensors, other high temperature parts [to 300C], radiation hardened parts including SRAMs [using silicon on insulator technology] PALs, PROMs, logic circuits, microcontrollers, and electronics speciality ASICs. Parts available for Space and Military applications.)

HONEYWELL — HUALON MICROELECTRONICS

Honeywell, Inc.
9201 San Mateo Blvd., NE
Albuqerque, NM 87113-2227
505-828-5000

(Parts used in Honeywell systems)

Hong Kong Crystals Company (HKC)
Unit 1105, 11/F
Golden Industrial Building
16-26 Kwai Tak Street
Kwai Chung, New Territories
Hong Kong
852 428 7183
Fax: 852 489 1137

(Crystals, crystal clock oscillators, ceramic resonators)

HP5 prefix
See *Opti, Inc.*

HPC prefix
See *National Semiconductor*

HPMX prefix
See *Hewlett-Packard Company*

(WLAN chip set)

HR prefix
See *Honeywell Solid State Electronics Center*, *Hitachi Semiconductor*

(Photo detectors)

HRD prefix
See *Natel Engineering Co., Inc.*

HRSD prefix
See *Natel Engineering Co., Inc.*

HS prefix
See *Harris Semiconductor* (HSX-XXXX-X part number), *Sipex* (Hybrid Systems Division)

HSB prefix
See *Hi-Sincerity Microelectronics Corp.*

HSC logic
See *GEC Plessy*

HSC prefix
Harris Semiconductor

HSCF prefix
See *Signal Processing Technologies, Inc.*

(Filters)

HSD prefix
See *Hi-Sincerity Microelectronics Corp.*

HSDC-XXXX part number
See *ILC DDC*

HSE
See *Hybrid Semiconductors & Electronics, Inc.*

HSMS prefix
See *Hewlett-Packard*

HSN prefix
See *IRT Corporation*

HSP prefix
See *Harris Semiconductor*

HSR-XXX part number
See *ILC Data Device Corp.*

HSSR-XXXX part number
See *Hewlett-Packard, Optical Communications Division*

HT logo (Within a circle on top of a triangle with the apex pointing to the left)
See *Hi-Tron Semiconductor Corp.*

HT prefix
See *Headland Technology* (Headland Products Division, of LSI Logic), *LSI Logic*, *Smar Research Corp.*

(Modem ICs)

HT-XXXX part number
See *Holtek, Holmate Technology Corp.*

HT Communications, Inc.
4480 Shopping Lane
Simi Valley, CA 93063
805-579-1700
Fax: 805-522-5295

(Chip set for telecommunication frame relay and DSU/CSU capabilities)

HTIU prefix
See *Honeywell*

HTK prefix
See *Headland Technology* (Headland Products Division, of LSI Logic), *LSI Logic*

Hua Ko Electronics
9 Dai Shun Str., Tai Po
New Territories 32901
Hong Kong
001 407 724 3000

(HC and HCT logic, SRAMs)

Hualon Microelectronics Corp. (HMC)
No. 1, R&D Road IV
Hsin-Chu Science Based Industrial Park
Hsinchu, Taiwan R.O.C.
886 35 774 945 8
Fax: 886 35 774 305

HUALON MICROELECTRONICS — HW

9-11 Shelter Street, Parkview
Causeway Bay
Hong Kong
852 577 5998
Fax: 852 5895 6659

6th Floor, Chung Shan N. Road
Taipei, Taiwan
886 2 536 2151
Fax: 886 2 531 3241

42 Chung Shan N. Road
Taipei, Taiwan
886 2 562 8913
Fax: 886 2 531 3214

HMC Philippines Trade Corporation
Corier Pela Roas Esteban
Makati, Metro Manila
Philippines
63 2 888 222
Fax: 63 2 815 4825

U.S. Parts:
The Summa Group, Ltd.
One California Street
19th Floor
San Francisco, CA 94111
415-288-0390
Fax: 415-288-0399

(Consumer ICs; such as melody generators, voice ICs, speech recognition ICs, voice/music ICs, sound effect ICs, calculator, time keeping, microcontrollers; telecommunications ICs such as tone dialers, pulse dialers, LCD phone, two line and cordless phone, call progress tone decoder and dial controllers, DTMF receivers, encoders and decoders; Computer ICs, display, I/O, mouse, floppy disk drive controllers, PC chip sets; ROMs, SRAMs, ASICs, Gate Arrays CCD image sensors and foundry services available.)

Huayue Electronic Devices Industry Co.
Shenzhen in Guangdong province of southern Peoples Republic of China. (Note: products only available through a state owned trading company, such as China National Electronics I/E Corp. [CNEIEC] which exports products outside of the Peoples Republic of China.

(Transistors, LEDs, power transistors, discrete ICs. Note: products only available through a state owned trading company, such as China National Electronics I/E Corp. [CNEIEC] which exports products outside of the Peoples Republic of China.)

Hughes Semiconductor Products Center
(Solid State Products)
500 Superior Ave.
P.O. Box H
Newport Beach, CA 92658-8903
714-759-2900, 2942, 2411, 2934
Fax: 714-759-2913
TWX: 910-596-1374/HACSSPD NPBH

59 Glover Avenue
Suite 21
Norwalk, CT 06850
203-846-7780
Telex: HUGHES AIR NLK

Hughes International Service Corporation
Schmaedel Str. 22
8000 Munich, Germany
49 89 834 7088
Telex: 5213856/HSPD

Queensway Industrial Estate
Glenrothes, Fife
KY7 5PY, Scotland
0592 754311
Fax: 0592 759775

(CMOS VLSI military gate arrays, LCD drivers, memories, microprocessors, mixed signal ASICs, diodes, transistors, and special purpose and custom products and HCMOS silicon foundry work)

Hunter Components
24800 Chagrin Blvd., Suite 101
Beachwood, OH 44122
216-831-1464
Fax: 216-831-1463

(LCD displays)

Hutson Industries
1000 Hutson Drive
P.O. Box 90
Frisco, TX 75034
214-377-2402
Fax: 214-377 2197
TLX: 910-860-5051

(Diodes, thyristors, SCRs)

HV prefix
See *Harris Semiconductor*, *Supertex, Inc.*

HV Component Associates, Inc.
P.O. Box 2484
Farmingdale, NJ 07727
800-548-0344
908-938-4499
Fax: 908-938-4451, 4499

(Diodes, bridges, and semiconductors)

HVCA marking
See *High Voltage Component Associates*

HVU prefix
See *Hitachi Semiconductor*

(Variable capacitance diodes)

HW prefix
See *Hitachi Semiconductor*

HX prefix
See *Honeywell Solid State Electronics Center*
(The HX indicates a Honeywell CMOS SOI [Silicon On Insulator] process), *RF Monolithics, Inc.*

HXO prefix
See *STC Defense Systems*

HY prefix
See *Hytek Microsystems, Inc.*

HY Electronics Corporation
Paralight (USA) Corporation
545 West Allen Avenue, #6
San Dimas, CA 91773
909-305-2393
Fax: 9130-2389
E-mail: HYLEDS@aol.com

(LEDs: discrete, matrix and numeric)

Hyashi Denkoh Co., Ltd.
U.S., AC Interface, Inc.
22391 Gilberto
Rancho Santa Margarita, CA 92688
714-858-1866
Fax: 714-858-1073

(Thin film platinum RTDs)

Hybrid Memory Products, PLC
Div. Implex PLC
Elm Road, West Chirton Ind. Est.
North Shields, Tyne & Wear
NE29 8SE
091 258 0690, Fax: 091 259 0997
TLX: 53206

(Memory products, memory modules, they have licensed the 3-D component packaging scheme, Trimrod, from Thomson-CSF.)

Hybrid Semiconductors & Electronics, Inc. (HSE)
87-45 Van Wyck Expressway
Jamaica, NY 11418
718-739-9000
Fax: 718-297-2633

(Diodes, transistors, thyristors, digital ICs)

Hybrid Systems, Inc.
See *SIPEX Corp.*

Hybrids
See *Hybrids International, Ltd.*

Hybrids International, Ltd.
311 N. Lindenwood Drive
Olanthe, KS 66062
913-764-6400
Fax: 913-764-6409

(VCXOs, crystal clock oscillators)

Hybridyne, Inc.
2155 North Forbes Blvd.
Tucson, AZ 85745
602-629-0001
(520 area code after 12/31/96)
Fax: 602-791-3899
(520 area code after 12/31/96)

(Thick film substrates and high density hybrids)

HYC prefix
See *Standard Microsystems Corp.*

(ARCNET transceiver)

Hy-Cal Engineering
9650 Telstar Avenue
El Monte. CA 91731
818-444-4000
Fax: 818-444-1314

P.O. Box 5291
Oak Brook, IL 60522-996
http://www.hycalnet.com/sensors/

(RTD, flow rate sensors, temperature humidity sensors. This company was part of the General Signal's Leeds and Northrup division before L&N was sold to Honeywell in 1996. It now reports to Honeywell Microswitch.)

HyComp, Inc.
165 Cedar Hill Street
Marlborough, MA 01752
508-485-6300
Fax: 508-481-1547

(Interface, linear ICs)

Hydra Chip Set
See *LSI Logic, Inc.*

Hydra trademark
See *Headland Technology*

HYF prefix
See *Hyundai Electronics*

HyperSPARC
See *Cypress Semiconductor*

Hyperstone Electronics, GmbH
AM Seerheim 8
Konstanz 78467
Germany
49 7531 980 30, 49 7531 677 89
Fax: 49 7531 517 25

Vallco Financial Center
10050 N. Wolfe Road
Suite SW1-276
Cupertino, CA 95014
408-253-0283
Fax: 408-446-5139

(Combination RISC and DSP microprocessor)

Hypres
175 Clearbrook Road
Elmsford, NY 10523
914-592-1190
Fax: 914-347-2239

(Joseph Voltage Standard ICs, the technology was transferred from the National Institute of Standards in 1994, and voltage standard ICs)

Hy-Q International (USA), Inc.
1438 Cox Avenue
Erlanger, Kentucky 41018
606-283-5000, Fax: 606-283-0883

(Crystals, crystal/LC filters, oscillators)

Hytek Microsystems, Inc.
400 Hot Springs Road
Carson City, NV 89706
702-883-0820
Fax: 702-883-0827

(Delay lines, hybrid microcircuits, subminiature thermoelectric coolers that work in conjunction with thermistor bridge, high speed laser diode drivers)

Hyundai Electronics America (Industries Co., Ltd.)
(Note the Digital Media Division is now called *Odeum Microsystems, Inc.*)
Worldwide Hyundai Electronics Industries Companies, Ltd.
Memory Business Div.
Hyundai Building
140-2, Kye-Dong, Chrongro-Ku
Seoul, Korea
http://www.hea.com

(Memory products, DRAMs, SRAMs. For MPEG-2 ICs, DVD, digital video disk ICs. See Odeum Microsystems, Inc.)

Hyundai Assembly and Test
510 Cottonwood Drive
Milpitas, CA 95035
408-232-8610
Fax: 408-232-8126

Dallas Office: 214-770-1930
Fax: 214-770-1933

(Subcontract assembly; can package devices in the MQUAD [tm Olin Corp.] package and the BGA package [licensed by Motorola])

San Jose, CA: 408-473-9229
Fax: 408-473-9370

Atlanta, GA: 404-242-6464
Fax: 404-242-6554

New Jersey: 201-262-7770
Fax: 201-261-2967

Dallas, TX: 214-770-1930
Fax: 214-770-1931

London, England: 44-81-741-8634
Fax: 44-81-563-0179

Paris, France: 33 1 6929 8319
Fax: 33 1 6929 8315

Frankfurt, Germany: 49-6142-9210
Fax: 49-6142-9212/14

Hong Kong: 852-596-1276/1282
Fax: 852-596-0815

Tokyo, Japan: 3-3211-4071/8
Fax: 3-3211-5447/8

Scotland: 44 786 449 386
Fax: 44 786 449 387

Singapore: 65-270-6300
Fax: 65-270-6102

Taipei, Taiwan: 8862-568-1134
Fax: 8862-543-3433

LCD Division:
San 136-1, Ami-Ri, Bubal-Eup
Ichon-Kun, Kyounggi-Do, Korea
0336-39-6404/6595;
0336-30-2084/3235
Fax: 0336-39-6415/6416;
Fax: 0336-30-3210

San Jose, CA: 408-232-8641/4
Fax: 408-232-8140

New Jersey: 201-262-7770
Fax: 201-261-2967

Dallas, TX: 214-770-1930
Fax: 214-770-1931

Atlanta, GA: 404-242-6464
Fax: 404-242-6554

Tokyo, Japan: 3 3211 4701/8
Fax: 3 3211 5447/8

Taipei, Taiwan:
886 2 716 9636/9637/9638
Fax: 886 2 713 4299

Singapore: 65 270 6300
Fax: 65 270 6102

Hong Kong: 852 2 971 1600
Fax: 852 2 971 1622

Frankfurt, Germany:
49 6142 921229
Fax: 49 6142 921225

London, England: 44 81 741 8634
Fax: 44 81 563 0179

Paris, France: 33 1 6929 8319
Fax: 33 1 6929 8315

Scotland: 44 786 449 386
Fax: 44 786 449 387

i logo and prefix
See *Intel Corp.*

I128-2 part number
See *Number Nine Visual Technologies*

(3D graphics IC)

IAC-XXXX part number
See *ILC-DDC*

IAM Electronics Co., Ltd.
9847, Akaho, Komagane-Shi,
Nagano 399-41
0265 82 5191
Fax; 0265 82 6039

(Hybrid ICs)

Manufacturers, Prefixes, Part Number Types, Logo Descriptions & Family Types

IBM prefix
See *IBM Corp.*

IBM Corp.
1000 River Street
Essex Junction VT 05452
1-800-769-ASIC (3754),
1-800-769-6984, 1-800-426-0181
802-769-6189
http://www.chips.ibm.com
http://power.globalnews.com/
(PowerPC News Home Page)

IBM Technology Products
Route 100
P.O. Box 100
Somers, NY 10589
914-766-1900

IBM Microelectronics
1580 Rte. 52, MS-B
Hopewell Junction, NY 12533-6531
1-800-IBM-0181 ext 904
(For SIMMs with error correction, ECC)
1-800-POWERPC, ext. 1420
(On the IBM PowerPC 603)
914-892-5384

Route 100
P.O. Box 100
Somers, NY 10589

P.O. Box 7932
Mt. Prospect, IL 60056-7932
800-IBM-0181

Literature:
IBM
Fulfillment Center
1000 Business Center Drive
Mt. Prospect, IL 60056
1-800-IBM-0181

Europe:
IBM
la Pompignane BP 1021
34006 Montpellier
France 33 6713 5757 (Francais)
33 6713 5756 (Italiano)

IBM
Postfach 72 12 80
30539 Hannover, Germany
49 511 516 3444 (English)
49 511 516 3555 (Deutsche)

Japan:
800, Ichimiyake, Yasu-Cho, Yasu-gun
Shiga-ken, Japan 520-23
81 775 87 4745

(RAMDACs. Note: IBM Technology Products changed its name to IBM Microelectronics in late 1993, 1-800-IBM-0181. ASICs including mixed signal types, microcontrollers, DC/DC converters, data compression ICs, MCM modules, SRAMs, VRAMs, EDRAMS [second sourced by Ramtron], FPGAs, etc. Many ICs are sold in modules using advanced packaging and interconnect technology to which adds resale value to the devices but IBM is also selling microprocessor "Power PC" ICs., and microcontrollers [with parts compatible to Intel devices, such as their 8096])

ICXX part number
See *Lucent Technologies* (formerly AT&T Microelectronics)

(Reconfigurable cell arrays FPGAs)

ICD prefix
See *IC Designs*

ICS
See *Integrated Circuit Systems, Integrated Component Systems, Inc.*

(Crystals and crystal oscillators)

IC Designs
12020 113th Avenue, N.E.
Kirkland, WA 98034
1-800-669-0557, 1-800-858-1810
Fax: 1-206-820-8959
Europe: 32 2 652 0270,
Fax: 32 2 652 1504
Asia: Fax: 1-415-940-4343

(Frequency synthesis ICs [clock generators] and ASSP programmable clock oscillator chips [QuiXTAL]. This company was acquired by Cypress Semiconductor Corporation in late 1993)

iC Haus GmbH
Am Kunmerling 18
Bodenheim 55294
Germany
6135 92 92 0
Fax: 6135 92 92 192

(Input ports, laser diode controllers, line drivers, encoders, switching regulators, microprocessor controllers)

IC-XX prefix
See *iC haus GmbH*

IC Works (ICW)
(A Samsung Semiconductor spin off)
3275 North First Street
San Jose, CA 95134
408-922-0202
Fax: 408-922-0833

(Low power or BiCMOS process technology foundry, SRAMs and mixed signal devices, cache memory devices, RAMDACs, FIFOs, IC clock synthesis [clock generator] chips, mixed signal and RF products.)

IC Sensors
See *EG&G IC Sensors*

Ichips Corporation
See *New Logic*

ICI
See *Integrated Circuits Incorporated*

ICL prefix
Formerly Intersil
See *Harris Semiconductor*

ICM prefix
See *International Crystal Mfg. Co., Inc.*, *Harris Semiconductor*

ICM International Controls & Measurement
P.O. Box 2819
Syracuse, NY 13220
800-365-5525
315-699-5266
Fax: 315-699-5260

(Thyristors)

ICO-Rally Corporation
2575 E. Bayshore Road
P.O. Box 51350
Palo Alto, CA 94303
415-856-9900
Fax: 415-856-8378

(Optoelectronics)

ICT, Inc.
(Formerly International CMOS Technology)

Headquarters and Western Area Sales:
2123 Ringwood Avenue
San Jose, CA 95131
1-800-729-7335
(1-800-SAY-PEEL)
408-434-0678
Fax: 408-432-0815
TWX: 910-997-1531
http: www.ictpid.com

320 New Crossing Trail
Kennesaw, GA 30144
404-516-0194

Central Area Sales:
1821 Walden Office Sq., Suite 400
Schaumburg, IL 60173
708-303-0012
Fax: 708-303-0284

(User programmable [PEEL] ICs, stereo enhancement processors)

I-Cube, Inc.
2328-C Walsh Avenue
Santa Clara, CA 95051
408-986-1077
Fax: 408-986-1629
BBS: 408-986-1652
Internet: marketing@icube.com
http://www.icube.com

(Programmable switches, interconnection devices which can be dynamically reconfigured, crossbar switches)

ICW prefix
See *IC (Integrated Circuit) Works*

ICX prefix
See *Sony*

(CCDs)

IDEA, Inc.
1300-B Pioneer Street
Brea, CA 92621-3728
310-697-4332
Fax: 310-690-1352

(LED displays including dot matrix, clock, counter, bar graph, bar module and lamp cluster arrays)

Ideal Semiconductor Inc. (ISI)
46721 Freemont Blvd.
Fremont, CA 94538
510-226-7000
Fax: 510-226-1564

(A manufacturer of obsolete parts from a variety of original manufacturers, including AMD, Harris, National Semiconductor, IDT, Signetics, Quality, Samsung and Zytrex. Military specification processing can be supplied)

IDEC Corporation
1213 Elko Drive
Sunnyvale, CA 94089
800-262-4332
Fax: 408-744-9055

(Sensors including photoelectric, fiber optic, laser, and ultrasonic types)

IDI
See *Industrial Power, Industrial Devices, Inc.*

(LEDs, lamps)

IDT (Also part prefix)
See *Integrated Device Technology, Inc.*

idv Solutions
1165 Walnut Avenue
Suite 4
Chula Vista, CA 91911
619-424-6630
Fax: 619-424-6673

(Battery charger modules)

IEE, Industrial Electronic Engineers, Inc.
Industrial Products Division
7740 Lemona Avenue
Van Nuys, CA 91409-9234
818-787-0311
Fax: 818-901-9046

(Vacuum fluorescent and other displays, keyboard systems)

IEP
See *International Electronic Products of America*

IEPC
See *International Electronic Products of America*

IEPC (International Electronic Products of America)
6-8 North Williams Street
Whitehall, NY 12887-1119
518-499-2435
Fax: 518-499-2437

(Silicon high voltage diodes)

IF prefix
See *InterFET Corporation*
(If FET devices)

IFN prefix
See *InterFET Corporation*
(If FET devices)

IFD prefix
See *Hewlett-Packard Co.*

IFP prefix
See *InterFET Corporation*
(If FET devices)

iFX prefix
See *Altera Corp.*

(Flexlogic FPGA)

IgT
See *Integrated Telcom Technology, Inc.*

IH prefix
See *Hy-Cal Engineering*

(Temperature humidity sensors)

II Stanley Electric Co.
See *Stanley Electric Company*

IIT
See *8x8 (Eight-by-Eight) Inc.*
(Formerly Integrated Information Technology, Inc.)

IL prefix
See *Siemens*

(Optocoupler)

ILC-DDC Data Device Corporation
105 Wilbur PL.
Bohemia, N.Y. 11716
1-800-332-1772
1-800-332-5757 customer service
516-567-5600
(International direct customer service)
Fax: 516-567-7358, 563-5208
TLX: 310-685-2203

12832 Valley View St., Suite 213
Farden Grove, CA 92645
714-895-9777
Fax: 714-895-4988
818-992-1772
Fax: 818-887-1372

21400 Ridgetop Circle
Suite 110
Sterling, VA 22170
703-450-7900
Fax: 703-450-6610

12 Furler St.
Totawa, NJ 07512
201-785-1734
Fax: 201-785-4132
Canada: 1-800-245-3413

DDC United Kingdom, Ltd.
Mill Reef House
9-14 Cheap Street
Newbury, Berkshire RG14 5DD
England
44 635 40158
Fax: 44 635 32264

DDC Ireland Ltd.
Cork Business and Technology Park
Model Farm Road
Cork, Ireland
353 21 341065
Fax: 353 21 341568

ILC-DDC Electronique
10 Rue Carle-Hebert
92400 Courbevoie
France
33 1 41 16 34 24
Fax: 33 1 41 16 34 25

ILC-DDC Electronik GmbH
P.O. Box 1212
D-86882 Landsburg am Lech
Germany
49 8191 3105
Fax: 49 8191 47433

DDC Electronik AB
P.O. Box 18 191 21 Sollentuna
Sweden
46 8 92 06 35
Fax: 46 35 31 81

DDC Electronics KK
Koraku 1-Chome Bldg 7F 2-8
Koraku 1-Chome
Bunkyo-Ku, Tokyo 112
Japan
81 33 814-7688
Fax: 81 33 814-7689

(MIL-STD-1553 devices, convertors, motor drivers, DC/DC converters, solid state power controllers. Note: ILC-DDC acquired the synchro-conversion line of Natel Engineering Company in 1995 [this includes resolver to digital, digital-synchro, synchro-digital, SINE to COSINE; angle simulators]. Radiation shielded versions of ILC DDC's hybrid and monolithic products, suitable for space applications, are available from Space Electronics, Inc. [SEI]).

ILCP prefix
See *Siemens*

ILD prefix
See *Siemens*

(Optocoupler)

ILH prefix
See *Siemens*

(Optocoupler)

ILQ — INFINITE TECHNOLOGY

ILQ prefix
See *Siemens*

(Optocoupler)

iLSI
See *Integrated Logic Systems, Inc.*

ILX prefix
See *Sony*

(CCDs)

IM prefix
See *Intelligent Motion Systems*

Image and Signal Processing, Inc. (ISP)
120 Linden Avenue
Long Beach, CA 90802
213-495-9533
Fax: 213-495-1258

(ICs used on their line of digital signal processing peripherals)

ImageNation
P.O. Box 276
Beaverton, OR 97075-0276
1-800-366-9131, 503-641-7408
Fax: 503-643-2458
Email: info@ImageNation.com
http://www.ImageNation.com/~image

(Custom ICs on their frame grabber circuit boards)

Imagine (Graphics controller IC)
See *Number Nine Computer Corporation*

IMB Electronic Products, Inc.
15421 Carmentia Road
Santa Fe Spring, CA 90670-5607
714-523-2110

(Memory products)

imc (As in imc-1000) prefix
See *Dimolex*

IMI
See *International Microcircuits, Inc.*

IMO VARO ESD
See *Varo, Inc.*

IMO TransInstruments
One Cowles Road
Plainville, CT 06062-1198
203-793-4516
Fax: 203-793-4514

(Pressure transducers)

IMP prefix
See *IMP, Inc.* (International Microelectronic Products, Inc.)

IMP, Inc. - International Microelectronic Products
2830 N. First St.
San Jose, CA 95134-2071
408-432-9100
Fax: 408-434-0335
Faxback: 800-249-1614
http://www.impweb.com
E-mail: info@epac.com

NE Sales Office:
Pinckney, MI
313-878-1502
Fax: 313-878-4736

(EPAC [tm], electrically programmable analog monitoring and diagnostic circuit, disk drive ICs)

IMR2(tm)/**QuIRT**(tm) chip set
See *Advanced Micro Devices*

IMS
See *Intelligent Motion Systems, Inc.*

IMS prefix
See *Institute of Microelectronics Circuits and Systems*

IMST prefix
See *Inmos*

INA prefix
See *Burr-Brown*, *Hewlett-Packard Corp.*

(RF, Microwave amplifier)

INA-XXXXX part number
See *Hewlett-Packard Co.*

(RF ICs)

Industrial Devices, Inc. IDI
A Chicago Miniature Company
260 Railroad Avenue
Hackensack, NJ 07601
201-489-8989
Fax: 201-489-6911

(LEDs and lamp assemblies)

ines
Messtechnik Datenverarbeitung
6711 S. Ivy Way, Ste. B-4
80112 Englewood
303-779-8354
Fax: 303-770-6536

(GPIB [General Purpose Instrumentation Bus] ICs, with parts compatible with NEC 7210C [IEEE 488.2 Controller, Talker and Listener with FIFO Buffer, etc.])

Infinite Solutions
300 Miller Ct. W.
Norcross, GA 30071
800-541-9801
Fax: 404-449-8280

(DSP ICs)

Infinite Technology, Corp.
2425 N. N. Central Expressway
#323
Richardson, TX 75080
214-437-7800, 7822
Fax: 214-437-7810

Manufacturers, Prefixes, Part Number Types, Logo Descriptions & Family Types

(Reconfigurable arithmetic datapath device a high speed accelerator IC for Data Stream Algorithms, SRAM programmable logic, an IC containing four 16 bit arithmetic processors)

InFocus Systems
7770 SW Mohawk Street
Tulantin, OR 97062
800-294-6400
503-685-8888
Fax: 503-692-4476, 503-685-8631

27700B S.W. Parkway Avenue
Wilsonville, OR 97070-9215
503-685-8609, 503-685-8888
Europe: 31 0 2503 23200,
Fax: 31 0 2503 24388

(LCDs, LCD projection panels and LCD projectors)

Info-Lite
4422 Cliff St.
Fairview, NJ 07022
201-941-4455
Fax: 201-941-4919

(Large digital instruments, displays, lighted indicators and matrix pinboards)

Infrared Laboratories, Inc.
1808 E. 17th Street
Tucson, AZ 86719
602-622-7074
(520 area code after 12/31/96)
Fax: 602-623-0765
(520 area code after 12/31/96)

(Linear ICs)

Infrascan, Inc.
7080 River Road, #102
Richmond, B.C. Canada V6X 1X5
604-723-8655
Fax: 604-278-3423

(Image processing ICs)

Information Storage Devices, Inc. (ISD)
2841 Junction Avenue, Suite 204
San Jose, CA 95134
800-825-4473, 408-428-1400
Fax: 408-428-1422
http://www.isd.com/

(A fabless company supplying single EEPROM ICs that can be used as voice record/playback devices, used in Voiceprint [tm] talking photographs. Mass storage memory.)

INI prefix
See *Initio Corporation*

Initio Corporation
2901 Tasman Drive
Suite 201
Santa Clara, CA 95054
1-800-99-INITIO
408-988-1919
Fax: 408-988-3254
sales@initio.com

(I/O ICs including second sources for Adaptec products, PCI to Fast/Wide SCSI adapter cards and chip sets)

INMOS Corporation
(Division of SGS-Thomson Microelectronics)
P.O. Box 23058
Colorado Springs, CO 80935
1000 Aztec West
Almondsbury
Bristol, United Kingdom BSAS 4SQ
44 01454 616815
Fax: 44 01454 617810

Sales Offices:
Santa Clara, CA: 408-727-7771
Torrance, CA: 213-530-7764
Denver, CO: 303-252-4100
Norcross, GA: 404-242-7444
Westborough, MA: 617-366-4020
Columbia (Baltimore), MD: 301-995-6952
Minneapolis, MN: 612-831-5626
Dallas, TX: 214-490-9522

INFINITE TECHNOLOGY — INTEGRATED CIRCUIT SYSTEMS

INMOS SARL
Paris, France
1 4687 22 01

INMOS Ltd.
Bristol, United Kingdom
454 616616

INMOS GmbH
Germany
089 319 1028

INMOS International
Tokyo, Japan
03 505 2840

Inova Microelectronics
(This company declared bankruptcy in 1991 and ceased producing material.)

Instrument Design Engineering Associates
1300 Pioneer Street, Suite B
Brea, CA 92621-3728
310-697-4332
Fax: 310-690-1352

(Optoelectronics)

Instruments, S.A.
See *ASA Jobin Yvon/Spex Division*

Intech Microcircuits Division
See *Lambda Advanced Analog Integran, Inc., JT PowerCraft, Inc.*, (Part of JT Electronics Corporation, Tokyo, Japan).

Integrated Circuit Systems, Inc. (ICS)
(Corporate Headquarters)
P.O. Box 968
2435 Blvd. of the Generals
Valley Forge, PA 19482-0968
1-800-220-3366
800-220-3366, 610-630-5300
Fax: 610-630-5399
http://www.icsinc.com

INTEGRATED CIRCUIT — INTEGRATED DEVICE TECHNOLOGY

1271 Parkmore Avenue
San Jose, CA 95126-3448
408-297-1201

(A fabless company supplying mixed signal ICs including multimedia ICs for audio and video, graphics ICs, power management ICs, frequency synthesizers, clock ICs, network ICs [for LAN, WAN, SONET/SDH, ATM, fast Ethernet], ASICs, etc.)

Integrated Circuits Inc.
10301 Willows Road
Redmond, WA 98052
206-882-3100, Fax: 206-882-1990

(Custom thick film hybrids)

Integrated CMOS Systems
(Became Vertex Semiconductor and now merged into Toshiba.)

Integrated Component Systems, Inc. (ICS)
5440 NW 55th Blvd, #11-105
Coconut Creek, FL 33073
1-800-396-5185
Fax: 305-480-6950

(Voltage controlled oscillators, wide band VCOs)

Integrated Device Technology, Inc. (IDT)
Corporate Headquarters
P.O. Box 58015, 2975 Stender Way
Santa Clara, CA 95054-3090
1-800-345-7015 then press 2
for nearest sales representatives in
the U.S. and Canada
408-727-6116, Fax: 408-492-8674
TWX: 910-338-2070
http://www.idt.com
FTP Server: Acrobat index:
<ftf:ftp.idt.com/pubs/docs/1000.pdf>
Text Index:<ftp:ftp.idt.com/pub/index>
E-mail: info@idt.com

Data Sheets/Application Notes/News Releases:
Fax-On-Demand:
1-800-9IDT-FAX,
1-800-943-8329, 408-492-8391
And follow the menu directions.
(For an index of current documentation, request document #1000)

Data CD-ROMs:
1-800-345-7015 x1

Central Headquarters:
(And Nebraska)
1375 E. Woodfield Road
Ste. 380
Schaumburg, IL 60173
708-517-1262

Eastern Headquarters:
(For Maine, Massachusetts, New Hampshire, Rhode Island, Vermont)
#2 Westboro Business Park
200 Friberg Parkway
Ste. 4002
Westboro, MA 01581
508-898-9266
(This is also a Technical Center)

Western Headquarters:
(For Hawaii, Northern Nevada))
2975 Stender Way
Santa Clara, CA 95054-3090
800-345-7015
408-492-8350
Fax: 408-492-8674
(This is also a Technical Center)

NE Regional Offices:
(For Delaware, Maryland, Virginia)
Horn Point Harbor
105 Eastern Avenue, Ste. 201
Annapolis, MD 21403
301-858-5423

One Greentree Centre
Ste. 202
Marlton, NJ 08053
609-596-8668

250 Mill Street
Ste. 107
Rochester, NY 14614
716-777-4040

1160 Pittsford Victor Road
Bldg. E
Pittsford, NY 14534

SE Regional Offices:
(For Georgia, Mississippi, South Carolina)
1413 S. Patrick Drive, Ste. 10
Indian Harbor Beach FL 32937
407-773-3412

18167 U.S. 19 North, Ste. 455
Clearwater, FL 34624
813-532-9988

1500 N.W. 49th Street, Ste. 500
Ft. Lauderdale, FL 33309
305-776-5431

NW Regional Offices:
1616 17th St.
Ste. 370
Denver, CO 80202
303-628-5494

7981 168th Avenue N.E.
Ste. 32
Redmond, WA 98052
206-881-5966

15455 NW Greenbriar Pkwy
Suite 210
Beaverton, OR 97006

SW Regional Offices:
6 Jenner Dr.
Ste. 100
Irvine, CA 92718
714-727-4438
(This is also a Technical Center)

16130 Ventura Blvd.
Ste. 370
Encino, CA 91436
818-981-4438

Manufacturers, Prefixes, Part Number Types, Logo Descriptions & Family Types

N. Central Regional Office
1650 W. 82nd St.
Ste. 1040
Minneapolis, MN 55431
612-885-5777

S. Central Regional Offices:
(For Arkansas, Louisiana, Oklahoma)
14285 Midway Road
Ste. 100
Dallas, TX 75244
214-490-6167
(This is also a Technical Center)

17314 State Highway 19
Ste. 242
Houston, TX 77064
713-890-0014

Other Offices:
(For Tennessee)
555 Sparkman Dr.
Ste. 1200-D
Huntsville, AL 35816
205-721-0211

11782 Jollyville Road
Suite 204B
Austin, TX 78759

European Headquarters/Northern Europe:
Integrated Device Technology, Ltd.
Prime House
Barnett Wood Lane
Leatherhead, Surrey KT22 7DG
44 372 363339
Fax: 44 372 378851

Central Europe Regional Office:
Gottfried Von-Cramm-Str.1
8056 Neufahrn
Germany
49 8165 5024

Southern Europe Regional Office:
15 Rue du Brisson aux Fraises
91300 Massy, France
33 1 69 30 89 00

Hong Kong Regional Office:
IDT Asia Ltd.
Suite 1003
China Hong Kong City
Tower 6
33 Canton Road Tsimshatsui
Hong Kong
852 736 0122
Fax: 852 375 2677

Japan Headquarters:
Nippon IDT KK
Sumitomo Fudosan Sanbancho Bldg.
6-26 Sambancho
Chiyoda-Ku
Tokyo 102
Japan
813 2221 9821
Fax: 813 3221 7372/9824

(MIPS controllers, memory products [SRAMs and DRAMs], EDC [Error Detection and Correction ICs], graphics [ADCs], logic products [decoders, multiplexers, counters, line drivers, transceivers, pipeline registers, buffers, latches, flip flops, etc.] memory modules, FIFO memories, hybrid ICs, RISC microprocessor components and subsystems, radiation hardened, enhanced and tolerant products, etc.)

Integrated Digital Products Corp.
1809 National Street
Anaheim, CA 92801
714-879-0810

(Microprocessor support ICs)

Integrated Information Technology, Inc. (IIT)
See *8x8 (Eight-by-Eight), Inc.*

Integrated Logic Systems, Inc.
4750 Edison Avenue
Colorado Springs, CO 80915
719-574-1688
Fax: 719-574-1993

(Gate arrays, programmable logic)

INTEGRATED DEVICE TECHNOLOGY — INTEGRATED TELECOM

Integrated Microelectronic Products
See *IMP, Inc.*

Integrated Microwave
11353 Sorrento Valley Road
San Diego, CA 92121-1009
619-259-2600
Fax: 619-755-8679

(600 to 2500 MHz ceramic filters, diplexers, lumped constant filters)

Integrated Optical Components, Ltd.
3 Waterside Business Park
Eastways, Witham
Essex CM8 3YQ, England
44 376 502110, Fax: 44 376 502125

(Integrated optical switches and modulators)

Integrated Power Semiconductors, Ltd.
See *Seagate Microelectronics*

Integrated Silicon Solutions, Inc. (ISSI)
836 N. St. Bldg. 5
Tewksbury, MA 01876
508-640-1400
Fax: 508-640-2355
Integrated Silicon Systems, Ltd.
United Kingdom

(A fabless company supplying affiliated with GEC Plessy, programmable IIR filter IC, SRAM memory ICs)

Integrated Telecom Technology, Inc. (IgT)
18310 Montgomery Village Avenue, Suite 300
Gaithersburg, MD 20879
301-990-9890
Fax: 301-990-9893
http://www.igt.com

INTEGRATED TELECOM — INTEL

(Telecommunications ICs including SONET processors [Async DS3 to SONET Mapper], CEPT ICs, gate arrays, and asynchronous transfer mode [ATM] ICs, gate arrays may also be second sourced by Altera Corporation)

Integrity Technology Corp.
Integrity Electronics
1400 Coleman Ave. #E-15
Santa Clara, CA 95050-4358
408-262-8640
Fax: 408-862-1680
Email: MCUclock@msn.com

(Ceramic resonators, Fax/modem coupling transformers)

Intel Corporation
2625 Walsh Ave.
Santa Clara, CA 95051
1-800-548-4725, 1-800-468-8118
(Literature distribution, see
1-800-879-4683 below)
1-800-628-8686
(Customer support)
916-356-3551
(Customer support outside U.S.)
408-765-4302

Fax: 1-800-628-2283 (Faxback), 916-356-3105 (U.S. or Canada)
44 0 793 496646 (Faxback Europe)
To download the FaxBACK catalogs call 1-800-897-2536 (U.S. or Canada).
For Europe, call the BBS number and look under FaxBACK area in the File Locator.

Internet: http://www.intel.com
http://www.intel.com/embedded/i960/ (for the i960 processor)
http://www.intel.com/design/mcs251
(MCS microcontroller design site)

BSDL (boundary scan)
BBS 916-356-3600
(U.S. or Canada)
44 0 793 496340 (Europe)

Literature Hotline:
1-800-548-4725 (U.S. or Canada)
In Europe, call an Intel office.

Microcontroller Code Software Updates (ApBuilder):
916-356-3600;
Europe: 44 0 793 496340:

Internet:
BUILDER_Software@ccm.hf.intel.com,
Fax: 602-554-7436
Address: ApBUILDER Software Feedback Mail Stop C3-64

Intel Corporation
5000 W. Chandler Blvd.
Chandler, AZ 85226

2200 Mission College Boulevard
P.O. Box 58119
Santa Clara, CA 95052-8119

3065 Bowers Avenue
Santa Clara, CA 95052
800-548-4725
Fax: 609-936-7619

5000 W. Chandler Blvd.
Chandler, AZ 85226

Literature Distribution Center:
1000 E. Business Center Drive
P.O. Box 7641
Mt Prospect, IL 60056-6090

P.O. Box 7641
Mt Prospect, IL 60056-7641
1-800-548-4725
1-800-548-4725

4024 Medford Drive
Huntsville, AL 35802
205-883-6137
Fax: 205-883-4826

410 North 44th St.
Suite 500
Phoenix, AZ 85008
800-628-8686
Fax: 602-244-0446

3550 Watt Avenue
Suite 140
Sacramento, CA 95821
800-628-8686
Fax: 916-488-1473

1900 Prairie City Road
Folsom, CA
916-356-6050

(Miniature Card [minicard] compact flash cards)

5-6, Tokodai, Tsukuba-shi
Ibarki 300-26
Japan
0298 47 8511

Intel Corporation S.A.R.L.
1, Rue Edison, BP 303
78054 Saint-Quentin-en-Yvelines
Cedex

Intel Corporation U.K. Ltd.
Pipers Way
Swindon
Wildtshire, England SN3 1RJ

Intel GmbH
Dornacher Strasse 1
85622 Feldkirchen/Muenchen
Germany

Intel Semiconductor Ltd.
32/F Two Pacific Place
88 Queensway, Central
Hong Kong

Intel Semiconductor of Canada, Ltd.
190 Attwell Drive, Suite 500
Rexdale, Ontario M9W 6H8
Canada

(Microprocessors, coprocessors, image processing, video processing and timing ICs, flash memory, microcontrollers [including parts for automotive transmission and engine control], board level products, modem ICs, FPGAs, 10base-T Ethernet ICs, PCMCIA controllers, etc. Military qualified parts available. Their neural networks IC business is now handled by Nestor, Inc. The Intel programmable logic line [5CXXX, 5ACXXX, iFXXXX, 85CXXX, and iPLD devices such as the 22V10] was acquired by the Altera Corporation on October 1, 1994.)

Intelligent Motion Systems, Inc. (IMS)
511 Norwich Ave.
Taftville, CT 06380
203-889-8353
Fax: 203-889-8720

(Microstepper motor controller IC)

Intellon Corporation
5100 W. Silver Springs Blvd
Ocala, FL 34482
904-237-7416
Fax: 904-237-7616
BBS: 904-237-8841 (9600,8,N,1)

(Power line transceiver IC and RF transceiver IC, SIMM interface modules for the CEbus and other wireless communications technology items)

InterChip Systems, Inc.
30 Massachusetts Avenue
North Andover, MA 01845-3413
508-687-8644

(MCMs)

Interdesign (Custom Arrays Corporation), Inc.
A Ferranti Company
525 Del Ray Ave #A
Sunnyvale, CA 94086
408-749-1166, Fax: 408-749-1718

(Gate Arrays)

InterFET Corporation
See *Interfet-ITAC*

InterFET-ITAC
322 Gold Street
Garland, TX 75042
214-487-1287
Fax: 214-276-3375

(High temperature parts such as microprocessors, memory, amplifiers, scintillation neutron detectors, high frequency oscillators and multichannel A/D converters. Military screening is available)

Intergraph Corp.
Mail Stop IW17D1
Huntsville, AL 35894-0001
1-800-763-0242
205-730-2700
205-730-5441 (International)
Fax: 205-730-9550
http://www.intergraph.com/

(Image processing ICs, graphics accelerators, workstations, interactive computer graphics systems, scanners, web servers, networking products, mass storage and backup products)

Intergraph
Advanced Processors Division
2400 Geng Road
Palo, Alto, CA 94303
415-494-8800
Fax: 415-856-0224

(Formerly the Fairchild Clipper [UNIX] processor division.)

Intermedics, Inc.
Company of Sulzermedica
4000 Technology Drive
Angleton, TX 77515-2523
1-800-231-2330
409-848-4578
Fax: 409-848-4580

(Custom microprocessors [Omega (tm)] with analog and digital circuitry for their products)

International CMOS Technology
See *ICT*

International Components Corporation
Intervox
105 Maxess Road
Melville, NY 11747
800-645-9154
516-293-1500
Fax: 516-293-4983

(Alarm systems, speakers, assemblies)

International Light
17 Graf Road
Newburtport, MA 01950-4092
508-465-5923
Fax: 508-462-0759

(High gain infrared detectors, spectradiometer, radiometers, photometers)

International Microcircuits, Inc. (IMI)
3350 Scott Blvd.
Building 36
Santa Clara, CA 95054
408-727-2280
TWX: 910-338-2032

(A fabless company supplying CMOS gate arrays)

International Crystal Mfg. Co., Inc. (ICM)
P.O. Box 26330
729 W. Sheridan
Oklahoma City, OK 73126-0330
1-800-725-1426
Fax: 1-800-322-9426 (24 hour line)
405-236-3741, Fax: 405-235-1904

INTERNATIONAL CRYSTAL — INTERSIL

(Crystals and crystal oscillators)

International Diode Corporation
229 Cleveland Avenue
Harrison, NJ 07029
201-482-6518
Fax: 201-482-4958

(Diodes, transistors, thyristors)

International Microcircuits, Inc. (IMI)
525 Los Coches Street
Milpitas, CA 95035
408-263-6300, Fax: 408-263-6571

(Semicustom CMOS gate arrays and standard cell custom ICs)

International Microelectronic Products (IMP), Inc.
2830 N. First Street
San Jose, CA 95134
408-432-9100
TWX: 910-338-2274
TLX: 4991041
Fax: 408-434-0335

(Custom and semicustom cell based ASICs, mixed and analog signal types)

International Microtronics Corp.
4016 E. Tennessee Street
Tucson, AZ 85714
602-748-8900
(520 area code after 12/31/96)
Fax: 602-790-2808
(520 area code after 12/31/96)

(Linear ICs)

International Power Devices, Inc. (IPD)
200 Butterfield Drive
Ashland, MA 01721
508-881-7434
Fax: 800-226-2100

155 North Beacon St.
Brighton, MA 01235
617-782-3331
Fax: 617-782-7416
ipdsales@ipd-hq.ccmail.compuserve.com

(DC/DC converters, switching power supplies)

International Power Semiconductor
7SDF-Seepz Anderi East
Bombay 400 096
India
6328534
TLX: 2840 GNO

(Transistors, thyristors)

International Power Sources, Inc.
200 Butterfield Drive
Ashland, MA 01721
508-881-7434
Fax: 800-226-2100, 508-879-8669

(DC/DC converters)

International Radiation Detectors (IRD)
2545 West 237th Street
Suite I
Torrance, CA 90505-5229
310-534-3 61
Fax: 310-534-3665

(Silicon photodiodes

International Rectifier
233 Kansas St.
El Segundo, CA 90245
310-322-3331, 310-607-8953
310-252-7105
(Product Information Center)
Fax: 310-322-3332
310-252-7100 (Fax on demand)
TLX: 66-4464
http://www.irf.com/

Europe:
Hurst Green
Oxted, Surrey RH8 9BB, England
(0883)713215
Fax: (0883) 714234
Telex: 95219

(HEXFET power MOSFETs, IGBTs, Schottkys, drivers, SCRs, rectifiers, SSRs [solid state relays, including photovoltaic types], modules, bridges, diodes)

International Semiconductor, Inc.
252 Cox St.
Roselle, NJ 07203
800-392-2474
908-245-2233
Fax: 908-245-5541

(Zener diodes, current regulator diodes, temperature- compensated reference diodes, transient voltage suppressors, zener and regulator diodes, switching diodes, variable capacitance [VVC] tuning diodes, silicon rectifier diodes and bridges, high voltage rectifier assemblies, diode arrays, surface mount parts available.)

International Sensor Systems, Inc.
Industrial Park
103 Grant Street
P.O. Box 345
Aurora, NE 68818
402-694-6111
Fax: 402-694-6180
Wats: 800-260-6287

(Custom thick film hybrid circuits)

International Thermoelectric, Inc.
See *ITI FerroTec*

Intersil
See *Harris Corporation*

Interpoint Corp.
10301 Willows Road
Redmond, WA 98052

P.O. Box 97005
Redmond, WA 98073-9705
1-800-822-8782
206-882-3100
Fax: 206-882-1990
E-mail: power@intp.com

12 Main Street
Garfield, NJ 07026
201-340-8277
Fax: 201-340-8286

Fleet, Hampshire, UK
44 1252 815511
Fax: 44 1252 815577
E-mail: poweruk@intp.com

Saint Gratien, France
33 1 34285455
Fax: 33 1 34282387
E-mail: powerfr@intp.com

Darmstadt, Germany
49 6151 177 629
Fax: 49 6151 177 654
E-mail: powergr@intp.com

(DC/DC converters, brushless motor controllers, custom hybrid circuits, units classified to Class K for spacecraft applications available.)

Inter-Technical Group, Inc.
2629 Saw Mill River Road
Bldg, 4C
P.O. Box 217
Elmsford, NY 10523
914-347-2474
Fax: 914-347-7230

(Photo detectors and receivers, opto couplers/interrupters, rotary encoders, solar cells, opto electronic ICs, resistors, thermistors, EMI/EMC coils, beads and cores, connectors, aluminum capacitors)

Intronics, Inc.
150 Dan Road
Canton, MA 02021
617-828-4992
Fax: 617-828-5050

(DC/DC converters, CRT distortion correction ICs)

Introtek
150 Executive Drive
Edgewood, NY 11717
516-242-5425
Fax: 516-242-5260

(Non-invasive air bubble and fluid detectors)

IOtech
25971 Cannon Road
Bedford Heights, OH 44146-1833
216-439-4091

(Functional replacement for the NEC uPD7210, an IEEE 488 based controller IC)

IP prefix
See *SemeLab Plc*,
Sumitomo Metals, *Siemens*

IPD
See *International Power Devices, Inc.*, *Siemens*

IPEC RG series
See *California Micro Devices, Inc.*

(Resistor networks)

IPS
See *Integrated Power Semiconductor, Ltd.*
International Power Semiconductors

IQ prefix
See *I-Cube, Inc.*

IQ Systems, Inc.
20 Church Hill Road
Newtown, CT 06470
203-270-8667
Fax: 203-270-9064

(Universal front panel controller ICs, speech messaging/DTMF controllers and other hardware and software products for real time I/O control)

IQC prefix
See *IQ Systems, Inc.*

IQS prefix
See *IQ Systems, Inc.*

IR, with the I followed by a diode symbol then an R, logo
See *International Rectifier*

IR-B prefix
See *Siemens*

IRC
A TT Group Company
Advanced Film Operations
4222 South Staples Street
Corpus Christi, TX 78411
512-992-7900
Fax: 512-992-3377

(Discrete and surface mount resistive products such as resistor networks)

IRD
See *International Radiation Detectors*

IRF prefix
See *International Rectifier*,
AVG Semiconductors
(Second source some devices),
Harris Semiconductors
(Second sources some parts)

IRFD — ISOTEMP RESEARCH

IRFD prefix
See *International Rectifier*, *Motorola*

IRFZ prefix
See *International Rectifier*

IRL prefix
See *Siemens*

Irvine Sensors Corp. (ISC)
3001 Redhill Ave.
Building III
Costa Mesa, CA 92626
714-549-8211
Fax: 714-557-1260, 714-549-5711
http;//www.irvine-sensors.com

Computer/DRAM Products:
21 Gregory Drive
South Burlington, VT 05403
802 860 6800
Fax: 802 860 4959

(A fabless company supplying memory stacks [Short Stack], a three dimensional chip stacking technology, infrared receiver IC)

ISA, Integration System Assemblies
600 W. Cummings Park
Woburn, MA 01801-6369
617-937-0177

(Founded in 1990 for thin film MCM manufacturing)

ISA Jobin Yvon/Spex Division
(Spex Industries)
An Instruments S.A. Company
3880 Park Avenue
Edison, NJ 08820
1-800-438-7739
(1-800-GET-SPEX),
800-782-0043
908-494-8660
Fax: 908-549-5125, 908-494-8796

16-18 rue du Canal
91165 Longjumeau cedex
33 1/64 54 13 00
Fax: 33 1/69 09 93 19

Benelux: 1720 24270
Netherlands: 0/1720 33323
Germany: 89/46 23 17 0
Italy: 2/57 60 30 50
United Kingdom: 181/204 81 42

(CCDs, [spectroscophic CCD detectors])

ISB prefix
See *SGS-Thomson*

(ASIC family)

ISC
See *Irvine Sensors Corp.*

ISD
See *Information Storage Devices, Inc.*, *Siemens*

Ishizuka Electronics Corp.
7-7 Kinshi 1-Chome
Sumida-Ku
Tokyo 130
Japan
81 03 3621 2704
Fax: 81 03 2623 6100, 7776
TLX: J33324 IZECCOJ

ISI
See *Ideal Semiconductor, Inc.*

ISO prefix
See *Burr-Brown*

Isocom PLC
Div. Implex PLC
Elm Road
West Chirton Ind. Est.
North Shields
Tyne & Wear
NE29 8SE
091 258 0690
Fax: 091 259 0997

Prospect Way
Park View Industrial Estate
Brenda Rd.
Hartlepool, Cleveland, England
0429 863609

256E. Hamilton Avenue, Ste. H
Campbell, CA 95008
408-370-2212

720 E. Park Blvd.
Plano, TX 75074
214-423-5521 #209
Fax: 214-422-4549

(Optoswitches, opto emitters and detectors, optocouplers including triac, SCR, transistor, darlington, multichannel, schmitt trigger, FET, high speed, AC input, and AC/DC logic types)

Isolink Inc. (ISO LINK)
501 Valley Way
Milpitas, CA 95035
408-946-1968
Fax: 408-946-1960

(Optocoupler products)

Isotek Corporation
566 Wilbur Avenue
Swansea, MA 02777
508-673-2900
Fax: 508-676-0888

(Resistor networks)

Isotemp Research Inc.
P.O. Box 3389
Charlottesville, VA 22903
804-295-3101
Fax: 804-977-1849

1750 Broadway St.
Charlottsville, VA 22901

(Crystal oscillators, crystal ovens)

ISP logo
See *Image and Signal Processing, Inc.*

ISP prefix
See *Emulex Micro Devices*

ispLSI
See *Lattice Semiconductor*

Isthmus Technology Group
8717 West 110th Street
Overland park, KS 66210
913-696-0011
Fax: 913-696-0012

(Sound and video over copper vire, versus coaxial cable IC. This IC supports transmission distances up to 3500 feet.)

ITAC Hybrid Technology
See *InterFet-ITAC*

ITC prefix
See *CP Clare*,
ITC Microcomponents, Inc.

ITC Microcomponents, Inc.
206-814 West 15th St.
North Vancouver B.C. V7P 1M6
604-985-6461
Fax: 604-985-6417

(Data acquisition and control products including a serial interface port for computer control of I/O ports, etc.)

ITI FerroTec
International Thermoelectric, Inc.
131 Stedman St.
Chelmsford, MA 01824
800-929-5665
508-452-0212
Fax: 508-452-0104

(Thermoelectric materials and modules)

itron
See *Noritake Co. Ltd.*

ITS prefix
See *Harris Semiconductor*
(Custom part)

ITS Electronics, Inc.
200 Edgeley Blvd., Unit #24
Concord, Ontario L4K 3Y8
Canada
416-660-0405
Fax: 416-660-0406

(Microwave components including dielectric resonator oscillators, voltage controlled oscillators, phase-locked loop oscillators, low noise amplifiers, linear power amplifiers, pulsed power amplifiers, diode limiters, precision power dividers, up/down converters, interference cancelling systems)

ITT prefix
See *ITT GTC*

ITT Components
1851 E. Deere Avenue
Santa Ana, CA 92705
714-261-5300

(Capacitors, crystals, crystal oscillators)

ITT GTC
7670 Enon Drive
Roanoke, VA 24019
703-563-8600, 3920
Fax: 703-563-3935

(RF and power amplifiers)

ITT Semiconductors (Far East), **Ltd.**
P.O. Box 47
Shinjuku Sumitomo Bldg
2-6-1, Nishi-Shinjuku
Shinjuku-Ku, Tokyo 163-02
Japan
81 3 3347 8881
Fax: 03 3347 8844

(Available from Taitron Components, Inc., 25202 Anza Drive, Santa Clarita, CA 91355-3496, 800-824-8766, 800-247-2232, 805-257-6060; Fax: 800-824-8329, 805-257-6415)

ITT Semiconductors
ITT-Intermetall
Hans Bunte Strasse 19
Freiburg 79108
Germany
761 517 0
Fax: 761 517 2174

(Note: The company has a discrete assembly facility in Colmar, France. ASIC audio and speech processing, digital TV, audio, video, automotive, telecommunications and television ICs, and hall effect sensors)

ITT Shadow, Inc.
8081 Wallace Road
Eden Prairie, MN 55344
612-934-4400
Fax: 612-934-9121

(Ultraminiature, ultra low profile surface mount side actuated or pushbutton switches)

IV prefix (Intersil VMOS part)
See *Harris Semiconductor*

Iwaki Musen Kenkyusho Co., Ltd.
485, Futako, Takatsu-ku
Kawasaki-shi
Kangawa 213
Japan
044 833 4311
Fax: 044 833 6605

(Hybrid ICs)

I - J

IWATA — JEROME INDUSTRIES

Iwata Electric Co., Ltd.
2-23-15, Kanda-Suda-cho,
Chiyoda-ku
Tokyo 101, Japan
03 3253 2365
Fax: 03 3255 3850

(Hybrid ICs)

Iwatsu Precision Co., Ltd.
1-7-41, Kugayama, Suginami-ku
Tokyo 168
03 3247 2211
Fax: 03 3247 2225

(Hybrid ICs)

IXR prefix
See *Burr-Brown*

IXYS Corp.
3540 Bassett St.
Santa Clara, CA 95054
408-982-0700
Fax: 408-496-0670

Europe:
49 6206 5030
Fax: 49 6206 503627

(MOSFETs, thyristors, IGBTs and diodes, military qualified devices are available.)

J prefix
(A JXXX part number, not to be confused with J for JAN or military qualified device)
See *InterFET Corporation* (FET devices), *Motorola*

James Electronics (Equipment and Supplies), Inc.
4050 N. Rockwell Street
Chicago, IL 60618-3796
800-438-1400
312-463-6500

(Telecommunications transformers)

JAN, JANTX, JANTXV prefix
(This denotes a semiconductor qualified to U.S. military standard MIL-S-19500.)

JAN Crystals
P.O. Box 60017
2341 Crystal Drive
Fort Meyers, FL 33906-4099
1-800-JAN-XTAL
941-936-2397

(Quartz crystals and oscillators)

Janos Technology, Inc.
Box 25 Route 35
Townshend, VT 05353
802-365-7714
Fax: 800-338-9764
802-365-4596

(Detectors, avalanche diodes, wavelength detector diodes, UV and blue optimized diodes)

Japan ACM Corporation
23-3 Daikyo-cho
Shinjuku-ku
Tokyo 160, Japan
813 3 3351 8231
Fax: 813 3 3351 7831

(Infrared transmission ICs)

Japan Minicomputer Systems Co.Ltd.
33-18, Takaido-Higashi 3-Chome,
Suginami-Ku, Tokyo 169
81 3 3333 3475

(Logic ICs)

Japan Pulse Laboratories, Inc.
425-3, Inari-Cho,
Isesaki-Shi, Gunma 372
Japan
0270 26 3369, 23-1031

(Hybrid ICs)

Japan Resistor Mfg. Co., Ltd.
2315, Kitano, Jyohano-Machi,
Higashi-Tonami-gun
Toyama 939-18
Japan
0763 62 1180
Fax: 0763 62 2848

(Hybrid ICs)

Japanese Products Corp.
272 Main Ave.
Norwalk, CT 06851
203-840-1590
Fax: 203-840-1601

(Linear ICs)

jbm logo
See *JBM Electronics, Inc.*

JBM Electronics, Inc.
One Commerce Drive
Bedford, NH 03110
603-623-0222
Fax: 603-623-0446

(Electromagnetic delay lines including active and passive types, DC/DC converters and pulse transformers)

JCA Technology, Inc.
3529 Old Conejo Road, Suite 118
Newbury Park, CA 91320-2155
805-498-6794
Fax: 805-499-2402
TLX: 510-601-6386 JCA TECH

1090 Avenida Acaso
Camarillo, CA 93012-8275
805-498-6794

(Hybrid microwave amplifiers, switches and subassemblies)

Jerome Industries Corporation
730 Division St.
Elizabeth, NJ 07201
908-753-5700, Fax: 908-353-1021
E-mail: jeromeIND@aol.com

Manufacturers, Prefixes, Part Number Types, Logo Descriptions & Family Types

JEROME INDUSTRIES — KAF-XXXX J - K

(Wall plug in, desktop power supplies, battery chargers and transformers)

Jetta Power Systems, Inc.
5252 Bolsa Avenue
Huntington Beach, CA 92649
714-899-6800
Fax: 714-899-6833 (sales)

(Power supplies)

Jewell Electrical instruments
Allard Industries, Inc.
An Allard/Nazarian Company
124 Jollette Street
Manchester, NH 03102
800-227-5955
603-669-6400
Fax: 603-669-5962

(Meters, indicators, solenoids, sensors, meter-relays, accelerometers, inclinometers)

JFU marking
See *Seiko Instruments, Inc.*

JFW
See *JFW Industries, Inc.*

JFW Industries Inc.
5134 Commerce Square Drive
Indianapolis, IN 46237
317-887-1340
Fax: 317-881-6790

(Fixed, rotary and programmable attenuators, matrix switches, switches for microwave applications)

Jiann WA Electronics Co., Ltd.
844-846, Sec. 1, Kong Fu Rd.
Wu-Chi
Taichung Hsien
Taiwan
886 46 561 373, 374
Fax: 886 46 566 418

(Optoelectronics)

JKL Components Corporation
13343 Paxton St.
Pacoima, CA 91331
800-421-7244
Fax: 818-897-3056

(Miniature fluorescent lamps)

JM prefix
See *GEC Plessy*
(Marconi Circuit Technology, Inc.)

JM38510 prefix
(This denotes a part qualified to U.S. military standard MIL-M-38510.)

Joelmaster Systems, Ltd.
4Fl, No. 46, Lane 184
Yung Chi Road
P.O. Box 51-29
Taipei, Taiwan
886 2 765 5605, 5606,
886 2 767 2137
Fax: 886 2 767 0540

(Diodes)

J P Technologies, Inc.
A TT Group Company
P.O. Box 1168
42 North Benson Avenue
Upland, CA 91786
909-946-1000
Fax: 909-946-6267
TLX (USA): 704980

(Sensors, strain gauges, RTDs)

JRC
See *NJR Corporation*
(A subsidiary of New Japan Radio Company, Ltd.)

JT PowerCraft, Inc.
(Part of JT Electronics Corporation Tokyo, Japan. This company is licensed to manufacture and sell dc/dc converters and power supplies based upon Vicor Corporation power supply technology.)

JTDA prefix
See *GHz Technology, Inc.*

JTDB prefix
See *GHz Technology, Inc.*

J.W. Miller Magnetics
P.O. Box 2859
306 E. Alondra Blvd.
Gardenia, CA 90248-1059
800-723-8588 (Fax-Back)
310-515-1720
Fax: 310-515-1962

(Chokes, inductors, transformers)

K prefix
See *Silicon Systems* (K-XXXX part numbers) *Optek Technology, Inc.*, *Champion Technologies* (Oscillators), *AVX Corporation* (Oscillators), *OKI Semiconductor*, *Krypton Isolation, Inc.*, *AVG Semiconductor*
(Second source gate arrays).

(Note: "K" prefix on a semiconductor, such as KN2222, indicates part manufactured in Korea.)

K68EC prefix
See *AVX Corp.*

(Clock oscillators)

KA prefix
See *Samsung Semiconductor*, *AVG Semiconductor*

(Second source gate arrays, microwave diodes, etc.)

KAF-XXXX part number
See *Eastman Kodak Co.*

(CCD devices)

117

KAI-XXXX — KELTRON POWER SYSTEMS

KAI-XXXX part number
See *Eastman Kodak Co.*

(CCD devices)

Kaman Instrumentation Corporation
P.O. Box 7463
1500 Garden of the Gods Road
Colorado Springs, CO 80907
800-552-6267
719-599-1132
Fax: 719-599-1823

(Sensors and measurement/sensing equipment)

Kanematsu USA, Inc.
543 W. Algonquin Rd.
Arlington Heights, IL 60005
Phone/Fax: 708-981-5655
TLX: 025 4306

(Optoelectronics, linear ICs)

KAOS Semiconductor
Sunnyvale, CA

(Mixed signal ASICs. In June 1995 it became a wholly owned subsidiary of Orbit Semiconductor, Inc.)

**Kappa Technologies, Inc.
(Kappa Networks, Inc.)**
1443 Pinewood St.
Rahway, N.Y. 07065
1-800-223-0603
908-396-9400
Fax: 908-388-4786

(Delay lines, DC/DC converters, isolation transformers, filters and chokes)

KASP-XXXX part number
See *Eastman Kodak Co.*

(ASICs for CCD devices)

Kawasaki LSI, U.S.A., Inc. (Steel Corporation)
4655 Old Ironsides Drive, Suite 265
Santa Clara, CA 95054
408-654-0180
Fax: 408-654-0198

501 Edgewater Drive, Suite 510
Wakefield, MA 01880
617-224-4201
617-224-2503

Kawasaki Steel Corporation
Makuhari Techno Garden B5
1-3 Nakase Mihama-ku
Chiba 261-01, Japan
81 43 296 7412
Fax: 81-43-296-7419
klsi@lsidiv.kawasaki-steel.co.jp

(Content addressable memories, microprocessors, Zilog Inc. is licensed to manufacture the KC80, an 8 bit microprocessor)

Kay Elemetrics, Corp.
12 Maple Avenue
P.O. Box 2025
Pine Brook, NJ 07058-2025
201-227-2000
Fax: 201-227-7760

(Attenuators)

KC prefix
See *Kawasaki Steel Corp.*

KCB prefix
See *Kyocera industrial Ceramics*

(LCD displays)

KDI/Triangle Electronics, Inc.
31 Farinella Drive
East Hanover, NJ 07936
201-884-1423
Fax: 201-884-0445

(RF/microwave thin/thick film/ microstrip resistors, amplifiers, attenuators, limiters, logarithmic amplifiers, switches, etc. Note: in early 1992 they acquired the assets of Optimax, Inc.)

KDS America
10910 Granada Lane
Overland Park, KS 66211
913-491-6825
Fax: 913-491-6812

(Crystals and crystal piezo oscillators)

KDSP-XXXX part number
See *Eastman Kodak Co.*

(ASICs for CCD devices)

KE prefix,
See *Optek Technology, Inc.*,
Kawasaki (Steel) LSI, U.S.A., Inc.

Keeper
(A trademark of Eagle Picher Electronics Div. for their lithium battery line)

KEFL
See *Khandelwal Electronics and Finance, Ltd.*

Kel-Com
Division of K&L Microwave, Inc.
408 Coles Circle
Salisbury, Maryland 21801
410-749-6774, Fax: 410-749-6887

(Filters)

Kelly Micro Systems
Irvine, CA

(This manufacturer of microprocessors and microprocessor memory upgrade modules, was acquired by Simple Technology, Inc. in late 1995.)

Keltron Power Systems, Inc.
225 Crescent Street
Waltham, MA 02154-3420
617-894-8700
Fax: 617-894-8710

KELTRON POWER SYSTEMS — KINGBRIGHT

(Power supplies)

Kenwood Corp.
2-17-5 Shibuya
Shibuya-ku
Tokyo 150, Japan
81 3 3488 5511
Fax: 81 3 3406 1680

(Voice recognition ICs)

Kepco Inc.
131-38 Sanford Ave.
Flushing, NY 11352
718-461-7000, Fax: 718-767-1102
TWX: 710-582-2631
http://www.kepcopower.com

(Power supplies, DC/DC converters)

Kern Co. Ltd.
5-50-8, Higashi-Nippori
Arakawa-ku
Tokyo 116, Japan
81 3 3806 7371, Fax: 03 3801 1857

(Linear, analog, logic, communication, telephone ICs, etc.)

Ketema
Rodan Division
2900 Blue Star St.
Anaheim, CA 92806
714-630-0081, Fax: 714-630-4131

(Inrush current limiting devices, temperature sensors and temperature ICs)

Keyence Corp. of America
17-17 Rte. 208 N.
Fair Lawn, NJ 07410
201-791-8811
Fax: 201-791-5791
Chicago, IL: 708-775-1110,
Fax: 708-775-1169

Los Angeles, CA: 310-540-2254, Fax: 310-316-1032

(Photoelectric sensors)

Keyrin Electronics Co., Ltd.
544-2, Guro-Dong
Guro-Gu
Seoul, Korea
02 858-3381
Fax: 02 864-2810

(Micro speakers)

Keystone Carbon Company
See *Keystone Thermometrics*

Keystone Thermometrics Corporation
808 US Highway 1
Edison, NJ 08817
908-287-2870
Fax: 908-287-8847

(Thermistors)

KF prefix
See *AVX Corp.*

(Clock oscillators)

KGF prefix
See *Oki Semiconductor*

(Parts usually marked with the "K" only because of device physical size limitations.)

KH prefix
See *Kulite Semiconductor Products*

KHO-CT, **HC** or **AC** prefix,
See *AVX Corp.*

(Clock oscillators)

KIC
See *KIC Corporation*

KIC Corporation
Division of Kimura International, Inc.
15 Trade Zone Court
Ronkonkoma, NY 11779
516-981-8788
Fax: 516-981-8826

(Thermal cutoffs, diodes)

Kilovac Corporation
P.O. Box 4422
Santa Barbara, CA 93140
800-253-4560
805-684-4560
Fax: 805-684-9679

(High voltage relays. This company became an independent division of Communications instruments, Inc. [CII], Fairview, NC in 1995.)

Kinematics & Controls Corp.
14 Burt Drive
Deer Park, NY 11729
800-833-8103
516-595-1803
Fax: 516-595-1523
TLX: 178 243 KACC UT

(Optoelectronics)

Kingbright Electronic Co. Ltd.
3Fl, No. 317-1
Chung Shan Road, Sec. 2
Chung Ho City
Taipei Hsien, Taiwan
886 2 249 9224
Fax: 886 2 240 3981
TLX: 22249 TWNKIN

225 Brea Canyon Road
City of Industry, CA 91789-3049
909-468-0500
Fax: 909-468-0505

(Optoelectronics including LEDs, [including surface mount and high brightness (to 4000 mcd), tricolor and blue LEDs], lamps, arrays, bargraphs, lightbars and displays [numeric and alphanumeric], backlights, phototransistor infrared diodes)

KINGSTON — KOEP PRECISION STANDARDS

Kingston Technology Corporation
17600 Newhope Street
Fountain Valley, CA 92708
800-259-9342
714-435-2600
Fax: 714-435-2699
http://www.kingston.com

(Memory modules)

Kinseki, Ltd.
Shin-Yuri 21 Bldg.
1-2-2, Manpukuji Asao-Ku,
Kawasaki-shi, 215 Japan
044 952 8141
Fax: 044 952 8140

American KSS Inc.
1735 Technology Drive, Suite 790
San Jose, CA 95110
408 437 9577
Fax: 408 437 1717

SchirmerstraBe 76
4000 Dusseldorf 1 Germany
0211 36815 0
Fax: 0211 36815 10

KSS Electronics Singapore PTE, LTD.
1 Selegie Road
#09-01/02 Paradiz Centre
Singapore 0718
338 8577
Fax: 338 8573

Suite 43, Carlton Offices
Aztec Centre, Aztec West
Almondsbury, Bristol BS12 4TD
United Kingdom
0454 614638
Fax: 0454 614700

(Synthetic quartz crystals, quartz crystals, crystal oscillators, monolithic crystal filters, ultrasonic delay lines, CCD delay lines)

KIR-XXXX part number
See *Eastman Kodak Co.*

(CCD devices)

Kistler Instrument Corporation
75 John Glenn Drive
Amherst, NY 14288-2171
716-691-5100
Fax: 716-691-5226
E-mail: kicsales@kistler.com

(Piezo-instrumentation products including accelerometers)

KIT-XX part number
See *Dimolex*

K&L Microwave, Inc.
408 Coles Circle
Salisbury, MD 21801
410-749-2424
Fax: 410-749-5725

(Filters and coaxial switches)

KLI-XXXX part number
See *Eastman Kodak Co.*

(CCD devices)

KM prefix
See *Samsung Semiconductor*

KMOS Semi-Custom Designs
1012 Stewart Drive
Sunnyvale, CA 94086
408-730-5944
Fax: 408-730-0176

(Analog and digital ASICS)

Knox Semiconductor, Inc.
Limerock St.
Camden, ME 04843
13 1/2 Quarry Road
P.O. Box 609
Rockport, ME 04856-0609
207-236-6076
Fax: 207-236-9558

(Semiconductors including rectifiers, zener and varactor diodes, etc., including military qualified parts)

KMZ prefix
See *Philips Semiconductor*

KOA Corp.
(Overseas Division)
3-2-3, Ebisu, Shibuya-ku,
Tokyo 150
Japan
81 3 34444 5151
Fax: 81 03 3444 5136

KOA Speer Electronics, Inc.
Box 547
Bradford, PA 16701
814-362-5536
Fax: 814-362-8883

(Hybrid ICs, chip resistors, resistor networks, chip capacitors, chip inductors, chip tantalum capacitors, MELF resistors, cermet trimmer potentiometers, chip circuit protectors [fuses], thermistors)

Kodak
See *Eastman Kodak Company*

Kodak Bekley Research
Berkley CA

(Reed Solomon error detection and correction ASIC)

Kodenshi Corp.
24-52 Makishima-Cho
Uji, Kyoto 611, Japan
0774 24 1121
Fax: 0774 24 1031
TLX: 54533650

(Optoelectronics)

Koep Precision Standards, Inc.
500-B Great Bay Blvd
Tuckerton, NJ 08087
609-296-9686
Fax: 609-296-9733

(Diodes, linear ICs)

Kokusai Electric Company, Ltd.
P'S Higashinakano Bldg., 3-14-20,
Higashi-Nakano
Nakano-Ku, Tokyo 164
03 3368 6111
Fax: 03 3365 9119

(Memory and hybrid ICs)

Kollmorgen Motion Technologies Group
Inland Motor
501 First St.
Radford, VA 24141
540-639-9045

(Motor control modules)

KOM prefix
See *Siemens*

Kopin Corporation, Inc.
695 Myles Standish Blvd
Taunton, MA 02780
508-824-6696
Fax: 508-822-1381

(Digital imaging ICs and display products)

Korer Electronics Company (KEC)
(Semiconductors, available from Taitron Components, Inc., 25202 Anza Drive, Santa Clarita, CA 91355-3496, 800-824-8766, 800-247-2232, 805-257-6060; Fax: 800-824-8329, 805-257-6415)

KR Electronics, Inc.
91 Avenel St.
Avenel, NJ 07001
908-636-1900, Fax: 908-636-1982

(Custom and standard filters including lowpass, bandpass, highpass, notch, band reject, matched phase and attenuation and group delay and attenuation equalizer networks, etc.)

KRC prefix
(A prefix typical of semiconductors made in Taiwan vs. the 2SC part made by Japanese manufacturers.)

Krypton Isolation, Inc.
39111 Paseo Parade Parkway
Suite 202
Fremont, CA 94538
510-713-9100
Fax: 510-713-9188

(DAA, Digital Access Arrangement IC for Fax Modems)

KS prefix
See *Samsung Semiconductor*,
Optek Technology, Inc.,
AVG Semiconductors
(Second source telephone ICs)

KSA prefix (Transistors)
See *AVG Semiconductor*
(Second source)

KSD prefix (Transistors)
See *AVG Semiconductor*
(Second source)

KSO-EL prefix
See *AVX Corp.*

(Clock oscillators)

KSS logo
See *Kinseki, Ltd.*

KT prefix
See *Optek Technology, Inc.*,
AVG Semiconductors
(Second source telephone ICs)

KTS Electronics
16406 N. Cave Creek Road, #5
Phoenix, AZ 85032-2919
Phone/Fax: 602-971-3301

(Crystals, clock oscillators, crystal filters, LC filters)

Kulite Semiconductor Products
1 Willow Tree Road
Leonia, NJ 07605
201-461-0900
Fax: 201-461-0990

1039 Hoyt Avenue
Ridgefield, NJ 07657
201-945-3000
Cable: Kultung
TLX: 135 458

(Transducer amplifiers)

Kung Dar Electronics Co. Ltd.
No. 292, Chung-Shan Road
Kuan-Miao Shian
Tainan Hsien
Taiwan
886 6 595 9763
Fax: 886 6 595 8697

(Diodes)

Kurta Corp.
3007 East Chambers Street
Phoenix, AZ 85040-3796
602-276-5533

(ASIC for a pen based system that transforms analog signals to digital information)

KVG GmbH
P.O. Box 61
D-6924 Neckarbischofsheim
Germany
07263/648-0
Fax: 07263/6196

KVG North America Inc.
2240 Woolbright Rd.
Suite 320 A
Boynton Beach, FL 33426-6325
407-734-9007
Fax: 407-734-9008

K - L
KVG — LAMBDA

(Quartz crystals, crystal oscillators, crystal filters)

KWMT prefix
See *Gentron Corp.*

KXO-CT, EH, HC or AC prefix
See *AVX Corp.*

(Clock oscillators)

Kyocera Corp.
5-22, Kita-Inoue-Cho
Higashino, Yamashina-Ku
Kyoto-Shi, Kyoro 607
Japan
075 592 3851

(Crystals, oscillators, hybrid ICs)

Kyocera Industrial Ceramics
5317 E. 4th Plain Bl.
Vancouver, WA 98661
206-750-6107

(Passive color LCD displays)

Kyodo Denki Co., Ltd.
1-9, Shimo-Maruko 4-Chome
Ohta-ku, Tokyo 146
Japan
81 3 3750 3311

(Hybrid ICs)

Kyopal C. Ltd.
2-83 Shurishike
Mukuoshi, Kyoto 617
Japan
81 75 935 5455
Fax: 81 75 935 5456

(Motor control ICs)

Kyowa Crystal Co. Ltd.
33-21 Nukui 3-Chome
Nerima-ku, Tokyo 176, Japan
81 3 3999 6166

(HMETS, High Electron Mobility Transistors)

L prefix
See *LSI Logic Corp.*,
SGS-Thomson Microelectronics,
Logic Devices, Inc.,
Lucent Technologies

L4 prefix
See *Logic Devices, Inc.*

L4C prefix
See *Logic Devices, Inc.*

L7C prefix
See *Logic Devices, Inc.*

L10 prefix
See *Logic Devices, Inc.*

L21 prefix
See *Logic Devices, Inc.*

L29 prefix
See *Logic Devices, Inc.*

L41 series
See *Frederick Components International Ltd.*

L7C prefix
See *Logic Devices, Inc.*

LA prefix
See *Semicoa* (Optoelectronics),
Gennum Corporation
(Linear arrays)

LA-XXX part number
See *Siemens*

LAH prefix
See *KDI/triangle Electronics, Inc.*

Lake Shore Cryotronics, Inc.
64 East Walnut Street
Westerville, OH 43081-2399
614-891-2243
Fax: 614-891-1392
TLX: 24-5415 CRYOTRON WTVL

(Cryogenic temperature sensors, measurement and control instrumentation and magnetic measurement and test systems)

Lambda Electronics, Inc.
Lambda Group of Unitech plc
515 Broad Hollow Road
Melville, NY 11747-3700
1-800-LAMBDA 4 (Eastern U.S.)
1-800-LAMBDA-5 (Western U.S.)
516-694-4200
Fax: 516-293-0519

Canada: 1-800-361-2578
(Montreal call 514-337-0311),
514-695- 8330

England, Ilfracombe (Coutant-Lambda Ltd.): 0271-865656

France, Orsay: 33 1 60 12 14 87
(Coutant-Lambda, S.A.)

Germany, Achern:
49 07841 6806 0

Israel, Tel Aviv: 972 3 648 7655

Italy, Milan: 39 2-660 40540

Japan, Tokyo: (Nemic-Lambda K.K.) 81-3-3447-4411

Korea, Seoul: (Nemic-Lambda) 02-556-1171

Malaysia: (Nemic-Lambda[M] Sdn. Bhd.) 60 03-756-6119 or 0739

Singapore: (Nemic-Lambda[S] Pte.Ltd.) 65-251-7211

Taiwan: 886 2 834 5151 (Nemic-Lambda K.K.)

(Power supplies, DC/DC converters)

Manufacturers, Prefixes, Part Number Types, Logo Descriptions & Family Types

LAMBDA — LATTICE SEMICONDUCTOR

Lambda Advanced Analog
2270 Martin Ave.
Santa Clara, CA 95050-2781
408-988-4930
TWX: 910-338-2213
Fax: 408-988-2702

(DC/DC converters [military units and devices qualified for space applications available], video DACs)

Lambda Novatronics, Inc.
500 S.W. 12th Avenue, P.O. Box 878
Pompano Beach, FL 33061-0878
800-952-6909, 305-784-2474
Fax: 305-783-4963

(Military power supplies, custom and off the shelf)

Lambda Semiconductor
See *Semtech Corpus Christie*

Lamp Technology, Inc.
1645 Sycamore Ave.
Bohemia, NY 11716
1-800-KEEP-LIT, 516-567-1800
Fax: 516-567-1806

(LED replacement lamps for incandescent bulbs, speciality lamps)

Language Systems Design, Inc.
P.O. Box 52026
Edmonton, Alberta T6G 2T5
Canada
403-448-1720, 403-489-9621
Fax: 403-963-1165, 403-489-3012
sales@lossless.com
http://www.ccinet.ab.ca/lsdi.html

(This company sells the lossless data compression IC developed by DCP Research Corporation)

Lansdale Semiconductor, Inc.
2929 S. 48th St., Suite #2
Tempe, AZ 85282
602-438-0123
Fax: 602-438-0138
http://ssi.syspac.com/~lansdale/

(A manufacturer or older technology products such as RTL, DTL, TTL and memory devices. Their product line includes devices formerly manufactured at the Signetics company closed bipolar wafer fabrication line in Orem, Utah, and Intel M82XX type bus controller and clock generator drivers. They also have Motorola Semiconductor discontinued military CMOS, MECL and Schottky logic lines. They have the ability to process parts to MIL-STD-883. They have a minimum order policy of $500.)

Lap-Tech, Inc.
230 Simpson Ave.
Bowmanville, Ontario
Canada L1C 2J3
416-623-4101
Fax: 416-623-3886

(Quartz crystals and crystal oscillators)

Lark Engineering Company
27151 Calle Delgado
San Juan Capistrano, CA 92675
714-240-1233
Fax: 714-240-7910

(Filters, including surface mount types)

LAS prefix
See Semtech Corpus Christie
(Formerly Lambda Semiconductor)

Lasar prefix
See *PMC-Sierra*

LaserMax, Inc.
3495 Winston Place, Building B
Rochester, NY 14623
716-272-5420
Fax: 716-272-5427

(Miniature diode lasers and electrical optical systems for scientific, medical, industrial and military applications)

Lattice Semiconductor Corp.
(GAL, ispLSI, pLSI prefixes)
5555 NE Moore Ct.
Hillsboro, OR 97124
800-327-8425
503-681-0118
Fax: 503-681-3037
TLX: 277338LSC UR
1-800-327-8425
(Literature Distribution, Application Help)

pLSI and ispLSI Applications:
1-800-Lattice
GAL Applications Hotline:
1-800-FASTGAL

1820 McCarthy Blvd
Milpitas, CA 95035
408-428-6400
Fax: 408-944-8450, 8444
http://www.latticesemi.com
http://memec.com/DataSheets/lattice/lattice.html

15707 Rockland Plaza
Ste. 110
Irvine, CA 92718
714-580-3880
Fax: 714-580-3888

12424 Research parkway
Suite 101
Orlando, FL 32826
407-281-6500
Fax: 407-658-0208

3091 Governors Lake Dr.
Bldg 100, Suite 500
Norcross GA 30071
404-446-2930
Fax: 404-416-7404
Tel: FAE-404-416-0679

1 Pierce Place, Suite 500-E
Itasca, IL 60143
708-250-3118
Fax: 708-250-3119

L | LATTICE SEMICONDUCTOR — LCD LIGHTING

67 S. Bedford St.
Suite 400 West
Burlington, MA 01803
617-229-5819
Fax: 617-272-3213

3445 Washington Dr.
Suite 105
Eagan, MN 55122
612-686-8747
Fax: 612-686-8746

115 Route 46
Suite F-1000
Mountain Lakes, NJ 07046
201-316-6024
Fax: 201-316-6619

Linden Oaks Park
70 Linden Oaks
3rd Floor
Rochester, NY 14625
716-383-5320
Fax: 716-383-5321

5555 N.E. Moore Ct.
Hillsboro, OR 97124
503-656-4808, 503-780-6771
Fax: 503-656-6541, 503-681-3037

100 Decker Ct.
Suite 280
Irving, TX 75062
214-650-1236
Fax: 214-650-1237

9430 Research Blvd.
Echelon IV, Suite 428
Austin, TX 78759
512-343-4506
Fax: 512-343-7309

France:
Les Algorithmes
Batiment Homere
91190 - Saint Aubin, France
33 1 69 33 22 77
Fax: 33 1 60 19 05 21

Germany:
Hanns-Braun-Strasse 50

85375 Neufahrn bei Munchen
Germany
49 8165 9516 0
Fax: 49 8165 9516 33

Hong Kong:
2802 Admiralty Centre, Tower 1
18 Harcourt Road
Hong Kong
852 2319 2929, 852 529-0356
Fax: 852 2319 2750, 852 866-2315

Japan:
N Bldg 9F
2-28-3 Higashi-Nihonbashi
Chuo-ku, Tokyo 103
81 3-5820-3533
81 3-5820-3531

United Kingdom:
Castle Hill House
Castle Hill
Windsor
Berkshire SL4 1PD
44 753-830-842
Fax: 44 753-833-457

(A fabless company supplying programmable logic, PLDs and Generic Array Logic-GALs, Note: they have a newsletter, Lattice-NEWS)

Laser Diode Products, Inc.
(Formerly Laser Diode, Inc.)
1130 Somerset St.
New Brunswick, NJ 08901
908-249-7000
Fax: 908-249-9165

(SONET link modules, high power emitter CW laser diodes)

Laser Optics, Inc.
P.O. Box 127
Danbury, CT 06813
203-744-4160
Fax: 203-798-7941

(Acousto optic modulators, Q-switches, beam deflectors)

Lasertron
37 North Avenue
Burlington, MA 01803
617-272-6462
Fax: 617-273-2694

(Lasers, detectors, transmitters and receivers. This company was acquired by Oak Industries in August 1995.)

LB-XXXX part number
See *Keyence Corp. of America*

(Laser displacement sensors)

LB prefix
See *Hitachi Semiconductor* (SONET devices), *Siemens*

LBC logic family
See *Texas Instruments*

LBB prefix
See *Siemens*

LC prefix
See *Gennum Corporation, Semtech Corp.*

(Transient voltage suppression diodes)

LCA prefix
See *LSI Logic*
(Programmable ASICs)

LCB prefix
See *LSI Logic*

(Programmable ASICs)

LCD Lighting, Inc.
11 Cascade Blvd
Milford, CT 06460-9998
203-882-5572
Fax: 203-882-5580

(Miniature fluorescent lamps for LCD backlights)

LCF Enterprises
651 Via Alondra, #710
Camarillo, CA 93012
805-388-8454
Fax: 805-389-5393

(RF power amplifier assemblies)

LCP prefix
See *SGS-Thomson Microelectronics*

LCX logic
(3.3 volts CMOS low voltage logic with tolerance to 5 volts on inputs and outputs)
See *Motorola, National Semiconductor, Toshiba*

LCX prefix, LCD displays
See *Sony Electronics, Inc.*

LD prefix
See *Gennum Corporation, Siemens, Silicon Logic, Inc.*

LDP
See *Laser Diode Products, Inc.*

LEA prefix
See *LSI Logic*

(Programmable ASICs)

Leadtron Enterprise Corp.
6F-6, No. 288
Chung Cheng Road, Shih Lin
Taipei, Taiwan
886 2 832 3430
Fax: 886 2 832 3433

(Diodes)

LeCroy Corporation
700 Chestnut Ridge Road
Chestnut Ridge, NY 10977-6437
914-425-2000

(Proprietary sampler and analog storage device for its DSO, digital storage oscilloscopes)

LED Technology, Ltd.
Unit 8 Pool Industrial Estate
Druids Road Redruth
Cornwall, TR15 3RH
United Kingdom
44 209 215 424
Fax: 44 209 215 197
TLX: 45273 FIBDAT G

(LEDs, optoelectronics)

Ledex Inc./Lucas
P.O. Box 427
801 Scholz Drive
Vandalia. Ohio 45377
800-282-3616
Fax: 941-665-4410
TWX: 810-450-2526
TLX: 288228

(Silicon surge suppressor diodes)

Ledtech Electronics Corp.
5th Floor, No. 5, lane 560
Chung Cheng Road
Hsin Tien City
P.O. Box 29-228
Taipei, Taipei Hsien, Taiwan
886 2 914 6891, 6895, 3241
886 2 914 3222
TLX: 31382 LEDTEDH

19209 Parphenia St., Suite D
Northridge, CA 91324
818-701-1341
Fax: 818-709-1012

(LED lamps and displays)

Ledtronics, Inc.
4009 Pacific Coast Highway
Torrance, CA 90505
310-534-1505
Fax: 310-534-1424

(LEDs [including surface mount types], multicolor bargraph LEDs, etc.)

Leecraft Manufacturing Co., Inc.
See *Lighting Components & Designs*

LEM U.S.A., Inc.
6663 W. Mill Road
Milwaukee, WI 53218
414-353-0711
Fax: 414-353-0733

(Power supply monitors, on-board current transducers)

Level One
See *Level One Communications, Inc.*

Level One Communications, Inc.
9750 Goethe Road
Sacramento, CA 95827
916-855-5000
Fax: 916-854-1101 (Main Fax)
916-855-1102
(Customer service Fax)
http://www.level1.com

(A fabless company supplying mixed signal ICs including 10Base T LAN devices, wide area networking devices, T1/E1 clock adapters, T1/E1 frame formatters, T1/E1 repeaters, hub ICs, and short-haul transceivers)

LF prefix
See *National Semiconductor, Logic Devices, Inc.*

LG prefix
See *Siemens*

LG Semicon
(Formerly Goldstar Electron, Lucky-Goldstar)
3003 North First Street
San Jose, CA 95134-2004
408-432-1331, 408-432-5000
Fax: 408-432-6067
http://www.goldstar.co.kr
(LG Electronics web page)

LG SEMICON — LINEAR SYSTEMS

Korea:
942, Daechi-Dong Kangnam-Gu
Seoul 135-280 Korea
822 3777 1114, 822 528 2884
Fax: 822 3459 3535,
822 528 2800/2880

Hong Kong:
8522 810 9209
Fax: 8522 810 9209

Japan:
81 33-224-0123
Fax: 81 33-224 0692

Singapore:
65-226-1191
Fax: 65-221-8575

Taiwan:
886 2 757 7022
Fax: 886 2 757 7013

Germany:
Jacob-Kaiser Strasse 12
4156 Willich 1
49-2154-492172
Fax: 49-2154-429424

(DRAMs, SRAMs, Video Drams, Mask ROMs, Linear ICs [timers, UARTs, floppy disk controllers, multiplexers, CRT controllers], earth leakage circuit breaker, PWM control circuits, operational and audio amplifiers, comparators, voltage regulators, CMOS [HCT and 4000B types] logic, LS TTL, STTL, line drivers and receivers, cross point switches, Ethernet controllers, etc.)

LGA prefix
See *Siemens*

LGB prefix
See *Siemens*

LGG prefix
See *Siemens*

LGH prefix
See *KDI/triangle Electronics, Inc.*, *Siemens*

LGK prefix
See *Siemens*

LGS prefix
See *Siemens*

LGS logo
See *LG Semicon*

LGT prefix
See *Siemens*

LGZ prefix
See *Siemens*

LH
See *Lik Hang Electronic Co. Ltd.*, *Lucent Technologies* (Formerly AT&T Microelectronics), *Sharp* (Solid state relays)

LH prefix
See *National Semiconductor*

(Note an ELH prefix is used by Elantec', who second sources some National Semiconductor parts. See also MICRA Corp. for another second source of some National Semiconductor parts, Sharp [used for SRAMs], and for LH-X part numbers see KDI/triangle Electronics, Inc., Siemens)

LHT prefix
See *Siemens*

LiD prefix
See *Vector Technology, Ltd.*

LIF prefix
See *KDI/triangle Electronics, Inc.*

Lighting Components & Designs
692 S. Military Trail
Deerfield Beach, FL 33442
305-425-0123
Fax: 305-425-0110

(Formerly Leecraft Manufacturing Company. Panel lights, lampholders, and lamps)

Lik Hang Electronic Co. Ltd. (LH)
Unit 1, Block 4, 13/R
Tai Ping Industrial Centre
51A, Ting Kok Road
Tai Po Market N.T. Hong Kong
852 664-6393
Fax: 852 665-0945

(IF transformers and coils)

LIM prefix
See *KDI/triangle Electronics, Inc.*

Linear Integrated Systems, Inc.
310 South Milpitas Blvd
Milpitas, CA 95035
408-263-8401
Fax: 408-263-7280

(JFETs, NPN and PNP transistors and dual transistors)

Linear Systems
See *Linear Integrated Systems*

Linear Systems, Inc.
4042 Clipper Court
Freemont, CA 94538
510-490-9160
Fax: 510-353-0261

(Linear ICs and transistors including replacements for discontinued Motorola transistors)

Manufacturers, Prefixes, Part Number Types, Logo Descriptions & Family Types

LINEAR TECHNOLOGY — LINFINITY

Linear Technology Corporation
1630 McCarthy Blvd.
Milpitas, CA 95035-7417
408-432-1900
1-800-637-5545, 1-800-4-LINEAR
(546327, for literature only)
Fax: 408-434-0507
Telex: 499-3977
http://www.linear-tech.com/

Central Region:
Chesapeake Square
229 Mitchell Court, Suite A-25
Addison, IL 60101
708-620-6910
Fax: 708-620-6977

Northeast Region:
3220 Tillman Drive
Suite 120
Bensalem, PA 19020
215-638-9667
Fax: 215-638-9764

266 Lowell Street
Suite B-8
Wilmington, MA 01887
508-658-3881
Fax: 508-658-2701

Northwest Region:
782 Sycamore Drive
Milpitas, CA 95035
408-428-2050
Fax: 408-432-6331

Southeast Region:
17060 Dallas Parkway
Suite 208
Dallas, TX 75248
214-733-3071
Fax: 214-380-5138

Southwest Region:
22141 Ventura Boulevard
Suite 206
Woodland Hills, CA 91364
818-703-0835
Fax: 818-703-0517

France:
Linear Technology S.A.R.L.
Immeuble "Le Quartz"
58 Chemin de la Justice
92290 Chatenay Malabry
France
33 1 41079555
Fax: 33 1 46314613

Germany:
Linear Technology GMBH
Untere Hauptstr. 9
D-8057 Eching
Germany
49 89 319741 0
Fax: 49 89 319482 1

Japan:
Linear Technology KK
5F YZ Building
4-4-12 Iidabashi Chiyoda-Ku
Tokyo, 102 Japan
81 3 3237 7891
Fax: 81 3 3237 8010

Korea:
Linear Technology Korean Branch
Nansong Building #505
Itaewon-Dong 260-199
Yongsan-Ku, Seoul
Korea
82 2 792 1617
Fax: 82 2 792 1619

Singapore:
Linear Technology Pte. Ltd.
101 Boon Keng Road
#02-15 Kallang Ind. Estates
Singapore 1233
65 293 5322
Fax: 65 292 0398

Taiwan:
Linear Technology Corp.
Rm. 801, No. 46, Sec. 2
Chung Shan N. Road
Taipei, Taiwan, R.O.C.
886 2 521 7575
Fax: 886 2 562 2285

United Kingdom:
Linear Technology (UK) Ltd.
The Coliseum, Riverside Way
Camberley, Surrey GU15 3YL
United Kingdom
44 276 677676
Fax: 44 276 64851

(Amplifiers [including voltage feedback op amps, instrumentation amps, comparators], references, ADCs, interface ICs, DC/DC converters, voltage regulators, power management ICs, pulse width modulators, MOSFET drivers, switching regulators, RS232 and RS485 transceivers and drivers, data acquisition ICs, voltage references, switched capacitance filters, lowpass filters, microprocessor supervisory ICs, comparators, power factor correction devices, Thermocouple cold junction compensator, analog switches, RMS-DC converter, etc.)

Linfinity Microelectronics, Inc.
(Formerly Silicon General, Inc.)
11861 Western Avenue
1-800-LMI-7011
Garden Grove, CA 92641
714-898-8121
Fax: 714-893-2570

Southeast Technical Group:
101 Washington Street
Suite 6
Huntsville, AL 35801
205-534-2376
Fax: 205-534-2384

(Voltage regulators, including adjustable types, military parts available, power supply, motion control, driver, interface, operational amplifiers, comparator, core memory, automotive ICs, modules including backlights for LCD displays, GTL and bus terminators, SCSI bus transceivers, and voltage regulators for pentium pro

LINFINITY — LOCKHEED MARTIN

computers. Note: Linfinity may have inventory of their obsolete product in stock [which is nonreturnable]. Contact them at 714-898-8121.)

Lintel NV/SA
Avenue de Jettelaan 32
Brussel 1980
Belgium
32 2 425 7767
Fax: 32 2425 3722

Lintel Sarl
7 Rue du Comm Riveree
Paris 75008
France
33 1 4562 0046
Fax: 33 1 4525 9544

(Encryption ICs)

Lion Enterprises Corp.
Room 137, Goldenhill Plaza
11th Floor, No. 25
Min Sheng E. Road
Taipei 10448 Taiwan
886 2 542 1287, 1289
Fax: 886 2 561 9250
TLX: 25964 LIONEC

(Diodes)

Lite-On, Inc.
720 S. Hillview Drive
Milpitas, CA 95035
408-946-4873, 408-956-8016
Fax: 408-942-1527, 408-942-1527

Taiwan Liton Electric Co., Ltd.
12th Fl. 25 Tunghua South Bend,
Taipei
Taiwan, R.O.C.
886 2 771 4321
Fax: 886 2 7511962
TLX: 24514 TWLITON
Cable: TWLITON Taipei

90, Chien I Rd., Chung Ho
Taipei Hsien, Taiwan R.O.C.
886 2 222 6181
Fax: 886 2 221 0660

(LEDs IR and visible, ethernet adaptors containing their own ASIC ICs, this company is the sister company of Diodes, Inc.)

Litronix
See *Siemens Components Inc., Optoelectronics Division*.

Littlefuse, Inc.
100 Johnson Street
Centralia, IL 62801-2251
1-800-999-9445
618-532-1926
Fax: 1-847-391-0894
Telnet: industry.net
http://www.industry.net/littlefuse.electronics
E-mail: lfelectr@interaccess.com

(Fuses [surface mount, clip mount, microminiature, etc] and fuse-holders)

Litton Solid State
3251 Olcott Street
Santa Clara, CA 95054-3095
408-988-1845
Fax: 408-970-9950

(High reliability components including MMIC amplifiers, FETs, diodes, switches and attenuators)

Litton Electron Devices
1215 South 52nd Street
Tempe, AZ
602-968-4471
Fax: 602-966-9055

(Infrared detectors and dewars)

LK prefix
See *SGS-Thomson*

LLIF prefix
See *KDI/triangle Electronics, Inc.*

LM prefix
See *National Semiconductor, Linear Technology Corp., Texas Instruments, Micrel Semiconductor, Motorola, Calogic Corp., Semiconductors, Inc., Samsung Semiconductor*

(Calogic, Semiconductors, Inc., and Samsung have National Semiconductor equivalents. Note: Silicon General also second sources parts with an SG prefix. Some devices also second sourced by Astec Semiconductor, and AVG Semiconductors. See Hitachi and Densitron if an LCD Display, ProSemi GmbH.)

L.M. Ericsson Components
See *Ericsson Components, Inc.*

LMA prefix
See *Logic Devices, Inc.*

LMC prefix
See *National Semiconductor*

LMD prefix
See *National Semiconductor*

LMU prefix
See *Logic Devices, Inc.*

LMS prefix
See *Logic Devices, Inc.*

LMX prefix
See *National Semiconductor*

LO prefix
See *Siemens*

LOA prefix
See *Siemens*

Lockheed Martin Information Systems
5600 Sand Lake Road
Orlando, FL 32819
407-826-3356, 407-356-2000
Fax: 407-356-2080

Manufacturers, Prefixes, Part Number Types, Logo Descriptions & Family Types

LOCKHEED MARTIN — LORAL MICROWAVE-NARDA — L

(3D chip sets for real time image generation)

Lockheed Martin Fairchild Imaging Sensors
1801 McCarthy Blvd
Milpital, CA 95035
1-800-325-6975
408-433-2500
TWX: 910-373-2110
Fax: 408-433-2508

(CCD image sensors)

Lockheed Martin Federal Systems
9500 Godwin Drive
Manassas, VA 22110-4198
703-367-3458
(QML contact Dave Polak)

1801 Route 17, #C
Oswego, NY 13827
800-426-3741
Fax: 607-751-6054

(ASICs including mixed signal types, microcontrollers, DC/DC converters, data compression ICs, MCM modules, SRAMs, RAD hard 32 bit chip sets, and military qualified parts, etc.)

Lockheed Sanders Corporation
(Military MMICs)
See *Lockheed Martin*
(For internal use)

Loctite Luminescent Systems, Inc.
See *Luminescent Systems, Inc.*

LOG prefix
See *Siemens*

Logic Devices, Inc.
628 E. Evelyn St.
Sunnyvale, CA 94086
800-851-0767
(Inside U.S. except CA)
800-233-2518 (Inside CA)

408-720-8630
408-737-3346
(Applications/literature hotline)
408-737-3300
Fax: 408-733-7690 (Fax Hotline), 408-732-9158
Telex: 172387
http://www.logicdevices.com
E-mail: litreq@logicd.mhs.compuserve.com

Southeastern U.S. Sales Office:
9700 Koger Blvd, Suite 204
St. Petersburg, FL 33702
813-579-9992
Fax: 813-576-5643

Northeast Region:
112 Meister Avenue
Somerville, NJ 08876
908-707-0033
Fax: 908-707-8574

Europe:
Warminster, Wiltshire
United Kingdom
44 01985 218888
Fax: 44 01985 218699

(A fabless company supplying CMOS devices, SCSI Bus controllers, pipeline registers, ALUs [Arithmetic Logic Units] barrel shifters, correlators, multipliers [digital-mixers], multiplier-accumulators, multiplier-summers, digital filters and image filters, frame buffers, FIFOs synchronous and asynchronous, SRAMs)

Logitek, Inc.
101 Christopher Street
Ronkonkoma, NY 11779
516-476-4200
Fax: 516-467-4090

(Power supplies, some similar to Abbott Electronics units)

Lohuis International
John Lijsenstraat 66
Meer-Hoogstraten B2321
Belgium
32 3 315 9228
Fax: 32 3 315 7913
TLX: 35661 LOHUIS B

(Optoelectronics)

LOK prefix
See *Siemens*

LOR prefix
See *Lockheed Martin Federal Systems* (Manassas, VA)

Lorain Products
1122 F St.
Lorain, OH 44052-2293
216-288-1122

(Power supplies)

Loral Fairchild Imaging Sensors
See *Lockheed Martin Fairchild Imaging Sensors*

Loral Federal Systems
(Formerly owned by IBM)
See *Lockheed Martin Federal Systems*

Loral Microwave-FSI
16 Maple Road
Chelmsford, MA 01824
508-256-4113
Fax: 508-937-3748

(Microwave switch modules, RF limiter modules, ECL VCOs, Comb generator modules, modular Gunn VCOs, etc.)

Loral Microwave-Narda
435 Moreland Road
Hauppauge, NY 11788
516-231-1700
516-231-1390 (International)
Fax: 516-231-1711

LORAL MICROWAVE-NARDA — LSI LOGIC

(Electromagnetic radiation measurement instruments, induced current products, EMC test equipment, Power/VSWR Monitors, integrated thermocouple power monitors, call through test set, isolators and circulators, mixers, terminations and phase shifters, mechanical switches, hybrids, couplers/detectors, attenuators, adaptors, multicouplers, etc.)

Loral ROLM Computer Systems
Division Loral Corporation
3151 Zanker Road
San Jose, CA 95134
800-321-7672
408-432-8000
Fax: 408-432-9496

(DSP MCM products)

Loral Semiconductor Division
75 Technology Drive
Lowell, MA 01851
508-256-4113

(Microwave diodes)

LOP prefix
See *Siemens*

Loras Industries, Inc.
2640 Freewood
Dallas, TX 75220
214-351-1234
Fax: 214-351-5628

(Transistors, thyristors, digital ICs)

LP prefix
See *National Semiconductor*, *Siemens*, *Supertex* (P Channel MOSFETs), *ProSemiGmbH*

LPD prefix
See *Semicon Components*

(Transient voltage stressor)

LPP prefix
See *Siemens*

LPR prefix
See *Logic Devices, Inc.*

LPT prefix
See *Siemens LPT*
(Low power CMOS logic family), *Pericom Semiconductor*, *Harris Semiconductor*

LR prefix
See *Supertex, Inc.*, *LSI Logic* (Microprocessors), *Siemens*

LR7 part number
See *Supertex, Inc.*

LRFMN-XX part number
See *RF Prime*

(Mixers)

LRFMS-X or **-XX** part number
See *RF Prime*

(Mixers)

LRFPS-X-X or **-X-XX** part number
See *RF Prime*

(Splitters)

LRFPX-XX part number
See *RF prime*

(Splitters)

LRFPQ-X part number
See *RF Prime*

(Splitters)

LRH prefix
See *Siemens*

LRZ prefix
See *Siemens*

LS logic family
See *Hitachi*, *Motorola, Inc.*, *National Semiconductor*, *SGS-Thomson, Lansdale*, *Goldstar Technology, Inc.*, *Rochester Electronics*, *AVG Semiconductors*, *Philips*

(Signetics)

LS prefix
See *Siemens*, *LSI Computer Systems, Inc.*, *Thin-Film Technology Corporation*

(PECL differential delay lines)

LSA prefix
See *Siemens*

LSC prefix
See *Motorola, Inc.*

LSG prefix
See *Siemens*

LSH prefix
See *Logic Devices, Inc.*, *Siemens*

LSI Computer Systems, Inc. (LSI/CSI)
1235 Walt Whitman Road
Melville, NY 11747
516-271-0400
Fax: 516-271-0405

(Custom and standard ICs including programmable digital delay timers, dividers, counters, quadrature pulse detectors, lighting controls, brushless DC motor controls, LCD display drivers, mask programmable melody generators, programmable keyless digital locks, telephone line controls)

LSI Logic Corporation
1551 McCarthy Road
Milpitas, CA 95035
1-800-574-4286

(Literature distribution)
408-433-8000
408-433-4288
(Literature distribution)
Fax: 408-434-6457??,
408-433-8989
http://www.lsilogic.com/
E-mail: info@lsil.com

Product Literature:
1-800-574-4286
415-940-6877, Dept JDS

1525 McCarthy Blvd #6760
Milpitas, CA 95035
408-433-7348
Fax: 408-433-6965

California:
Irvine, CA: 714-553—5600
Fax: 714-474-8101
San Diego, CA: 619-635-1300
Fax: 619-635-1350

Florida:
407-728-9481
Fax: 407-728-9587
Boca Raton, FL: 407-395-6200
Fax: 407-394-2865

Georgia: 404-395-3800
Fax: 404-395-3811
Illinois: 708-995-1600
Fax: 708-995-1622
Maryland: 301-897-5800
Fax: 301-897-8389
Columbia, MD: 410-740-9191
Fax: 410-740-5587
Massachusetts: 617-890-0180
Fax: 617-890-6158
Minnesota: 612-921-8300
Fax: 612-921-8399
New Jersey, Edison: 908-549-4500
Fax: 908-549-4802
New York: 914-226-1620
Fax: 914-226-1315
Victor, NY: 716-225-8820
Fax: 716-223-8822
North Carolina: 919-783-8833
Fax: 919-783-8909

Oregon: 503-645-9882
Fax: 503-645-6612
Texas: 512-388-7294
Fax: 512-388-4171
Dallas, TX: 214-788-2966
Fax: 214-233-9234
Washington: 206-822-4384
Fax: 206-827-2884

Canada:
Calgary: 403-262-9292,
Fax: 403-262-9494
Kanata: 613-592-1263,
Fax: 613-592-3253
Montreal (Pointe Claire):
514-694-2417,
Fax: 514-694-2699
Toronto (Etobicoke):
416-620-7400, Fax: 416-620-5005

Germany:
Stuttgart: 49 711 139690,
Fax: 49 711 8661428

Israel: 972 3 5403741,
Fax: 972 3 5403747
Italy: 39 39 6056881,
Fax: 39 39 6057867

Japan:
Tokyo: 81 3 5463 7811,
Fax: 81 3 5463 7825
Osaka: 81 6 947 5281,
Fax: 81 6 947 5287

Korea: 82 2 561 2921,
Fax: 82 2 554 9327
Spain: 34 1 3672200,
Fax: 34 1 3673151
Sweden: 46 8 703 4680,
Fax: 46 8 7506647
Switzerland: 41 32 536363,
Fax: 41 32 536367
Taipei, Taiwan: 886 2 755 3433, Fax: 886 2 755 5176
United Kingdom, Bracknell:
44 344 426544, Fax: 44 344 481039

LSI Logic Europe plc:
44 1753 680009
Fax: 44 1753 680179

(ASICs, including radiation hardened versions, MIPS, SPARC and RISC microprocessors, Digital Signal Processing DSP products, gate arrays, image and video compression products [such as a JPEG processor], graphics accelerator and controller ICs, Reed-Solomon Codecs, DBS [Direct Broadcast Satellite] receiver ICs, Internet ASICs. Note: LSI Logic no longer supplies military qualified devices)

LSK prefix
See *Siemens*

LSS prefix
See *Siemens*

LST prefix
See *Siemens*

LT or LTC prefix
See *Linear Technology Corp.*, *Motorola, Inc.* (Transistors), *Astec Semiconductor*, *Texas Instruments* (Second source), *Melcher, Inc.*

(For AC/DC converters = LT prefix)

LTE logo
See *Lansdale Semiconductor, Inc.*

LTI
See *Gennum Corporation*

LTI
182-6, Sadang-5
Dongjak, Seoul, Korea
822 587 4086, Fax: 822 587 4088

(Power amplifiers for cellular telephones)

LTL prefix
See *Lite-On*

(LED)

L LU — LUCKY-GOLDSTAR

LU prefix
See *Siemens*

Lucas Deeco
See *Deeco*

Lucas Electronics and Systems
Lucas NovaSensor
1055 Mission Ct.
Freemont, CA 94539
510-490-9100
Fax: 510-770-0645

(Accelerometers and pressure sensors, including surface mount types and micromachined sensors)

Lucas Schaevitz
7905 N. Route 130
Pennsauken, NJ 08110
609-662-8000
Fax: 609-662-6281

543 Ipswich Road
Slough, Berks SL1 4EG
England
0753 537622
Fax: 0753 823563

(Displacement transducers, position transmitters, pressure sensors, transducers and transmitters, inclinometers, and other instrumentation)

Lucas Stability Electronics, Ltd.
63 Greystone Road
Antrim, BT41 2QN
United Kingdom
0849 46 30 35
Fax: 0849 42 8094

(Diodes, transistors)

Lucas Weinschel
P.O. Box 6001
Gaithersburg, MD 20884-6001
800-638-2048, 301-948-3434
Fax: 301-948-3625

(RF products including: fixed and variable attenuators, terminations, loads, attenuators, phase shifters, power splitters and dividers, etc.)

Lucent Technologies
555 Union Blvd.
Allentown, PA 18103
1-800-372-2447
215-439-5237
Fax: 1-215-778-4106

Headquarters:
295 N. Maple Drive
Basking Ridge, NJ 07920
908-221-2000
Fax: 908-221-1211

1090 E. Duane Ave.
Sunnyvale, CA 94086
1-800-372-2447, 408-522-5555
Fax: 408-522-4401

4995 Patrick Henry Drive
Santa Clara, CA 95054
800-372-2447

Canada:
1-800-553-2448
Fax: 215-778-4106

Europe:
AT&T Deutschland GmbH
Bahnhofstr 27A
D-8043 Unterfoehring, Germany
089/950 86-0
Fax: 089/950 86-111

Asia Pacific:
AT&T Microelectronics, Asia/Pacific
14 Science Park Drive
#03-02A/04 The Maxwell
Singapore 0511
65 778-8833, Fax: 65 777-7495
Telex: RS 42898 ATTM

Japan:
AT&T Japan Ltd.
7-18, Higashi-Gotanda 2-Chome
Shinagawa-Ku, Tokyo 141
81 3 5421 1600
Fax: 81 3 5421 1700

Data requests in Europe:
44 732 742 999, Fax:
44 732 741 221

Technical Inquiries in Europe:
Central Europe:
49 89 95086 0 (Munich)
Northern Europe:
44 344 487 111 (Bracknell U.K.)
France: 33 1 47 67 47 67
Southern Europe:
39 2 6601 1800 (Milan)
34 1 807 1 1700 (Madrid)

(Formerly AT&T Microelectronics, Bipolar ICs, GaAs ICs for fiber optics [GaAs discontinued in mid 1995], reprogrammable FPGAs, ASICs, fast SRAMs [discontinued in mid 1995 along with flash memories but modules are available from Rumarson Technologies, Inc. in Kenilworth, NJ], dual port RAMs, solid state relays [the solid state relay line was sold to the Optoelectronics division of Siemens Components, Inc. in Cupertino, CA. in mid 1995], laser diodes, digital signal processors [DSPs], microprocessors, coprocessors [microprocessors including the Hobbit line discontinued in mid 1995], communication ICs; high voltage relays, solid state switches, transistor arrays, video coder/decoder chip sets, video compression chip sets including MPEG-2 decoders, fax and modem chip sets, ATM ICs, SONET/ATM [Synchronous Digital Hierarchy] ICs, [military qualified manufacturer listed] and power supplies)

Lucky-Goldstar
See *LG Semicon*

LUH — M100

LUH prefix
See *Siemens*

Lumex Opto/Components, Inc.
292 E. Hellen Rd.
Palatine, IL 60067
800-616-4444
708-359-2790
Fax: 800-944-2790, 708-359-8904

(Gas discharge tubes, lamps, LEDs and LCDs, including LED light bars and blue LEDs, photo resistors)

Luminescent Systems, Inc. (LSI)
101 Etna Road
P.O. Box 9004
Lebanon, NH 07366-9004
603-448-3444
Fax: 603-448-3452
TWX: 710-366-0607

77 Olean Road
East Aurora, NY 14052-2593
716-655-0800
Fax: 716-655-0309

37 Inwood Road
Rocky Hill, CT 06067
203-721-9500
Fax: 203-529-8163

Mechelsesteenweg 313
2550 Kontich, Belgium
32 3 458 38 52
Fax: 32 457 79 60

(Formerly Loctite Luminescent Systems, Inc., electroluminescent lamps [MaxEL] for backlighting)

LVC logic
See *Philips Semiconductor*

LVQ logic
(Low voltage quiet CMOS logic introduced in December 1995)
See *Motorola Semiconductor*

LVT logic
See *Texas Instruments*

Lutze, Inc.
1911 Associates lane
Charlotte, NC 28217
800-447-2371

(Power supplies)

LX prefix
See *SenSym, Inc.*,
Linfinity Microelectronics, Inc.

LXD Inc.
7650 First Pl.
Oakwood Village, OH 44146
216-786-8700
Fax: 216-786-8711

(LCD displays, including one for the discontinued Doric 130A meter)

LXM prefix
See *Linfinity Microelectronics, Inc.*

LXT prefix
See *Level One Communications (Technology), Inc.*

LY prefix
See *Siemens*

LYB prefix
See *Siemens*

LYK prefix
See *Siemens*

LYT prefix
See *Siemens*

LYU prefix
See *Siemens*

LYZ prefix
See *Siemens*

LZR Electronics, Inc.
8051 Cessna Avenue
Gaithersburg, MD 20879
301-921-4600
Fax: 301-670-0436

(Instant data transmission system 301-921-4607. External power supplies, DC/DC converters, DC to AC power inverters, jacks-plugs-connectors, cord sets, custom cable assemblies and miniature speakers)

M prefix
See *Mitsubishi Electronics America, Inc., Hughes Semiconductor Products Center, Valtronic USA, Inc.*
(CMOS SRAM modules),
*TranSwitch Corp.,
SGS-Thomson Microelectronics, ALI - Acer Laboratories, Inc.*
(Pacific Technologies Group),
Micronetics (VCOs),
AVG Semiconductors
(Second source some ICs),
Optical Imaging Systems, Inc.
(LCD displays),
EM Microelectronic
(Display drivers, real time clocks)

M underlined logo
See *Monolithic Sensors, Inc.*

M over "PC" logo
See *Monitor Products Company*

M logo, looks like two pulses
See *Mallory*

(Capacitors)

M100 series
See *Micro Networks*

(Clock oscillators)

M198 — M/A COM SEMICONDUCTOR

M198 part number
See *Daico Industries, Inc.*

M2B part number
See *VideoLogic*

M38510/XXXXXXXX part number
(Indicates a part qualified to MIL-M-38510 military specification. All the slash sheets [i.e., individual devices covered by slash sheets to this drawing] are inactive and are not to be used for new designs. MIL-M-38510 has been replaced by MIL-PRF-38535.)

M5M prefix
See *Mitsubishi*

M6XXX series
See *Mitsubishi*

(Programmable ASICs)

MA prefix
See *Matsushita* (Panasonic), *GEC Plessy* (Marconi Circuit Technology, Inc.), *M/A-Com*, *Analog Systems*

MA-XXXX-XXX part number
See *Harris Semiconductor*
(Custom part)

M/A (Microwave Associates) Com Semiconductor (Silicon) Products, Inc.

M/A-COM Gallium Arsenide Products, Inc.
an Amp Company
100 Chelmsford St.
Lowell, MA 01853-3294
1-800-366-2266
508-453-3100
Fax: 508-656-2740
617-821-9198
(LOUD&Clear newsletter)

43 South Ave.
Burlington, MA 01803
1-800-366-2266
617-272-3000
TWX: 710-332-6789
TLX: 94-9464

New England Field Office:
M/A-COM Components Marketing, Inc.
Suite 182
155 New Boston Road
Woburn, MA 01801
617-938-8600
TWX: 710-348-1339
Modem: 617-933-0882

Upper New York State:
Syracuse Field Office
115 Twin Oaks Drive
Syracuse, NY 13206
315-437-0315
TWX: 710-541-0494
Modem: 315-437-4980

Philadelphia Area:
1800 Byberry Road
Suite 1002
Huntingdon Valley, PA 19006
215-938-0550
TWX: 510-665-6533
Modem: 215-947-2058

Chicago Area:
380 West Palatine Road, Suite 2
Wheeling, IL 60090
312-459-8440
TWX: 910-651-3026
Modem: 312-459-8380

Indiana/Ohio Area:
3230 Mallard Cove Lane
Ft. Wayne, IN 46804
219-432-8424
TWX: 810-332-1513
Modem: 219-432-3609

Southeastern Region:
M/A-COM Components Marketing, Inc.
19640 Club Horse Road, Suite 425
Gaithersburg, MD 20879
301-670-1700
TWX: 710-828-9819
Modem: 301-670-1703

Atlanta Area:
1640 Powers Ferry Road
Suite 160, Bldg 2
Marietta, GA 30067
404-956-0351
TWX: 810-751-0244
Modem: 404-956-0354

Warner Robbins, GA:
1532 Watson Blvd.
Suites 4 & 6
Warner Robbins, GA 31093
912-929-1000

North Carolina Area:
4915 Waters Edge Drive
Suite 265
Raleigh, NC 27606
919-851-6220/1
TWX: 510-928-1835

Florida Area:
P.O. Box 3147
Indiatlantic, FL 32903
305-729-6400
TWX: 810-848-0008

Dallas Area:
1143 Rockingham Drive
Suite 110
Richardson, TX 75080
214-234-2463
TWX: 910-867-4769
Modem: 214-644-8259

Phoenix Area:
3260 No. Hayden Road, Suite 115
Scottsdale, AZ 85251
602-949-1642
TWX: 910-950-1281
Modem: 602-994-8543

Manufacturers, Prefixes, Part Number Types, Logo Descriptions & Family Types

M/A COM SEMICONDUCTOR

Denver Area:
14 Inverness Drive - E
Bldg. F, Suite 236
Englewood, CO 80112
303-790-0538, 39, 42
TWX: 910-935-0156

California:
9800 Sepulveda Blvd, Suite 610
Los Angeles, CA 90045
213-641-5311
TWX: 910-328-6132
Modem: 213-645-2220

280 Newport Center Drive
Suite 200
Newport Beach, CA 92660
714-720-0355
Modem: 714-720-9154

5553 Hollister Ave.
Suite F
Goleta, CA 93117
805-964-4844
TWX: 910-334-4840

3650 Clairemont Drive
Suite #5
San Diego, CA 92117
619-483-5099
TWX: 910-335-1737

421 N. Brookhurst, Suite 108
Anaheim, CA 92801
714-758-1921
TWX: 910-591-1921

1754 Technology Drive
Suite 129
San Jose, CA 95110
408-298-2525
TWX: 910-338-2140
Modem: 408-298-2120

Seattle/Western Canada Area:
1495 Gilman Blvd.
NW Suite 17
Issaquah, WA 98027
206-392-4990
TWX: 910-443-3038
Modem: 206-392-8497

Asia/Pacific: 81 03 3226 1671

Toronto Canada Area:
MA Electronics Canada, Ltd.
6547 Mississauga Rd.
Mississauga, Ontario L5N 1A6
416-821-3548
TWX: 610-492-2999

Ottawa Canada Area:
M/A-COM Electronics Canada
190 Colonnade Rd.
Suite 204
Nepean, Ontario K2E 7J5
613-727-9800
514-848-7081 (Montreal)
Telex: 053-3507

Asia: 81 03 3226 1671

Australia:
Microwave Associates
Australia Pty Ltd.
2nd Floor, 20 Clark Street
Crows Nest NSW 2065
Australia
61 2 438 1299
Telex: 71464

Belgium, Luxembourg:
Microwave Associates International
92-94 Sq. E. Plasky
1040 Bruselles
Belgium
32 2 735 01 95
Fax: 736 53 72
Telex: 846223281
Cable: MELABSA BRUSSELES

Europe/Middle East/Africa:
44 0344 869 595

East Europe:
Microwave Associates International
19a Creffield Road
Ealing
London, W5, 3RR
United Kingdom
44 1 992-6396
Telex: 85182295
Cable: MUWAVE LONDON W5

United Kingdom:
Microwave Associates, Ltd.
Humphreys Road
Dunstable LU5 4SX
Bedfordshire, United Kingdom
44 582 60 14 41
Telex: 85182295
Cable: Microwave Dunstable

France:
M/A-COM France S.A.
6-8 Rue Du 4 Septembre
92130 Issy Les Moulineaux
France
33 1 45 54 97 58
Telex: 842202100

Southern Italy:
M/A-COM Italia, S.P.A.
Components Division
Via M. Macchi 28
20124, Milan, Italy
2 669 6368
Telex: 340564

Netherlands:
Microwave Associates
Components BV
Lelystraat 2
3601 BV Maarssen
The Netherlands
P.O. Box 277
31 3465 66024
Telex: 84447808

Sweden:
M/A-COM A/B
Wallingatan 38
S-111 24 Stockholm, Sweden
46 8 14 03 50
Telex: 10540 MACOM S

Germany/Austria:
M/A-COM GmbH
Fasanweg 4
8016 Feldkirchen, Germany
49 89 903 80 34
Telex: 841529103

M/A COM SEMICONDUCTOR — MAGNECRAFT S-D

(Silicon and GaAs components including: microwave components, subsystems, varactor diodes [abrupt, hyperabrupt and controlled gamma devices], power generation and amplification diodes, bipolar transistors, switching diodes [PINs], limiter diodes, power Schottky rectifiers, FETs, connectors, isolators, circulators, attenuators, couplers, filters, phase shifters, E/M switches, I/F amplifier assemblies, RF/IF, DLVA and IFMS subsystems, antennas and antenna arrays, GPS, horns and arrays, etc., for wireless, cellular, commercial, telecommunication and military applications.)

MAAM prefix
See *M/A-Com*

Mabuchi Motor Co. Ltd.
430 Matsuhidai Matsudo-shi
Chiba-ken, 270 Japan
473 84 1111
Fax: 473 85 2026
TLX: 2993 218 TYOMAB J
Cable: TYOMAB MATSUDO

475 Park Avenue South
New York, NY 10016
212-686-3622

111 North Bridge Road
#21-02 Peninsula Plaza
Singapore 0617
339 9991

19 Sam Chuk Street
San Po Kong
Kowloon, Hong Kong
3285575

6/F No. 3 Sec.1 Tun Hwa
South Road
Taipei, Taiwan
2 7510155

Zeil 65-69
60313 Frankfurt/Main 1
Germany
69 282252

(Motors)

MAC prefix
See *Mitech Corporation*

MACH prefix
See *Advanced Micro Devices* (AMD)

Macrochip Research
1301 N. Denton Drive
Carrolton, TX 75006
214-242-0450
Fax: 214-245-1005

(Power driver ICs)

Macronix Inc.
No. 3 Creation Road III
Science-Based Industrial park
Hsin-Chu, Taiwan R.O.C.
886 35 788 888
Fax: 886 35 788 887, 886

12F, 4, Sec. 3, Min-Chuan E. Rd.
Taipei, Taiwan R.O.C.
886 02 509 3300
Fax: 886 02 509 2200

Macronix International Co. Ltd.
3F, No. 56,
Park Avenue 2
Science Based Industrial Park
Hsin Chu, Taiwan R.O.C.
886 35 783-333, 886 35 788 888
Fax: 886 35 778 689, 886 35 788 886

Room 223, 2F, No. 602
Min-Chuan E. Rd.
Taipei, Taiwan, R.O.C.
Macronix America, Inc.
1338 Ridder Park Drive
San Jose, CA 95131
800-432-1621
408-453-8088, 408-451-0888
Fax: 408-453-8488, 408-451-0889

Macronix Japan K.K.
Sekine Build,
3-1-1 Amanuma,
Suginami-Ku, Tokyo, Japan 167
03 3393-4645
Fax: 03 3398-5163

Grote Winkellaan 95, Bus 1
1853 Strombeck
Belgium
32 2 267 7050
Fax: 32 2 267 9700

(Data communication ICs [including UARTs, DUARTs], fax/modem ICs, PC Chip sets, Video and Graphics ICs [including 2-D graphics controllers], EPROMs, ROMs, flash memory, masked ROM, TELE-COM board level products, flash memory and Ethernet ICs)

MacroTek
Emil-Figge-Str. 76
D-4600 Dortmund 50
Germany
49 231 9742 151
Fax: 49 231 9742 120

(Chip sets for the PowerPC processor)

MAD prefix
See *Motorola, Inc.*

Magnecraft/Struthers-Dunn (MSD)
700 Orange St.
Darlington, SC 29532
803-393-5421
Fax: 803-393-4123

211 Waukegan Rd. #210
Northfield, IL 60093
708-441-2525
Fax: 708-441-2520

(Relays)

MAGNETEK — MARSHALL

MagneTek
112 East Union Street
Goodland, Indiana 47978
219-297-3111
Fax: 800-795-7020 (fast fax),
219-297-3554

(Transformers, chokes, inductors, current sensor transformers)

Magnetic Circuit Elements, Inc.
1540 Moffett St.
Salinas, CA
408-757-8752
Fax: 408-757-5478

(Magnetics such as telephone transformers)

Magnetico, Inc.
182 Morris Avenue
Holtsville, NY 11742
516-654-1166
516-654-1167
(Engineering hotline)
Fax: 516-758-7408

(Magnetics for military and aerospace applications)

MAH
See *M-tron Industries, Inc.*

(Oscillators)

Maida Development Company
20 Libby Street
P.O. Box 3529
Hampton, VA 23663
804-723-0785
Fax: 804-772-1194
TLX: 82 3443

(Diodes, thermistors, metal [zinc] oxide varistors)

Mako
(An R4000 chip set for PCI bus)
See *Toshiba America*

Mallory
North American Capacitor Company
P.O. Box 1284
7545 Rockville Road
Indianapolis, IN 46214-1284
317-273-0090
Fax: 317-273-2400

(Capacitors)

Mammoth Memory, Inc.
2640A 118th Avenue N
St. Petersburg, FL
800-808-5070
Fax: 813-532-2015

(Supplier of memory modules made by Q-1 Technologies, St. Petersburg, FL)

Manutech, Inc.
8181 N.W. 91 Terrace, Bldg. 10
Miami, FL 33166
305-888-2800
Fax: 305-888-2628

(Magnetic components including transformers and filter inductors)

MAP prefix
See *WSI* (Waferscale Integration), *National Semiconductor*

(Programmable logic)

Marcon Electronics Co., Ltd.
1-1, Saiwai-Cho, Nagai-Shi
Yamagata 993
Japan
0238 84 2131
Fax: 0238 88 3694

(Hybrid ICs, MOVs)

Marconi Electronic Devices, LTD.
I.C. Division,
See *GEC Plessy Semiconductors*

(Marconi Electronic Devices, LTD. and Plessy Semiconductors LTD. have been grouped together to form GEC Plessy Semiconductors)

Marktech International Corp.
5 Hemlock St.
Latham, NY 12110
518-786-6591
Fax: 518-786-6559

(LEDs, LED displays including dot matrix types, dip and miniature switches, IC sockets and connectors)

Marl International Ltd.
Ulverton
Cumbria, LA12 7RY
United Kingdom
44 229 582430,
Fax: 44 229 585155

(Optoelectronics)

Marlow Industries, Inc.
10451 Vista Park Road
Dallas, TX 75238-1645
214-340-4900
Fax: 214-341-5212

(Thermoelectric coolers)

Mars (Chip set)
See *Intel Corp.*

Marshall Thermocouples
L.H. Marshall Company
Box 02226
Columbus, OH 43202
800-THERMOC
614-294-6433
Fax: 614-294-0297

(Thermocouples)

MARTEL ELECTRONICS — MAXIM

Martel Electronics
P.O. Box 897
Windham, NH 03087
800-821-0023
603-893-0886
Fax: 603-898-6820

(Analog to RS232 converter IC)

Master Instrument Corp.
No. 252, Yuan Huan E. Road
Feng Yuan
Taichung 42023
Taiwan
886 4 525 1600
Fax: 886 4 523 5787
TLX: 58492 MICCORP

(Diodes, optoelectronics)

MAS prefix
See *Micronas, Inc.*

Mass Memory Technology, Inc.
See *WaferDrive*

Matra Harris
See *Siliconix* (Temic)

(Matra split from Harris in 1988)

Matra MHS Electronics Corp., see Siliconix

Matrox Electronic Systems, Ltd.
Matrox Graphics
1055 St. Regis Blvd
Dorval, PQ H9P 2T4
Quebec, Canada
514-685-2630
Fax: 514-685-2853
TLX: 05 822 798
(ASIC Design)

1075 Broken Sound Parkway
Boca Raton, FL 33487

(Graphic accelerator chip sets, multimedia [GUI, Video 3-D] graphics controller)

Matsushita Electric Industrial Co. Ltd.
1006 Kadoma
Kadoma-shi
Osaka 571, Japan
81 6 908 1121
Fax: 81 6 906 1507

Matsushita Electronics Corporation
1-1 Saiwai-cho
Takatsuki-shi
Osaka, 569 Japan
0726 82 5521
Fax: 0726 82 3093

Matsushita Electric Corporation
of America
1 Panasonic Way
Secaucus, NJ 07094
800-344-2112
201-348-7000

(Includes Panasonic parts, ICs for VCRs, televisions, CD players, speech synthesis and recognition, VCR, watch, clock, disk drive controller, hybrid, logic, microcontrollers, voltage comparators and regulator ICs, Hall effect sensors, GaAs cellular phone ICs, ferroelectronic [FRAM] memory cards, plasma displays, etc.)

Matsusada Precision Devices, Inc.
2570 W. El Camino Real
Suite 306
Mountain View, CA 94040-9852
1-800-422-HVPS
Fax: 415-949-1217
Europe: 49 89 4991691
Fax: 49 89 403602

(High voltage power supplies for PMTs and APDs, etc.)

Matthey Electronics
Burslem, Stoke-On-Trent
England
0 782 577588
Fax: 0 782 838558
TLX: 36341

U.S. Agent:
Television Equipment Associates, Inc.
Box 393
South Salem, NY 10590
914-763-8893
Fax: 914-763-9158

(Delay lines, transmission line and video filters)

MAX prefix
See *Maxim Integrated Products, Altera Corp.*

(Alteras "MAX" EPLDs are second sourced by Cypress Semiconductor Corp.)

MaxEL
See Luminescent Systems, Inc. (LSI)

(Electroluminescent lamps)

Max-Lion Corp.
8th-1, 285 Roosevelt Road
P.O. Box 84-539 Taipei
Taiwan
886 2 363 4133
Fax: 886 2 362 6803

(Diodes, thyristors, optoelectronics)

Maxim Integrated Products
120 San Gabriel
Sunnyvale, CA 94086
1-800-998-8800
408-737-7600
Fax: 408-737-7194
http://www.maxim-ic.com

Manufacturers, Prefixes, Part Number Types, Logo Descriptions & Family Types

MAXIM — MCE

Germany:
Maxim GmbH 089 898 13 70
Distributors: 0130 82 79 25

Japan:
Maxim Japan Co. Ltd.
03 3232 6141

United Kingdom:
Maxim Integrated Products
0734 845255

France: 1 30 60 91 60
Distributors: 1 30 60 93 39
Taiwan: 35 777659

(Various ICs including linear devices, op amps, comparators, ADCs, references, interface ICs, microprocessor supervisors, DC/DC converters, wideband and current feedback amplifiers, switching regulators, voltage regulators, voltage monitors, power management ICs, analog multiplexers, switched capacitance filters, etc. Note: Samples of up to 2 pieces are available from Sunnyvale CA at Ext 6215. The company will also take small orders of 100 pieces or less if there is a long delivery schedule and the need is urgent.)

Maxtek Components Corporation
13335 Southwest Terman Road
P.O. Box 1480
Beaverton, OR 97075-1480
1-800-4MAXTEK
503-627-4133
Fax: 503-627-4651
technology@maxtek.com
Email: technology

8425 Heather Place
Boynton Beach, FL 33437-2929
800-694-6166
407-736-8004
Fax: 407-736-2060

Maxtek Components Corporation U.K.
44 0 1734 303434
Fax: 44 0 1934 520385
Email:
100756.2565@compuserve.com

(Incorporated in early 1995 by Tektronix and Maxim Integrated products to develop high frequency MCMs, ASICs, and hybrid devices. The company evolved from Tektronix's MCM and hybrid facility. RF power amplifiers, active filters, fiber optic drivers/receivers, A/D converters and ATE components [pin drivers and attenuators])

MB prefix
See *Fujitsu*, *Mitel Semiconductor*, *Analog Systems*

MBR prefix
See *Motorola*

(Schottky barrier rectifiers)

MBM prefix
See *Fujitsu*

MC logo
See *Micro Crystal*

(Crystals and oscillators)

MC prefix
See *Motorola, Inc.*,
Performance Motion Devices,
MICRA Corporation
(Second sources some National Semiconductor parts),
IBM (Microcontrollers),
Texas Instruments (Second source),
Intel, *Watkins-Johnson*,
Analog Systems,
Standard Microsystems Corp.
(IBM PS/2 microchannel bus interface), *AVG Semiconductors* (Second source telephone and interface, etc. ICs)

MC-42 prefix
See *NEC*

(DIMM memory modules)

MCA prefix
See *Siemens*

MCA4 prefix
See *Motorola*

(Programmable ASICs)

McCoy Electronics, Inc.
100 Watts Street
Mt Holly Spring, PA 17065
717-486-8294
Fax: 717-486-5920

(Crystals)

MCC
See *Microelectronics,*
Computer Technology Corp.

MCCF prefix
See *Motorola*

MCD prefix
See *Motorola*

McDonnell Douglas
Headquarters:
Airport Road & McDonnell Blvd.
P.O. Box 516
St. Louis, MO 63166
314-232-7503
Fax: 314-777-1096

(ICs for their own products)

MCE logo
See *Magnetic Circuit Elements, Inc.*

(Small magnetics)

MCE SEMICONDUCTORS — MEDIAVISION

MCE Semiconductors, Inc.
111 Fairfield Drive
West Palm Beach, FL

(This company declared bankruptcy in mid 1993 and is no longer in operation)

MCF prefix
See *Motorola*

MCG Electronics, Inc.
12 Burt Drive
Deer Park, NY 11729-5778
516-586-5125

McGuirk Electronics Co., Inc.
212 S. Main Street
P.O. Box 853
Middleton, MA 01949
617-933-3500
Fax: 508-774-8748

(Digital, interface, linear and memory ICs)

MCH prefix
See *Interpoint*

(DC/DC converters)

MCL
See *Mini-Circuits*

MCM prefix
See *Motorola, Inc.*, *National Semiconductor*

MCR prefix
See *Instruments SA*

(CCD Detectors)

MCS prefix
See *Intel*

MCT prefix
See *Harris Corp.*
(MOS-controlled thyristors), *Siemens*

MD prefix
See *Intel Corp.*, *Mitel Semiconductor*, *Harris Semiconductor*
(Usually a custom part)

MD54-XXXX part number
See *M/A-Com*

(GaAs MMIC mixers)

MDC prefix
See *Motorola Inc.*

MDC logo
See *Maida Development Corporation*

MDJ, Inc.
P.O. Box 1019
Glen Alpine, NC 28628
704-584-4007
Fax: 704-584-7000

(Fan controller IC)

MDL prefix
See *Siemens*

MDL-XXXXXX-X part number
See *Varitronix*

(LCD displays)

MDPLDXXX part number
See *Intel*

(PLD device)

MDS prefix
See *GHz Technology, Inc.*

MDSP prefix
See *IBM Microelectronics*

(DSP ICs)

MDW prefix
See *Japan ACM Corporation*

Mean Well Enterprises Co., Ltd.
4F-2, No. 26,
Wu-Chuan 2nd Road
Hsin Chuang,
Taipei, Taiwan R.O.C.
886 2 299 6100
Fax: 886 2 299 6200
http://www.shinestar.com/www/newera
E-mail: mw@tpts1.seed.net.tw

(U.S. Representative, Jameco Electronics, 800-831-4242, Fax: 800-237-6948. Power supplies)

MEC logo
See *Mercury Electronic Ind. Co. Ltd.*, *Martel Electronics* (MEC232)

MEC prefix
See *Harris Semiconductor*

Meder Electronic GmbH
Robert-Bosch-StraBe 4
D-W-77— Singen
49 0 77 31 or 6 20 02
Fax: 49 0 77 31 or 6 65 24

In the U.S.:
Gem Electronics
160 Fernwood Drive
East Greenwich, RI 02818
401-885-8454
Fax: 401-885-1741

(Molded case reed relays, reed sensors, and reed switches)

MediaChips, Inc.
(Freemont CA)
Purchased by Opti, Inc. in late 1993
Media Computer Technologies
(MCT)
Santa Clara, CA

(This company that supplies video processing IC's, was acquired by C-Cube Microsystems in late 1995.)

MediaVision Technology, Inc.
See *Aureal Semiconductor*

MEDL — MERCURY UNITED ELECTRONICS

MEDL
See *Marconi Electronic Devices*

MEDM prefix
See *Technitrol, Inc.*

Medtronic Micro-Rel
2343 West 10th Place
Tempe, AZ 85281-5164
602-968-6411
Fax: 602-968-9691
TWX: 910-950-1941

(Driver/receiver ICs, full custom and semi-custom ICs and hybrid devices, for military [including MIL-STD-1772 certification, and MIL-STD-1553 products], medical, industrial and commercial applications)

Megadyne Corp.
8718 Arlington Blvd
Fairfax, VA 22031
703-820-5232
TLX: 4931706 MEGAUI

(Linear ICs)

Megapower
508 Division St.
Campbell, CA 95008
408-376-3500
Fax: 408-376-3672

(Power supplies)

MEH
See *M-tron Industries, Inc.*

(Oscillators)

Melcher Inc.
Uster Switzerland
187 Billerica Road
Chelmsford, MA 01824
800-828-9712
Fax: 508-256-4642

Germany, Melcher GmbH:
040 89 68 27, Fax: 040 89 83 59
France, Melcher SA:
1 69 05 99 11 Fax: 1 69 96 04 54
Great Britain, Melcher Ltd.:
0425 47 47 52 Fax: 0425 47 47 68
Ireland, Melcher S.r.l.:
0266 10 10 63 Fax: 0266 120 10 62
CH: Melcher AG, Ackerstrasse 56,
CH-8610 Uster, 01 944 81
Fax: 01 940 98 58

(DC/DC converters and power supplies)

Melcor
Materials Electronic Products Corporation
1040 Spruce Street
Trenton, NJ 08648
609-393-4178
Fax: 609-393-9461

(Thermoelectric coolers)

Melles Griot
4665 Nautilus Court South
Boulder, CO 80301
303-581-0337
Fax: 303-581-0960

Canada: 613-226-5880
Denmark: 5361 5049
France: 01 3460 5252
Germany: 06151 86331
Hong Kong: 724 5023
Japan: 03 3407 3614
Netherlands: 08360 33041
Singapore: 743 5884
Sweden: 08 630 1040
Taiwan: 035 775 111
United Kingdom: 223 420071

(Diode laser assemblies, high speed silicon photodiodes, laser diode drivers and controllers, optics and optical components, optical mechanical items such as positioners and isolation systems, and instruments including laser beam measuring instruments and detectors and amplifiers.)

MEM prefix
See *Advanced RISC Machines*

Memory X
3914 Murphy Canyon Road
San Diego, CA 92123-4413
619-292-1151

(Memory SIMM modules)

Memtech Technology Corporation
1257 Tasman Drive, #A
Sunnyvale, CA 94089-2229
800-445-5511
408-745-1600
Fax: 408-745-1733

(Bubble memory controllers, sense amplifiers, coil predrivers, etc.)

MemTech
1259 Birchwood Drive
Sunnyvale, CA 94089
1-800-445-5511

(Solid state drives)

Mentor GmbH & Co.
Otto-Hahn Stass 1
D 4006
Erkrath-Unterfeldhaus
Germany
021 000 0000
Fax: 021 120 0241
TLX: 8586734MENT

(Optoelectronics)

Mercury United Electronics, Inc.
10822 Edison Court
Rancho Cucamonga, CA 91730
714-466-0427
Fax: 714-466-0762

MERCURY ELECTRONIC — MHPM

Mercury Electronic Ind. Co., Ltd.
No. 506 Chung Yang Road
Nan King District
Taipei, Taiwan ROC
02 783 4202
Fax: 886 2 782 7230

(Quartz crystals, crystal oscillators, crystal filters)

Meredith Instruments
5035 N 55th Avenue
Glendale, AZ 85301-7500
602-934-9387

(Ultraminiature diode lasers)

Merrimac Industries, Inc.
41 Fairfield Place
West Caldwell, NJ 07006
201-575-1300
Fax: 201-575-0531

(Couplers, dividers, wideband digital attenuators, hybrids and mixers, mmic integration)

Mesostate LCD Industries Co., Ltd.
No. 154, Ping Ho Road
Chung Ho City
Taipei Hsien
Taiwan
886 2 221 4402
Fax: 886 2 221 2386

(LCDs)

MetaDesign Semiconductor
2895 Northwestern Parkway
Santa Clara, CA 95051
408-986-9000
Fax: 408-748-1038

(PCMCIA functionality IC that supports FAX/MODEM chip sets)

Metaflow Technologies, Inc.
4250 Executive Square, Suite 300
La Jolla, CA 92037
619-452-6608
Fax: 619-452-0401

(RISC microprocessors for workstations)

Metalink Transmission Devices, Ltd.
8 Elkahi Street
Tel Aviv, Israel
972 3 642 0006
Fax: 972 3 642 0070

325 E. Elliot Road
Chandler, AZ
602-926-0797

(Chip sets for high data-rate subscriber loop [HDSL] telecommunications [for E1/T1 lines])

Metelics Corp.
975 Stewart Avenue
Sunnyvale, CA 94086
408-737-8181
Fax: 408-737-7645

(Diodes)

Metolius, Inc.
14127 125th Ave., NE
Kirkland, WA 98034
206-820-0760
Fax: 206-820-6976

(Image processing IC's)

MF prefix
See *Intel, Analog Systems*

MF Electronics Corp.
10 Commerce Drive
New Rochelle, NY 10801
914-576-6570
Fax: 914-576-6204

(Crystal oscillators, VCXOs, etc.)

MFF prefix
See *Motorola, Inc.*

MFL prefix
See *Interpoint Corporation*

MFT prefix
See *MF Electronics*

(Crystal oscillators)

MG prefix
See *Intel Corp.,
Harris Semiconductor*
(Usually a custom part)

MG Electronics
32 Ranick Road
Happauge, NY 11788
516-582-3400
Fax: 516-582-3229

(Speakers, buzzers, transducers)

MGA-XXXXX part number
See *Hewlett-Packard Company*
(Low noise RF amplifier),
Matrox Electronic Systems, Ltd.
(Matrox Graphics)

MGF prefix
See *Mitsubishi Electronics America, Inc.*

MH prefix
See *Mitel Semiconductor*

MHF prefix
See *Interpoint*
(DC/DC converters)

MHP prefix
See *Micro-Switch, Motorola, Inc.*
(IGBT modules),
Interpoint (DC/DC converters)

MHPM prefix
See *Motorola, Inc.*
(IGBT modules)

Manufacturers, Prefixes, Part Number Types, Logo Descriptions & Family Types

MHS
See *Siliconix*
(Matra MHS, formerly Matra Harris MHS)

MHV prefix
See *Interpoint*

(DC/DC converters)

MHW prefix
See *Motorola, Inc.*

MIC prefix
See *Micrel Semiconductor, Inc.*, *Macrochip Research*

MICRA Corp.
10301 Willows Road
P.O. Box 97005
Redmond, WA 98073-9705
800-822-8782
206-882-3100
Fax: 206-882-1990

120 Ricefield Lane
Hauppauge, NY 11788
516-231-8600
Fax: 516-231-8951

(Operational amplifiers [high speed, high slew rate, etc.], current dividers, level shifters, power drivers, active filters, etc. This company second sources some National Semiconductor parts.)

Micracor
43 Nagog Park
Acton, MA 01720
508-263-1080
Fax: 508-263-1448

(Diode lasers and tunable external cavity semiconductor diode lasers)

Micrel Semiconductor, Inc.
1849 Fortune Drive
San Jose, CA 95131
408-944-0800
Fax: 408-944-0970

(PCMCIA Vpp controllers, voltage regulators [including low voltage dropout regulators], switching regulators, references, MOSFET drivers, MOSFET and PCMCIA power switches, display switches and drivers, power management ICs, proprietary ICs and second source products from National Semiconductor, Teledyne, etc. They have been qualified to produce some SMDs, such as dual power MOSFET drivers. This company also produces some components obsoleted by other manufacturers.)

Micro Circuit Engineering Ltd.
Alexandra Way
Ashchurch, Tewksbury Gloucester Shire GL2 8TB
United Kingdom
011 44 684 297777

(MIL-STD-1553 circuits)

Micro-C Corporation
11085 Sorrento Valley Court
San Diego, CA 92121
619-552-1213
1-800-723-1357
FAX: 619-552-1219

Micro-C/I.I. Ltd.
1 Whittle Place
South Newmoor Industrial Estate
Irvine, Scotland KA11 4HR
011-44-294-221836
011-44-294-2211837
Fax: 011-294-221838

(Recycled or "reconditioned" ICs, parts pulled off of circuit assemblies, they also sell, with their logo, flash EPROMs, SRAM's and some processor ICs)

Microchip Technology, Inc.
2355 W. Chandler Blvd.
Chandler, AZ 85224-6199
602-963-7373
602-786-7200
(520 area code after 10/21/95)

MHS — MICROCHIP TECHNOLOGY

Fax: 602-345-3390, 899-9210
http://www.ultranet.com/biz/mchip/
www.mchip.com/microchip

ASIC Product Division
2674 North First Street
San Jose, CA 95134
408-943-1332
Fax: 408-943-1335

(Custom devices used to replace high speed [22V10] PALs, known as HAL products. This company was formerly known as ASIC Technical Solutions, Inc. [ATS] before being acquired by Microchip Technology in 1996.).

North CA, AK, CO, HI, ID, MT, NV (except Clark County), OR, UT, WA, WY, Canada: Saskatchewan, Manitoba, Alberta, British Columbia:
2107 N. First St.
Suite 410
San Jose, CA 95131
408-436-7950
Fax: 408-436-7955

South CA, AZ, NM, NV (Clark County) and TX (El Paso):
1411 W. 190th Street
Suite 230
Gardenia, CA 90248-4307
213-323-1888
Fax: 213-323-1424

Georgia, AL, FL, MS, Western TN:
1513 Johnson Ferry Road, NE
Suite B-16
Marietta, GA 30062
404-509-8800
Fax: 404-509-8600

1521 Johnson Ferry Rd. NE
Suite 170
Marietta, GA 30062

MICROCHIP TECHNOLOGY — MICRO LINEAR

Illinois, IN, KY, MI, MN, ND, OH, Western PA, SD, WI, WV:
665 Tollgate Road, Unit C
Elgin, IL 60123
708-741-0171, Fax: 708-741-0638

Massachusetts, CT, ME, NH, RI, VT, Canada: Manitoba, Ontario, Quebec and Eastern Provinces:
Five The Mountain Road, Suite 120
Framingham, MA 01701
508-820-3334, Fax: 508-820-4326

New York, DC, DE, MD, NC, NJ, Eastern PA, SC, VA, Eastern TN:
300 Wheeler Rd, Suite 206
Hauppauge, NY 11788
516-232-1930
Fax: 516-232-1935

150 Motor Parkway, Suite 416
Hauppauge, NY 11788
516-273-5305
Fax: 516-273-5335

Texas, AR, IA, KS, LA, MO, NE, OK:
17480 N. Dallas Parkway, Suite 114
Dallas, TX 75287
214-733-0391
Fax: 214-250-4631

France, Italy, Spain, Portugal:
Arizona Microchip Technology SARL
2, Rue Du Buisson aux Fraises
F-91300 Massy, France
1 69 30 90 90
Fax: 1 69 30 90 79

United Kingdom, Ireland, Scandinavia:
Arizona Microchip Technology Ltd.
Unit 3, Meadow Bank
Furlong Road
Bourne End, Bucks SL8 5AJ
England
0628 850303
Fax: 0628 850178

Germany, Africa, Austria, Benelux, Eastern Europe, Israel, Switzerland, Turkey, The Middle East:
Arizona Microchip Technology GMBH
Alte Landstrasse 12-14
D-8012 Ottobrunn
Germany
089 609 6072
Fax: 089 609 1997
TLX: 524518

Asia/Pacific:
Microchip Technology Inc.
4F Madre Matsuda Building
4-13, Kioi-Cho, Chiyoda-Ku
Tokyo 102 Japan
3 234 8774
Fax: 3 234 8549

Microchip Technology International, Inc.
Shinyokohama Gotoh Bldg.
8F, 3-22-4
Shinyokohama, Kohoku-Ku
Yokohama-Shi
Kanagawa 222 Japan
45 471-6166
Fax: 45 471-6122

(Microcontrollers, EEPROMs, ASICs for energy management applications, wireless security products [based on Keeloq code-hopping technology], etc.)

MicroClock
274 E. Hamilton Avenue
Campbell, CA 95008-0240
408-364-4900

(CMOS clock and timing control devices including clock synthesizers and generators)

Micro Crystal
Division of SMH (US) Inc.
702 W. Algonquin Road
Arlington Heights, IL 60005
708-806-1485
Fax: 708-593-5062

Micro Crystal
Div. of ETA
CH-2540 Grenchen
Switzerland
065 51 82 82
Fax: 065 51 82 83

(Oscillators, VCOs, etc.)

Micro Electronics, Ltd.
38 Hung To Road
Kwun Tong P.O. Box 69477
Kowloon, Hong Kong
3 430181 6
Fax: 3 410321
TLX: 43510 MICROHX

(Diodes, thyristors, transistors, LEDs, linear ICs)

Micro International, Inc.
P.O. Box 218018
Nashville, TN 37221-8018
615-352-0700
Fax: 615-352-0780

(Multi-chip lid packages)

Micro Linear Corp.
2092 Concourse Dr.
San Jose, CA 95131
408-433-5200
Fax: 408-432-0295, 1627
E-mail: info@ulinear.com

(A fabless company supplying A/D and D/A converters, telecommunications ICs [such as token ring transceivers], data communications ICs [transceivers, ethernet, fiber optic and 10Base-T parts], SCSI terminators, disk drive ICs [such as servo chip sets, read data processors, disk voice coil servo driver, sensorless spindle motor controller, area and servo burst area detectors], power supply ICs [pulse width modulators, electronic ballast controllers, power factor controllers, DC/DC converters, voltage regulators], BiCMOS arrays, programmable filters, etc.)

Micro Networks
324 Clark Street
Worcester, MA 01606
508-852-5400
Fax: 508-853-8296, 508-852-8456
http://www.mnc.com
Email: sales@mnc.com

9330 L.B.J. Freeway, Suite 900
Dallas, TX 75243
214-437-1800
Fax: 214-680-3410
Fax: 714-261-6761

Unitrode S.R.L.
Via Del Carracci 5
20149 Milano, Italy
39 2 4800-7831
Fax: 39 2 4800-8014

Unitrode Electronics GmbH
Hauptstrasse 68
8025 Unterhaching
Germany
49 89 619-004
Fax: 49 89 617984

Unitrode (U.K.) Ltd.
6 Cresswell Park
Blackheath, London
SE3 9RD
United Kingdom
44 1 318 1431
Fax: 44 1 318-1431

Unitrode Electronics Asia Ltd.
Suite 1221 East Wing
New World Center
24 Salisbury Road
Kowloon, Hong Kong
852 3 722-1101
Fax: 852 369-7596

(A/D, D/A converters [including pin for pin replacements for some Burr-Brown military devices], track and hold amplifiers, data acquisition systems, amplifiers, V/F [voltage-frequency] converters, clock oscillators)

Micro Power Systems
See *Exar*

MicroLithics Corp.
See *Visicom Laboratories, Inc.*

MicroModule Systems, Inc.
10500-A Ridgeview Ct.
Cupertino, CA 95014-0736
408-864-7437
Fax: 408-864-5950

Eastern US and European Business office:
8 Stream Mill Hill Road
Brookline, NH 03033
603-672-8301
Fax: 603-673-4965

9 Paradise Road
Ipswich, MA 01938-1220
508-356-3845
Fax: 508-356-4719

(MCM-D interconnect substrates through standard and custom multichip modules and subsystems, including specific cache SRAMs, 80486 IC to upgrade 80386SX systems [clips on top of the 386 part], and a 80486 upgrade for a 80286 system [replaces the 286 microprocessor.] This company was formed in 1992 to acquire Digital Equipment Corporation's multichip module plant in Cupertino CA.)

Micro Oscillator, Inc.
1030 E. El Camino Real, Suite 209
Sunnyvale, CA 94087
408-984-8139
Fax: 408-984-8139

(Bare chip oscillators, and CMOS oscillators, such as the Micro Oscillator [tm])

Micron Semiconductor, Inc.
(An operating unit of Micron Technology Inc.)
8000 S. Federal Way
Boise, ID 83707-0006

MICRO NETWORKS — MICRON SEMICONDUCTOR

800-932-4992
(Customer comment line)
208-368-3900
Fax: 208-368-3342
(Customer comments), 4431

Internet: http//www.micron.com
E-mail: prodmktg@micron.com
(For an individual "username" @micron.com)
Data Fax: 208-368-5800, fax on demand that can be used to get data sheets and literature

Micron Quantum Devices, Inc.
Santa Clara, CA 208-368-3950

International Phone:
01 208-368-3410

Micron Europe Ltd.
Micron House
Wellington Business Park
Dukes Ride
Crowthorn, Berkshire RG45 6LS
44 1344 750750
Fax: 44 1344 750710

European International Response
Centre Fax: 44-732-741221

Micron Semiconductor
(Deutchland) GmbH
Sternstrasse 20, D-85609
Aschheim, Germany
49 89 904 8720, 49 89 9030021
Fax: 49 89 904 87250,
49 89 9043114

Micron Semiconductor Asia Pacific
Pte Ltd.
629 Aljunied Road #07-21
Cititech Industrial Building
Singapore 1438
65 841 4066
Fax: 65 841 4166

Syute 1010, 10th Floor

MICRON SEMICONDUCTOR — MICROSEMI

333 Keeling Road, Sec. 1
Taipei, 110 Taiwan ROC
886 2 757 6622
Fax: 886 2 757 6656

(Memory products such as flash memory, DRAMs and VRAMs, note: Military parts are now sold by Austin Semiconductor).

Micron Communications
Boise, ID
208-368-4000

(RF transceiver ICs)

microParts
Gesellschaft fur
Mikrostrukturtechnik mbH
Hauert 7 D-44227
Dortmund, Germany
49 231 9799 0
Fax: 49 231 9799 100

American Laubscher Corporation
516-694-5900
Fax: 516-293-0935

(Microspectrometer)

Micro-Rel
See *Medtronic Micro-Rel*

Micro-Precision Technologies, Inc.
12B Manor Parkway
Salem, NH 03079
603-893-7600
Fax: 603-893-9110

(Interface, linear ICs)

Micropride Ltd.
Unit 16 Shipyard International
Estate Brightlingsea
Essex C07 0AR, United Kingdom
020 630 4957
Fax: 020 630 4830

(Optoelectronics)

Micro Switch
(A Honeywell Div.)
11 W. Spring Street
Freeport, IL 61032
800-537-6945
815-235-6600
Fax: 815-235-6545

300 Winding Brook Drive
P.O. Box 5930
Glastonbury CT 06003
203-659-6472

574 Springfield Ave.
Westfield, NJ 07090
201-233-9200

(Various sensors [such as current sensors], industrial control switches and controls including Hall Effect devices, surface mount optoelectronic emitters and detectors for optical encoders.)

Micronas Oy Ltd.
Micronas Inc.
Kamreerintie 2
P.O. Box 51
Espoo SF02771, Finland
358 0 80521
Fax: 358 0 805 3213
TLX: 1000691

2, Stirling Road
Glenrothes, Fife KY66 2ST
England
44 1592 744 755
Fax: 44 1592 774 200

(Interface, linear, microprocessor support ICs, DACs, asynchronous to synchronous converters)

Micronetics (Wireless)
26 Hampshire Drive
Hudson, NH 03051
603-883-2900
Fax: 603-882-8987
E-mail: micrnet@aol.com

(Built in calibration modules [BICM] for wireless receivers and Noizeg [tm] series of broadband noise sources, diodes, VCOs [some parts developed by Qualcomm, Inc.])

Micropac Industries, Inc.
905 E. Walnut Street
Garland, TX 75040
214-272-3571
Fax: 214-494-2281

Optoelectronic Products Division
725 E. Walnut Street
Garland TX 75040
214-272-3571
Fax: 214-487-6918

Micropac Bremen
Ronzelestr .57
2800 Bremen-Horn
Germany
011 49 421 239716
Fax: 011 49 421 237637

(Optoelectronic products, LEDs, small signal hybrids, power hybrids, power op amps, solid state relays, DC/DC converters, microwave VCO's)

Microsemi Corporation
2830 S. Fairview Street
Santa Ana, CA 92704
714-979-8220
Fax: 714-557-5989
Telex: 4720306

8700 East Thomas Road
P.O. Box 1390
Scottsdale, AZ 85252
602-941-6300
Fax: 602-947-1503
TWX: 910-950-1320
(Previously Siemens, then Coors Components)

800 Hoyt St.
Broomfield, CO 80020
303-469-2161
Fax: 303-469-2161
(Formerly Coors Components)

MICROSEMI — MIKRON INSTRUMENT

580 Pleasant Street
Watertown, MA 02172
617-926-0404
Fax: 617-924-1235

(Semiconductor devices and LEDs, including military qualified parts. They also manufacturer transistor products previously provided by Raytheon Semiconductor)

MicroSignals Inc.
38-42 9th Street
Long Island City, NY 11101
718-937-7955
Fax: 718-729-9140

(RF components and subsystems)

Microtemp
(A trademark of Therm-O-Disk, a thermal cutoff)

MicroTouch Systems, Inc.
300 Griffin Park
Wilmington, MA 01844
508-694-9900
Fax: 508-659-9100

(Touch screen ASICs)

Microwave Diode Corp.
US Route 3
P.O. Box 250
West Stewartstown, NH 03597
603-246-3362
Fax: 603-246-8161

(Diodes, optoelectronics, thyristors)

Microwave Oscillator Corporation
1671 Dell Avenue
Suite 206
Campbell, CA 95008
408-379-7980
Fax: 408-379-7982

(Cavity oscillator/multiplier products including free running resonator oscillators, voltage tunable dielectric resonator oscillators, phase locked dielectric resonator oscillators, phase locked oscillators, etc. to GHz frequencies.)

Microwave Semiconductor Corp.
See *SGS-Thomson*

Microwave Signal, Inc. (MiSig)
22300 Comsat Drive
Clarksburg, MD 20871
301-428-4500
Fax: 301-540-8512

(This subsidiary of AMP, Inc. can design/supply MMIC chips, components and subsystems)

Microwave Solutions, Inc.
3200 Highlane Avenue
Suite 3A
National City, CA 91950
800-967-4267
Fax: 619-474-7003

(Microwave amplifiers, FET doublers, oscillators both free running and crystal types, solid state TWT replacements, down converters)

Microwave Technology, Inc.
4268 Solar Way
Freemont, CA 94538
510-651-6700
Fax: 510-651-2208

(Thyristors, linear ICs, microwave JFETs [trademarked SSTs, solid state triodes])

Midcom
P.O. Box 1330
121 Airport Drive
Watertown, SD 57201
800-643-2661
Fax: 605-886-4486
BBS: 605-882-0349

(Low profile data communication transformers)

Mietec Alcatel
10, Rue Latecoere
P.O. Box 57
Velizy, Cedex 78140
France
33 1 46 32 53 86
Fax: 33 1 46 32 55 68
TLX: 85739

(Interface, linear ICs)

Migatron Corporation
935 Dieckman Road
Woodstock, IL 60098
815-338-5800
Fax: 815-338-5803

(Ultrasonic sensors)

Mikroelektronik Dresden GmbH
Grenzstrasse 28
Dresden 01109
Germany
351 588 464
Fax: 351 593 3123

5020 Erfert
Juri-Gagarin Ring 154
Germany
057 0900
TLX: 61433

(SRAMs and microcontrollers)

Mikron Instrument Company, Inc.
16 Thornton Road
Oakland, NJ 07436
1-800-631-0176
201-405-0900
Fax: 201-405-0090

(Non-contact temperature measuring equipment, infrared temperature sensors)

MIL ELECTRONICS — MISTRAL SPA

Mil Electronics, Inc.
1 Chestnut Road
Nashua, NH 03060
603-882-3200
Fax: 603-881-8661

(Linear ICs)

Miller
See *J.W. Miller Magnetics*

Milliren Technologies, Inc. (MTI)
Two New Pasture Road
Newburyport, MA 01950
508-465-6064
Fax: 508-465-6637

(Precision crystal oscillators including VXCOs, TCXOs, OCXOs and miniature types)

MIM Electronics, Ltd.
Broadway, Dukinfield
Cheshire, SK16 4UU
United Kingdom
061 339 6028
Fax: 061 330 0944

(Microprocessor support ICs)

MIMA prefix
See *Motorola*

Minco Technology Labs, Inc.
1805 Rutherford Lane
Austin, TX 78754
512-834-2022
Fax: 512-837-6285

(This company packages and screens IC die, they also have die to support discontinued Motorola military products.)

Minelco Inc.
See *Electrodynamics, Inc.*

Mini-Circuits
P.O. Box 350166
Brooklyn, N.Y. 11235-0003
718-934-4500
Fax: 718-332-4661
http://www.minicircuits.com

North American Distribution Center:
800-654-7949
Fax: 417-335-5945

European Distribution Center:
Dale House, Wharf Road
Frimley Green
Camberley, Surrey GU16 6LF
England
44-252-835094
Fax: 44-252-837010

Germany:
MUNICOM GmbH
D-8217 Grassau, Germany
49 8641 5036

Italy:
Microelit S.P.A.
20146 Milan, Italy
39 2 4817900
00141 Rome, Italy
39 6 86894323

Japan:
Mini-Circuits Yokohama
Yokohama, 222 Japan
81 4 5545 1673

(Spdt switches, splitters, mixers, bias tees, VCOs, attenuators, transformers, etc.)

Mini-Systems, Inc. (MSI)
237 Enterprise Road
Deltona, FL 32725
407-574-0208
Fax: 407-574-8262

(Sunbelt Microelectronics Division, custom hybrid circuits, thick film resistors, and packaging)

168 E. Bacon Street
Plainville, MA 02762
508-695-2000
Fax: 508-695-8758

(Electronic Package Division, flatpack packages, chip carriers, and microwave packages)

45 Frank Mossberg Drive
P.O. Box 1659
Attleboro, MA 02703-0028
508-226-2111
Fax: 508-226-2211
E-mail; msithin@mini-systemsinc.com

(Thick/Thin Film Division, chip resistors, meets military QPL requirements, custom networks, MOS-type chip capacitors)

Minolta Camera Co. Ltd.
3-13, Azuchi-Machi 2-Chome,
Chuo-ku, Osaka-shi, Osaka 541
06 271 2251

(Instrument, camera, VCR ICs)

MIPS Technologies, Inc.
Mail Stop 410
2011 N. Shoreline Blvd.
Mountain View, CA 94043-1389
415-390-2573
http://www.mips.com

(RISC microprocessors)

MiSig
See *Microwave Signal, Inc.*

Mistral SpA
Via Dell'Irto 32
04013 Latina
Scalo, Italy
0773 3194 52, 0773 3195 71 2 3
TLX: 680009 MSTR

(Diodes, transistors)

MITECH — MJD

Mitech Corporation
347 Spence Avenue
Milpitas, CA 95035
Fax: 408-262-8770

(Addressable controllers)

Mitac Japan Corporation
1A, 5-7-14, Kita-Shinagawa,
Shinagawa-ku
Tokyo 141, Japan
81 3 5420 2822

(Disk drive control ICs)

Mitel Semiconductor
360 Leggett Dr.
P.O. Box 13320, or P.O. Box 13089
Kanata, Ontario, Canada K2K 1X3
800-267-6244
613-592-2122
Fax: 613-592-4784
TLX: 053-3221, 053-4596
TWX: 610-562-8529

http://www.semicon.mitel.com
http://www.semicon.mitel.com/ic_datasheets.html
http://www.semicon.mitel.com/distributors.html

2321 Morena Blvd, Suite M
San Diego, CA 92110
619-276-3421
Fax: 619-276-7348
TWX: 910-335-1242
Easylink: 62011928

2255 Crescent Drive
Mt. Dora, FL 32757
904-383-8877
Fax: 904-383-8822

Mitel Telecom Ltd.
Fabrikstrasse 17
D-7024 Filderstadt-4
Germany
National: 0711-77-015-22
International: 49-711-7701-522
Fax: 49-711-7701-524
TLX: 41-7255637

Mitel Telecom ltd.
Portskewett, Caldicot,
Near Newport,
Gwent NP6 4YR
National: 0291-43-00-00
International: 44-291-43-00-00
Fax: 0291-436389
TWX: 497360

(Telecommunications ICs such as: analog switch arrays, DTMF signaling components, Line and Trunk interfaces, modems and support equipment, custom hybrid devices, Subscriber Line and Network interfaces, Codec rate adaptors, trunk PLLs, ISDN cards, etc.)

Miteq
(Microwave Information
Transmission Equipment)
100 Davids Drive
Hauppauge, NY 11788
516-436-7400
Fax: 516-436-7430

(Microwave control products such as: solid state switches, attenuators, solid state phase shifters, vector modulators)

Mitsubishi Electronics America, Inc.
1050 E. Arques, Ave.
Sunnyvale, CA 94086
408-730-5900, 408-774-3189
408-522-7498, 7499
(Eastern U.S. and Canada)

5665 Plaza Drive, P.O. Box 6007
Cypress, CA 90630-0007
714-220-2500

Repairs: 800-446-6866
Tech Support and Parts:
800-344-6352
Sales: 800-843-2515
AutoFax: 714-236-6453
BBS: 714-236-6286
Spare Parts: Fax: 800-556-5015
Tech Support Fax: 714-236-6418
TSUPPORT@MSM.MEA.COM

Mitsubishi Electric Corp.
2-2-3 Marunouchi, Chiyoda-ku
Tokyo 100, Japan
81 3 3218 2111
Fax: 981 3 3218 3686,
981 3 3218 2431

(Gate arrays, microcontrollers, laser diodes, SRAMs, flash memory [some parts jointly developed with Hitachi], ASICs, CCD linear image sensors, disk drive control, speech synthesis and recognition, television, VCR, hybrid, logic ICs, MPEG ICs and chipsets, AM-LCDs, VRAMs, etc.)

Mitsubishi Cable America, Inc.
520 Madison Avenue
New York, NY 10022
1-800-262-6200 (outside NYS)
212-888-2270
Fax: 212-888-2276

(LEDs, discrete and chip)

Mitsumi Electric Co. Ltd.
8-8-2, Koyuryo-Cho
Chofu-Shi
Tokyo 182
Japan
03 3489 5333
Fax: 03 3488 3031, 03 3488 1228

(Hybrid and linear ICs; camera, car, CD player, television, VCR, and telecommunications ICs, voltage regulators)

MJ prefix
See *Harris Semiconductor*

(PNP transistors)

MJD prefix
See *Motorola* (Transistors),
Samsung (Transistors)

MJE — MOLECTRON DETECTOR

MJE prefix (Transistors)
See *Motorola, AVG Semiconductor* (Second source),
Advanced Semiconductor,
Central Semiconductor,
NAS Electronische Halbleiter GmbH, SGS-Thomson, Samsung, Selelab PLC, Space Power Electronics, Solid State, Inc.,
Semiconductor Technology, Inc.

MK prefix
(For parts formerly made by formerly Mostek)
See *SGS-Thomsom,*
AVG Semiconductors
(Second source some ICs),
Interpoint (DC/DC converters),
MicroClock (Timing products)

MKT prefix
(For parts formerly made by Mostek)
See *SGS-Thomson*
(Siemens for metallized film capacitors)

ML prefix
See *Micro Linear Corp., Motorola*

MLT prefix
See *Analog Devices*

MN prefix
See *Micro Networks* (A division of Unitrode), *Matsushita* (Panasonic), *Micronetics* (Crystal oscillators. Note: devices not equivalent)

MM prefix
See *National Semiconductor*
(If MM[number][letter]-XXXXX-XXXXX see Matra-MHS),
Analog Systems, ATI Technologies (MPEG ICs)

MMAD prefix
See *Motorola*

MMBT prefix (Transistors)
See *ROHM Electronics Division, Motorola Semiconductor Products*

MMBTA prefix (Transistors)
See *ROHM Electronics Division, Motorola Semiconductor Products*

MMBTH prefix (Transistors)
See *ROHM Electronics Division, Motorola Semiconductor Products*

MMBV prefix (Transistors)
See *ROHM Electronics Division, Motorola Semiconductor Products*

MMC Electronics America, Inc.
(Subsidiary of Mitsubishi Materials)
4080 Winnetka Avenue
Rolling Meadows, IL 60008
708-577-0200
Fax: 708-577-0201

(Surge absorbers)

MMFT prefix
See *Motorola*

MMG-North America
126 Pennsylvania Avenue
Patterson, NJ 07053-251'2
201-345-8900

(Ferrite products including beads, cores and surface mount ferrites)

MMI (Monolithic Memories, Inc.)
See *AMD*
(Advanced Micro Devices)

MMPQ prefix
See *Motorola*

MMS-part number prefix
See *MicroModule Systems, Inc.*

MMS prefix
See *Motorola*

MMST prefix (Transistors)
See *Motorola Semiconductor Products*

MMST prefix (Transistors)
See *ROHM Electronics Division*

MMSV prefix
See *Motorola*

MMUN prefix
See *Motorola*

MOC prefix
See *Motorola* (Solid state relay), *Siemens*

Modco
55 Freeport Blvd. #28
Sparks, NV 89431
1-800-OSCILLATOR
702-331-2442
Fax: 702-331-6266

(Miniature surface mount VCOs)

Modular Devices, Inc. (MDI)
One Roned Road
Shirley, N.Y. 11967
1-800-333-7697
(1-800-333-POWR)
516-345-3100, Fax: 516-345-3106

(DC/DC converters, military, space and high reliability units available)

Modular Power Systems
See *Transistor Devices, Inc.*

Modupower
See *Semtech*

MOI
See *Micro Oscillator, Inc.*

Molectron Detector, Inc.
7470 S. W. Bridgeport Road
Portland, OR 97224
800-336-4340, 503-620-9069
Fax: 503 620-8964
TLX: 5106002976

MOLECTRON DETECTOR — MOSPEC

(Pyroelectric hybrid detectors)

Monitor Products Co., Inc.
502 Via Del Monte, P.O. Box 92828
Oceanside, CA 92054
619-433-4150
Fax: 800-434-0255

(Crystals, crystal oscillators, VCXOs, OCXSs, TCXOs, TCVCXOs)

Monolithic Memories, Inc.
See *AMD, Advanced Micro Devices*

Monolithic Sensors, Inc.
2800 W. Golf Road
Rolling Meadows, IL 60008
800-860-1116 ext. 816
708-871-3917, 708-437-8090
Fax: 708-437-8144

(Silicon capacitive pressure sensors)

Monitor Products Company, Inc.
502 Via Del Monte
Oceanside, CA 92054
619-433-4510
Fax: 619-434-0255

(Frequency control devices, oscillators)

Monsanto Optoelectronics
See *Quality Technologies*

Morgan Matroc, Inc.
Electro Ceramics Division
232 Forbes Road
Bedford, OH 44146
216-232-8600
Fax: 216-232-8731

(Piezoelectric sensors and assemblies)

Morrihan International Corp.
8-5 Fl., No. 57
Fou Shin North Road
Taipei, Taiwan
886 2 752 2200
Fax: 886 2 741 4690
TLX: 20422 MORRIHAN

(Optoelectronics, transistors, interface, linear ICs)

Mosaic Semiconductor, Inc.
7420 Carroll Road
San Diego, CA 92121
619-271-4565
Fax: 619-271-6058

(Memory products, note: military qualified product no longer available)

Mosaid Technologies, Inc.
2171 McGee Side Road
P.O. Box 13579
Kanata, Ontario K2K 1X6, Canada
613-836-3134
Fax: 613-831-0796

(A fabless company supplying advanced chips for memory and ASIC applications and manufactures test systems for memory ICs.

Mosel-Vitelic, Inc.
1 R&D Road 1
(No. 1, Creation Rd. 1)
Science based Industrial Park
Hsin Chu, Taiwan, R.O.C.
886 35 770055, 886 35 783 344,
886 2 545 1213
Fax: 886 35 772788,
886 35 792 860

3910 North First Street
San Jose, CA 95134
408-433-6000
Fax: 408-433-0952
http://www.moselvitelic.com

Northeastern Area:
71 Splitbrook Road, #306
Nashua, NH 03062
603-891-2007
Fax: 603-891-3597

Central and Southeastern Area:
604 Fieldwood Circle
Richardson, TX 75081
214-690-1402
Fax: 214-690-0341

Southwestern Area:
5150 E. Pacific Coast Highway, #200
Long Beach, CA 90904
310-498-3314
Fax: 310-597-2174

Hong Kong:
19 Dai Fu Street
Taipo Industrial Estate
Taipo, NT, Hong Kong
011 852 665 4883
Fax: 011 852 664 7535

Japan:
Rm. 302, Annex-G
Higashi-Nakano, Nakano-ku
Tokyo 164, Japan
011 81 3 3365 2851
Fax: 011 81 3 3365 2836

Europe:
C/O AMT
Saville Court
11 Saville Place
Clifton, BS8 4EJ, England
44 272-237594
Fax: 44 272-237598

(Memory ICs including VRAMs and DRAMs)

Mosis
4676 Admiralty Way
Marina del Rey, CA 90262
310-822-1511, Fax: 310-823-5624
URL: http:/www.isi.edu/mosis
E-mail: mosis-help@misis.edu

(IC and MCM fabrication and prototypes)

Mospec Semiconductor Corp.
2Fl-5, No. 620, Tun Hwa S. Road
Taipei 10661, Taiwan
886 2 741 6464
Fax: 886 2 741 6460
TLX: 12200 PECORPTE

MOSPEC — MOTOROLA

(Diodes, transistors)

Mostek
(Formerly Thomson-Mostek and now SGS-Thomsom)

MOST
See *MoSys*

Motorola, Inc.
Semiconductor Products Sector
3102 N. 56th Street
Phoenix, AZ 85018

Motorola Design-
NET FAX Services
5005 East McDowell
Phoenix, AZ 85008
602-244-6591
(Automated fax data)
Fax: 602-244-6693

Technical Resources Hotline:
1-800-521-6274
1-800-845-MOTO in the U.S.
(Power PC products)
512-343-8940 outside the U.S.
(Power PC products)
1-800-441-2447 (Literature only)
FaxBack: 1-800-774-1848
(For data sheets, see below)

Databook sheets and application notes: http://design-net.com

68K, PowerPC, Coldfire and FlexCore products:
1-800-248-8567
(Embedded systems hotline)
512-891-8775 (Aesop questions)

Email:
aesop_support@pirs.aus.sps.mot.com
Modem: 512-891-3650
(Embedded systems division)
1-800-843-3451

n/8/1/f
Data rate less than 14,400 bps
Terminal emulation VT100
Login: pirs

Mfax (tm), to order any of over 25k documents via Fax call:
1- 800-774-1848
(U.S. and Canada), 602-244-6609
Mfax assistance: 602-244-6591
Mfax (tm) Email:
RMFAX0@email.sps.mot.com

If have Internet and graphic server (like netscape): URL:
http://motserv.indirect.com
http://part.net/partnet
http://freeware.aus.sps.mot.com.freeweb/index.html
http://motserv.indirect.com/home2/sales_office.html
http://www.mot.com/
http://design-net.com
(OEM price Book and master Selection Guide)

http://design-net.com/cisc/CISC_home.html (68HC05 family of Customer Specified Integrated Circuits)
RMFAXO@email.sps.mot.com
(Static rams)
http://www.mot.com/FastSRAMs
(static rams)
Internet Access:
Telnet: pirs,aus.sps.mot.com
(login: pirs)
WWW: http://pirs,aus.sps.mot.com/aesop/hmpg.html

Query by Email:
aesop_query@pirs.aus.sps.mot.com
(Type "HELP" in the text body.)

Literature Distribution Centers:
Motorola Literature Distribution
P.O. Box 20912
Phoenix, AZ 85036
800-441-2447
602-303-5454, 602-994-6561
Fax: 602-994-6430

European Literature Distribution Centre:
Motorola Ltd.
88 Tanners Drive
Blakelands, Milton Keynes
MK14 5BP, England

Japan Literature Distribution Center:
Nippon Motorola Ltd.
4-32-1 Nishi-Gotanda,
Shinagawa-ku,
Tokyo 141 Japan
81 3 3440 3311

Asia Pacific Literature Distribution Center:
Motorola Semiconductors H.K. Ltd.
Silicon Harbour Center
No. 2, Dai King Street
Tai Po Industrial Estate
Tai Po
N.T. Hong Kong

Other Offices:
Chandler, AZ
602-814-4157
Motorola DSP Operation
Austin, TX
512-891-2030

Motorola Microprocessor and Memory Technologies Group
6501 William Cannon Dr. W.
Austin, TX 78735-8598
800-845-6686

CISC Division
P.O. Box 52073
M/D 56-102
Phoenix, AZ 85072
512-891-2035

Analog IC Division
2100 E. Elliot Road
Tempe, AZ 85284
602-413-3615

MOTOROLA

Advanced Microcontroller Div.
6501 William Cannon Drive
W. Austin, TX 78735
512-891-3255, 2035
Fax: 512-891-2652

MOS Memory Products:
512-322-8832
SRAM's
Box 52073
MD 56-102
Phoenix, AZ 85072
1-800-347 MITO (6686)
512-928-7726

Gate Arrays:
Chandler AZ
602-821-4172

P.O. Box 1466
Austin, TX 78767
Logic IC Division
Mesa, AZ
602-962-3410

Sales Offices:
Alabama, Huntsville:
205-464-6800
Alaska: 800-635-8291
Arizona, Tempe: 602-897-5056

California:
Agoura Hills: 818-706-1929
Los Angeles: 310-417-8848
Irvine: 714-753-7360
Sacramento: 916-922-7152
San Diego: 619-541-2163
Sunnyvale: 408-749-0510

Colorado:
Colorado Springs: 719-599-7497
Denver: 303-337-3434

Connecticut, Wallingford:
203-949-4100

Florida:
Maitland: 407-628-2636
Pompano Beach/Ft. Lauderdale:
305-486-9776
Clearwater: 813-538-7750

Georgia, Atlanta: 404-729-7100
Idaho, Boise: 208-323-9413

Illinois:
Chicago/Hoffman Estates:
708-490-9500
Shaumburg: 708-413-2500

Indiana:
Ft. Wayne: 219-436-5818
Indianapolis: 317-571-0400
Kokomo: 317-457-6634

Iowa, Cedar Rapids: 319-373-1328
Kansas, Kansas City/Mission:
913-451-8555
Maryland, Columbia:
410-381-1570
Massachusetts, Marlborough:
508-481-8100
Woburn: 617-932-9700
Michigan, Detroit: 313-347-6800
Minnesota, Minnetonka:
612-932-1500
Missouri, St. Louis: 314-275-7380

New Jersey:
(Fairfield)
100 Passaic Ave
Fairfield NJ 07006
201-808-2400,
Fax: 201-808-2411

New York:
Fairport: 716-425-4000
Hauppauge: 516-361-7000
Fishkill: 914-896-0511

North Carolina, Raleigh:
919-870-4355

Ohio:
Cleveland: 216-349-3100
Columbus/Worthington:
614-431-8492
Dayton: 513-495-6800

Oklahoma, Tulsa: 800-544-9496
Oregon, Portland: 503-641-3681
Pennsylvania, Colmar:
215-997-1020

Philadelphia/Horsham:
215-957-4100
Tennessee, Knoxville:
615-690-5593

Texas:
Austin: 512-873-2000
Houston: 800-343-2692
Plano: 214-516-5100
Seguin: 210-372-7620

Virginia, Richmond: 804-285-2100
Washington, Bellevue:
206-454-4160
Seattle Access: 206-622-9960
Wisconsin, Milwaukee/Brookfield:
414-792-0122

Canada:
British Columbia, Vancouver:
604-293-7605
Ontario, Toronto: 416-497-8181
Ottawa: 613-226-3491
Quebec, Montreal: 514-731-6881

International Sales Offices:
Australia:
Melbourne: 61-3 887-0711
Sydney: 61 2 906-3855

Brazil, Sao Paulo: 55 11 815-4200
China, Beijing: 86 505 2180

Finland:
Helsinki: 358 0 693 58 40
Car phone: 35 8 49 211501

France Paris/Vanves:
33 1 40 955 900

Germany:
Langenhagen/Hannover:
49 511 789911
Munich: 49 89 92103-0
Nurenberg: 49 911 64-3044
Sindelfingen: 49 7031 69 910
Wiesbaden: 49 611 761921

MOTOROLA — MPSH

Hong Kong:
Kwai Fong: 852 6106888
Tai Po: 852 6668333

Motorola Semiconductors H.K., Ltd.
8B Tai Ping Industrial Park,
51 Ting Kok Road, Tai Po, N.T.
Hong Kong
852 26629298

India, Bangalore 91-812 627094
Israel, Herzlia: 972 9 590222
Italy, Milan 39 2 82201

Japan:
Fukuoka: 81 92 725 7583
Gotanda: 81 3 5487 8311
Nagoya: 81 52 232 3500
Osaka: 81 6 305 1802
Sendai 81 22 268-4333
Takamatsu: 81 878 37 9972
Tokyo: 81 03 3440-3311

Nippon Motorola Ltd.
3-20-1, Minami-Azabu,
Minato-ku, Tokyo 106
03 3440 3311
Fax: 03 3440 3505

Korea:
Pusan 82 51 4635-035
Seoul 82 2 554-5118

Malaysia, Penang 60 4 374514

Mexico:
Mexico City 52 5 207-7880, 7727
Guadalajara: 52 36 21-89-77,
21-91-29
Marketing: 52 36 21 8977
Customer Service: 52 36 669 9160

Netherlands, Best: 31 4998 612 11
Puerto Rico, San Juan:
809-793-2170
Singapore 65 2945438
Spain, Madrid 34 1 457-8204, 8254
Sweden, Solna 46 8 734-8800

Switzerland:
Geneva 41 22 799 11 11
Zurich 41 1 730-4074

Taiwan, Taipei: 886 2 717-7089
Thailand, Bangkok: 66-2 254-4910
United Kingdom, Aylesbury:
44 296 395-252

(This vendor offers a large variety of products including: mixed signal and programmable ASICs, FPGAs, cross point switches, SRAMs [including burst types], CATV amplifiers, digital signal processor ICs [DSPs], Full-Motion Video Decoders, set-top video ICs, microcontrollers, microcomputer products; TTL, ECL and CMOS circuits; analog and interface ICs including op amps, comparators, voltage regulators and references, voltage monitors, crowbar circuits, zero voltage switches, motor controller ICs, A/D and D/A converters, Sigma-Delta converters, ethernet transceivers, line drivers and receivers, telecommunication ICs, power and signal ICs, GaAs rectifiers, crystal oscillators, AM stereo decoders, video circuit ICs [including set-top box ICs], automotive electronic circuits, timing circuits, smoke detector ICs; bipolar JFET, and TMOS FET transistors; tuning [abrupt and hyper abrupt], signal, hot carrier, multiple switching, rectifier, and zener diodes; thyristors, SIDACs, PUTs, optoelectronics, pressure sensors, RF ICs, etc. They also do some foundry work and MCM modules. Note: An announcement that the Military Qualified Product line was being discontinued was made in October 1994, phasing out in 18 months. The Military Bipolar power transistor line was sold to Omnirel in mid 1995.)

MP prefix
See *Micro Power Systems, Inc.*

MPA10XX part number
See *Motorola, Inc.*

(FPGAs)

MPC prefix
See *Burr-Brown* (Multiplexer),
Motorola (Microprocessors),
Monitor Products Company, Inc.
(Crystals and crystal oscillators)

MPD prefix
See *Micrel, Inc., Siemens*

MPF prefix
See *Motorola*

MPIC prefix
See *Motorola Semiconductor*

MPL
50 W. Hoover Ave.
Mesa, AZ 85210
800-443-7722
Fax: 602-962-5750

Tafernstrasse 20
5405 Dattwil
Switzerland
41-56 83 30 80
Fax: 41-56 83 30 20

(PCMCIA memory card readers, uses their custom devices)

MPN prefix
See *Motorola*

MPQ prefix
See *Motorola,
Central Semiconductor*

MPS prefix
See *MOS Technology, Motorola, Zetex, AVG Semiconductor*
(Second source)

MPSH prefix
See *Motorola*

MPSW — MSK CORPORATION

MPSW prefix
See *Motorola*

M-Pulse Microwave
576 Charcot Avenue
San Jose, CA 95131
408-432-1480
Fax: 408-432-3440

(Microwave semiconductors including diodes, transistors, etc.)

MPX prefix
See *Motorola*
(Sensor Products Division)

MQ prefix
See *Intel Corp.*

MR prefix
See *Medtronic Micro-Rel*,
Harris Semiconductor
(Custom part), *Intel*, *LSI Logic*
(Microprocessors),
Analog Systems,
Motorola Semiconductor (Standard and fast recovery rectifiers)

MRFA prefix
See *Motorola*

MRFIC prefix
See *Motorola*

MRG prefix
See *Micro-Rel*

MRV Communications, Inc.
8943 Fullbright Ave.
Chatsworth, CA 91311
818-773-9044
Fax: 818-773-0261

Eastern Region:
7 Laurel Street
Woburn, MA 01801
617-932-6260
Fax: 617-932-6260

Western Regional Office:
11160 Jolyville Rd. #1603
Austin, TX 78759
512-338-0066
Fax: 512-338-0067

European Sales Office:
Weinstrasse 33/35, D-722202
Nagold, Germany
49 7452 61226
Fax: 49 7452 61200

(Uncooled DFB lasers)

MS prefix
(MOS Electronic Corporation)
See *Mosel-Vitelic*,
Mitel Semiconductor, *Zetex*,
Systronix, Inc.

M.S. Kennedy Corporation (MSK)
8170 Thompson Road
Cicero, NY 13039-9393
315-699-9201
Fax: 315-699-8023

(Power operational amplifiers, video amplifiers, motor control/bridge drivers, MIL-STD-1772 qualified hybrids)

MSA prefix
See *Motorola*

MSB prefix
See *Motorola*

MSC prefix
See *Motorola*, *GEC Plessy*

(Smart power IC to meet standards for electrical equipment drawing near sinusoidal current in phase with the line voltage)

MSD prefix
See *Motorola*, *Siemens*

MSH prefix
See *Datel, Inc.*

MSI logo
See *MSI Electronics Inc.*,
Microwave Solutions, Inc.
(Also see Mini-Systems, Inc. for custom devices)

MSI Electronics, Inc.
3100 47th Avenue
LIC, NY 11101-3068
718-937-3330

(Varactor diodes)

MSIS Semiconductor
(Formerly MSI/Scorpion Semiconductor)
1999 Concourse Drive
San Jose, CA 95131
408-944-6270, 6271
Fax: 408-944-6272

(This company produces the full line of P-Channel Silicon gate MOS technology products formerly supplied by AMD [Advanced Micro Devices]. They offer products and design services in N-Channel and CMOS process technology. There is a minimum order policy of $500 for commercial parts and $1000 for military parts)

MSK logo
See *M.S. Kennedy Corp.*

MSK prefix
See *M.S. Kennedy Corp.*

MSK Corporation
To-kan No. 2 Castle, Rm 1011
5-12, Nishi-Shinjuku 3-Chome
Shinjuku-ku, Tokyo 160
Japan
81 3 3342 3838

(Logic, memory, linear, hybrid, digital video, watch and clock ICs)

MSM — MURATA

MSM prefix
See *Oki Semiconductor*

MSP prefix
See *Texas Instruments*

M-Systems
4655 Old Ironsides Drive
Santa Clara, CA 95054-1808
408-654-5820, Fax: 408-654-9107
Europe: 31 10 262 1144
Fax: 31 10 462 4855
Israel: 972 3 647 7776
Fax: 972 3 647 6668

(High density flash memory and flash disks)

MT prefix
See *Micron Technology* (and Austin Semiconductor who now manufactures the military product line of Micron Technology), *Mitsubishi*, *Mitel*, *Intel*

MTA prefix
See *Microchip Technology*

MTB prefix
See *Motorola Inc.*

MTE prefix
See *Microchip Technology*

MtH prefix
See *Metalink Transmission Devices, Ltd.*

MTI
See *Milliren Technologies, Inc.*

MTP prefix
See *AVG Semiconductors*

(Second source transistors)

MTR prefix
See *Interpoint Corporation*

M-tron Industries, Inc.
P.O. Box 630
Yankton, SD 57078
1-800-762-8800
Fax: 605-665-1709

(Quartz crystals, quartz crystal oscillators and special hybrid products)

MU prefix
See *Music Semiconductors*

Multice and Company, Ltd.
P.O. Box 47-91
Taipei, Taiwan, R.O.C.
886-2-812-3313-5
Fax: 886-2-812-3316
Fty Tel: 886-6-272-1725/6

(Speakers, buzzers, transducers)

Multichip Technology
(A subsidiary of Cypress Semiconductor, products marketed under the Cypress Semiconductor name)

MUN prefix
See *Motorola*

(Bias resistor transistors)

MuP prefix
See *Ultra Technology*

MUR prefix
See *Motorola Semiconductor*

(Ultra fast rectifiers)

Murata Mfg. Company
(Headquarters)
2-26-10, Tenjin
Nagaokakyo City,
Kyoto Prefecture 617,
Japan
81 075-951-9111
Fax: 81 075-954-7720

International Division
Yokohama-shi,
Kanagawa 226
Japan
81 45 931 7111
Fax: 81 45 931 7105

Murata/Erie North America
2200 Lake Park Drive
Smyrna GA 30080
1-800-831-9172
1-800-241-6574
(For technical questions on EMI filters)
404-436-1300
Fax: 404-436-3030
Telex: 26-1301

27 Concord Street
El Paso, TX 79906
915-778-6606
Fax: 915-778-0217

Al Santos
1,893-6 Andar-Conjunto 62
CEP 01419 Sao Paulo, SP Brazil
55 11 287 7188/4232
Fax: 55 11 289 8310

Av Buriti 7040 Distrito Industrial
Manaus-Am, Postal 413,
CEP 69075, Brazil
55 92 615 2112/2114/2145/2160
Fax: 55 92 615 1625

Holbeinstrasse 21-23,
8500 Nurnberg 70, Germany
49 911 66870
Fax: 49 911 6687193/6687288

Parc Technologique 18, 22,
Avenue
Edouard Herriot 92356
Le Plessis Robinson Cedex France
33 1 4094 8300
Fax 33 1 4094 0154

Via Sancarlo, 1 20040 Caponago,
Milano Italy
39 2 95743000
Fax: 39 2 95740168/95742292

MURATA — MX-COM

Oak House, Ancells Road,
Ancells Business Park,
Fleet Aldershot Hampshire,
GU13 8UN United Kingdom
44 252 811666
Fax: 44 252 811777

Plantenweg 69-71 2132 HM
Hoofdorp, N. Amsterdam,
the Netherlands
31 2503 21506
Fax: 31 2503 21306

Singapore:
65 758 4233, Fax: 65 753 6181
Malaysia:
60 3 293 5227, Fax: 60 3 293 8018
Taiwan:
886 2 562 4218,
Fax: 886 2 536 6721
Thailand: 66 2 236 3512/3491,
Fax: 66 2 238 4873
Hong Kong: 852 7822618,
Fax: 852 7821545
Korea: 82 2 730 7605/7321,
Fax: 82 2 739 5483
China: 86 1 465 1303,
Fax: 86 1 465 1305

(GaAs Hall Effect devices, infrared and semiconductive magnetic [angle, magnetorestive, rotation and rotational angle] sensors, ceramic capacitors, ceramic resonators, ceramic and EMI/RFI filters, crystal oscillators, miniature coaxial connectors, filtered "D" connectors, current detectors [for Fax and Modems], SCSI switching terminators, DC/DC converters, delay lines, piezoelectric elements/alarms, potentiometers, resistor networks, thermistors, hybrid ICs, pyroelectric infrared sensors, temperature sensors, airflow sensors, ultrasonic sensors, etc.)

MuRata Electronics
RF/Microwave Division
1900 West College Avenue
State College, PA 16801

(Ovenized crystal oscillators)

Music Semiconductors
Far East Headquarters:
12th Floor, First Bank Building
8737 Paseo de Roxas
Makati, Metro Manila
Philippines
63 2 815 8536 ext. 6201
Fax: 63 2 813 1744
http://www.music.com/music

European Headquarters:
Music Semiconductors, NV
Kleinstraat 1-D
6422 PS Heelen
The Netherlands
31 45 417580
Fax: 31 45 428391

Korea: 82 2 552 3773

USA Headquarters:
Music Semiconductors, Inc.
1150 Academy Park Loop, #202
Colorado Springs, CO 80910
800-933-1550, 800-788-6874
719-570-1550
Fax: 719-570-1555
E-mail: info@music.com

P.O. Box 415
234 West Mill Road
Long Valley, NJ 07853
908-876-9691
Fax: 908-876-9542

1136 E. Hamilton Ave.
Suite 203
Campbell, CA 95008
408-371-3993
Fax: 408-371-0878

(Video processing, timing and programmable logic ICs, content addressable memory [some designed for LAN applications], graphics and direct color palettes some with synthesizers, they also second source some Oak Technology, Inc. graphic chips, token ring source route [SRT], FDDI and Ethernet bridge interfaces)

MV prefix
See *Datel, Inc.*, *GEC Plessy*, *Motorola*

MVA prefix
See *Aureal Semiconductor* (MediaVision, Inc.)

MVAM prefix
See *Motorola*

MVD prefix
See *Aureal Semiconductor* (MediaVision, Inc.), *Datel, Inc.*

MVV Prefix
See *Aureal Semiconductor* (MediaVision, Inc.)

Mwave
See *IBM Microelectronics*

(DSP IC)

MWD prefix
See *LeCroy Corporation*

(Proprietary sampler and analog storage device in their DSO, digital storage oscilloscopes)

MX logo
See *Macronix, Inc.*

MX prefix
See *Maxim Integrated Products*, *MX-Com Inc.*, *Macronix Inc.*, *Datel, Inc.*, *SZE Microelectronics GmbH*

MX-Com Inc.
4800 Bethania Station
Winston-Salem, NC 27105-1201
800-638-5577
910-744-5050
Fax: 910-744-5054
TLX: 62892598, 5101012852

M - N   MX-COM — NATIONAL HYBRID

(Telecommunication ICs including MODEM ICs and mixed signal ICs for tone signaling, secure speech, voice storage, data transmission and voice filtering, GMSK wireless data modems, CVSD codecs, digital control amplifiers, telephone and signal processing ICs, etc.)

MXD prefix
See *Maxim Integrated Products, Inc.*

(Silicon delay lines)

MYE prefix (Transistors)
See *AVG Semiconductor*
(Second source)

Mykotronx, Inc.
357 Van Ness Way
Torrence, CA 90501-1483
310-533-8100

(Clipper encryption ICs and electronic security hardware was acquired by Rainbow Technologies, Inc. [Irvine, CA] in early 1995)

MZ prefix
See *Intel*

N logo (for National)
See *Matsushita* (Panasonic)

N prefix
See *Philips Semiconductors*

(Signetics)

NAC prefix
See *Lucas NovaSensor*

NAiS
(Trademark of Aromat Corporation)

Nan Ya Technology Corporation
Nan Ya Plastics Corporation
(Subsidiary Formosa Plastics Group)
3 Fl., No. 201
Tung Hwa N. Rd.
Taipei, Taiwan R.O.C.
886 2 712 2211
Fax:
886 2 712 9211, 886 2 719 7413

(A 1995 start-up company that manufactures LCD displays, and DRAMS with technology licensed from Oki Semiconductor)

Nanmac Corporation
9-11 Mayhew Street
Framingham Ctr., MA 01701
800-758-3932
508-872-4811
Fax: 508-879-5450

(Thermocouples)

Nano-Pulse Industries
440 Nibus Street
P.O. Box 9398
Brea, CA 92622
714-529-2600
Fax: 714-671-7919
Hong Kong: 852-329-7688,
Fax: 852-352-3272

(DC/DC converters and data networking transformers and filter products)

Nanotec Pty Ltd.
P.O. Box 7991
Hennopsneer 0046
South Africa
Fax: 2712 665 1343

Nanotec Europe
5a Southlands
Aston OX18 2DA
England
Fax: 44 1993 851 923

(Encoder/decoders, token ring controllers)

NAP-XX part number
See *F.W. Bell*

(Hall effect current sensor)

Natel Engineering Co.
9340 Owensmouth Avenue
Chattsworth, CA 91311
818-734-6500

(High power density DC/DC converters. Note: Natel sold their converters including resolver to digital, digital-synchro, synchro-digital, SINE to COSINE; angle simulators to ILC Data Device Corporation in 1995)

National
See *Matsushita* (Panasonic)

National Electronics Institute
Bandung, Indonesia

(CMOS wafer fabrication plant started operation in 1994)

National Hybrid, Inc.
2200 Smithtown Avenue
Ronkonkoma, NY 11779
516-981-2400
Fax: 516-981-2445, 8888
(Sales Fax)
TWX: 910-380-4391

(Custom thick and thin film hybrid circuits, power hybrids, power GaAs FET amplifiers, active filters, dual transceivers, dual redundant remote transceivers, Enhanced controllers with bus controllers, remote terminal, bus monitor, SRAM and transceivers; Advanced terminals with dual redundant RISC processor based bus controller, remote terminal, bus monitor and internal RAM, Devices for MIL-STD-1553/1760 [MAC AIR data bus systems], this company is qualified to MIL-STD-1772)

NATIONAL INSTRUMENTS — NATIONAL SEMI.

National Instruments
6504 Bridge Point Parkway
Austin, TX 78730-5039
800-433-3488
512-794-0100
Fax: 512-794-8411
E-mail: info@natinst.com
http://www.natinst.com/

Australia: 03 879 9422
Austria: 0662 435986
Belgium: 02 757 00 20
Canada: 800-433-3488,
519 622 9310
Denmark: 45 76 26 00
Finland: 90 527 2321
France: 1 48 65 33 70
Germany: 089 714 50 93
Italy: 02 48301892
Japan: 03 3788 1921
Mexico: 95 800 010 0793
Netherlands: 01720 45761
Norway: 32 848400
Spain: 91 640 0085
Sweden: 08 730 49 70
Switzerland: 056 27 00 20
United Kingdom: 0635 523545

(GPIB [General Purpose Instrumentation Bus] ICs [IEEE 488.2 Talker/Listener Interface] and interface circuit cards)

National Semiconductor Corporation
2900 Semiconductor Drive
P.O. Box 58090
Santa Clara, CA 95052-8090
1-800-272-9959
(Hotline, Arlington, TX)
  (Extension 626 for MCMs)

1-800-642-3827 (Service center)
408-721-5000
408-721-3884
408-721-2636 (DSP IC's)
Cable: NATSEMICON
TLX: 346353
TWX: 910-339-9240
http://www.national.com
http://www.national.com/see/PS/PowerSOT.html
(DMOS power MOSFETs)
E-mail: cjgfsc@tevm2.nsc.com

1090 Kiefer Road
Sunnyvale, CA 94086-3737
408-721-5000
Fax: 408-245-9655

MultiChip Module Business Unit:
1090 Kiefer Road
M/S 16-184
P.O. Box 58090
Santa Clara, CA 95052-8090
408-721-7297
Fax: 408-721-8078

Customer Support Center-Information Hotline:
800-272-9959
Fax: 800-462-9672

1111 W. Bardin Road
Arlington, TX 76017

Hard copy data mailed from:
319 Glenn Street
Crawfordsville, IN 47933
317-364-3934

Australia:
Suite #4, Level 5
3 Thomas Holt Drive
North Ryde, N.S.W. 2113
Sydney, Australia
02-887-4355
Telex: AA 27173
Fax: 02-805-0298

Bldg 16, Business Park Dr.
Melbourne, VIC 3168
Australia
3-558-9999
Fax: 3-558-9998

Belgium:
Vorstiaan 100
B-1170 Brussels, Belgium
02-661-0680
Fax: 02-660-2395

Brazil:
Av. Brig. Faria Lima, 1409
6 Andar
Cep-014515, Paulistano
Sao Paulo, SP Brasil
55-11 212-5066
Telex: 391-1131931 NSBR BR
Fax: 55-11 212 1181

Canada:
5925 Airport Road
Suite 615
Mississauga, Ontario L4V 1W1
416-678-2920
Fax: 416-678-2535

Denmark:
Ringager 4A, 3
DK-2605 Brandby
Denmark
43 43-32-11
Telex: 15179
Fax: 43 43-31-11

Finland:
Kaupparkartanonkatu 7 A22
SF-00930 Helsinki
Finland
90 33-80-33
Telex: 126116
Fax: 90 33-81-30

France:
Centre d'Affaires La Boursidiere
Batiment Champagne
BP 90
Route Nationale 186
F-92357 Le Plessis Robinson
Paris, France
01 40-94-88-88
Telex: 631065
Fax: 01 40-94-88-11

Hong Kong:
852 2737 1600

NATIONAL SEMICONDUCTOR

Ireland:
Unit 2A
Clonskeagh Square
Clonskeagh Road
Dublin 14 Ireland
01 69-55-89
Telex: 91047
Fax: 01 830650

Italy:
Strada 7, Pallazo R/3
I-20089 Rozzano-Milanofiori
Italy
02 57500300
Telex: 352647
Fax: 02 57500400

Japan:
Sanseido Bldg 5F
4-15, Nishi-Shinjuku, Shinjuku-ku
Tokyo, Japan 160
81 43 299-7001, 81 43 299 2300
Fax: 81 43 299-7000

Korea:
13th Floor, Dai Han
Liffe Insurance 63 Building
60, Yoido-Dong, Youngdeungpo-Ku
Seoul, Korea 150-763
02 7848051/3
Fax: 02 784-8054
TLX: 24942 NSRKLO

Mexico:
Electronica NSC de Mexico SA
Juventino Rosas No. 118-2
Col Guadalupe Inn
Mexico, 01020 D.F. Mexico
905-524-9402
Fax: 905-524-9342

Netherlands:
Pampus Alaan 1
1382 JM Weesp, The Netherlands
0 29 40 3-04-48
Fax: 0 29 40 3-04-30
Telex: 10956

Norway:
National Semiconductor U.K. Ltd.
P.O. Box 57, Isveien 45
N-1393 Ostenstad, Norway
2 79-6500
Fax: 2 79-6040

Singapore:
National Semiconductor Asia
Pacific Pte. Ltd.
Southpoint 200
200 Cantonment Road #13-01
Singapore 0208, Singapore
65 225-2226
Fax: 65 225-7080
Telex: NATSEMI RS 50808

Spain:
Calle Agustin de Foxa, 27 (9oD)
28036 Madrid
Spain
01 7-33-29-58
Fax: 01 7-33-80-18
TLX: 46133

Sweden:
National Semiconductor AB
P.O. Box 1009
Grosshandiarvaegen 7
S-121 23 Johanneshov
Sweden
08 7228050
Fax: 08 7229095
TLX: 10731 NSCS

Switzerland:
Alte Winterhurerstrabe 53
Postfach 567
CH-8304 Wallisellen-Zurich
Switzerland
01 8-30-27-27
Fax: 01 8-30-19-00

Taiwan:
National Semiconductor (Far East),Ltd.
Taiwan Branch
9th Floor, No. 18
Sec. 1, Chang An East Road
Taipei, Taiwan, R.O.C.
02 521-3288
Fax: 02 561-3054

United Kingdom:
National Semiconductor (UK) Ltd.
The Maples, Kembrey Park
Swindon, Wiltshire SN2 6UT
United Kingdom
07-93 49-81-41-103
Fax: 07-93 10-35-06
TLX: 444674

Germany:
National Semiconductor GmbH
IndustrestraBe 10
D-8080 Furstenfeldbruck
Germany
49 81-41 1-3-0
49 0 180 532 78 32 (English)
49 0 180 530 85 85 (German)
Fax: 49 81-41 10-35-54
TLX: 527649

MisburgerstraBe 81D
D-3000 Hannover 61
Germany
49 08 575-072
Fax: 49 08 561-740
TLX: 527649

FlughafenstraBe 21
D-6078 Neu-Isenburg/
Zeppelinheim
Germany
49 69 69-45-12
Fax: 49 69 69-40-85
TLX: 4185397

Untere Waldplatze 37
D-7000 Stuttgart 80
Germany
49 07-11 68-65-11
Fax: 49 07-11 68-65-260
TLX: 7255993

Puerto Rico:
La Electronica Bldg.
Suite 312, R.D. #1 KM 14.5
Rio Piedias, Puerto Rico 00927
809-758-9211
Fax: 809-763-6959

Manufacturers, Prefixes, Part Number Types, Logo Descriptions & Family Types

(Besides offering a large variety of products, such as logic ICs [including FCT logic], op amps, ADCs, references, data acquisition ICs, voltage regulators, semiconductors, microcontrollers, futurebus controllers and interfaces, JTAG boundary scan bus devices, DECT [Digital European Cordless Telecommunications] ICs, 10base-T Ethernet ICs, disk drive ICs, silicon temperature sensors, and military qualified devices, mixed signal ASICs, radiation hardened parts, flash memory storage subsystems [hybrid MCM devices], DC/DC converters in a 24 pin DIP package, etc. They also do some foundry work and do custom MCM module designs.)

NC prefix
See *Noise/COM*

NChip, Inc.
A subsidiary of Flextronics International, Ltd.
1971 N. Capitol Avenue
San Jose, CA 95132
408-945-9991, Fax: 408-945-0151
http://www.nchip.com

(Multichip modules and various IC packaging)

NCR Microelectronic Products Div.
See *Symbios Logic, Inc.*

nCUBE Corp.
919 East Hillside Blvd.
Foster City, CA 94404
1-800-654-CUBE
415-593-9000 (International)

(Custom processor IC used in their parallel processor computer systems)

NDK America, Inc.
47671 Westinghouse Drive
Fremont, CA 94539
800-635-9825
Fax: 510-623-6590

(Crystals, oscillators and filters)

NE prefix
See *Philips* (Signetics), *AVG Semiconductors* (Second source NE558 quad timer, etc.), *NEC Electronics, Inc.*

NEC Electronics, Inc.
475 Ellis Street
Mountain View, CA 94039-7241
415-960-6000
Fax: 415-965-6077, 6130
1-800-366-9782 for literature
(8am-4pm pacific time)
415-965-6776 (Technical literature)
http://www.nec.com

401 Ellis St.
Mountain View, CA 94039
1-800-366-9782 (Literature line)
415-960-6000
415-965-6620 (DSP ICs)
Fax: 1-800-729-9288 (Literature)

California Eastern Laboratories
4590 Patrick Henry Drive
Santa Clara, CA 95056-0964
408-988-3500, 988-7846
Fax: 800-390-3232, 408-988-0279

Los Angeles, CA: 310-645-0985
San Diego, CA: 619-467-6727
Bellevue, WA: 206-455-1101
Richardson, TX: 214-437-5487
Olathe, KS: 913-780-1380
Woodridge, IL: 708-241-3040
Timonium, MD: 410-453-6600
Middleton, MA: 508-762-7400
Hackensack, NJ: 201-487-1155
or 487-1160
Altamonet Springs, FL:
407-774-7682
Snellville, GA: 404-978-4443
Nepean, Ontario, Canada:
613-726-0626

NEC Corp.
5-7-1 Shiba, Minato-ku
Tokyo 108-01, Japan
81 3 3454 1111
Fax: 81 3 3798 1510, 1509

NATIONAL SEMICONDUCTOR — NEL

Germany: 0211-650302
Fax: 0211-6503490
The Netherlands: 040-445-845
Fax: 040-444-580
Sweden: 08-753-6020
Fax: 08-755-3506
France: 1-3067-5800
Fax: 1-3946-3663
Spain: 1-504-2787
Fax: 1-504-2860
Italy: 02-6709108
Fax: 02-66981329
United Kingdom: 0908-691133
Fax: 0908-670290
Ireland: 01-6794200
Fax: 01-6794081
Hong Kong: 755-9008, 886-9318
Fax: 796-2404, 886-9022
Taiwan: 02-719-2377
Fax: 02-719-5951
Korea: 02-551-0450
Fax: 02-551-0451
Singapore: 253-8311
Fax: 250-3583
Australia: 03-8878012
Fax: 03-8878014
Japan: 03-3454-1111,
03-3798-6148
Fax: 03-3798-6059, 03-3798-6149

(Various products including RISC microprocessors, microcontrollers, SRAMs, DRAMs, VRAMs, ASICs, hybrid, logic voltage comparators and regulator ICs, disk drive controller ICs, LCD displays, Graphics ICs [including 3D graphics controller ICs], AM-LCDs, watches, clocks, telecommunications, VCR and television ICs, relays, etc.)

NEL Frequency Controls, Inc.
P.O. Box 457
357 Beloit Street
Burlington, WI 53105-0457
414-763-3591
Fax: 414-763-2881

NEL — NEW JERSEY SEMICONDUCTOR

(Crystal filters, quartz crystals and crystal oscillators)

NeoCad
2585 Central Avenue
Boulder, CO 80301
303-442-9121
Fax: 303-442-9124

(FPGA foundry)

NeoMagic Corporation
Santa Clara, CA
408-988-7020, Fax: 408-988-7032

(Mobile computing systems ICs, graphics ICs and devices with embedded DRAMs, Magic-Graphic IC for notebook computers)

NES
See *New England Semiconductor*

Neural Semiconductor, Inc.
P.O. Box 16877
Irvine, CA 92713
714-753-7500
Fax: 714-753-7537

(Interface ICs)

Neuralogix, Inc.
See *Adaptive Logic, Inc.*

New England Photoconductor (NES)
253 Mansfield Avenue, P.O. Box M
Norton, MA 02766
508-285-5561, Fax: 508-285-6480

(Diodes, optoelectronics)

New England Semiconductor (NES)
6 Lake Street
Lawrence, MA 01841
1-800-446-1158, 508-794-1666
Fax: 508-689-0803

(Military qualified semiconductors including bipolar [EPI- Base, MESA and Planar process] devices and ultrafast recovery rectifiers, and voltage regulators. The company can also package die in standard or custom packages.)

New Japan Radio Company, Ltd.
17th Floor Arco Tower
8-1 Shimomeguro 1-Chome
Meguro-ku
Tokyo 153 Japan
81 3 5434 8335
81 3 5434 8222
Fax: 81 3 5434 8261,
81 3 5434 8257

6001 Beach Road #16-09
Golden Mile Tower
Singapore 0719
65 2972102, 2972567
Fax: 65 2982019

NJR Corporation
A Subsidiary of New Japan Radio Company, Ltd.
340-B East Middlefield Rd.
Mountain View, CA 94043
415-961-3901
Fax: 415-969-1409

(Linear, digital ICs, hybrid, logic and disk drive ICs, diodes, transistors, op amps, voltage comparators and voltage regulators, disk drive controllers, speech synthesis and recognition ICs Hall effect sensors, telecommunications, television, VCR, watch and clock ICs, and second source for National Semiconductor [LM devices] and Texas Instruments [TLO devices])

NewLogic Corporation
Vancouver, WA

(Logic and memory on a single chip designer was acquired by Paradigm Technology [San Jose, CA] in April 1996).

NTE prefix
NTE

NTE (New Tone) Electronics, Inc.
44 Farrand Street
Bloomfield, N.J. 07003
1-800-631-1250
201-748-5089
Fax: 201-748-6224, 201-732-1326
QuickFacts (Fax): 1-800-683-3292
Telex 333226
http://www.nteinc.com

(A manufacturer of generic replacement semiconductors, and supplier of capacitors, resistors, relays, etc. Note: Cross references are available on computer disk.)

Newbridge Microsystems, Inc.
Division of Newbridge Networks Corporation
603 March Road
Kanata Ontario, Canada K2K 1X3
1-800-267-7231
613-592-0714
Fax: 613-592-1320
http://www.newbridge.com

(Data encryption products, VMEbus circuits [some parts jointly developed with DY 4 Systems, Inc.], Digital Audio Companding filter, ESM Window filter, communications devices such as SCSI Controller and Serial Communications controller, RF communications circuits, DC/DC converters and RF support circuits, PCI bridge ICs, Microprocessors)

New Jersey Semiconductor Products Co., Inc.
20 Stern Avenue
Springfield, NJ 07081
201-376-2922
Fax: 201-376-8960
TLX: 138720

(Diodes, thyristors, transistors)

Manufacturers, Prefixes, Part Number Types, Logo Descriptions & Family Types

NEWPORT COMPONENTS — NIC COMPONENTS

Newport Components, Ltd.
4 Tanners Drive
Blakelands North
Milton Keynes, MK14 5NA
United Kingdom
44 0908 615232
Fax: 44 0908 617545

(Digital, interface, linear ICs)

Newport Technology, Inc.
6321 Angus Drive
Raleigh, NC 27613
919-571-9405
Fax: 919-571-9262

Newport Components, Ltd.
4 Tanners Drive
Blakelands North
Milton Keynes MK14 5NA
England
44 01 908 615232
Fax: 44 01 908 617545

Newport Electronics Guangzhou, Ltd.
5th Floor, Building A1
Bei Wei, No. 1 District
Guangzhou, Guangdong 510730
Peoples's Republic of China
86 20 221 8066
Fax: 86 20 221 5902

(DC/DC converters, inductors, delay lines, interfaces)

Newport Wafer Fab Ltd.
Newport, South Wales
United Kingdom
44 1633 811313
Fax: 44 1633 816910

(Formerly an Inmos Ltd. facility sold by SGS-Thomson to QPL International Holdings Ltd. This is a wafer foundry.)

New Tone Electronics, Inc.
See *NTE*

Nexcom Technology, Inc.
532 Mercury Drive
Sunnyvale, CA 94086
408-730-3690
Fax: 408-720-9258

(Flash EEPROM and flash memory products including solid state disks and application specific modules)

NexGen Microproducts, Inc.
1623 Buckeye Drive
Milpitas, CA 95035
800-8NEXGEN (800-863-9436)
408-435-0202
Fax: 408-435-0262

18 Bis Rue du Montceau
77133 Fericy (France)
33 1 64 23 68 65
Fax: 33 1 64 23 61 91

(Microprocessor ICs. This company was acquired by Advanced Micro Devices [AMD], Inc. in 1995 and is now a wholly owned subsidiary of AMD.)

NeXt Sensors
1253 Reamwood Avenue
Sunnyvale Avenue
Sunnyvale, CA 94089
408-744-0452
Fax: 408-744-0392

Sarstedt, Germany
0 5066/65 6 48
Fax: 0 5066/65 6 75

(Silicon micromachined pressure sensors)

NEZ prefix
See *NEC Electronics*

NF Corporation
3-20, Tunashima-Higashi, 6-Chome,
Kohoku-ku
Yokohama-shi, Kanagawa 223
Japan
045 545 8111

(Linear ICs)

NFCSXX prefix
See *Computer Products*

(DC/DC converters)

NGK Insulators, Ltd.
2-56, Suda-Cho, Mizuho-Ku
Nagoya-Shi, Aichi 467
Japan
056 876 3119
Fax: 056 877 2901
Was: 052 872 7171

(Hybrid ICs)

nh logo
(Stylized to look like a waveform)
See *North Hills Signal Processing*

NH prefix
See *Intel Corp.*

NHi logo
See *National Hybrid, Inc.*

NHI prefix
See *National Hybrid, Inc.*

Ni prefix
See *Nestor, Inc.*

NIC Components Corp.
6000 New Horizons Blvd.
Amityville, NY 11707
516-226-7500
Fax: 516-226-6262

2070 Ringwood Avenue
San Jose, CA 95131
408-954-8470
Fax: 408-954-0349

(Rectifier diodes; ceramic, tantalum and aluminum capacitors; resistors and networks; thermistors; inductors; ferrite beads; varistors)

NICHICON — NIPPON STEEL SEMICONDUCTOR

Nichicon Corporation
Uehara Bldg, Karasumahigashi-Iru,
Oike-Dori, Nakagyo-Ku
Kyoto-Shi, Kyoto 604
Japan
075 231 8461
Fax: 075256 4158

(Hybrid ICs, capacitors)

Nidec/Power General
152 Will Drive
P.O. Box 189
Canton, MA 02021
617-828-6216
Fax: 617-825-3215

(Power supplies, DC/DC converters)

Nihon Inter Electronics Corp.
Nihon Semiconductor, Inc.
T.O.C. Bldg No. 22-17
Nishi-Gotanda, 7-Chome,
 Shinagawa-Ku
Tokyo 141 Japan
81 3 494 7411
Fax: 81 3 494 7414
TLX: 232 2994

1204, Soya, Hatano-Shi,
Kanagawa 257
Japan
0463 82 1111
Fax: 0463 81 2709

(Diodes, transistors, thyristors, hybrid ICs)

Nikko Denshi Company Ltd.
5-1, Tobitakyu 1-Chome
Chiyoda-Ku,
Tokyo 100 Japan
81 424 87 1321

(HEMTS - High Electron Mobility Transistors)

Nikkohm Co., Ltd.
3-31-2640, Minami-Cho
Misawa-Shi
Aomori 033 Japan
0176 53 2105
Fax: 0176 53 2106

(Hybrid ICs)

Nikkoshi Co. Ltd.
Nagai Bldg, 4-11,
Nihombashi-Honcho 4-Chome
Chuo-Ku, Tokyo 103
Japan
81 3 3245 0804

(Converter ICs, audio ICs, CD player, digital video, speech synthesis and recognition, VCR, hybrid and logic ICs)

Nimbus Technology
See *Alliance Semiconductor*

Nippon Avionics Co., Ltd.
3-20-1, Nishi-Shimbashi,
Minato-Ku,
Tokyo 105 Japan
03 5401 7351
Fax: 03 5401 7341

(Hybrid ICs)

Nippon Chemi-Con Corporation
1-167-1, Higashi-Ohme
Ohme-Shi, Tokyo 198
Japan
0428 22 1251
Fax: 0428 24 5890

(Hybrid ICs)

Nippon Electric Co. Ltd.
Nippon Industries Co. Ltd.
20 2-404, Honcho 1-Chome,
Shibuya-Ku
Tokyo 151
81 3 3374 7871

(HMETS - High Electron Mobility Transistors)

Nippon Gakki Co. Ltd.
Electronic Equipment Business
Section 203
Matsunokijima, Toyooka-Mura
Iwata-Gun
Shizuoka-Ken, 438-01
Japan
053962 3215

(Microprocessor support ICs)

Nippon Precision Circuits Ltd.
4-3-9, Taihei
Sumida-Ku
Tokyo 130, Japan
81 3 5608 5577
Fax: 81 3 5608 5566

(Hybrid, converter, watch and clock, audio, CD player, com-munication, television, VCR and digital video ICs)

Nippon Seiki Co., Ltd.
2-2-34, Higashizaoh
Nagakoa-Shi, Niigata 940
Japan
0258 24 3311
Fax: 0258 21 2151

(Hybrid ICs)

Nippon Steel Semiconductor Corp. (NPNX)
1580, Yamamato, Tateyama-Shi
Chiba, 294
Japan
0470 23 3121
Fax: 0470 23 2171

Tokyo Sales Office:
4F 25 Chuo Bldg
2-8-3 Kandatsukasa-Cho
Chiyoda-Ku
Tokyo 101, Japan
03 5294 2701
Fax: 03 5294 2707

Manufacturers, Prefixes, Part Number Types, Logo Descriptions & Family Types

NIPPON STEEL SEMICONDUCTOR — NMH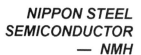

Nippon Steel Semiconductor
U.S.A. Corporation
2900 Gordon Avenue, Suite 206
Santa Clara, CA 95051
408-524-8000
Fax: 408-524-8040

Nippon Steel Semiconductor
Singapore Pte. Ltd.
16 Raffles Quay #11-01,
Hond Leong Building
Singapore 048581
65 2255088
Fax: 65 2253088

Logic Products and ASIC Design Service:
Semiconductor Division
Product Development Group
5-10-1, Fuchinobe, Sagamihara-Shi
Kanagawa 229 Japan
0427 68 5727
Fax: 0427 685750

(Note: NMB Semiconductor Co. Ltd. was purchased by Nippon Steel Company in early 1993, from Minebea Co. Ltd. and the company was renamed in March 1993. DRAMs, memory modules, and serial EEPROM.)

Nissei
See *Nissei Electric Co., Ltd.*

Nissei Electric Co., Ltd.
Hiroo Office Bldg. No. 3-18,
1-Chome, Hiroo Shibuya-Ku
Tokyo T 150 Japan
033 442 8151
Fax: 033 442 8692
TLX: 34156
Cable Address: CONDENISSEI

Nissei Denki America, Inc.
Nissei-Arcotronics
7873 S.W. Nimbus Avenue
Beaverton, OR 97005
503-646-1434
Fax: 503-641-9594

(EMI-RFI suppressors)

Nissei Denki Pvt., Ltd.
11 Tuas Avenue 7
Singapore 2263
861 1488
Fax: 861 4469

(Film capacitors)

Nitsuko Corp.
2-6-1, Kita-Mikata, Takatsu-Ku,
Kawasaki-Shi
Kangawa 213
Japan
044 811 1111
Fax: 044 822 1455

(Hybrid ICs)

NJ prefix
See *GEC Plessy*

NJL prefix
See *New Japan Radio Company, Ltd., NJR Corporation*

NJM prefix
See *New Japan Radio Company, Ltd., NJR Corporation*

NJU prefix
See *New Japan Radio Company, Ltd., NJR Corporation*

NJR Corporation
See *New Japan Radio Company, Ltd.*

NJS
See *New Jersey Semiconductor Products Co., Inc.*

NJU prefix
See *New Japan Radio Company, Ltd., NJR Corporation*

NL prefix
NEC Corp.

(LCD displays)

NLX prefix
See *Adaptive Logic, Inc.*

NM prefix
See *National Semiconductor, NeoMagic Corporation*

NMA prefix
See *Newport Technology*

(DC/DC converters)

NMB prefix
See *Newport Technology*

(DC/DC converters)

NMB Technologies, Inc.
See *Nippon Steel Semiconductor Corporation*

NMCM prefix
See *Newport Technology*

(DC/DC converters)

NMCS prefix
See *Newport Technology*

(DC/DC converters)

NMD prefix
See *Newport Technology*

(DC/DC converters)

NME prefix
See *Newport Technology*

(DC/DC converters)

NMF prefix
See *Newport Technology*

(DC/DC converters)

NMH prefix
See *Newport Technology*

(DC/DC converters)

NMP prefix
See *Newport Technology*

(DC/DC converters)

NMS prefix
See *Newport Technology*

(DC/DC converters)

NMV prefix
See *Newport Technology*

(DC/DC converters)

NMX prefix
See *Newport Technology*

(DC/DC converters)

NMYD, NMYS prefix
See *Newport Technology*

(DC/DC converters)

Noge Electric Ind. Co. Ltd.
10-1, Fukura 2-Chome
Kanazawa-Ku, Tokyo 106
Japan
81 3 3440 3311

(Memory)

Noise/COM
E. 49 Midland Avenue
Paramus, NJ 07652
201-261-8797
Fax: 201-261-8339

(Noise source instruments and references)

Noizeg (tm)
See *Micronetics*

Nonvolatile Electronics, Inc.
11409 Valley View Road
Eden Prairie, MN 55344
612-829-9217
Fax: 612-829-9241
http://www.nve.com
email: info@nve.com

(GMR magnetic sensors)

Noritake Co., Ltd.
1-36 Noritake Shinmachi 3 Chome
Nishi-Ku, Nagoya 451, Japan
052 562-0336
Fax: 052 581 1679

2635 Clearbrook Dr.
Arlington, Heights, IL 60005
1-800-837-4727
708-439-9020
Fax: 708-593-2285

919 Atlanta Gift Mart
230 Spring St., N.W.
Atlanta, GA 30303
404-529-9568
Fax: 404-529-9455

Noritake, Co., Inc.
945 Concord St., Suite 118
Framingham, MA 01701
508-626-0811
Fax: 508-626-0429

2454 Dallas Trade Mart
Dallas, TX 75207
214-742-9389
Fax: 214-747-5065

2050 E. Vista Bella Way
Compton, CA 90220
310-603-9770
Fax: 310-603-9810

75 Seaview Dr.
Secaucus, NJ 07094
201-319-0600
Fax: 201-319-1962

Noritake Canada, Ltd.
90 Nuggett Avenue
Againcourt, Ontario M1S 3A7
Canada
416-291-2946
Fax: 416-292-0239

Noritake Europa GmbH
Frankfurter Strasse 97-99
W-6096 Raunheim, Germany
06142-43095/96/97
Fax: 06142-22799

(Vacuum fluorescent displays)

North Hills Signal Processing
A Porta Systems Company
575 Underhill Boulevard
Syosset, NY 11791
516-682-7740
Fax: 516-682-7704

(MIL-STD-1553 data bus couplers and accessories)

Nortel (Northern Telecom)
200 Athens Way
Nashville, TN 37228
615-734-4000
Fax: 615-734-5190

8200 Dixie Road
P.O. Box 3000
Brampton, ON L6V 2M6
Canada
905-452-2000
Fax: 905-452-5901

3 Robert Speck parkway
P.O. Box 458, Station A
Mississauga, Ontario
Canada L42 3C8

9701 Data park
P.O. Box 1222
Minneapolis, MN 55440
800-328-6760
Fax: 612-932-8235

NORTEL — NVE

685 E. Middlefield Road
P.O. Box 7277
Mountain View, CA 94039
415-969-9170
Fax: 415-966-1098

97 Humbolt Street
Rochester, NY 14609
800-833-7477
Fax: 716-654-2180

England:
44 1803 662875

(ICs for their own products, telecommunication components. Note: The miniature telephone relay line was sold to Communications Instruments, Inc. [CII] in 1985.)

Northrop-Grumman Corporation
1840 Century Park East
Los Angeles, CA 90067
310-553-6262

Northrop Electronics Systems Division
2301 W. 120th Street
P.O. Box 5032
Hawthorne, CA 90251
213-600-3000
Fax: 213-600-4099

Northrop Corporation
600 Hicks Road
Rolling Meadows, IL 60008
708-259-9600
Fax: 708-870-5706

(GaAs MMICs, in-house capabilities for military applications are available for commercial uses)

Northwest Microcircuits, Inc.
211 S. 9th St, P.O. Box 1479
Philomath, OR 97370
503-929-6424
Fax: 503-929-6428

(Linear, microprocessor support ICs)

NOV prefix
See *Novacom*

Nova Engineering, Inc.
4747 Devitt Drive
Cincinnati, OH 45246-1105
513-860-3456
Fax: 513-860-3535

(Low frequency synthesizer modules [containing TCXO, prescaler, VCO and loop filter])

Novacom Technology, Inc.
5 Aldrin Road
Plymouth Industrial Park
Plymouth, MA 02360
508-747-6811
Fax: 508-747-6844

68 Hashoftin Street
Ramat-Hasharon 47113
Israel
972 3 540 7474
Fax: 972 549 6602

(Six-lobe active retiming concentrator for token ring networks)

NovaSensor
See *Lucas NovaSensor*

NPC
See *Nucleonic Products Company*

NPI
See *Nano-Pulse Industries*, *Lucas NovaSensor*

NPNX Corporation
See *Nippon Steel Semiconductor Corporation*

NS prefix
See *National Semiconductor*, *Nova Engineering, Inc.*

(Frequency synthesizer modules)

NSC prefix
See *National Semiconductor*

NT prefix
See *NUMA Technologies*

NTC/American Bright Optoelectronics Corporation
460 W. Lambert Rd., Suite H
Brea, CA 92621
714-257-0800
Fax: 714-257-1310

(LED lamps, infrared diodes, phototransistors, digital displays, dot matrix displays, clock displays, cluster LED displays)

NTQ prefix
See *Nanotec*

Numa Technologies
5551 Ridgewood Dr.
Suite 303
Naples, FL 33963
813-591-8008
Fax: 813-591-8704

(Digital frequency discriminator, demodulates and measures frequency; period to digital converters)

Number Nine (Computer) Visual Technology Corporation
18 Hartwell Avenue
Lexington, MA 02173
800-438-6463 (1-800-GET-NINE)
617-674-0009
Fax: 617-674-2129
http://www.nine.com

(Graphics accelerator circuit boards with their custom graphics controller IC, the Imagine)

NV1 part number
See *NVidia Corporation*

NVE logo
See *Nonvolatile Electronics, Inc.*

N - O
NVIDIA — OHMTEK

NVidia Corporation
1226 Tiros Way
Sunnyvale CA 94086
408-720-7132, 0204
Fax: 408-438-0204
http://www.nvidia.com

(A fabless company with an integrated multimedia accelerator IC licensed to SGS-Thomson, graphics controllers)

Nx prefix
See *NexGen Microproducts, Inc.*

NX prefix
See *Nexcom Technology, Inc.*

NXX part number
See *Hewlett-Packard Co.*

(RF ICs)

NXXXXHx part number
See *Westcode Semiconductors*

Nymph (tm)
See *SaRonix*

O2Micro
Santa Clara, CA

(Battery management, DC power plane ICs, finger driver cursor devices, PC card host adaptors)

Oak
See *OTI, Oak Technology, Inc.*

OAK
(Tradename of a DSP Semiconductor digital signal processor IC)

Oak Frequency Control Group (OFC)
100 Watts Street, P.O. Box B
Mt. Holly Springs, PA 17065
717-486-3411, Fax: 717-486-5920
http://www.ofc.com

(This company consists of McCoy Electronics, Ovenaire, Croven Crystals and Spectrum Technology, manufacturers of quartz crystal frequency control products such as high performance, surface mount OCXOs, VCXOs and clocks.)

Oak-Grigsby
88 N. Dugan Rd.
P.O. Box 890
Sugar Grove, IL 60554-0890
708-556-4200
Fax: 708-556-4216

(Optical encoders, switches)

Oak Technology, Inc. (OTI)
139 Kifer Ct.
Sunnyvale, CA 94086
408-737-0888
Fax: 408-737-3838

(A fabless company supplying PC chip sets, PC [digital] audio ICs, graphics accelerator and controller chip sets, CD-ROM ICs)

OBG prefix
See *Siemens*

(Bar graph displays; discontinued)

OCP prefix
See *Matsushita* (Panasonic)

Odeum Microsystems, Inc.
166 Baypoint Parkway
San Jose, CA 95134
408-473-9274, 408-232-8342
Fax: 408-473-9370, 408-232-8125

(Formerly Hyundai Digital Media Division prior to 1996. This company is focused on silicon and software solutions for home theatre and home entertainment systems including DVD, digital satellite and set-top boxes, MPEG ICs, etc.).

OEi
See *Optical Electronics, Inc.*

OFC logo
(The "O" has 3 wavy lines through it)
See *Oak Frequency Control Group*

Ohio Semitronics, Inc.
4242 Reynolds Drive
Hilliard, OH 43026
614-777-1005
Fax: 614-777-4511

(Hall Effect sensor probes)

Ohkura Electric Co., Ltd.
20-8, Narita-Nishi 3-Chome,
Suginami-Ku
Tokyo 166 Japan
81 3 3398 5111

(Hybrid ICs)

Ohmcraft, Inc.
3800 Monroe Avenue
Pittsford, NY 14534
716-586-0823
Fax: 716-586-0015

(Resistor networks)

Ohmite Manufacturing Co.
Headquarters:
3601 Howard Street
Skokie, IL 60076
708-675-2600
708-675-1505

(Hybrid microcircuits, resistors)

Ohmtek
A company of Vishay
2160 Liberty Drive
Niagra Falls, NY 14304
716-283-4025
Fax: 716-283-5932

(Resistor networks, surface mount resistors)

OHN prefix
See *Optek Technology, Inc.*

OHS prefix
See *Optek Technology, Inc.*

OIS
See *Optical Imaging Systems, Inc.*

Okaya Electric Industries Co., Ltd.
1-8-3, Shibuyaka
Tokyo 150 Japan
03 3400 8511
Fax: 03 3499 3437
TLX: 2423489

Okaya Electric America
503 Wall Street
Valpariso, IN 46383
219-477-4488, Fax: 219-477-4856

(Transient protection devices for power and communications lines)

Oki Semiconductor
785 North Mary Avenue
Sunnyvale, CA 94086-2909
1-800-OKI-6388
408-720-1900
Fax: 408-720-1918, 408-737-6503
http://www.oki.co.jp/
http://www.okisemi.com

Oki Electric Industry Co., Ltd.
1-7-12, Toranomon,
Minato-Ku, Tokyo 105
81 3 3501 3111
Fax: 81 3 3581 5522

(HMETS - High Electron Mobility Transistors, microcontrollers, ASICs, gate arrays, hybrid ICs, logic, memory, disk drive, speech synthesis and recognition, watch and clock, VCR, television, and telecommunications ICs [including ICs for the PCS or personal-communications services market such as codecs, modems and echo canceller ICs], SAW filters, GaAs RF devices for wireless communications)

Okita Works Co. Ltd.
14Fl Sintaiso Building 2gokan
2-10-7 Dogenzaka, Sibuya-Ku,
Tokyo 150 Japan
03 3464 3561
Fax: 03 3464 3482

(Photo DMOS-FET relays and reed relays)

Olektron Corp.
See *ST Olektron Corp.*

OM prefix
See *Matsushita* (Panasonic),
Philips Semiconductors,
Omnirel Corp.

(Power MOSFETS, voltage regulators)

OMA prefix
See *Omnirel Corp.*

(Operational amplifiers, similar to OPA prefixed devices)

OMC prefix
See *Omnirel Corp.*

OMD prefix
See *Omnirel Corp.*

Omega
See *Omega Micro, Inc.*

Omega symbol with a "T" inside it or followed by "TEK"
See *Ohmtek*

Omega symbol followed by a Micro symbol
See *Omega Micro*

Omega Micro, Inc.
155 North Wolf Road
Sunnyvale, CA 94086
408-522-8895
Fax: 408-774-9330

440 Oakmead Parkway
Sunnyvale, CA 94086
408-992-1100
Fax: 408-774-9330

(PCMCIA controller ICs and related items.)

Omega Power Systems, Inc.
8968 Fullbright Avenue
Chatsworth, CA 91311
818-727-2216
Fax: 818-727-2219
http://www.primenet.com/~omega/
sales1@megapower.com

(Switching power supplies, dc/dc converters)

Omega Research Limited
Dagety Bay
Scotland
Phone/Fax: 01383 820583

In the U.S. contact:
Saelig Company
1193 Moseley Road
Victor, NY 14564
716-425-3753
Fax: 716-425-3835
http://www.memo.com/saelig

(DC/DC converter modules)

OMH prefix
See *Omnirel Corp.*

Omnirel Corp.
205 Crawford Street
Leominster, MA 01453
508-534-5776
Fax: 508-537-4246

Field Sales Offices:
Southwestern Area:
818-889-3838

O

OMNIREL — OPAMP LABS

Northwestern Area:
408-629-4789

Eastern Area:
301-317-6979

Northeastern Area:
508-534-5776

(Power MOSFETs, IGBTs, Rectifiers, DC/DC converters, power hybrids including amplifiers, voltage regulators, and contract manufacturing. Military [DESC] qualified parts are available. In mid 1995 they acquired Motorola's military bipolar transistor line.)

Omni-Wave Semiconductor
See *Advanced Data Technology, Inc.*

Omniyig, Inc.
3350 Scott Blvd, Bldg. #66
Santa Clara, CA 95054-3125
408-988-0843
Fax: 408-727-1373

(Miniature YIG oscillator, etc.)

Omron Corporation
Karasumaru Nanajo, Shimogyo-Ku
Kyoto-Shi, Kyoto 600
075 344 7000
Fax: 075 344 7001

Omron Electronics, Inc.
Control Components, Inc.
1 E. Commerce Drive
Schaumburg, IL 60173
800-62-OMRON
708-843-5339 (in IL),
708-843-7900
Fax: 708-843-7787

(Primarily a supplier of industrial control systems and relays this American division of a Japanese electronics conglomerate is deeply involved in Fuzzy logic, including ICs, the company also manufacturers optical sensors including units that detect displacement changes as small as 10 micrometers).

Omron Corp.
Nanajo-Saguru
Karasuma-Dori, Shimogyo-Ku
Kyoto 600 Japan
075 344 7000
Fax: 075 344 7131

(Reflection type wafer sensors)

OMS prefix
See *Omnirel Corp.*

ON prefix
See *Harris Semiconductor*
(Custom part)

On Chip Systems
1190 Coleman Avenue
San Jose, CA 95110
408-988-5400
Fax: 408-988-5488

(Frequency synthesizers, sample and hold amplifiers, active filters, analog signal conditioners, phase locked loops, Modem ICs, music synthesizer ICs, analog multiplexer and squarer ICs. Attenuator and electronic volume control IC, waveform and function generator ICs, CD sampler and synthesizer ICs)

Oneida Electric Mfg. Inc.
503 Randolph St, P.O. Box 678
Meadville, PA 16335
814-336-2125
Fax: 814-336-2126

(Diodes)

Ono Sokki Co., Ltd.
1-16-1, Hakusan, Midori-Ku,
Yokohama-Shi
Kangawa 226
045 935 3888
Fax: 045 390 1303

(Measuring instrument, hybrid ICs)

OnSpec Electronic, Inc.
3056A Scott Blvd
Santa Clara, CA 94054
408-727-1819
Fax: 408-727-2219

(Enhanced parallel port ICs. They also do ASICs and software for data acquisition, multimedia and MPEG1 video CD products)

Onset Computer
Box 3450
Pocasset, MA 02559
508-563-9000
Fax: 508-563-9477

("Tattletale" data logger/controller engine IC used in their data logging/control card)

On-Trak Photonics, Inc.
22471 Aspan St.
Lake Forest, CA 92630
714-587-0769
Fax: 714-587-9524

(Position sensing detectors for non-contact measurement of position, distance, motion and vibration)

OP prefix
See *Analog Devices*
(Linear Technology, Maxim, National Semiconductor, Second source some parts).
Also see *Optek Technology, Inc.*

OPA prefix
See *Burr-Brown Corp.*

Opamp Labs, Inc.
1033 N. Sycamore Ave.
Los Angeles, CA 90038
213-934-3566
Fax: 213-462-6490

(Operational amplifiers)

OPB prefix
See *Optek Technology, Inc.*

OPC prefix
See *Optek Technology, Inc.*

OPC-XXXXXX part number
See *Opto Power Corp.*

Opcoa
Div. Refrac Electronics Corp
12881 Knotts St. #109
Garden Grove, CA 92641
714-898-9333

(LEDs)

Operating Technical Electronics, Inc. (OTE)
850 Greenview Drive
Grand Prairie, TX 75050
214-988-6828
Fax: 214-641-7089

7F, Block B, No. 58, Sec. 1
Min Sheng E. Rd.
Taipei, Taiwan R.O.C.
011 886 2 551 2834
Fax: 011 886 2 511 2473

Rm. 1413-14,
New World Office Bldg.
(West Wing)
20 Salisbury Rd., T.S.T.
Kowloon, Hong Kong
011 852 2722 0888
Fax: 011 852 2722 6116

(Intelligent battery guard [prevents discharge of vehicle battery by a load], switching power supplies [Anoma Electric Company, Ltd.], wall mounted power supplies, miniature brushless DC fans)

OPF prefix
See *Optek Technology, Inc.*

OPI prefix
See *Optek Technology, Inc.*

OPL prefix
See *Optek Technology, Inc., Yamaha*
(If OPLX part number, sound synthesizer IC)

OPM prefix
See *Optek Technology, Inc.*

OPR prefix
See *Optek Technology, Inc.*

OPS prefix
See *Optek Technology, Inc.*

OPT logo
See *OPT Industries, Inc.*

OPT prefix
See *Burr-Brown, Optimum Semiconductor*

OPT Industries, Inc.
300 Red School Lane
Phillipsburg, NJ 08865
908-454-2600
Fax: 908-454-3742

(Transformers, power supplies, coils, filters, inductors, reactors)

Optek Technology, Inc.
1215 West Crosby Road
Carrollton, TX 75006
1-800-481-7820
214-323-2200, 214-323-7035
Fax: 214-323-2392, 2396

(Formerly TRW Optron. Small signal transistors, optical emitter and detectors and Hall Effect sensors, military approved parts available)

Opti, Inc.
2525 Walsh Ave.
Santa Clara, CA 95051
408-980-8178
Fax: 408-980-8860

888 Tasman Drive
Milpitas, CA 95035
http://www.opti.com

OPB — OPTICAL IMAGING SYSTEMS

9th Fl., No. 303, Sec. 4, H
Taipei, Taiwan
886 2 325 8520, Fax: 886 2 325 6520

(A fabless company supplying PC Chip sets, including audio ICs [sound controllers], and SCSI to VL-bus interface IC, graphics ICs, Video Electronic Standards Association United Memory Architecture chip sets)

Optical Communication Products, Inc.
9736 Eaton Avenue
Chatsworth, CA 91311
818-701-0164
Fax: 818-701-1468

(PIN photodiode transmitter/receiver modules, laser diodes, optical amplifiers, photodiodes, fiber optic receivers and transmitters)

Optical Electronics, Inc. (OEi)
P.O. Box 11140
3939 S Park Avenue, Suite 180
Tucson, AZ 85714
602-889-8811
(520 area code after 12/31/96)
Fax: 602-889-8575
(520 area code after 12/31/96)

(Amplifiers and analog function modules, D/A converters, sample and hold and track and hold amplifiers, polar to Cartesian converter, real time imaging processing IC)

Optical Imaging Systems, Inc. (OIS)
47050 Five Mile Road
Northville, MI 48167
313-454-5560
Fax: 313-416-3318

(Active matrix full color and monochrome LCDs for aircraft cockpit applications)

OPTIMUM — ORIGIN ELECTRIC

Optimum Semiconductor
5201 Great America parkway
Suite 219
Santa Clara, CA 95054
408-748-1111
Fax: 408-980-8209

(D/A converters)

Opto 22
43044 Business Park Drive
Temecula, CA 92590-3665
1-800-321-OPTO, 1-800-321-6786
909-695-9299, 909-695-3000
1-800-HLP-OPTO,
1-800-457-6786
Technical Assistance
(24 hour hotline)
Fax: 909-695-2712
http://www.opto22.com

(Solid state relays)

Opto Diode Corp
914 Tourmaline Drive
Newbury Park, CA 91320
805-499-0335
Fax: 805-499-8108

(IR emitters, including high power and rad hard units)

Opto Power Corp.
3321 E. Global Loop
Tuscon, AZ 85706-5008
520-746-1234
Fax: 520-234-3300
http://www.optopower.com

(Diode laser drivers, diode laser arrays, optical re-imaging units)

Optoram Electronic Co. Ltd.
Nichimen Nihombashi Heights 503
28-1,Hakazaki-Cho, Nihombashi,
Chuo-Ku
Tokyo 103 Japan
81 3 5965 6021

(Memory, hybrid, logic and linear ICs)

OptoSwitch, Inc.
See *Clarostat Sensors and Controls*

Opto Technology, Inc.
562 Chaddick Dr.
Wheeling, IL 60090
708-537-4277
Fax: 708-537-4785

(Optoelectronics including optical sensors)

Optotek Ltd.
62 Steacie Drive
Kanata, ON K2K 2A9
Canada
613-591-0336
Fax: 613-591-0584

(LED displays, photoemitter displays, digital status indicator)

Optrex Corp.
3-14-9 Yushima
Bunkyo-Ku
Tokyo 113 Japan
03 832 5357

Asahi Glass/Optrex
23399 Commerce Drive
Farmington Hills, MI 48335
313-471-6220
Fax: 313-471-4767
44160 Plymouth Oaks Drive
Plymouth MI 48170
313-416-8500
Fax: 313-416-8520

Optrex-Satorl
3830 Del Amo Blvd. #101
Torrance, CA 90503
310-214-1791

(LCD displays)

Optron Products (or TRW Optron)
See *Optek Technology*

OR prefix
See *Oren Semiconductor*

OR5-X part number
See *Saelig Company*

(DC/DC converters)

Orbit Semiconductor, Inc.
1215 Bordeaux Drive
Sunnyvale, CA 94089
1-800-331-4617
1-800-647-0222 (in CA)
408-744-1800
Fax: 408-747-1263
http://www.orbitsemi.com

(A foundry and a service for customers that require small to medium volume of chips produced on a fast turnaround schedule. They can convert FPGAs, that are used in higher production volume designs into standard ICs and can second source Gate Arrays from other vendors [with their Encore netlist conversion service]. The company became a unit of DII Group, Inc. in 1996.).

ORCA
See *Lucent Technologies*
(Formerly AT&T Microelectronics)

Oriel Corp.
250 long Beach Blvd
P.O. Box 872
Stratford, CT 06497-9988
203-377-8282
Fax: 203-378-2457

(Optoelectronics)

Origin Electric Co., Ltd.
1-18-1 Takada
Toshima-Ku,
Tokyo 171 Japan
81 3 3983 7111
Fax: 03 3988 6369
TLX: 2722068

(Diodes, diode bridges, hybrids, and arrays, etc.)

Orion
(A MIPS processor by Integrated Device Technology, also the name of a chip set by Intel Corp. for P6 servers and workstations)

Orion Electric Co., Ltd.
257 Kongdan-Dong
Kumi-Shi, Kyongsandbuk-Do
Korea
82 546 467 5420, 3
Fax: 82 546 462 2496, 7
Sales: 82 546 467 5490-5
Fax: 82 546 467 5488
See *Hantronix, Inc.*,
Franchised distributor in the U.S.

(TN, HTN, STN LCD modules)

Orion Industries, Inc.
6 Industrial Drive
Windham, NH 03087
603-437-4446, 603-894-4242
Fax: 603-437-3260

(DC/DC converters)

Ortel Corp.
2015 W. Chestnut Ave.
Alhambra, CA 91803-1542
818-281-3636
Fax: 818-281-8232
TLX: 752 434

(Optoelectronics)

Oryx Power Products
Division Oryx Technology Corporation
1000 Milwaukee Avenue
Glenview, IL 60025
800-827-8720
Fax: 708-391-7531

(Switching power supplies)

OS prefix
See *KDI/triangle Electronics, Inc.*

OSC-XXXX part number
See *ILC-DDC*

Oscillatek Corporation
625 North Lindenwood
Olathe, KS 66062
913-829-1777
Fax: 913-829-3505
TLX: 437045

(Oscillators including OXCOs, OVCXOs, TCXOs, VCXOs, hybrid clocks, military, aerospace devices available)

OSD prefix
See *Centronic, Inc.*

OTE
See *Operating Technical Electronics, Inc.*

OTI prefix
See *Oak Technology, Inc.*

OTL prefix
See *Optotek Ltd.*

OTR prefix
See *Opto Technology, Inc.*

OW prefix
See *Advanced Data Technology*

Oxford Computer, Inc.
39 Old Good Hill Road
Oxford, CT 06478
203-881-0891
Fax: 203-888-1146 (14,400 8N1)
E-mail Compuserve 70714,2166
Internet:oxfordcomput@delphi.com

(Electronic Image processing, optical memory, pattern recognition ICs and foundry services. ICs include intelligent image sensor, neural network module, image sensor and optical character recognition, etc.)

Oxley, Inc.
Priory Park, Ulverston
Cumbria, LA12 9QG
United Kingdom
44 0 0229 52621
Fax: 44 0 0229 55090
TLX: 65141

25 Business Park Drive
Branford, CT 06405
203-488-4135

(Industrial LED indicators, EMI filters)

OZ prefix
See *O2Micro*

P prefix
See *GEC Plessy, Intel Corp., Weitek, Philips Semiconductors, Performance Semiconductor Corp., Phase IV Systems, Inc.*

P logo
With a lightning bolt through it
See *Pericom Semiconductor Corporation*

P2NXXXXX
(A plastic encapsulated 2NXXXXX transistor EIA designated, manufactured by a variety of vendors)

P35-XXXX-X part number
See *Daico Industries, Inc.*

P4 prefix
See *Performance Semiconductor Corp.*

P54C
(A 3.3 version of the Intel Pentium Microprocessor)

P82 prefix
See *Chips and Technologies*

PA — PARAMETRIC INDUSTRIES

PA prefix
See *Apex Microtechnology Corp.*, *Hewlett Packard* (Microprocessor ICs), *Hitachi* (RISC microprocessors), *Unisys Corporation* (ASICs), *ICT* (PLDs)

PAC prefix
Trademark of *Waferscale Integration, Inc.*

(Programmable stand-alone microcontroller)

Paccom Electronics
3928 148th Avenue NE
Redmone, WA 98052-5350
800-426-6252, 206-883-9200
Fax: 206-881-6959

(MOVs)

Pacific Communications Sciences, Inc. (PCSI)
9645 Scranton Road
San Diego, CA 92121
619-535-9500

(This company is owned by Cirrus Logic. Wireless communications ICs including 2 way paging [pACT] chip sets)

Pacific Hybrid Microelectronics
See *Pacific Microelectronics Corporation*.

Pacific Microelectronics Corporation (PMC)
10575 Southwest Cascade Blvd.
Portland, OR 97223
1-800-622-5574
503-684-5657
Fax: 503-620-8051
http://www.pmcnet.com/~pmc
Email: pmc@pmc@pmcnet.com

(Design services for design, assembly, testing of multichip modules, electronic circuit boards and IC packages including BGA packages)

Pacific Semiconductor
(Acquired by ASIC Semiconductor in January 1996)

Pacific Silicon Technologies dba
373 River Oaks Circle, Suite 1004
San Jose, CA 95134
408-955-9020

(ASIC service and wafer foundry; See also PSD)

PAL, PALCE prefix
See *AMD*

Pal-Tech Electronics
See *Fujitsu-Towa*

Panametrics, Inc.
221 Crescent St.
Waltham, MA 02154-3497
1-800-833-9438
617-899-2719, Fax: 617-894-8582

(Hybrid circuit humidity sensors)

Panasonic
Panasonic Industrial Co.
One Panasonic Way
Panazip 2F-3
Secaucus, N.J. 07094
201-348-7000, Fax: 201-348-8164

(Various ICs used in their consumer products)

Milpitas, CA (gate arrays)
408-945-5672
Panjit Semiconductor

(Semiconductors, available from Taitron Components, Inc., 25202 Anza Drive, Santa Clarita, CA 91355-3496, 800-824-8766, 800-247-2232, 805-257-6060; Fax: 800-824-8329, 805-257-6415)

Paradigm Technology, Inc.
71 Vista Montana
San Jose CA 95134
1-800-767-4530
1-800-767-9530
408-954-0500
Fax: 408-954-8913, 408-954-0664

(A fabless company supplying SRAM devices, FIFO buffer memory ICs. While Paradign does not supply military qualified parts, National Semiconductor is buying their dice and supplying military product.)

Paradise Systems, Inc.
A Western Digital Company
99 South Hill Drive
Brisbane, CA 94005
415-468-7300
Fax: 415-468-7323, 415-468-7324
Telex: 709368

(ICs generally on their computer graphics cards)

Paralight (USA) Corporation
See *HY Electronics Corporation*
545 West Allen Avenue, #6
San Dimas, CA 91773
909-305-2393
Fax: 909-305-2389
E-mail: HYLEDS@aol.com

International Paramount Light Electronics Company, Ltd.
4F, No. 1, Chien Yi Road,
Chung Ho City
Taipei, Taiwan R.O.C.
02 225 3733
Fax: 02 225 4800

(LEDs: discrete, matrix and numeric)

Parametric Industries, Inc.
742 Main St.
Winchester, MA 01890
617-729-7333

(Diodes)

PASCALL ELECTRONICS — PE-XXXXX

Pascall Electronics
P.O. Box 844
Frank Ewing, TN 38459
1-800-484-8639 (ext 4353)
615-424-8844
Fax: 616-424-1187

(DC/DC converters, aerospace, defense and commercial use)

pASIC prefix
See *Cypress Semiconductor*

(FPGAs)

Patriot Scientific Computing Corporation
12875 Brookpointer Place
Suite 300
Poway, CA 92064
619-679-4428
Fax: 619-679-4429
http://www.PTSC.com

(RISC microprocessors, including the ShBoom [originally developed by nanoTronics] for JAVA, modem/interface use and the U.S. Automotive industry, microwave ground penetrating radar products.)

PBF prefix
See *Motorola*

PC prefix
See *National Semiconductor*, *Harris Semiconductor* (Custom part), *CMD Technology, Inc.* (PCI to IDE controller IC), *ICT, Inc.* (Stereo enhancement processor), *Powertip Technology Corporation* (LCDs)

PCA
See *PCA Electronics, Inc.*, *AVG Semiconductors* (Second source TV ICs)

PCA, PCD, PCF prefix
See *Philips Semiconductor* (Formerly Signetics), *AVG Semiconductors* (Second source some ICs)

PCA Electronics, Inc.
16799 Schoenborn Street
Sepulveda, CA 91343
818-892-0761
Fax: 818-894-5791
TLX: 67-4681
TWX: 910-496-1525

(Delay lines, toroid power inductors, pulse transformers, current sensing transformers, RF inductors, telecommunication transformers [LAN, Ethernet, etc.])

PCB Piezotronics, Inc.
3425 Walden Avenue
Depew, NY 14043-2495
716-684-0001
Fax: 716-684-0987
TWX: 710-263-1371

(Accelerometers)

PCD prefix
See *Philips Semiconductor*

PCF prefix
See *Philips Semiconductor*

PCI prefix
See *PLX Technology, Inc.*, *CMD Technology* (Controller ICs), *Wolfson Microelectronics* (PCI interface ICs)

PCI Inc.
1530 Parkmore Avenue
Suite A
San Jose, CA 95128
408-282-0200
Fax: 408-282-0216

(Optoelectronics)

PCM prefix
See *Burr-Brown*, *National Semiconductor*

(PCMCIA interface ICs)

PCMC
See *MetaDesign Semiconductor*

PCSI
See *Pacific Communications Sciences, Inc.*

PCT prefix
See *On Chip Systems*

PCTech
Subsidiary of Zeos International Ltd.
980 W. Lakewood Avenue
Lake City, MN 55041-2101
612-332-4111, 612-345-4555

(PCI bus interface ICs to disk drives)

PD prefix
See *Siemens*

PD&E, Inc.
Newington Industrial park
Newington, NH 03801
608-436-8121

(Diodes)

PDI
See *Precision Devices, Inc.*

PDM prefix
See *SenSym Inc.*, *Paradigm Technology, Inc.*

PDSP prefix
See *GEC Plessy, Siemens*

PE-XXXXX part number
See *Pulse Engineering, Inc.*

PE-XXXXX — PG

(DC/DC converters)

PEB prefix
See *Siemens*

PED Ltd.
Exning Road, Newmarket
Suffolk, CB8 OAX
United Kingdom
44 638 665 161
Fax: 44 638 660 718

(Optoelectronics)

PEEL prefix
See *ICT, Inc.*

Penny Technologies, Inc.
78 Kingsland Ave.
Greenpoint, NY 11222
718-782-4149
Fax: 718-963-2930

(Delay modules, delay lines, cellular-pcs-pcn filters, microwave filters)

Penstock
520 Mercury Drive
Sunnyvale, CA 94086
1-800-Penstock
800-736-7862
Fax: 408-730-4782
Northern CA: 408-730-0300
Canada: 613-592-6088

(This value added distributor is part of the Avnet Electronic Marketing Group. Avnet Corp. is based in Great Neck, NY.)

Pentastar Electronics, Inc.
110 Wynn Dr.
Huntsville. AL 35807
205-895-2076, 2006
Fax: 205-895-2356

(DC/DC converters)

Pentium (Processor)
(Trademark of the Intel Corp. for the P5 microprocessor. The Pentium Pro is the name of the P6 version.)

Pepi logo
See *Portage Electric Products, Inc.*

Pepperl + Fuchs, Inc.
1600 Enterprise Parkway
Twinsburg, OH 44087
216-425-3555
Fax: 216-425-4607

(Photoelectric and inductive proximity sensors, etc.)

Peregrine Semiconductor Corporation
2909 Canon Street
San Diego, CA 92106
300 Orchard City Drive, Suite 142
Campbell, CA 95008
Fax: 619-523-2655

(A 1995 start-up that manufactures ultra thin single crystal silicon on sapphire substrate technology for wireless communication ICs)

Performance Electronics Pkg. Svcs., Inc.
1565 Creek St.
San Marcos, CA 92069
800-255-8607
Fax: 619-471-9691

(Memory modules including custom modules and replacements for those obsoleted by the original manufacturer [TI, Toshiba, Mitsubishi, Hitachi, etc.])

Performance Motion Devices (PMD)
97 Lowell Road
Concord, MA
508-369-3302
Fax: 508-369-3819

(4 axis motion control ICs)

Performance Semiconductor Corporation
610 E. Weddell Drive
Sunnyvale, CA 94089
408-734-9000, 8200
Fax: 408-734-0258, 0962
TLX: 6502715784

(Digital, interface and microprocessor ICs. Note: FCT logic line sold to Cypress Semiconductor in late 1993)

Pericom Semiconductor Corp.
2380 Bering Drive
San Jose, CA 95131
1-800-435-2336
Fax: 408-321-0933
Fax on Demand: 1-800-609-4695, 415-688-4354
http://www.pericom.com
E-Mail: nolimits@pericom.com

(Formerly Pioneer Semiconductor Corp., FCT and clock generation ICs, IEEE 802.12 transceivers, DRAMs, token ring hub ICs, graphics ICs.)

Petrond Microwave
2628 Bayshore Parkway
Mountain View, CA 94043-1013
415-968-5986

(Microwave power amplifiers)

PF prefix
See *Hitachi Semiconductor*

(RF power amplifier modules)

PFC prefix
See *Philips Semiconductors*

PG prefix
See *Powertip Technology Corporation*

(LCDs)

PGT prefix
See *Philips Semiconductor*

(Programmable ASICs)

Philbrick, Teledyne-Philbrick
See *TelCom Semiconductor, Inc.*
(Formerly Teledyne Components),
Philco Radio Television Ltd.

PH prefix
See *Philips Semiconductors*

(Signetics)

Phase IV Systems, Inc.
3405 Triana Blvd
Huntsville, AL 35805-4695
205-535-2100
Fax: 205-535-2110

(Microwave and high speed digital components including GaAs multiplexers, 64 bit parallel to serial converters)

PHD prefix
See *Philips Semiconductors*

(Signetics)

Phihong Enterprise Company, Ltd.
No. 16, Lane 530
Chung Cheng N. Rd.
Sanchung City, Taipei,
Taiwan R.O.C.
011 886 2 988 2126
Fax: 011 866 2 981 7086

(Switching power supplies including external units for portable electronics.)

Phihong USA
1585 McCandless Drive
Milipitas CA 95035-8001
408-946-7888
Fax: 408-946-8098
http://www.phihongusa.com

Philips Key Modules
2029 Gateway Place
Suite 100
San Jose, CA 95110-1-17
800-235-7373
408-453-7373

(Laser diodes and modules used in Philips devices)

Philips Semiconductors
440 Wolfe Road
Sunnyvale, CA 94088-3409
1-800-227-1817
1-800-447-3762
(Literature distribution center),
Ext. 1500
400-991-2000
408-991-2339 (PLDs and FPGAs)
http://www.semiconductors.philips.com/ps/

Literature Distribution Center:
Philips Electronics
North America Corporation
1000 Business Center Drive
Mt. Prospect, IL 60056-6091
1-800-447-3762

5600 MD Eindhoven
The Netherlands
31 40 722091, Fax: 31 40 724825

Discrete Semiconductors Group
100 Providence Pike
Slatersville, RI 02876-2078
401-762-3800
Fax: 401-767-4497, 4493

Philips Components/Discrete
Products Division
Riviera Beach, FL
800-447-3762

Philips Components
1440 W. Indiantown Rd.
P.O. Box 689605
Jupiter, FL 33468

PGT — PHILIPS SEMICONDUCTOR

Amperex Electronic Co.
Optoelectronics Business Group
George Washington Hwy.
Smithfield, RI 02917
401-232-0500

(LEDs, CCDs, and other optoelectronics)

Philips Semiconductors-Signetics
A North American Philips Company
811 E. Arques Ave.
Sunnyvale, CA 94088
408-991-2000, 991-4577,
991-4566
1-800-227-1817 ext 761D
Fax: 408-991-2133

(Microcontrollers)

9201 Pan American Freeway N.E.
MS 08
Albuqerque, NM 87113
505-822-7629

(Complex Programmable Logic Device, CPLD business unit)

1144 East Arques Ave. MS 45
Sunnyvale, CA 94088
408-991-4577

Alabama, Huntsville: 205-464-0111

California:
Calabasas: 818-880-6304
Irvine: 714-833-8980,
714-752-2780
San Diego: 619-560-0242
Sunnyvale: 408-991-3737

Colorado, Englewood:
303-792-9011
Georgia, Atlanta: 404-594-1392
Illinois, Itasca: 708-250-0050
Indiana, Kokomo: 317-459-5355
Massachusetts, Westford:
508-692-6211

PHILIPS SEMICONDUCTOR

Michigan, Novi: 313-347-1400
New Jersey, Toms River: 908-505-1200
New York, Wappingers Falls: 914-297-4074
Ohio, Columbus: 614-888-7143
Oregon, Beaverton: 503-627-0110
Pennsylvania, Plymouth Meeting: 215-825-4404
Tennessee, Greenville: 615-639-0251

Texas:
Austin: 512-339-9945
Richardson: 214-644-1610

Philips Semiconductors
Box 218, Bldg BAF-1
5600 MD Eindhoven
The Netherlands
31 40 722091, 31 40 783749
Fax: 31 40 724825

Argentina, Buenos Aries: 1 541-4106
Australia, North Ryde: 2 805-4455
Austria, Wein: 0222 60-101-820
Belgium, Brussels: 02 525-61-11
Brazil, Sao Paulo: 011 829-1166

Canada:
Etobicoke, Ontario: 416 626-6676
Nepean Ontario: 613 225-5467
Scarborough, Ontario: 416 292-5161

Chile, Santiago: 02 077-3816
Columbia, Bogota (Iprelenso Ltd.): 01 2497624
Denmark, Copenhagen S: 01 883333
Finland, Esposo: 358-0-502-6508
France: Issy-les-Moulineaux, Cedex: 01 40-93-80-00
Germany, Hamburg: 40 3-296-1
Greece, Tavros: 01 4984-339/4894911
Hong Kong, Kwai Chung, Kowloon: 0 424-5121
India, Bombay (Pelco Electronics Ltd.): 022 493-8541
Indonesia, Jakarta Selatan: 021 515690
Ireland, Dublin: 01 69-33-55
Italy, Milano: 02 67-52-1
Japan, Tokyo: 03 740-5029
Korea, Seoul: 02 794-5011
Malaysia, Petaling Java: 03 757-5511

Mexico:
Juarez, Chihuahua: 16 18-6701/02
Mexico City: 52-55-33-3858/59

New Zealand, Auckland: 09 894-160
Norway, Oslo: 02 74-10-10
Peru, Cadesa, San Isidro: 14 707-080
Philippines, Makati-Rizal: 63-2-810-01-61
Portugal, Lisbon: 019 68-31-21
Singapore, Singapore: 65 350-2000
South Africa, Martindale: 27-11-470-5911
Spain, Barcelona: 03 301-63-12
Sweden, Stockholm: 0 782-10-00
Switzerland, Zuerich: 01 488-2211
Taiwan, Taipei: 886-2-500-5842
Thailand, Bangkok: 02 233-6330-9
Turkey, Ticaret A.S., Istanbul: 01 179-27-70
United Kingdom, London: 01 580-6633
Uruguay, Montevideo: 91 56-41/42/43/44
Venezuela, Caracas (Magnetica S.A.): 02 241-7509
Zimbabwe, Harare: 4 47211

(Op Amps, microcontrollers, programmable ASICs, RF/wireless communication devices [LNAs (low noise amplifiers) FM/IF demodulators, audio processors, prescalers, synthesizers, cellular radio/paging/cordless phone chipsets], programmable logic [CPLDs, PALs, FPGAs, address decoders, sequencers], BiCMOS, CMOS and Bipolar ICs, data communications ICs [serial data controllers, multi-protocol converters, controllers, I/O processors, 10base-T Ethernet ICs], Video and Audio ICs [encoders, decoders and image resizing ICs, MPEG parts including MPEG-1 and MPEG-2 ICs], UARTs [Universal Asynchronous Receiver/Transmitter ICs], GPS [Global Positioning System] chipsets, fluorescent starter IC, television, and military qualified parts, surface PTC thermistors. Note: Philips discontinued manufacturing EPROMs in 1994.)

Philips Components, Discrete Products Division

(Ferrites, diodes, transistors, resistors, capacitors)
2001 W. Blue Heron Blvd.
Riviera Beach, FL 33404
407-881-3200
1-800-447-3762

(Magnetoresistive sensors and other automotive and industrial products, including video decoders and scalers)

Philips Photonics

100 Providence Pike
P.O. Box 278
Slatersville, RI 02876
401-762-3800, 401-769-8304
Fax: 401-767-4493, 401-769-4335

3 Baldwin Green Common, Suite 201
Woburn, MA 01801
617-932-4748
Fax: 617-932-4612

France: 33 1 40996069
Fax: 33 1 40996402
Germany: 49 40 3296640
Fax: 49 40 3296917
Japan: 81 3 37405031
Fax: 81 3 37405035
United Kingdom: 44 1306512121
Fax: 44 1306512122

PHILIPS SEMICONDUCTOR — PHOTRON

Philips Components:
Asia: 65 350 2553
Fax: 65 252 3068
Australia: 0 2 8054403
Fax: 0 2 8054466
Austria: 0 1 60101 1220
Fax: 0 60101 1975
Belgium: 0 2 7418211
Fax: 0 2 7358667
Bulgaria: 32 2 7625396
Fax: 32 2 7625862
Brazil: 0 11 4598211
Fax: 0 11 4598282
China: 852 4245121
Fax: 852 4806960
CIS: 31 40 723217
Fax: 31 40 723085
Czech/Slovakia: 49 40 3296640
Fax: 49 40 3296917
Denmark: 0 86 848484
Fax: 0 86 848244
Finland: 9 050261
Fax: 9 0520971
Greece: 0 1 4894911
Fax: 0 1 4815180
Hong Kong: 4245121
Fax: 4806960
Hungary: 49 40 3296640
Fax: 49 40 3296917
India: 0 3 22 4921550
Fax: 0 22 4938722
Ireland: 0 1 640203
Fax: 0 1 640210
Italy: 5843883398
Fax: 584388959
Netherlands: 0 20 5696611
Fax: 0 20 6953941
New Zealand: 0 9 8494160
Fax: 0 9 8497811
Norway: 0 22 748000
Fax: 0 22 748341
Pakistan: 0 21 577031 to 35
Fax: 0 21 5691832
Poland: 49 40 3296640
Fax: 49 40 3296917
Portugal: 0 1 3883121
Fax: 0 1 3883208
Romania: 32 2 7625396
Fax: 32 2 7625862
South Africa: 0 11 4705434
Fax: 0 11 475494
Spain: 0 1 4042200
Fax: 0 1 3265453
Sweden: 0 8 6322403
Fax: 0 8 6322745
Switzerland: 0 1 4882211
Fax: 0 1 4817730
Taiwan: 0 2 5097666
Fax: 0 2 5005827
Turkey: 0 1 2792770
Fax: 0 1 2693094

Other Countries:
33 55 863757
Fax: 33 55 863773

(Image intensifier tubes, photomultiplier tubes, streak tubes, image detector tubes, microchannel plates, neutron detectors, electron multipliers)

Phoenix Contact, Inc.
P.O. Box 4100
1 Phoenix Plaza
Harrisburg, PA 17111
800-888-7388
Fax: 717-944-1625
http://www.phoenixcontact.com
http://industry.net/phoenix.contact

W160 N10174 Cherokee Court
Germantown, WI 53022
414-251-0059
Fax: 414-251-0597

20115 Redwood Road #12
Castro Valley, CA 94546
303-438-0458

1811 Santa Rita Road, #222
Pleasantron, CA 94566
510-426-1515

(Electronic and electrical sockets, contacts, connectors; including DIN rail types; terminal blocks, interface and relay systems, signal condition modules, I/O modules, transient protection devices, DIN rail mounted power supplies, distributed I/O systems)

Phoenix Logistics, Inc.
2202 East Magnolia
Phoenix, AZ 85034
602-231-8616

(MIL-STD-1553 data bus couplers and RF cables assemblies for nuclear, military and aerospace applications)

Phoenix Microwave Corp.
100 Emlen Way
Telford, PA 18969
215-723-6011
Fax: 215-723-6015

(Microwave frequency amplifiers)

Photek
26 Castleham Road
St. Leonards on Sea
East Sussex TN38 9NS
44 424 850555
Fax: 44 424 850051
In the U.S.: 602-967-0602,
Fax: 602-967-7770

(Biplanar photodiodes)

Photonics Imaging
6967 Wales Road
Norwood, OH 43619
419-666-6325
Fax: 419-666-0785

(AC gas [color plasma] discharge displays)

Photron Semiconductor Corp.
No. 45, Kuang Fu S. Road, Hsinchu Industrial Park
Hukou, Hsinchu Hsien
Taipei, Taiwan R.O.C.
886 35 985 889
Fax: 886 35 982 700

(Diodes [including schottky types], thyristors, transistors, DIACs, SCRs and SIDACs, are available in the U.S.

PHOTRON — PIXEL VISION

from Taitron Components, Inc., 25202 Anza Drive, Santa Clarita, CA 91355-3496, 800-824-8766, 800-247-2232, 805-257-6060; Fax: 800-824-8329, 805-257-6415)

Photonics Research, Inc. (PRI)
350 Interlochen Blvd.
Broomfield, CO 80021-3400
303-465-6491

(Vertical cavity surface emitting lasers)

PHY prefix
See *Phylon, Inc.*

Phy-Chem Scientific Corporation
36 West 20th Street
New York, NY 10011
212-924-2070
Fax: 212-243-7352

(Humidity sensors)

Phylon Inc.
4027 Clipper Ct.
Fremont, CA 94538
510-656-2606, Fax: 510-656-0902

(Fax/Voice/Modem chip sets, DSVD products for networks and PCs, etc.)

PI
See *Polytronix, Inc.*

(Displays)

PI prefix
See *Pericom Semiconductor Corporation*

PIC prefix
See *MicroChip Technology, Pericom, Inc.*

(Transceivers for token ring networks)

Piccolo
(DSP coprocessor cores by Advanced RISC Machines, Ltd., England)

PICO Electronics, Inc.
453 N. MacQuesten Parkway
Mt. Vernon, N.Y. 10552
800-431-1064, 914-699-5514
Fax: 914-699-5565

(DC-DC Converters, Transformers, Power Inductors)

PicoPower Inc.
2680 N. First Street
San Jose, CA 95134
408-954-8880

(80X86 core logic chip sets; a subsidiary of Cirrus Logic)

Piezo Crystal Co.
100 K Street
Carlisle, PA 17013
717-249-2151
Fax: 717-249-7861

(Crystals, crystal oscillators, commercial, military and high reliability applications, discrete component type units.)

Piezo Technology, Inc. (PTI)
P.O. Box 547859
Orlando, FL 32854

2525 Shader Road
Orlando, FL 32804
407-298-2000
Fax: 407-293-2979

(TCXOs temperature controlled crystal oscillators, switched filters)

Pilkington Microelectronics Ltd.
Sherwood House
Gadbrook Business Centre
Rudheath Northwich
Chesire, United Kingdom
CW9 7TN
0606 49582

Fax: 0606 49615
E-mail: info@pmel.com
Software: ftp.cica.indiana.edu
micros.hensa.ac.uk
WWW sites thru Mosaic
(through Demon Systems):
http://demon.co.uk/pmel

(FPGAs, PLDs, programmable video DSPs)

PINE
(Trade name of a DSP Semiconductor digital signal processor IC)

Pioneer Semiconductor Corp.
See *Pericom Semiconductor Corp.*

Pixel Magic, Inc.
138 River Road
Andover, MA 01810
Fax: 508-688-5475
E-mail: hr@pixelmagic.com

(Digital signal processors and image processing ICs. This company acquired by Oak Technology in 1995)

Pixel Semiconductor
3460 Lotus Drive
Plano, TX 75075
214-985-2345

(Programmable digital video processors, image resizing ICs, etc., became a subsidiary of Cirrus Logic in 1991)

Pixel Vision
Advanced Sensors Division
Huntington Marina Executive Center
4952 Warner Avenue, Suite 300
Huntington Beach, CA 92649
714-840-9778
Fax: 714-840-2627

(CCD and imaging sensors for scientific, space, military, medical and commercial applications)

PIXTECH — PMB

PixTech, Inc.
3350 Scott Blvd.
Building #37
Santa Clara, CA 95054
408-986-8868
Fax: 408-986-9896
http://www.Pixtech.com

PixTech SA
Montpellier, France

(FED [field emission] flat panel displays)

PKC prefix
See *On Chip Systems*

Planar Systems, Inc.
1400 N.W. Compton Dr.
Beaverton, OR 97006
503-690-1100, 503-863-3213,
503-690-6967
Fax: 503-690-1493

Planar International. Ltd.
P.O. Box 46
FIN-02201
ESPOO, Finland
358 0 420 01
Fax: 358 0 422 143

(Flat panel [EL] displays)

Plasmaco, Inc.
South Street
Highland, NY 12528
914-883-6800

(Plasma display panels, this company was acquired by Matsushita Industrial Co., Ltd. in January 1996)

Platan
Fryazino, (near) Moscow

(LCDs for portable computer applications)

PLED prefix
See *Optical Communication Products, Inc.*

PLD prefix
See *Optical Communication Products, Inc.*

Pletronics, Inc.
19015 36th Avenue W
Suite H
Lynnwood, WA 98036-5762
206-776-1880
Fax: 206-776-2760

(Crystals and oscillators)

PLC prefix
See *Philips Semiconductors* (Signetics),
Adaptive Networks, Inc.

PLCD prefix
See *Siemens*

PLD prefix
See *Cypress Semiconductor*

Plessy Semiconductor Corporation
P.O. Box 660017
1500 Green Hills Road
Scotts Valley, CA 95066
408-438-2900

(FPGAs, licensed from Pilkington Technology, etc. NOTE: Marconi Electronic Devices, LTD. and Plessy Semiconductors LTD. have been grouped together to form GEC Plessy Semiconductors, GEC purchased Plessy plc in 1989)

Plessy Tellumat
South Africa

In the U.S. contact:
LNY Sales, Inc.
548 Sunrise Highway
West Babylon, NY 11704-6003
516-661-8900

(Low noise wireless communications amplifiers)

PLH prefix
See *Philips Semiconductors*

(Signetics)

PLH-XX part number
See *KDI/triangle Electronics, Inc.*

PLHS prefix
See *Philips Semiconductors*

(Signetics)

pLSI prefix
See *Lattice Semiconductor Corporation*

PLUS prefix
See *Philips Semiconductors*

(Signetics)

Plus Logic
(Acquired by Xlinx in 1992)

PLX Technology
390 Potrero Avenue
Sunnyvale, CA 94086
408-774-9060
Fax: 408-774-2169
http://www.plxtech.com

(PCI bus interface, microprocessor support, programmable logic, and clock distribution ICs)

PM prefix
See *PMC-Sierra, Inc.*,
Analog Devices
(Precision Monolithics Division)

(If P subscript M [PM], see Premier Magnetics, Inc., Pixel Magic, Inc.)

PMB prefix
See *Siemens Components*

PMC-SIERRA — POLYTRON DEVICES

PMC-Sierra, Inc.
8501 Commerce Ct.
Burnaby, BC, Canada V5A 4N3
604-668-7300
Fax: 604-668-7301
e-mail: info@pmc-sierra.bc.ca

30 March Road
Suite 430
Kanata, Ontario
Canada K2K 2E2
613-599-7270
Fax: 613-599-8067

One Apple Hill
Suite 316
Natick, MA 01760-5659
508-650-3431
Fax: 508-650-3434

2116 Walsh Avenue
Suite C2
Santa Clara, CA 95050
408-986-1222
Fax: 408-986-1224

European Head Office:
St. James Court
Wilderspool Causeway
Warrington Chesire WA4 6PS, England
44 19 25 65 11 12
Fax: 44 19 25 65 00 33
E-mail: livesey@pmc-sierra.bc.ca

Germany and Central Europe:
Sierra Semiconductor GmbH
Rosenkavalierplatz 12
81925 Munchen
Germany
49 89 9101088
Fax: 49 89 9102232

Southern Europe:
Sierra Semiconductor SRL
Via Poli, 14
24081 Arona (Novara), Italy
39 322 45744
Fax: 39 232 48107

United Kingdom and Northern Europe:
Sierra Semiconductor Ltd.
Terminal 3-3B2
Stonehill Green
Westlea, Swindon SN5 7HB
England
44 793 618492
Fax: 44 793 618589

(This fabless company, spun off from MPR Teltech, Inc. was the research arm of the British Columbia phone company. Sierra took a majority stake in the company in 1990 and acquired it in September 1994. The company offers data communications chip sets including SONET based interface ICs for ATM local area networks [LANs], Chip Framer that can terminate 4 T1 digital telephone lines, and Asynchronous transfer mode ICs)

PMD
See *Performance Motion Devices*

PMI, Precision Monolithics, Inc.
See *Analog Devices*

PML prefix
See *Philips Semiconductors*

(Signetics)

PN prefix
See *Amplifonix*

PNY Electronics, Inc.
200 Anderson Ave.
Moonachie, NJ 07074
201-438-6300
Fax: 201-438-9144

P.O. Box 144
Aubudon, NJ 08106
609-547-1219
Fax: 609-854-5644
http://www.pny.com

(Memory modules)

PO prefix
See *Zetex*

Poach (Chip set)
See *Appian Technology, Inc.*

(ZyMOS)

Polara Engineering, Inc.
4115 W. Artesia Avenue
Fulerton, CA 92633
714-521-0900
Fax: 714-522-8001

(Delay lines)

Polaroid Corporation
549 Technology St.
Cambridge, MA 02139
119 Windsor Street
Cambridge, MA 02139
800-343-5000
617-577-4681
Fax: 617-577-3303

(Piezo transducers with asymmetrical beam angle and ability to withstand harsh environments)

Pole/Zero Corporation
5530 Union Center Drive
West Chester, OH 45069-4821
513-870-9060

(Digitally tuned notch filters)

Polyfet RF Devices
1110 Avenedia Acaso
Camarillo, CA 93012
805-484-4210
Fax: 805-484-3393

(Power MOSFET RF transistors)

Polytron Devices, Inc.
P.O. Box 398
River Street Station
Paterson, NJ 07544
201-345-5885
Fax: 201-345-1264

(DC/DC converters and power supplies)

Polytronix, Inc.
805 Alpha Drive
Richardson, TX 75081
214-238-7045
Fax: 214-644-0805

(LCDs, sister company is Picvue Electronics in Taiwan)

Portage Electric Products, Inc. (PEPI)
7700 Freedom Avenue, N.W.
Canton, OH 44720
1-888-GO-4-PEPI
216-499-2727, Fax: 216-499-1853

(Thermal protectors)

Potter & Brumfield
See *Siemens Electromechanical Components*
(Acquired by Siemens in 1986).

Powerchip Semiconductor Corporation
8 Fl., Sec. 3
Nanking E. Road
Taipei, Taiwan R.O.C.
886 2 517 0055, Fax: 886 2 517 2017
Taipei, Taiwan

(Mitsubishi Electric Corp., the Japanese Trading firm Kanematsu Corp. and UMAX of Taiwan established this new DRAM manufacturing company in late 1994. It started sample production of 16M DRAMs in mid 1996. Devices are sold under the Mitsubishi and UMAX logos. In the U.S. parts are sold directly to OEMs [original equipment manufacturers])

Power Components of Midwest
P.O. Box 1348
Mishawaka, IN 46545
800-2211-9257
Fax: 219-256-6643

(Tilt sensors that do not use mercury)

Powercube Corp.
8 Suburban Park Dr.
Billerica, MA 01821
508-667-9500

(Power supplies, DC/DC convertors)

Power Convertibles
3450 S. Broadmont Dr.
Suite 128
Tucson, AZ 85713
602-628-8292
(520 area code after 12/31/96)
Fax: 602-770-9369
(520 area code after 12/31/96)
Datafax: 520-628-8691
http://www.pcc1.com

(DC/DC converters)

Power General
Division of Nidec Corp.
Canton, MA
617-828-6216
Fax: 617-828-3215

(DC/DC converters)

Power Integrations, Inc.
411 Clyde Avenue
Mountain View, CA 94043
415-960-3572
Fax: 800-468-0809, 415-940-1226

(Power switching regulators [including TOPSwitch, a three terminal off line phase width modulator switch- a three phase modulator and a high voltage power MOSFET in a TO-220 or D pack package] and parts used in solid-state electronic ballasts)

Power-One
740 Calle Plano
Camarillo, CA 93012
805-987-8741
Fax: 805-388-0476
http://www.power-one.com

POLYTRON DEVICES — POWER SOLUTIONS

(Power supplies including power units)

Power Products International, Ltd.
Commerce Way
Edenbridge
Kent TN8 6ED
United Kingdom
44 732 866 424
Fax: 44 732 866 399

(Diodes)

Power Semiconductors, Inc.
6352-F Corte Del Abeto
P.O. Box 9000-365
Carlsbad, CA 92009
800-654-8721
619-438-7873
Fax: 619-438-0437

(Diodes, thyristors)

Power Sensors Corporation
1113 Tower Road
Schaumburg, IL 60173
1-800-9PLANAR
(1-800-975-2627)
847-884-5898
Fax: 847-884-5899

(High density switching power supplies, 33W/cubic inch and DC/DC converters)

Power Solutions, Inc.
4699 N. Federal Highway
Pompano Beach, FL 33064
305-943-4110
Fax: 305-943-6068

17 Hunters Lane
Roslyn, NY 11576
516-627-5628
Fax: 516-627-6032

(DC/DC converters, power supplies)

POWER SPECTRA — PRECISION DEVICES

Power Spectra, Inc.
919 Hermosa Ct.
Sunnyvale, CA 94086-4103
408-737-7977
Fax: 408-732-1832

(Diode lasers including pulsed quantum well types)

Power Switch Corporation
17 Vreeland Street
Lodi, NJ 07644
201-478-0300
Fax: 201-478-4650
E-mail: pwrswitch@aol.com

(Power supplies)

Power Systems, Inc.
Subsidiary of UMB (US), Inc.
45 Griffin Road South
Bloomfield, CT 06002
860-726-1300
Fax: 860-726-1495

(DC/DC converters [the VersaMod (tm)], units similar to Vicor Corporation designs)

Power Tech Inc.
0-02 Fairlawn Avenue
Fairlawn, N.J. 07410
201-791-5050
Fax: 201-791-6805
TLX: 353970

(High current transistors, military qualified semiconductors, etc.)

Power Technology Inc.
P.O. Box 191117
Little Rock, AR 72219-1117
Plant Location
7925 Mabelvale Cutoff
Mabelvale, AK 72103-2213
501-568-1995
Fax: 501-568-1994

(Laser diode systems, laser power supplies, laser diode pulse drivers)

Powertip Technology Corporation
Taiwan
In the U.S.
Okaya Electric America, Inc.
503 Wall Street
Valpariso, IN 46383
219-477-4488
Fax: 219-477-4856
Faxback: 1-800-755-8561

(LCDs)

Power Trends, Inc.
1101 North Raddant Road
Batavia, IL 60510
708-406-0900
Fax: 708-406-0901
http://www.powertrends.com/isr/

(Switching regulators and DC-DC converters)

Powerex
(A division of Westinghouse, General Electric and Mitsubishi)
200 Hillis Street
Youngwood, PA 15697
1-800-451-1415
Fax: 412-925-4393

(Power semiconductors including fast diodes, power transistors, bridge rectifiers, fast recovery rectifiers, IGBTs, thyristor and diode modules)

PowerPC
A trademark of *Motorola*

(RISC processors jointly developed by Apple Computer, IBM and Motorola in an alliance formed in October 1991, with the first IC the PowerPC 601 fabricated in October 1992 and released in April 1993)

PowerPlay64
See *Videologic, Inc.*

Powertech, Inc.
0-02 Fair Lawn Avenue
Fair Lawn, NJ 07410
201-791-5050, Fax: 201-791-6805
TLX: 353970 POWERTECH UD
http://www.power-tech.com

(Optoelectronics, high power silicon transistors, thyristors)

PPC Products Corporation
7516 Central Industrial Drive
Riviera Beach, FL 33404
407-848-9606, Fax: 407-848-1607

(Transistors, power transistors, rectifiers, including military qualified parts)

PPM prefix (PPM-XXXX)
See *On Chip Systems*

PPMC prefix
See *Dimolex Corp.*
EMBE Electronik

(Stepper motor controllers)

PQ prefix
See *Sharp*

(Voltage regulators for switching power supplies)

PQR prefix
See *Lintel NV/SA*

PR prefix
See *Philips Semiconductors*

PRCD prefix
See *California Micro Devices*

(Resistor-capacitor arrays)

Precision Devices, Inc. (PDI)
8840 N. Greenview Drive
Middletown, WI 53562
1-800-274-XTAL
Fax: 608-831-3343
Sales@PDIXTAL.Com

Manufacturers, Prefixes, Part Number Types, Logo Descriptions & Family Types

(Crystals, crystal oscillators, crystal filters)

Precision Monolithics, Inc.
See *Analog Devices*

Premier Magnetics, Inc.
27111 Aliso Creek Road, Suite 175
Aliso Viejo, CA 92656
714-362-4211, Fax: 714-362-4212

(Various magnetic products including transformers, inductors, saturable reactors, low power inverters, balun filters, floating switches, clock generators, delay lines, pulse discriminators, pulse width generators, DC/DC converters, networking products)

PREM Magnetics, Inc.
3521 N. Chapel Hill Road
McHenry, IL 60050
815-385-2700
Fax: 815-385-8578

(Transformers)

Prime View International Co. Ltd.
21 Prosperity Rd.
Science-Based Industrial Park
Hsinchu, Taiwan R.O.C.
886 35 773 233
Fax: 886 35 773 033

(Active matrix LCD panels)

Princeton Technology Corporation
Republic of China, Taiwan
886 2 916 2151
Fax: 886 2 917 4598

(Remote control encoder/decoders)

Projects Unlimited, Inc.
3680 Wyse Road
Dayton, OH 45414
513-890-1918
Fax: 513-890-4911
Internet: sales@pui.com

(Audio alert transducers and alarms)

ProSemi GmbH
Robert Bosch Strasse 28
Bensheim 64625
Germany
6251 63036
Fax: 6251 63 030

(Voltage regulators, switching transistors)

Protek Devices
2929 South Fair Lane
Tempe, AZ 85282
602-431-8101
Fax: 602-431-2288

(Diode arrays, transient suppressors including devices for I/O ports and data lines)

Provector AB
P.O. Box 402
Hagersten 12904, Sweden
46 8 880 511
Fax: 46 8 880 311

(Encoder ICs)

Proxim, Inc.
295 N. Bernardo Avenue
Mountain View, CA 94043
800-229-1630
415-960-1630
Fax: 415-964-5181

(Numeric controlled oscillators)

Proxitronic
Robert-Bosch Str. 34
D-64625 Bensheim, Germany
49 0 6251 17030
Fax: 49 0 6251 170390

Stanford Computer Optics, Inc.
415-494-7797
Fax: 415-494-6711

(Proximity focus optoelectronics for high resolution UV imaging)

PRECISION DEVICES — PUI

PS prefix
See *Hitachi Semiconductor* (SONET devices), *ProSemi GmbH*

PSXX (or **PSXXX**) part number
See *I-Cube, Inc.*

(Crossbar switch)

PSB prefix
See *Siemens*

PSD prefix
See *WaferScale Integration, Inc.* (WSI), *Philips Semiconductors* (Second source for these field programmable microcontroller peripheral ICs)

PSL prefix
See *Blue Sky Research*
(Point source laser diodes)

PSR prefix
See *Photron Semiconductor*

PSW-XXXX part number
See *Mini-Circuits*

PT prefix
See *On Chip Systems*,
Power Trends (Voltage regulators),
Princeton Technology Corporation

PTC
See *Princeton Technology Corporation*

PTI
See *Piezo Technology, Inc.*

PTM prefix
See *Opti, Inc.*

PUI logo
See *Projects Unlimited, Inc.*

P – Q

PULSAR — Q-BIT

Pulsar Microwave
10 Blanjen Terrace
Clifton, NJ 07014
201-667-0224
Fax: 201-677-0424

(Microwave hybrids)

Pulse Engineering, Inc.
See *Pulse*

Pulse
12220 World Trade Drive
San Diego, CA 92128
P.O. Box 12235
San Diego, CA 92112
619-674-8100, 8203, 8374
Fax: 619-674-8262
QuickFax: 619-674-9672
(Press 2 for information on use, enter 1001 for menu of data sheets)
http://www.pulseeng.com

Dunmore Road
Tuam, County Galway, Ireland
353 93 24107
Fax: 353 93 24883

England: 44 522 688699
Germany: 49 89 96 3046
France: 31 1 60 19 111
Hong Kong: 852 425 1651

(Token ring filters, delay lines, magnetics, PCMCIA wireless transceivers)

Fil-Mag locations:
9445 Farnham
San Diego, CA 92123
619-569-6577
Fax: 619-569-6073

FEE S.A.
Zone Industrielle
39270 Orgelet, France
33 84 35 53 83
Fax: 33 84 25 46 41

Fil-Mag GmbH
Raiffeisenstr.2
D6054 RODGAU
Germany
49 06106-2011
Fax: 49 06106-24286

FIL-Mag Taiwan Corp.
P.O. Box 26-11
6 Central Sixth Road
KEPZ, Kaohsiung, Taiwan
886 7-821-3141
Fax: 886 7-841-9707

Fil-Mag Singapore PTE LTD
Block 3027, UBI Road 1
#03-120
Singapore, 1440
65 7415277
Fax: 65 7413013

(Formerly FEE Fil-Mag, then Fil-Mag. In October Pulse Engineering was acquired by Fil-Mag and them became known as Pulse, a wholly owned subsidiary of Technitrol. This company manufacturers communication products such as data line transformers and 10Base-T filters, DC/DC converters, impedance matching transformers, RF transformers, current transformers, delay lines.)

Punjab Semiconductor Devices
Punjab Wireless Systems, Ltd.
B-53 Phase IV S.A.S.
Nager (Mohali)
Chandigarh, 160 055 India
91 172 87652, 87657, 87867
Fax: 91 172 0395 319

PV prefix
See *International Rectifier*

(Photovoltaic relay)

PVT prefix
See *International Rectifier*

(Photovoltaic relays)

PWR-XXXX part number
See *ILC-Data Device Corp.*

PWR-TOP part number
See *Power Integrations, Inc.*

PX prefix
See *Pixel Semiconductor*

PXB prefix
See *Siemens Components, Inc.*

PXXXXH part number
See *Westcode Semiconductors*

PYR part number
See *Sensar, Inc.*

Pyromation, Inc.
5211 Industrial Road
Fort Wayne, IN 46825
219-484-2580
Fax: 800-837-6805

(RTDs and thermocouples)

PZ prefix
See *Keyence Corp. of America, Photoelectric Sensors, Philips Semiconductors, Inc.*

(CPLDs)

PZT prefix
See *Motorola*

Q prefix
See *Quality Semiconductor, Inc., Qualcomm, Inc.*

QB prefix
See *QBAR Tech, Inc.*

Q-bit Corporation
2144 Franklin Drive NE
Palm Bay, FL 32905
800-236-1772
407-727-1838
Fax: 407-727-3729
TWX: 510-959-6257

(RF amplifiers)

QBAR
(A trademark of QBAR Tech Inc. for a non-inverting digital transistor, a solid-state normally closed switch.)

QBAR Tech, Inc.
264-260 Hearst Way
Kanata, ON K2L 3H1 Canada
613-592-3633, Fax: 613-592-2344

812 Proctor Avenue
Odgensburg, NY 13669
315-393-3793
Fax: 315-393-9017

QBH prefix
See *Q-bit Corporation*

QC prefix
See *GEC Plessy*

(Crystal oscillators)

QL prefix
See *QuickLogic Corp.*

QLogic Corp.
3545 Harbor Blvd.
Costa Mesa, CA 92626
800-662-4471
(1-800-ON-CHIP-1)
714-438-2200
Fax: 714-668-6950

(This company was formerly the Emulex Micro Devices [EMD] Division, renamed in May 1993. Interface ICs, SCSI processors for disk drive controllers)

QLS prefix
See *Triquint Semiconductor*

QR prefix
See *National Semiconductor*

QRB prefix
See *Quality Technologies Corporation*

QRM prefix
See *TRM, Inc.*

QSI prefix
See *Quality Semiconductor, Inc.*

Q Source, Inc.
2852 F Walnut Avenue
Tustin, CA 92680
714-832-3312
Fax: 714-832-5010

(Diodes, thyristors, transistors)

Q-Tech Corporation
10150 W. Jefferson Blvd.
Culver City, CA 90232-3510
213-836-7900
Fax: 213-836-2157
TLX: 69-6140

2201 Carmelina Ave.
Los Angeles, CA 90064
213-820-4921, Fax: 213-207-2248
TLX: 69-6140

(Crystal, clock oscillators and hybrid devices)

QT Optoelectronics
610 N. Mary Avenue
Sunnyvale, CA 94086
800-LED-OPTO

(Surface mount chip LEDs)

QTLP prefix
See *QT Optoelectronics*

QT prefix
(For crystal oscillators/crystals)
See *Q-Tech Corp.*,
Quality Thermistor, Inc.

(Thermistors)

Qualcomm, Inc.
10555 Sorrento Valley Road
San Diego, CA 92121
619-597-5005, Fax: 619-452-9096
WWW: http://www.qualcomm.com/

Q-BIT — QUALITY SEMICONDUCTOR

(Speech compression ICs, CDMA. Code Division Multiple Access) ASICs)

Quality Semiconductor, Inc. (QSI)
851 Martin Avenue
Santa Clara, CA 95050-2903
408-450-8000
408-450-8080
(Headquarters and bay area region)
408-450-8019, 986-1700,
986-8063 (Sales)
408-986-8326 (Headquarters)
408-450-8051 (Far East sales)
Fax: 408-496-0591 (Headquarters),
408-496-0773 (Sales)
1-800-609-3669 (For literature)
Fax: 818-365-5428
(For literature requests, fax requests to this number)
http://www.xmission.com/~qsi
http://www.qualitysemi.com

Western region USA:
6 Morgan, Suite 100
Irvine, CA 92718
714-587-3040, Fax: 714-587-1521

Central region USA:
555 Republic Drive, Ste. 200
Plano, TX 75054
214-516-0488, Fax: 214-516-0394

Eastern region USA:
945 Concord Street
Framingham, MA 01701
508-620-4725
Fax: 508-626-2601

European Headquarters:
Suite A, Unit 6
Mansfield Park,
Four Marks, Hampshire GU34 5PZ
United Kingdom
44 0 1420 563333
Fax: 44 0 1420 561142

QUALITY SEMICONDUCTOR — QUIXTAL

(A fabless company supplying logic [such as FCT logic], SRAMs, switching and FIFOs, Asynchronous Transfer Mode [ATM] network ICs, bus switches, bus exchange, multiplexers/demultiplexers, synchroswitches, crossbar switches, JTAG ICs)

Quality Technologies Corporation
610 North Mary Avenue
Sunnyvale, CA 94086
408-720-1440
Fax: 408-720-0848
TLX: 6731382
MCI Mail ID: 4286653

758 Highway 18, Suite 108
East Brunswick, NJ 08816
201-254-1588
Fax: 201-254-5875

3030 Holcomb Bridge Road, Suite F
Norcross, GA 30071
404-446-2939
Fax: 404-446-3805

Trinity Centre
2340 E. Trinity Mills Road, Suite 330
Carrollton, TX 75006
214-418-2953
Fax: 214-416-8361

Quality Technologies Europe N.V.
24A Stationsstraat
1720 Groot-Bijgaarden
Belgium
02 466 35 40
Fax: 02 466 67 89
TLX: 24157

Quality Technologies GmbH
Max-Huber-Strasse 8
8045 Ismaning West Germany
089 96 30 51
Fax: 089 96 54 74
TLX: 5214223

Quality Technologies Italia S.R.L.
Viale Campania 42
20133 Milano, Italy
02 73 06 69
Fax: 02 74 04 76

Quality Technologies UK Ltd.
10, Prebendal Court, Oxford Road
Aylesbury, Buckinghamshire
HP 19-13EY, United Kingdom
0296 39 44 99
Fax: 0296 39 24 32
TLX: 83192

Quality Technologies France S.A.
Immeuble La Pyramide
80 Avenue du General de Gaule
94009 Creteil Cedex France
01 43 99 25 12
Fax: 01 43 99 17 41
TLX: 262819

(Optical couplers, infrared sensors and emitters, displays [seven segment and multicolor digits], light bars and bargraph displays, solid state lamps, optical proximity switches)

Quality Thermistor, Inc.
2147 Centurion Place
Boise, ID 83709
800-554-4784
Fax: 208-376-4754

(Thermistors)

Quantum Effect Design, Inc. (QED)
3255-3 Scott Blvd, Suite 200
Santa Clara, CA 95054
408-565-0315
Fax: 408-565-0335
http://www.qedinc.com
E-mail: rkepple@qedinc.com

(Design of microprocessor solutions for customers, including designing specific microprocessors)

Qua-tron Electronics Industrial Co., Ltd.
No. 14-1 Taiji Road
Hsiao Kand District
Kaohsiung, Taiwan
886 7 802 3890, 886 7 803 4554
Fax: 886 7 802 3884

(Diodes)

QuickLogic Corporation
2933 Bunker Hill Lane
Santa Clara, CA 95054
1-800-842-3742
(1-800-842-FPGA)
408-987-2000
Fax: 408-987-2012
e-mail: info@qlogic.com
http://www.quicklogic.com

1 Batchelder Rd.
Marblehead, MA 09145
Phone/Fax: 617-631-4279

516 23rd St.
Manhattan Beach, CA 90266
310-546-5589
Fax: 310-546-5589

2613 Hampshire Drive
Garland, TX 75040
214-414-6908
Fax: 214-414-8374

Europe: 49 89 899 143 28
Fax: 49 89 857 77 16

(A fabless company supplying one time programmable antifuse FPGAs, ASICs, military screened parts available. Note: This company was acquired by Cypress Semiconductor in 1995).

QuikVoice
See *Eletech Electronics*, *Eletech Enterprise Co.*

QuiXTAL
See *IC Designs* (A subsidiary of Cypress Semiconductor)

R — RANCHO TECHNOLOGY

R prefix
See *Rockwell International Corp.*,
Raytheon Semiconductor
(PROMs),
Integrated Device Technology
(IDT, microprocessors),
Radisys Corporation,
Toshiba (RISC processors),
Isotek Corporation
(Surface mount resistors)

RXX part number
(On SOT transistor)
See *Rohm*

RXXXCHx part number
See *Westcode Semiconductors*

R4 part number
See *Toko America, Inc.*

R4000, R4400
See *MIPS Technologies, Inc., IDT*

R7H prefix
See *Optotek Ltd.*

R&E International, Inc.
210 Goddard Blvd
King of Prussia, PA 19406
800-253-7007
215-992-0727
Fax: 215-992-0734

(CMOS digital logic, including parts formerly made by SSS and Sprague Semiconductor)

RAB prefix
See *Ledtronics*

(LEDs)

RAD prefix
See *Federal Systems Company*
(RAD hard chip sets),
Infinite Technology Corporation
(An IC with four 16 BIT arithmetic processors)

Rad Data Communications
900 Corporate Drive
Mahwah, N.J. 07430
201-529-1100
Fax: 201-529-5777

8 Hanechoshet Street
Tel Aviv 69710 Israel
972-2-6458181
TLX: 371263 RADCO IL
Fax: 972-3-498250, 5447851

Radio Shack
Tandy Distributor
Sales/Tandy National Parts
900 E. Northside Drive
Fort Worth, TX 76102
817-870-5600
Fax: 817-332-4216
Fax: 817-332-4216
Sales: 1-800-THE-SHACK

(Service manuals and repair parts for Radio Shack merchandise, Micronta, Optimus, Radio Shack, etc.)

Radisys Corporation
15025 Koll Parkway South
Beaverton, OR 97006
503-646-1800
Fax: 503-646-1850

(Control logic chip sets [memory/bus] for microprocessors, embedded PCs)

Radstone Technology Corporation
20 Craig Road
Montvale, NJ 07645
800-368-2738
Fax: 201-391-2899

(Custom ICs are often found on their circuit cards)

RAH prefix
See *KDI/Triangle Electronics, Inc.*

Rakon Ltd.
Private Bag 99943, Newmarket
1 Pacific Rise, Mt. Wellington
Auckland, New Zealand
64 9 573 5554
Fax: 64 9 573 5559

(Crystals and TCXOs)

RAL logo
See *Raltron Electronics Corporation*

Raltron Electronics Corporation
2315 NW 107th Ave.
Miami FL 33172
305-593-6033
Fax: 305-594-3973
Telex: 441588 RALSEN
E-mail: sales@raltron.com

(Crystals, crystal filters, ceramic resonators, VCO Oscillators)

Ramtron International Corporation
1850 Ramtron Drive
Colorado Springs, CO 80921
800-545-3726
719-481-7000
Fax: 719-488-9095, 719-481-9170
e-mail: info@ramtron.com
web server: http//www.csn.net/ramtron

(Nonvolatile ferroelectric RAMs, known as FRAMs [This technology also licensed to Hitachi Ltd.], EDRAMS [second source of IBM]. See also *Enhanced memory Systems, Inc.*)

Rancho Technology, Inc.
10783 Bell Court
Rancho Cucamonga, CA 91730
909-987-3966
Fax: 909-989-2365

RANCHO TECHNOLOGY — RAYTHEON

Ranch International
CH-6900 Lugano
Via Sorengo, 6
Switzerland
41 91 568722
Fax: 41 91 569843

(SCSI interface products including: single IC enhanced parallel port SCSI adapter, CD ROM kits, SCSI host adapter kits, PCMCIA SCSI-2 adapter card, SCSI-2/3 con-verters)

Rawmat Electronics (S) Pte Ltd.
1160 #02-05/07 Depot Road
Singapore 109647
65 271 9927
Fax: 65 270 8636

(Thick film networks, chip resistors)

RAY prefix
See *Raytheon Semiconductor*

Raychem Corporation
Circuit Protection Division
300 Constitution Drive
Menlo Park, CA 94025-1164
415-361-6900

(Polyswitch [trademark] positive temperature coefficient circuit protective devices, used in NiCad battery packs and on circuit boards)

Raytek, Inc.
1201 Schaffer Road
Box 1820
Santa Cruz, CA 95061-9924
800-227-8074
408-458-1110
Fax: 408-458-1239

(Non-contact thermocouples)

Raytheon Company
Semiconductor Division
350 Ellis Street
Mountain View, CA 94039-7016
415-968-9211
800-722-7074 (For literature)
Fax: 415-969-8556
Internet: Applications:
@lj.sd.ray.com
Sales: @lj.sd.ray.com
TWX: 910-379-6481

10A Goodyear
Irvine, CA 92718
714-830-2808
Fax: 714-830-2607

760 N. Mary Avenue
Sunnyvale, CA 94086
408-522-7072
Fax: 408-522-7061

130 Ridgewood Drive
Longwood, FL 32779
407-682-6988
Fax: 407-682-6404

1430 Branding Lane, #129
Downers Grove, IL 60515
708-810-1577
Fax: 708-810-1683

63 Second Avenue
Burlington, MA 01803
617-272-1313
Fax: 617-270-7965

3600 Ridgeway Road
Dayton, OH 45419-1125
513-293-1888
Fax: 513-293-8943

116 Sussex Place
Berwyn, PA 19312
215-640-4057
Fax: 215-640-4058

354 Zion Road
Neshanic Station, NJ 08853
908-369-8502
Fax: 908-369-8780

59 Cardinal Street
Pearl River, NY 10965
914-620-1326
Fax: 914-620-0951

Raytheon Semiconductor
International Company

France:
LaBoursidiere, RN 186
F-92350 Plessis Robinson

Cedex, France
33-1-46310676
Fax: 33-1-46324608
01 46 3106 76

Germany:
Kathi-Kobus-Str.24/11
D-80797 Munich Germany
49 89 187053
Fax: 49 89 183758

Japan:
Matsukaze Building 5/F
4-1-1 Kitashinagawa Shinagawa-Ku
Tokyo 104 Japan
81-3 3280 4776
Fax: 81-3-3280-4156

England:
Elizabeth Way
The Pinnacles
Harlow, Essex CM19 5AZ
England
44 0279 421510
Fax: 44 0279-421316

Pelican House
83 New Street
Andover, Hampshire SP10 1DR
England
44 264-334616
Fax: 44 264-334620

(Bipolar PROMs, small signal transistors, gate arrays, and Linear ICs such as precision operational amplifiers, multimedia, ATE, A/D and D/A converters, mixed signal ASICs, ARINC line drivers and receivers, DSP,

Multimedia, High Speed Data Converters, Image manipulation, and second sources to Analog Devices parts. Military qualified devices available. Note: Transistor products discontinued by Raytheon are now manufactured by Microsemi Corporation.)

Quincy Operations
465 Centre Street
Quincy, MA 02169
617-984-8493, 617-984-8508, 4104
(MCM modules)
Fax: 617-984-8515
Fax: 617-984-4193, 8515
(MCM modules)

(Power supplies, DC/DC converters, optoelectronics, MCM modules, special purpose CRT tubes)

RB prefix
See *ILC Data Device Corp.*

RBC prefix
See *Ricoh Corp.*

RBG-XXX part number
See *Siemens*
(Bar graph displays, discontinued)

RBV prefix
See *HV Component Associates, Inc.*

RC prefix
See *Raytheon Co.*, *Rockwell International Corporation*

RCA
For OEM devices
See *Harris Corporation*
(For generic replacement devices [SK series], see Thomson)

RCC prefix
See *Raytheon Co.*

RCD prefix
See *SGS-Thomson Microelectronics*
(Resistor, capacitor and schottky diode termination network ASICs)

RCD Components, Inc.
520 E. Industrial Park
Manchester, NH 03109
603-669-0054
Fax: 603-669-5455

(Resistors [fusible and temperature sensitive], coils, delay lines)

RCS prefix
See *Third Domain, Inc.*

RCV prefix
See *Hitachi Semiconductor* (SONET devices),
Rockwell (Modem IC)

RCM prefix
See *Rohm Corp.*

(LCD displays)

RD prefix
See *Ricoh Corp.*

RDC-XXXXX part number
See *ILC Data Device Corp.*

RdF Corporation
23 Elm Avenue
P.O. 490
Hudson, NH 03051
800-445-8367
603-882-5195
Fax: 603-882-6925
E-mail: sensor@rdfcorp.com

(Thermocouple air sensors, temperature sensors)

RE logo
See *Rawmat Electronics*

Real Time Devices, Inc. (RTD)
200 Innovation Boulevard
P.O. Box 906
State College, PA 16804-0906
814-234-8087, Fax: 814-234-5218

RAYTHEON — RECTON ELECTRONIC ENTERPRISES

RTD Europa: 36 1 212 0260
RTD Scandinavia: 358 0 346 4539

FaxBack (R): 814-235-1260
BBS: 814-234-9427

(Computer modules, which contain their custom parts)

REALTEK Semi-Conductor Company, Ltd.
1F, No. 11
Industry E. Road IX
Science Based Industrial Park
Hsinchu, Taiwan R.O.C.
886 35 780 211
Fax: 886 35 776 047
http://www.realtek.com.tw
E-mail: cn@realtek.com.tw

(This fabless company offers LAN, Ethernet and data communication ICs)

Rectron Electronic Enterprises, Inc.
13252 E. Temple Avenue
Industry, CA 91746
818-333-3802
Fax: 818-330-6296

Acumer Micro Computer, Inc.
San Jose, CA
408-894-1888
Fax: 408-456-6967

Rectron, Ltd.
4F-2, No. 2, Lane 609, Section 5
Chung-Shin Rd., San Chung City
Taipei hsien, Taiwan R.O.C.
886 2 999 8525
Fax: 886 2 999 8818

2 Fl.-1, Lane 15, Tzu Chiang St.
Tu-Cheng
Taipei Hsien, 10508
Taiwan
Phone/Fax: 886 2 268 3701,
886 2 268 1314 to 1316

RECTRON ELECTRONIC ENTERPRISES — RF PRIME

(Low power rectifiers, diodes, and transistors)

Reeves-Hoffman
Division Dynamics Corporation of America
400 W. North Street
Carlisle, PA 17013
717-243-5929
Fax: 717-243-0079

(Oscillators, filter crystals)

REF prefix
See *Analog Devices*
(Also Linear Technology, Burr-Brown Corp., GEC Plessy, and Raytheon, that second sources some of these voltage references.)

REF-XXXXX part number
See *ILC Data Devices Corp.*

Refac Electronics Corp.
10230 S. 50th Pl
Phoenix, AZ 85044
602-496-0035
Fax: 602-496-0168

(Optoelectronics)

REG prefix
See *Burr-Brown*

Reliable Electronic Manufacturing (REM)
Division of Wyvern Technologies
1205 E. Warner Avenue
Santa Ana, CA 92705
1-800-962-1085
714-966-0710
Fax: 714-556-7014

(This company design and manufactures replacements for obsolete components, a variety of arrays, networks, piggyback boards and SIP and DIP packaged parts. This includes RC networks, field programmable modules and active modules. Military specification devices available.)

Reliability, Inc.
P.O. Box 218370
16400 Park Row
Houston, TX 77218
713-492-0550
Fax: 713-492-0615

(DC/DC converters)

Rel-Labs, Inc.
30 Midland Avenue
Hicksville, NY 11801
516-935-7272
Fax: 516-935-8568

(Transistors, digital, linear and interface ICs)

Relm Communications, Inc.
7707 Records Street
Indianapolis, IN 46226
800-228-8108
Fax: 317-545-2170

(Crystals)

REM
See *Reliable Electronic Manufacturing*

Renaissance Microsystems
See *Exponential Technology, Inc.*

Renard Manufacturing, Co., Inc.
3305 NW 79th Avenue
Miami, FL 33122-1015
800-327-7244
305-592-1500
Fax: 305-593-9990

(Diodes)

RF prefix
See *Ricoh Corp., RF Micro-Devices*

RF Electronics, Inc.
1091 Duryea Avenue
Irvine, CA 92714-5517
800-523-1094
714-263-0661
Fax: 714-263-0662

(Diodes)

RF Micro-Devices, Inc.
7341-D West Friendly Avenue
Greensboro, NC 27410
919-855-8085
Fax: 919-299-9809

(RF application ICs utilizing bipolar silicon, gallium arsenide [MESFET] and advanced heterojunction bipolar transistor technologies, RF power amplifiers, LNA/mixers, quadrature modulators/demodulators, IF amps, receivers and transmitters, etc.)

RF Monolithics
4441 Sigma Rd
Dallas, TX 75244
214-233-2903
Fax: 214-387-8148

(SAW, surface acoustic wave, devices including quartz resonators, filters, delay lines, clocks)

RF Power Components, Inc.
20 Peachtree Ct.
Holbrook, NY 11741
516-467-4230
Fax: 516-467-4324

(Directional couplers, resistors and terminations, 90 degree hybrids)

RF Prime, Inc.
10569 Old Placerville Road
Sacramento, CA 95827
800-878-4669
916-368-4400
Fax: 916-368-4401

(Surface mount RF mixers, power splitters and RF transformers)

RF Products, Inc.
Davis & Copewood Street
Camden, NJ 08103
609-365-5500
Fax: 609-342-9757

(Tunable filters, multicouplers, RF MOSFETs for communications systems)

RF5 prefix
See *Ricoh Corporation*

(DC/DC converters, power supplies)

RFM logo
See *RF Monolithics, Inc.*

RFMD
See *RF Micro-Devices, Inc.*

RFV
See *Harris Semiconductor*

RFX prefix
See *Rockwell International Corporation*

RGB prefix
See *IBM Microelectronics*

RH prefix
See *Actel Corporation*

(Radiation hardened ICs)

RH5 prefix
See *Ricoh Corporation*

(Switching regulators, DC/DC converters)

Rhombus Industries, Inc.
15801 Chemical Lane
Huntington Beach, CA 92649
714-898-0960, Fax: 714-896-0971
TWX: 910-596-1826
Telex: 62891137
RHOMBUS HTBH

(Delay lines and pulse transformers)

Rhythm Watch Company, Ltd.
2-27-7, Taito, Taito-Ku
Tokyo 110
Japan
81 3 3833 7311
Fax: 03 3831 6043

(Logic, camera ICs)

ri logo
See *Reliability, Inc.*

Richardson Electronics Ltd.
40W267 Keslinger Road
LaFox, IL 60147
708-208-2200
1-800-RF Power
Fax: 708-208-2550
http://www.rell.com
Canada: 800-348-5580
U.K.: 0522 542631
France: 1 34 26 4000
Italy: 055 420 10 30
Spain: 1 528 37 00
Germany: 089 80 02 13 1
Japan: 3 3874 9933
Singapore: 65 298 4974

(RF and microwave transistors, semiconductors, modules and tubes)

Ricoh Company Ltd.
1-15-5 Minamiaoyama,
Minato-Ku, Tokyo 107
Japan
03 3479 3111
Fax: 03 3403 1578, 03 401 8710

115 Passaic Ave
Fairfield, NJ 07006
201-882-2272
Fax: 201-808-7514

3001 Orchard Parkway
San Jose, 95134-2088
408-432-8800
Fax: 408-432-8375

(Logic, programmable logic, CMOS/BiCMOS gate arrays, ROM, SRAM, microprocessors, voltage regulators, thermal print heads, disk drive controllers, digital signal processors, communication products; such as digital filters, UARTs, real time clocks [RTCs], power supply ICs, disk drive ICs, audio and digital video ICs, speech synthesis and recognition, watch and clock, ICs)

RIM prefix
See *Optotek Ltd.*

Risho Kogyo Co., Ltd.
2-9-1, Dojima, Kita-Ku
Osaka-Shi, Osaka 530
Japan
06 345 8331
Fax: 06 345 1388

(Hybrid ICs)

RJ prefix
See *Rad Data Communications*

RL prefix
See *Siemens*

RLA prefix
See *Raytheon Semiconductor*

RLC Electronics
83 Radio Circle
Mt. Kisco, NY 10549
914-241-1334
Fax: 914-241-1753

(Coaxial switches)

RM prefix
See *Raytheon Semiconductor*, *Rockwell* (Device family), *Siemens*

RN5 prefix
See *Ricoh Corporation*

(Voltage detector, voltage regulators)

RO ASSOCIATES — ROCKWELL INTERNATIONAL

RO Associates
246 Caspian Dr.
P.O. Box 61419
Sunnyvale, CA 94088
800-443-1450
408-744-1450
Fax: 408-744-1521
http://www.roassoc.com/
E-mail: sales@roassoc.com

(DC/DC converters, switching power supplies, they also sell Tohritsu Co.[TRT] DC/DC converters)

Rochester Electronics, Inc.
10 Malcolm Hoyt Drive
Newburyport MA 01950-4018
508-462-9332
Fax: 508-462-9512

36 Evelyn Road
Dunstable, Beds
LU5 4NG, United Kingdom
582 603439
Fax: 582 476238

KH Electronics Corp.
Landic No. 2 Akasaka Bld. 3 Fl.
10-9 Adasaka 2-Chome
Minato-Ku, Tokyo 107, Japan
03 3587 1041
Fax: 03 3584 6394

(A supplier of discontinued and custom packaged military and commercial semiconductors that can manufacture parts from die masters, They can supply obsolete products from various original manufacturers including AMD, Harris Semiconductor, Intel, Texas Instruments, Microchip Technologies, Inc. and National [and Fairchild] Semiconductor)

Rockwell International
Semiconductor Systems
(Formerly called Rockwell Telecommunications)
4311 Jamboree Rd.
P.O. Box C, MS 501-300
Newport Beach, CA 92658-8902
714-833-6849
http://www.nb.rockwell.com

10700 W. Higgins Road
Rosemount, IL 60018
1842 Reynolds
Irvine, CA 92714
714-833-4655
ELS 62108710
TWX: 910-595-2518

3375 Scott Blvd., Suite 410
Santa Clara, CA 95051
408-980-1900
TLX: 756560

921 Bowser Road
Richardson, TX 75080
214-996-6500
TLX: 73-307

10700 West Higgins Road, Ste. 102
Rosemont, IL 60018
312-297-8862
TWX: 910-233-0179
(RI MED ROSM)

5001B Greentree Executive Campus
Route 73
Marlton, N.J. 08053
609-596-0090
Fax: 609-596-5681
TWX: 710-940-1337
COMNET: 662-1000

Europe:
Rockwell International GmbH
Fraunhoferstrasse 11
D-8033 Munchen-Martinaried
Germany
089 857 6016
TLX: 0521/2650 rimd d

Rockwell International (U.K.) Ltd.
Heathrow House, Bath Rd.
Cranford, Hounslow,
Middlesex, England TW5 9QW
01 759-9911

Rockwell Collins Italana S.P.A.
Via Boccaccio, 23
20123 Milano, Italy
39-2 498-7479
TLX: 316562 RCIMIL 1

Far East:
Rockwell International Overseas Corp.
Itohpia Hirakawa-cho Bldg.
7-6, 2-Chome, Hirakawa-Cho
Chiyoda-Ku, Tokyo 102, Japan
03 265 8806
TLX: J22198

Rockwell-Collins International
Tai Sang Commercial Bldg., 11th Floor
24-34 Hennessy Rd.
Hong Kong
5 274-321
TLX: 74071 HK

Rockwell International Corporation
Electro-Optical Center
P.O. Box 3150
3370 Miraloma Avenue
Anaheim, CA 92803
714-762-2718
Fax: 714-762-6105

(Focal plane arrays, telecommunication products such as Modem, Fax/data [modem] and digital cordless telephone ICs; microprocessors, microcomputers [AIM and RM families], memory products such as NVRAM, ROMs, display controllers, GaAs ASICs, etc.)

Rockwell International Corporation
Electro-Optical Center
P.O. Box 3150
3370 Miraloma Avenue
Anaheim, CA 92803
Fax: 714-762-1493

(Focal plane arrays)

Manufacturers, Prefixes, Part Number Types, Logo Descriptions & Family Types

ROHM Electronics
ROHM Co. Ltd.
21 Saiin Mizosaki-Cho, Ukyo-Ku
Kyoto-Shi, Kyoto 615, Japan
81 75 311 2121
Fax: 81 75 315 0172

ROHM Electronics Division
3034 Owen Drive
Antioch, TN 37013
1-800-955-ROHM
615-641-2020
Fax: 615-641-2022

ROHM Electronics GmbH
24 Rue Saarinen Silic 224
94528 Rungis Cedex, France
011 33-1-46-759051
Fax: 011 33-1-46-750047

Muhlenstrasse 70
D04052 Korschenbroich 1
Germany
011 49-2161-6101-35
Fax: 011 49-2161-6421-02

ROHM Electronics Co. Ltd.
30 Canton Road, Tsimsatsui Room
Tower 1, Silvercord
Kowloon, Hong Kong
011 852-3756262
Fax: 011 852-3758971

United Kingdom:
ROHM Electronics U.K.
15 Peverel Drive
Granby, Milton Keynes
MK1 1NN United Kingdom
011 44-908-271311
Fax: 011 44-908-270380

(Transistors, laser diodes, LEDs, LCD displays, logic and hybrid ICs, voltage comparators and voltage regulators, disk drive controller, audio and communication, television, VCR, and telecommunications ICs, IC protectors, resistor networks, ceramic chip capacitors, ceramic resonators, etc.)

ROJ prefix
See *Datel, Inc.*

ROLM Computer Systems
See *Loral ROLM Computer Systems*

Rosetta Technologies, Inc.
265 25th St.
West Vancouver, British Columbia
V7V 4H9
604-925-0820

(Chip sets for joysticks and game pads for user programmable buttons)

Ross Technology, Inc.
Subsidiary of Fujitsu Ltd.
5316 Highway 290 W. Ste. 500
Austin, TX 78735-8930
1-800-774-ROSS
512-349-3108, 512-919-5207
Fax: 512-349-3101, 512-919-5200

E-Mail:
ROSS_INFO+aAYN%AYN
@MCIMAIL.COM
info@ross.com
http://www.ross.com

P.O. Box 140407
Austin, TX 78714-9943

Avenue Ernest Solvay, 80
1310 LaHulpe
Belgium
32 2 652 1014
Fax: 32 2 652 1062
E-Mail: rosseurope@attmail.com

(Formerly a subsidiary of Cypress Semiconductor [with SPARC products marketed under the Cypress Semiconductor name] Purchased by Fujitsu Ltd. in 1993 and manufacturers HyperSparc microprocessor upgrades)

RP prefix
See *Ricoh Corp.*, *Siemens*,
ILC Data Device Corporation
(If RP-XXXXXXX part number)

ROHM ELECTRONICS — RTB

RPM prefix
See *LSI Logic Corp.*

RRL prefix
See *Siemens*

RR-XXX part number
(PNP transistors sold by Radio Receptor Co., Inc. NY in 1953.)

RS5 prefix
See *Ricoh Corporation*

(DC/DC converters, power supplies)

RSC prefix
See *Ricoh Corp.*,
Raytheon Semiconductor

(Mixed signal ASICs)

RSC-XXX part number
See *Sensory Circuits, Inc.*

(Voice recognition, speech synthesis and general control ICs)

RSD/100
See *Lockheed-Martin*

(3D graphics IC)

RSM
See *Sensitron Semiconductor* (A division of RSM Electron Power, Inc.)

RSS prefix
See *Third Domain, Inc.*

RSS-XXXX part number
See *Third Domain, Inc.*

RT prefix
See *Rancho Technology, Inc.*

RTB prefix
See *RTB Technology, Inc.*

R-S RTB — SAGE LABORATORIES

RTB Technology, Inc.
3500 West Balcones Center Drive
Austin, TX 78759-6509
512-338-3678
Fax: 512-338-3900

(Memory Cube, three dimensional memory modules)

R.T.C. LaRadiotechnique-Compelec (RTC logo)
RTC prefix
See *Fox Electronics*

(Real time clock module)

rtd logo
See *Real Time Devices, Inc.*

RTG, Inc.
2790 Skypark Drive
P.O. Box 3986
Torrance, CA 90510
310-534-3016
Fax: 310-534-3728
E-mail: sales@rtg.com

(This company designs and delivers analog and mixed signal ASICs)

RTL prefix
See *REALTEK Semiconductor Corporation*

RTO (A two-lead TO-220 package)
See *Vishay Resistors*

RTX prefix
See *Harris Semiconductor*

RTZ prefix
See *Optical Communication Products, Inc.*

Rubycon America, Inc.
4293 Lee Avenue
Gurnee, IL 60031
708-249-3450, Fax: 708-578-1300

(Inductors, choke coils)

RX prefix
See *RF Monolithics*

RX5 prefix
See *Ricoh Corp.*

(Voltage regulators)

RXT prefix, Transistors
See *ROHM Electronics Division*

S logo
See *Simtek Corp., Staktek Corp.*

(Memory stacked modules)

S prefix
See *American MicroSystems (AMI), AMCC Applied Micro Circuits Corp., Philips Semiconductor, ISA Jobin Yvon-Spex (Spex Industries) (CCDs), GHz Technology, Inc.*
(Microwave transistors)

S part number prefix
See *Seiko Instruments*

S, with a diode symbol in the middle of the S
See *Taiwan Semiconductor Co. Ltd.*

S (Schottky) logic family
See *Hitachi, Motorola, Inc., National Semiconductor, SGS-Thomson, Lansdale, LG Semicon* (Formerly Goldstar Technology, Inc.), *Rochester Electronics, Philips, Texas Instruments*

S3, Inc.
2770 San Tomas Expressway
Santa Clara, CA 95051-0968
408-980-5400
Fax: 408-980-5444

(A fabless company supplying video graphics ICs [including the "Verge"], MPEG ICs, graphics accelerator ICs including Windows accelerator IC's, Video Electronic Standards Association United Memory Architecture chip sets)

S3LV prefix
See *Applied Micro Circuits Corporation*

SA prefix
See *Philips Semiconductors, Harris Semiconductor*

SAA prefix
See *Philips Semiconductors (Signetics), AVG Semiconductors* (Second source some ICs)

SAB prefix
See *Philips Semiconductors*

SABC prefix
See *Siemens*

SAC-TEC, Inc.
23625 Madison St.
Torrance, CA 90505
310-375-5295
Fax: 310-375-9647

(Multichip modules incorporating EEPROM)

SAD prefix
See *EG&G Reticon*

Sage Active Microwave
26 Clinton Drive
Suite 114
Hollis, NH 03049-6521
603-598-6900
Fax: 603-598-9065

(Surface mount mixers)

Sage Laboratories, Inc.
11 Huron Drive
Natwick, MA 01760-1338
508-653-0844
Fax: 508-653-5671

(RF and Microwave components including combiner assemblies, antenna switching networks, beam forming networks, splitting and coupling networks, switches, rotary joints/couplers, phase shifters)

Sakugi Densi Seisakusho
472, Shimo-Sakugi, Sakugi-Mura
Futami-Gun
Hiroshima 728-01, Japan
082 455 2315
Fax: 082 455 3546

(Automotive ICs)

Samsung Display Devices
Daekyung Bldg., 12 Fl.
120-2-Ka, Taepyung-Ro
Choong-Ku
Seoul, Korea
822-727-3331-3, 3351-3
Fax: 822-7275949, 727-3389

Samtron Display, Inc.
14251 East Firestone Blvd.
Suite 101
La Mirada, CA 90638
310-802-8425
Fax: 310-802-8820

Samsung Display Devices:
18600 Broadwick Street
Rancho Dominguez, CA 90220
310-537-7000
Fax: 310-537-1033

European Office:
Samsung Haus AM
Unisys Park 1
65843 Salzback
Frankfurt, Germany
49 6196 74001 5
Fax: 49 6196 758149

Hong Kong:
66F Central Plaza
18 Harbor Road
Wanchai, Hong Kong
852 2862 6053
Fax: 852 2866 2548

Malaysia:
Lot 635 & 660
Kawasan Perindustria
Taunku Jaafgar,
71450 Sungai Gadut
Negeri Sembilan Darul Khusus
West Malaysia
696 776 160
Fax: 606 776 164

Japan:
Hamacho Center, Building 17F
31-1, Hamacho 2-Chome
Nihonbashi-Hamacho
Chuo-ku, Tokyo 103
Japan
813 564 19875
Fax: 813 564 19876

Crystal Tower 6F
2-27, 1-Chome, Shiromi
Chuo-ku
Osaka 540 Japan
816 949 4801
Fax: 816 949 5133

(LCD, including AM-LCD, displays, Vacuum Fluorescent Displays [VFD])

Saitama Parts Industrial Co., Ltd.
8-22-20, Tsuji
Urawa-Shi
Saitama 336 Japan
048 862 0074
Fax: 048 862 0080

(Hybrid ICs)

Sakata USA Corporation
651 Bonnie Lane
Elk Grove Village, IL 60007
800-323-6647
708-593-3211
Fax: 708-364-5290

Sakata INX Corp.
23-37, Edobori, 1-Chome, Nishi-Ku
Osaka-Shi
Osaka 550 Japan
06 447 5877

SAGE LABORATORIES — SAMSUNG SEMICONDUCTOR

(Diodes, transistors, logic, linear and hybrid ICs, voltage comparators, disk drive controller, audio, telecommunication, television, VCR, and Hall effect ICs, etc.)

Samsung Semiconductor, Inc.
8/10FL., Samsung Main Bldg.
250, 2-KA, Taepyung-Ro
Chung-Ku, Seoul, Korea
C.P.O. Box 8780
Seoul 727-7114
Fax: 753-0957
TLX: KORSST K27970

3655 No. First Street
San Jose, CA 95134
1-800-446-2760
408-954-7000

3725 N. First Street
San Jose, CA 95134
408-434-5400

(LCD displays)

24 Fl. Admiralty Center
Towr 1, 18 Harcourt Road
Queensway, Hong Kong
862-6900
Fax: 866-1343
TLX: 80303 SSTC HX

RM 2401, 24F, International Trade
Bldg. 333
Kee Lung Road, Sec. 1, Taipei
Taiwan R.O.C.
2 757-7292
Fax: 2 757 7311

8F Sudacho Verde Bldg.
2-3 Kanda-Sudacho Chiyoda-Ku
Tokyo, 101 Japan
03 3258 9501
Fax: 03 3258-9695
TLX: 2225206 SECJPN J

SAMSUNG SEMICONDUCTOR — SANREX

Mergenthaler Allee 38-40
D-6236 Eschborn, Germany
0-6196-900920, Fax: 0-6196-900989
Telex: 4072678 SSED
Frankfurt Germany:
0049-6196-90090,
Fax: 0049-6196-900989

10F., Collyer Quay #14-07
Ocean Building Singapore 0104
535-2808, Fax: 532-6452, 227-2792

(VRAMs, DRAMs, SGRAMs, flash memory and logic ICs, memory cards, LCD controllers, graphics ICs [including 3D graphics ICs])

Samsung Microwave Semiconductor, Inc.
(Formerly Harris Microwave Semiconductor, Inc. [HMS])
1530 McCarthy Blvd.
Milpitas, CA 95035
408-433-2222, Fax: 408-432-3268

Asia:
822 259 4574
Fax: 822 259 2469

(GaAs FETs, MMICs [amplifiers and attenuators, low noise gain blocks] and custom foundry services)

San Francisco Telecom, Inc. (SFT)
101 Townsend Street
San Francisco, CA 94107-1912
415-777-3439

(A custom IC design company whose products for QAM digital radios and telecommunications include: multiple line rate ASICs for T1/T3 and E1/E3 framing, high speed digital line access IC and cable TV modems and wireless data applications ICs. This company was acquired by Level One Communications, Inc. in June 1995.)

Sanchez
See *Sanchez Associates*

Sanchez Associates, Inc.
437 South Union Street
Lawrence, MA 01843
508-688-1299

(High voltage op amps)

Sand Microelectronics
Santa Clara, CA
408-235-8600
Fax: 408-235-8601

(PCI cores for IC designs including synthesizible Universal Serial Bus [USB] Cores)

SanDisk Corp.
140 Caspian Ct.
Sunnyvale, CA 94089
408-562-0500
Fax: 408-562-0503

(Flash chip technology mass storage devices [flash memory, flash PC cards] and flash chip sets. Note: this company is in partnership with Seagate Technology. This company was known as SunDisk prior to August 1995)

Sanken Electric Co. Ltd.
3-6-3 Kitano
Niiza-Shi, Saitama 352
048 472 1111
Fax: 048 471 6249

2500 W. Higgins Road
Hoffman Estates, IL 60195
708-519-1717
Fax: 708-884-1511

Sanken USA, Inc.
4333 Harbor Pointe Blvd. SW
Mukilteo, WA 98275
206-347-9000
Fax: 206-355-0195

(Linear ICs, hybrid ICs, voltage regulators, transistors, amplifiers, power hybrid ICs)

Sankyo Seiki Mfg. Co., Ltd.
1-17-2, Shinbashi
Minato-Ku
Tokyo 105 Japan
81 3 3502 3711
Fax: 03 3502 6228

(Hybrid ICs)

San-O Industrial Corporation
91-3 Colin Drive
Holbrook, NY 11741
516-472-6666
Fax: 516-472-6777

(Fuses, fuseholders, etc.)

SanRex Corporation
3000 Marcus Ave. #3-E8
New Hyde Park, NY 11042-1006
516-352-3800
Fax: 516-352-3956

SanRex Europe GmbH
Knorstrasse 142
80937 Munich, Germany
89 3112034
Fax: 89 3161636

SanRex Limited
Room 307
Kowloon Plaza

485 Castle Peak Road
Kowloon, Hong Kong
7441310, 7856313
Fax: 7856009, 7862165

SaxRex Welding Equipment Pte. Ltd.
Block 6015
Ang Mo kio Industrial Park 3
#01-332
Singapore 2056
4826576
Fax: 4824766

Manufacturers, Prefixes, Part Number Types, Logo Descriptions & Family Types

SANREX — SARATOGA SEMICONDUCTOR

(See Sansha Electric manufacturing Co., Ltd. for the Japan headquarters of this company. Power semiconductors including transistor modules, SCR/diode modules, triacs, IGBts, three phase bridge rectifiers, fast recovery diodes)

Sansha Electric Mfg. Co. Ltd.
2-14-3 AWAJI
Higashiyodogawa-Ku
Osaka 533 Japan
81 06 325 3863
Fax: 81 06 321 0355
TLX: 5233617 SANREXJ

(See SanRex for US and European offices. Power semiconductors including transistor modules, scr/diode modules, triacs, IGBts, three phase bridge rectifiers, fast recovery diodes)

Sanshin Electric Co., Ltd.
2-7-20 Shin-Yokohama,
Kohoku-Ku,
Yokohama-Shi, Yokohama 222
Japan
045 471 1011
Fax: 045 471 1010

(Hybrid ICs)

Sante Fe Laser Co.
2424 E. Aragon
Tucson, AZ 85706
602-889-9147
(520 area code after 12/31/96)
Fax: 602-889-0539
(520 area code after 12/31/96)

(Diode and lamp pumped lasers. This company was formerly a division of CVI Laser Corporation.)

Sanwa Denshi Co. Ltd.
58-5, Nakamaru-Cho, Itabashi-Ku
Tokyo 173 Japan
81 3 3959 6611

(Logic, memory ICs)

Sanyo Semiconductor Corp.
7 Pearl Court
Allendale, N.J. 07401
201-641-2333
201-641-3000
(Parts and service information, eastern region, for manufactured products)
201-825-8080
(Semiconductor parts)
201-784-0303
(Semiconductor parts, when going through distribution)
Fax: 201-641-0311

Sanyo Electric Co. Ltd.
2-5-5 Keihan-Hondori,
Moriguchi-Shi
Osaka 570, Japan
81 6 991 1181
Fax: 81 6 991 5411

(Has licensed the flash architecture designs of Silicon Storage Technology (SST), Sunnyvale, CA)

Chicago: 708-285-0333
Fax: 708-285-1133
Florida: 904-376-6711
Fax: 904-376-6722
Atlanta: 404-279-7377
Fax: 404-279-7280
Texas: 214-480-8345
Fax: 214-480-8351
Canada: 416-421-8344
Fax: 416-421-8827

Europe: 089 460095 0,
44 895 810018
Fax: 089 460095 90,
44 895 812050

Asia: 4891533
Fax: 4891726

Taiwan: 02 522 1311
Fax: 02 543 4659
Singapore: 7363100
Fax: 7361230
Australia: 02 763 3822
Fax: 02 746 1344

(Hall Effect sensors including InSb linear output types and GaAs types, hybrid and logic ICs, speech synthesis and recognition, television [including a NTSC Scan Converter IC], VCR, and telecommunications ICs, etc.)

Sanyo Energy (USA) Corp.
2001 Sanyo Ave.
San Diego, CA 92173
619-661-6620
Fax: 619-661-6743

(Battery charging hybrid IC for NiCd and NiMH batteries)

Sanyo Video Components
3333 Sanyo Road
Forrest City, AR 72335
501-633-5030
Fax: 501-633-6720

2001 Sanyo Avenue
San Diego, CA 92173
619-661-6835
Fax: 619-661-1055

(SAW and dielectric filters)

Sara chip set
See *TranSwitch, Corp.*

Saratoga Semiconductor
686 W. Maude Avenue
Sunnyvale, CA 94086
408-522-7500
Fax: 408-245-3713

9900 West Sample Road
Suite 300
Coral Springs, FL 33065
305-344-4477
Fax: 305-344-9808

3800 North Wilkie Road, Suite 300
Arlington Heights, IL 60004
312-506-9877
Fax: 312-506-9815

SARATOGA SEMICONDUCTOR — SITEQ

P.O. Box 189
Cold Springs Harbor, NY 11724
516-367-6110
Fax: 516-367-7643

15150 Preston Road
Suite 300
Dallas, TX 75248
214-404-9805
Fax: 214-404-0905

European Headquarters
9 Erlenstrasse
Langerbach, Germany
D8051
49 8761 60967
Fax: 49 8761 60780

(BiCMOS products including SRAMs, cache tags, FIFOs, logic circuits, etc.)

SaRonix
151 Laura La.
Palo Alto, CA
800-227-8974
415-856-6900
Fax: 415-856-4732
http://www.saronix.com

(Crystal oscillators and crystals [Nymph (tm)])

Savoy Electronics, Inc.
1175 NE 24th St.
P.O. Box 5727
Fort Lauderdale, FL 33310
800-359-5400
305-563-1333
Fax: 305-563-1378

Sawtek Inc.
P.O. Box 609501
Orlando, FL 32860-9501
407-886-8860
Fax: 407-886-7061

(VCSOs [Voltage Controlled SAW Oscillators])

SBA prefix
See *ILC Data Device Corp.*

SBC prefix
See *Carroll Touch, Inc.*

SBH prefix
See *Siemens*

SBL prefix
See *Siemens*

SBL prefix
See *Siemens*

SBM prefix
See *Siemens*

SBT prefix
See *Sicon, Inc.*

SBX prefix
See *Sony Semiconductor*

SC prefix
See *S-MOS Systems,
Philips Semiconductor,
Electronic Technology Corporation,
Motorola Semiconductor,
Applied Microcircuits Corp.,
Silicon Composers, Inc.*

(Microprocessors)

SCAN prefix
See *National Semiconductor*

SCB prefix
See *Philips Semiconductors*

(Signetics)

SCC prefix
See *Philips Semiconductors*
(Signetics)
(SenSym for sensor products)

SCCW prefix
See *Motorola*

SCD prefix
See *ILC Data Device Corporation,
Siemens*

SCDV prefix
See *Siemens*

SCE prefix
See *Siemens*

SCF prefix
See *Siemens*

Schaffner EMC, Inc.
9-B Fadem Rd.
Springfield, NJ 07081
201-379-7778
Fax: 201-379-1151

(Chokes for EMI suppression)

Schurter, Inc.
1016 Clegg Court
P.O. Box 750158
Petaluma, CA 94975-0158
1-800-848-2600, ext. 286
707-778-6311
Fax: 707-778-6401
E-mail: 73024.2314@compuserve.com
http://www.schurterinc.com

(Fuses and fuseholders including surface mount types, power entry modules)

SCI prefix
See *S-MOS Systems*

(Converter IC)

Sciteq Electronics, Inc.
4775 Viewridge Ave.
San Diego, CA 92123
619-292-0500
Fax: 619-292-9120

(Frequency synthesis products including a waveform IC)

SCM prefix
See *Sony*

(Memory modules)

SCN prefix
See *Philips Semiconductors*

SCOG
See *Video International Development Corporation*

Scorpion Technologies
See *MSIS*

SCO prefix
See *STC Defense Systems*

Scorpio Communications
575 5th Ave.
Suite 1103
New York, NY 10036
212-221-5998, 408-866-4450
Fax: 212-724-2220

(ATM switch ICs)

SCP
See *Sokol Crystal Products, Inc.*

SCV prefix
See *Newbridge Microcircuits*
(And DY 4)

SCX prefix
See *SenSym, Inc.*

SD prefix
See *Optek Technology, Inc.*, *SGS-Thomson Microelectronics, Inc.*, *SanDisk* (Flash memory products), *Calogic*, *Fincitec OY*

SD-XXXXXXX-XXX part number prefix
See *ILC Data Device Corporation*

SDC-XXXXXXX
(or XXXXX-XXX) part number
See *ILC Data Device Corporation*

SDL, Inc.
80 Rose Orchard Way
San Jose, CA 95134
408-943-9411
Fax: 408-943-1070

(Formerly known as Spectra Diode Laboratories, a manufacturer of visible laser diodes, CW linear arrays, fiber coupled laser diodes, etc.)

SDT, Inc.
See *Omega Micro, Inc.*

SDU prefix
See *S-MOS Systems*

Spectra Diode Laboratories
See *SDL, Inc.*

SE prefix
See *Seiko* (Seiko-Epson), *Philips* (Signetics), *International Light*

SEA prefix
See *S-MOS Systems*

(RAMDAC)

Seagate Microelectronics, Ltd.
MacIntosh Road
Kirkton Campus
Livingston EH54 7BW, Scotland
44 0506 416416
Fax: 44 0506 413526
TLX: 727360

2722 HWY 694 Frontage Road,
Suite 20
New Brighton, MN 55112
1-800-866-3034
Fax: 612-631-3540

900 Disc Drive
Scotts Valley, CA 95066

(Formerly Integrated Power Semiconductors, Ltd., power ICs)

SCM —
SEEQ

Seastar Optics, Inc.
P.O. Box 2219
2045 Mills Road
Sidney, B.C. Canada V8L 3S8
604-656-0891
Fax: 604-655-3435

316 Second Avenue South
Seattle, WA 98104
1-800-663-8375
206-623-2855

(Laser diodes and laser diode modules)

SEC logo
See *Samsung*

Secowest Italia SpA
Via Liguria 49
Borgaro, 10071
Italy
011 470 14 84, 011 470 37 03
Fax: 011 470 42 90
TLX: 2212527 RECTLT

(Diodes, thyristors)

SED prefix
See *S-MOS Systems*

(LCD, VFD, PDP display driver)

SEEQ Technology, Inc.
Corporate and International
Sales Office
(Also Australia and New Zealand)
47131 Bayside Parkway
Fremont, CA 94538
510-226-7400
Fax: 510-657-2837

Northwest Sales Office and Asia-Pacific Sales Office:
3945 Freedom Circle
Suite 650
Santa Clara, CA 95054
408-988-4900
Fax: 408-988-5506

SEEQ — SEIKO INSTRUMENTS

Southwest Sales Office:
Westlake Village, CA 91362
818-597-1020
Fax: 818-597-1121

Southeast Sales Office:
3450 East Lake Road
Suite 206-9
Palm Harbor, FL 34685
813-789-4195
Fax: 813-787-6290

East Sales Office:
24 New England
Executive Park
Burlington, MA 01803
617-229-6350
Fax: 617-273-0322

Mid-America Sales Office:
300 Martingale Road
Suite 605
Schaumberg, IL 60173
708-517-1515
Fax: 708-517-1519

Northern European Sales Office:
SEEQ International Ltd.
Dammas House
Dammas Lane, Old Town
Swindon SN1 3EF,
United Kingdom
44 793 694999
Fax: 447 93616201
Telex: 444588

South European Sales Office:
SEEQ International Sarl
4 Allee de Pomone, RN13
78100 Saint-Germaine-en-Laye
France
33 (1) 30 61 21 23
Fax: 33 (1) 30 61 21 92
Telex: 699912

(A fabless company supplying nonvolatile memory devices, EPROMs, communication ICs such as Ethernet LAN controllers, 10base-T Ethernet ICs, etc. Note: the EEPROM [or E2PROM] was sold to Atmel Corp. in February 1994. Military qualified products no longer available.)

SEG
See *Shenzhen Electronics Group*

Segor-Optoelectronique
7, Rue Du
Commandant Louis Bouchet
L hay-Les-Roses, 94240
France
1 46 86 14 71

(Optoelectronics)

Seiko-Epson
Seiko Epson Corp.
3-3-5 Owa, Suwa-Shi
Nagano 392, Japan
81 266 52 3131
Fax: 81 266 52 8775

Seiko Instruments, Inc.
Head Office
31-1, 6-Chome Kameido,
Koto-Ku, Tokyo 136, Japan
03-636-3821, 03-3682-5201
Fax: 03-638-0627, 03-3637-0117
Telex: 2622162 Dseiko J

IC Overseas Sales Branch:
563 Takatsuku-Shinden
Matsudoshi Chiba-Ken, 271 Japan
0473-92-0667
Fax: 0473-92-7112

Seikosha Co. Ltd.
4-1-1 Taikei, Sumida-Ku
Tokyo 130, Japan
81 3 5610 7067
Fax: 81 3 5610 7162

Seiko Instruments U.S.A., Inc.
Semiconductor Products Group
1150 Ringwood Court
San Jose, CA 95131-1726
408-433-3208
Fax: 408-433-3214, 3201

Epson America, Inc.
Component Sales Dept.
310-787-6300
Fax: 310-782-5320

Seiko Instruments, Inc.
Siemens Strasse 9
D-6078 Neu-Isenburg Germany
061 02 297-0
Fax: 061 02 297-222

Lyoner Strasse 36 D-6000
Frankfurt am Main 71
Germany
49-69-6630000
Fax: 49-69-6667003

Ivo-Hauptmann - Ring 1
Hamburg 22159
Germany
40 645 892 0
Fax: 40 645 892 29
TLX: 211 331

4-5/F, Wyler Centre
2 200 Tai Lin Pai Road
Kwai Chung,
N.T. Kowloon, Hong Kong
852-4218611
Fax: 852-4805479

5F-1 No. 99 SEC. 2
Chung Shan N. Rd.
Taipei 104, Taiwan R.O.C.
886-2-563-5001
Fax: 886-2-521-9519

(FCT CMOS logic devices, SRAMs, E2PROMs, Fuse ROM, mask ROMs, linear ICs [voltage detectors, voltage regulators, voltage comparators, battery backup ICs], LSI devices [color palette, speech synthesis and recognition ICs], microprocessors, telecommunication ICs, timer/watch/clock ICs, display drivers, sensors [Hall Effect], JEIDA/PCMCIA memory, speciality cards, LCD graphic display modules, etc.)

Manufacturers, Prefixes, Part Number Types, Logo Descriptions & Family Types

Seiko Instruments
Electronic Components Division
2990 West Lomita Blvd.
Torrance, CA 90505
310 517-7830
310 517-7833
Fax: 310 517-7792

(Batteries)

Seiko Instruments, USA Inc.
Liquid Crystal Display Dept.
2990 W. Lomita Blvd.
Torrance, CA 90505
310-517-7829

(LCD displays)

Seiwa Techno Systems, Corp.
Ichigaya White Bldg.
3-26 Ichigaya Honmura-Cho
Shinjuku-Ku, Tokyo 162
Japan
03 3267 2351
Fax: 03 3267 0810; 03 3267 4743

(Watch ICs)

SEK prefix
See *S-MOS Systems*

(Module)

SEL prefix
(On crystal oscillators)
See *SaRonix*

SemeLab Plc.
Coventry Road, Lutterworth
Leicestershire LE17 4JB, England
44 455 554711
Fax: 44 455 552612
TLX: 341927

2722 Highway 694, Suite 20
New Brighton, MN 55112

(Linear pulse width modulator ICs, voltage regulators [including adjustable types], military parts available etc., diodes, thyristors, transistors)

Semefab, Ltd.
United Kingdom
Ste 800
1901 N. Roselle Road
Schaumburg, IL 60195
708-490-6475

(ASICs and various ICs)

Semetex Corporation
3450 Fujita St.
Torrance, CA 90505
310-539-0201
Fax: 310-375-7420

(Transistors, optoelectronics)

Semi Dice, Inc.
Semi Dice SoCal
High Rel Division
10961 Bloomfield St.
P.O. Box 3002
Los Alamitos, CA 90720
310-594-4631, 714-952-2216
Fax: 310-430-5942

Semi Dice North
1600 Wyatt Dr. #14
Santa Clara, CA 95054
800-325-6030 (outside CA)
408-980-9777, Fax: 408-980-9499

Semi Dice East
24 Norfolk Ave.
So. Easton, MA 02375
508-238-8344
Fax: 508-238-9204

Semi Dice Long Island
12-5 Technology Dr.
East Setauket, NY 11733
516-941-3190, Fax: 516-941-3185
Note: Call 1-800-345-6633 outside CA and MA.

(This company packages and screens IC die; they also have die to support discontinued Motorola military products. This company is a franchised supplier of bare die from various companies including Allegro Microsystems, Inc. [bare die and MOS capacitors]; Analog Devices [PMI]; BKC Semiconductors; Calogic; Compensated Devices, Inc.; International Rectifier; Philips; Sipex; Semtech; Sussex Semiconductor, Inc. They also have passive components from Amitron; California Micro Devices; Johanson Dielectrics; Novacap; Semi Films Division; State of the Art, Inc.; Tansitor Electronics, Inc., and Vishay [Sprague])

Semicoa Optoelectronics
333 McCormick Ave.
Costa Mesa, CA 92626
800-854-3188
714-979-1900
Fax: 714-557-4541

(Optoelectronic products including silicon photodiodes, military qualified devices available)

Semicon Components, Inc.
10 North Avenue
Burlington, MA 01803
617-272-9015
Fax: 617-272-9109

(Diodes, thyristors, transient voltage suppressors)

Semicon, Inc.
ST Semiconductors of Indiana
415 N. College Avenue
P.O. Box 609
Bloomington, IN 47402-0609
812-332-1435
Fax: 812-331-8706
(Diodes, thyristors, optoelectronics)

Semiconductor Circuits, Inc.
Subsidiary Astec America, Inc.
49 Range Road
Windham, NH 03087
800-448-4724, 603-893-2330
Fax: 603-893-6280

SEIKO INSTRUMENTS — SEMICONDUCTOR CIRCUITS

SEMICONDUCTOR CIRCUITS — SEMTECH

(Power supplies and DC/DC converters)

Semiconductor Laser International Corporation (SLI)
148 Vestal Parkway East
Vestal, NY 13850
607-754-0112
Fax: 607-754-5974

(High power laser products including laser wafers, bars, arrays and single devices)

Semiconductor Technology, Inc. (STI)
3131 S.E. Jay Street
Stuart, FL 33497
305-283-4500
Fax: 407-286-8914
TWX: 510-953-7511

(Transistors, diodes, bridge rectifiers, including military qualified versions- high voltage and general purpose types)

Semiconductors, Inc.
3680 Investment Lane
Riviera Beach, FL 33404
800-327-6183
407-842-0305
Fax: 407-845-7813
TLX: 62014516

(Diodes, SCRs, triacs, power MOSFETs, voltage regulator ICs, germanium transistors)

Semicustom Logic, Inc.
1630 Oakland Road
Suite A103
San Jose, CA 95131
408-452-1766
Fax: 408-452-0563

(Microprocessor support ICs)

Semikron Inc.
11 Executive Drive
P.O. Box 66
Hudson, NH 03051
800-258-1308, 603-883-8102
Fax: 603-883-8021
TLX: 6711011 SEMKRO UW

(IGBT power modules and bridges)

Semitron Industries, Ltd.
Chelworth Industrial Estate
Cricklade Swindon
Wiltshire SN6 6HQ
United Kingdom
44 793 751 151
Fax: 44 793 751 808
TLX: 44848

(Diodes, thyristors, transistors)

Semitronics Corporation
Semitron Semiconductors
64 Commercial Way
Freeport, N.Y. 11520
800-645-3960, 516-693-9400
Fax: 516-623-6954
TLX: 298294 SEMT UR

(Diodes, thyristors, rectifiers, transistors)

Semtech
Corporate Headquarters/Western regional Sales:
652 Mitchell Road
Newbury Park, CA 91320
805-498-2111
Fax: 805-498-3804

Eastern Regional Sales:
387 Main Street
Ridgefield, CT 06877
203-438-8877
Fax: 203-438-9410

Central Regional Sales:
1701 Greenville Avenue, Suite 501
Richardson, TX 75081
214-793-9903
Fax: 214-437-6350

Semtech Corpus Christie
(Formerly Lambda Semiconductor)
121 International Blvd.
Corpus Christie, TX 78406
1-800-451-5883
512-289-7780
Fax: 512-289-0472
e-mail: NPSMTCHAD@AOL.COM

European Manufacturing and Sales Office:
Newark Road South, EastField
Glenrothes, FIFE KY7 4NF
Scotland
01592 773520
Fax: 01592 774781

France Sales Office:
17 Avenue Marc Sangnier
92398 Villeneuve la Garenne Cedex
France
1 40 85 90 91
Fax: 1 47 98 07 51

Germany Sales Office:
Weinstrasse 2
D-74172 Neckarsulm
Federal Republic of Germany
07132 37780
Fax: 07132 37775

(Linear voltage regulators, linear low dropout regulators, switching voltage regulators, voltage references, ultra fast rectifiers, voltage converters for battery applications, interface drivers, overvoltage protectors, voltage regulators, zeners, transient voltage suppressors, rectifiers, half current half wave assemblies, high voltage half wave assemblies, center taps and doublers, single phase full wave bridges, three phase bridges, high voltage capacitors [Notes (1) source for Modupower DC/DC converters, (2) Semtech Corporation purchased Gamma Inc.] [ECI Semiconductors, Santa Clara, CA], a supplier of foundry wafers and linear, custom linear and digital arrays, in 1995., Military qualified devices available.)

Sensar, Inc.
P.O. Box 3538
Princeton, NJ 08543-3538
609-951-9544
Fax: 609-951-9458

(A spin off from the David Sarnoff Research Center owned by SRL International. It manufacturers, markets and provides contract services for real time vision and recognition patented technology, including it's "PYR" IC for image processing and computer vision applications.)

SensArray
3 Ray Avenue
Burlington, MA 01803
617-273-7373
Fax: 617-273-2552

(Infrared sensing and imaging array modules)

Senisys
See *Clarostat Sensors and Controls Group*

Sensing Devices, Inc.
1809 Olde Homestead Lane
Lancaster, PA 17601
717-295-2311
Fax: 717-295-2314

(Temperature sensors)

Sensitron Semiconductor
(RSM Sensitron Semiconductor, a division of RSM Electron Power, Inc.)
221 West Industry Court
Deer Park, N.Y. 11729
516-586-7600
Fax: 516-242-9798
TLX: 96-7737

(High current, MELF and surface mount square packaged hermetic high current [and high voltage] rectifiers for military and high reliability applications. Custom assemblies available)

SensorTec, Incorporated
16335-7 Lima Road
P.O. Box 373
Huntertown, IN 46748
219-637-3442
Fax: 219-637-3172

(Thermocouples, RTDs, PRTs, thermistors)

Sensors Unlimited, Inc.
3490 U.S. Route 1
Princeton, NJ 08540
609-520-0610
Fax: 609-520-0638

(InGaAs photodiodes)

Sensory Circuits, Inc.
1735 N. First St., Suite 313
San Jose, CA 95112-4511
408-452-1000
Fax: 408-452-1025

(A fabless company supplying voice recognition, speech synthesis and general control ICs)

Sensotek, Inc.
1200 Chesapeake Avenue
Columbus, OH 43212
1-800-848-6564 (order line)
614-486-7723
TWX: 810-482-1188

(Gage and absolute pressure transducers, differential pressure transducers, load cells, piezoelectric and piezoresistive accelerometers, displacement transducers and associated instruments)

Sen-Sym Inc.
1244 Reamwood Avenue
Sunnyvale, CA 94089
1-800-45-SENSYM
(1-800-457-3679)
Fax: 408-734-0407
Milpitas, CA
800-573-6796
Fax: 408-954-9458

1512 Ridgeside Drive
Suite 203, Box 136
Mt. Airy, MD 21771
301-829-9603
Fax: 301-829-0970

(Pressure sensors and transducers)

SEP Electronic Corporation
3 FL, No. 25-4 Pao-Hsing Road,
Hsin Tien
Taipei Hsien, Taiwan
886 2 917 5364, 886 2 917 9954
Fax: 886 2 917 9954

(Diodes, optoelectronics)

Seponix Corporation
2151 O'Toole Avenue
Suite L O'Toole Business Center
San Jose, CA 95131
800-237-4590
408-922-0133
Fax: 408-922-0137

(Modem and telecommunications interface ICs)

Servoflo Corporation
75 Allen Street
Lexington, MA 02173
617-862-9572
Fax: 617-862-9244

(Pressure sensors by Fujikura, etc.)

Setra Systems, inc.
45 Nagog Park
Acton, MA 01720
800-257-3872
508-263-1400
Fax: 508-264-0292

(Accelerometers, pressure transducers, digital systems)

SFH prefix
See *Siemens Components*

SG — SHANGHAI RADIO FACTORY

SG prefix
See *Sony*, *Linfinity Microelectronics, Inc.*, *Motorola Semiconductor Products*

SGH prefix
See *Sony*

(GaAs transistor)

SGS
See *SGS-Thomson*

SGS-Thomson Microelectronics
(ST logo)
1000 E. Bell Road
Phoenix, AZ 85022
602-867-6100
602-867-6340 (DSP ICs)
Fax: 602-867-6102
http://www.st.com

303 Williams Avenue, Suite 1031
Huntsville, AL 35801-5104

2055 Gateway Place, Suite 300
San Jose, CA 95110

3 Hutton Centre Drive, Suite 850
Santa Ana, CA 92707

5619 Scotts Valley Drive, Suite 180
Scotts Valley, CA 95066

1860 Industrial Circle, Suite D
Longmont, CO 80501

902 Clint Moore Road,
Building 3, Suite 20
Boca Raton, FL 33487

1300 East Woodfield Road, Suite 410
Schaumburg, IL 60173-5444

404 Pennsylvania Parkway, Suite 360
Indianapolis, IN 46280

2745 Albright Road
Kokomo, IN 46902-3996

Lincoln North, 2nd Floor
55 Old Bedford Road
Lincoln, MA 01733
617-259-0300
Fax: 617-259-4420, 4421

17197 North Laurel Park Drive,
Suite 253
Livonia, MI 48152

8525 Edinbrook Crossing
Brooklyn Park, MN 55443

2000 Regency Parkway, Suite 535
Cary, NC 27511

170 Mt. Airy Road, Building A,
Suite AA3
Basking Ridge, NJ 07920

1307 White Horse Road,
Building F
Voorhees, NJ 08043
609-772-6222

2-4 Austin Court
Poughkeepsie, NY 12603-3633

4900 South West Meadows Road,
Suite 475
Lake Oswego, OR 97035

211 Commerce Road
Montgomeryville, PA 18936

406 Union Avenue, Suite 645
Knoxville, TN 37902

8911 Capital of Texas Hwy.,
Suite 1120
Austin, TX 78759

1310 Electronics Drive,
Mail Stop 764
Carrollton, TX 75006-5039
214-466-7404
Fax: 214-466-7352

20515 SH249 #498,
Building CCA3
Houston, TX 77070

International Sales Office:
2723 37th Avenue, N.E.
Suite 206
Calgary, Alberta, Canada T1Y 5R8

301 Moodie Drive, Suite 307
Nepean, Ontario, Canada K2H 9C4

900 The East Mall, 3rd Floor
Etobicoke, Ontario, Canada M9B6K2

R. Henrique Schaumann 286
05413 Sao Paolo, Brazil

Via C. Olivetti 2
Agrate Brianza, Italy
39 39 6355 021

(Microprocessors, microcontrollers, EPROMs, flash memory, discrete semiconductors, gate arrays, flash memory, MPEG encoders/decoders, disk drive ICs, lamp dimmer IC, DACs, solid state relays [surface mount types], video compression ICs including MPEG types, Improved Quality Television [IQTV] IC, .etc., SONET/SGH ICs, smart power ICs for battery charging, audio processor ICs [including devices for hi-fi systems and karoake machines] Note: See also Thomson Components and Tubes Corporation)

SGM prefix
See *Sony Semiconductor*

(GaAs MOSFETs)

SH prefix
See *Hitachi Semiconductor*

Shanghai Radio Factory Number 7
See *Advanced Semiconductor Manufacturing Company of Shanghai* (ASMC)

Sharc
(Super Harvard Architecture DSP)
See *Analog Devices*

Sharlight Electronics Co., Ltd.
No. 112, Section 3, An Ho Road
Hsin Tien City
Taipei Hsien, Taiwan
886 2 944 5973, 886 2 944 5975
Fax: 886 2 946 9732

(Optoelectronics, microprocessor support ICs)

Sharp Corporation
Japan, International Sales Department
Electronics Components Group
22-22, Nagaike-Cho, Abeno-Ku,
Osaka 545, Japan
81 06 621-1221, 06 625 3007
Fax: 81 6117-725300,
6117-725301, 6117-725302,
06 628 1653, 06 628 1667

IC Sales Department
International Sales and
Marketing Group
IC/Electronic Components
Integrated Circuits Group
2613-1 Ichinomoto-Cho
Tenri-City, Nara 632, Japan
07436 5-1321
TLX: LABOMETA-B J63428
Fax: 07436 5-1532

Sharp Electronics Corporation
North America
Microelectronics Division
Sharp Plaza
Mahwah, NJ 07430-2135
1-800-642-0261 ext 900
201-529-8757, Fax: 201-529-8759
1-800-833-9437 (FASTFAX)
Telex: 426903
(SHARPAM MAWA)
http://www.sharplabs.com (PCs, digital video and imaging)
1-800-237-4277
Information on Sharp Products:
1-800-642-0261 ext. 900
Fax: 206-834-8996

Microelectronics Group
(Including optoelectronics)
5700 NW Pacific Rim Blvd.
Camas, WA 98607
800-642-0262, 0261, 360-834-2500
206-824-8966 (gate arrays)
206-834-2500 (DSP IC's)
TLX: 49608472 (SHARPCAM)
Fax: 206-834-8903, 360-834-8903
FastFax: 800-833-9437
Literature: 800-642-0261

Sharp Electronics (Europe) GmbH
Microelectronics Division
SonninstraBe 3
2000 Hamburg 1, Germany
49 040 23775-0
Fax: 49 40 231480
Telex: 2161867 (HEEG D)

20097 Hamburg, Germany
49 40 2376-2286
Fax: 49 40 231480,
49 40 2376 2232

Sharp-Roxy (Hong Kong) Ltd.
3rd Business Division
Room 1701-1711
Admiralty Centre, Tower 1
18 Harcourt Road, Hong Kong
5 8229311/8229348
Fax: 5 297561/8660779
Telex: 5 297561, 74258 SRHL HX

Sharp-Roxy Sales (Singapore)
PTE, Ltd.
100G Pasir Panjang Road
Singapore 0511
4731911, Fax: 4794105
Telex: 55504 (SRSSIN RS)

Sharp-Roxy Sales
& Service Company
(M), SDN.BHD
IC/Electronics Component Department
No. 11B, Jalan 223, Section 51-A
46100 Pataling Jaya
Selangor, Malaysia
3 7571477
Fax: 3 7571736
TLX: RMKL MA37167

Sharp Electronics Components
(Korea) Corporation
RM 501 Geosung Bldg,
541 Dohwa-Dong
Mapo-Ku, Seoul, Korea
02 711 5813
Fax: 02 711 5819
TLX: SHARPCC K22080

Sharp Electronic Components
(Taiwan) Corporation
7F, No. 16, Sec. 4, Nanking E. Rd.
Taipei, Taiwan, Republic of China
02 741 7341
Fax: 02 741 7326, 02 741 7328
TLX: 10518 SECT

(Microprocessors [including RISC type], microcontrollers, SRAMs, fiber optics, various ICs, GaAs Hall Effect devices, gate arrays, LCDs, AM-LCDs, electroluminescent displays, infrared data links, etc.)

SHB prefix
See *S-MOS Systems*

(Hybrid IC)

SHC prefix
See *Burr-Brown*

Shelly Associates, Inc.
14811 Mayford Road
Tustin, CA 92680-7253
714-669-9850
Fax: 714-669-1081

(Graphic LCD modules-passive matrix color and monochrome, discrete LEDs)

Shian Yih Electronic Industry Co., Ltd.
No. 22 Kno Yea 24th road,
Taichung Industrial Estate
P.O. Box 46-13 Taichung, Taiwan
886 4 254 0111, Fax: 886 4 252 7646
TLX: 57405 SHIANYIH

SHIAN YIH — SIEMENS COMPONENTS

(Optoelectronics)

Shimadzu
See *Soltec*

Shindengen Electric Mfg. Co. Ltd.
New Ohte-Machi Bldg.
2-2-1, Ohte-Machi, Chiyoda-Ku
Tokyo 100
Japan
03 3279 4431
Fax: 03 3279 6478

Shindengen America, Inc.
2985 E. Hillcrest Drive
Westlake Village, CA 91362
800-634-3654 (Western US)
800-543-6525 (Eastern US)
Fax: 805-373-3710

5999 New Wilke Road
Suite 406
Rolling Meadows, IL 60008
800-543-6525
Fax: 708-593-8597

(DC/DC converters, hybrid ICs, MOSFETs, schottky diodes, regulators, and power supplies)

SHM prefix
See *Datel, Inc.*

Shogyo International Corp.
287 Northern Blvd.
Great Neck, NY 11021-4799
516-466-0911
Fax: 516-466-0922
Cable: SHOGYONEWYORK
TLX: 4758024

(Battery holders, buzzers, cable/cord assemblies, connectors, lamps/bulbs, speakers, switches, transformers [including miniature metal can units], adaptors)

Shokai Far East Ltd.
9 Elena Court
Peekskill, NY 10566
914-736-3500
Fax: 914-736-3656

(Optoelectronics)

Shoreline Electronics, Inc.
2098 B. Walsh Avenue
Santa Clara, CA 95050
408-987-7793
Fax: 408-987-7735

(Rechargeable battery charging ICs)

Si
See *Syntar Industries, Inc.*

Si prefix
See *Siliconix*

SI prefix
See *Sanyo Energy Corp.*

S.I. Diamond Technology, Inc.
2435 North Blvd.
Houston, TX 77098-5015
713-529-9040

(Field-emission displays)

SI Semiconductors Co. Ltd.
(Shenzhen, Guangdong province of southern Peoples Republic of China, a joint venture between SEG and America IBDT Co., a Hong Kong Based trading company. Bipolar power transistors and discrete ICs)

SICOM, Inc.
7585 East Redfield Road
Scottsdale, AZ 85260
602-483-2867
Fax: 602-483-7986
E-mail: info@SICOM.com

(Custom ASICs, including a radiation hardened CMOS digital sub-band tuner IC)

SID Microeletronica
R. Paes Leme, 524-80
05 424 Sao Paolo - SP
Brazil
011 813 0333

(Thyristors, transistors, digital and linear ICs)

Siemens Components, Inc.
Inquiry Center
1000 Business Center Drive
Mt. Prospect, IL 60056
800 77-SIEMENS
1-800-888-7729
(Literature and technical assistance)
Fax: 708-296-4805

10950 North Tantau Avenue
Cupertino, CA 95014
408-777-4500, 408-777-4968
Fax: 408-777-4987
(Also serves the Northwest USA)
http://www.sci.siemens.com
http://www.siemens.de/ec
(Europe, relays connectors, fiber optics, hybrids)

Suite 320
625 The City Drive South
Orange, CA 92668
714-971-1274
Fax: 714-971-1294
(Also services southern CA)

Integrated Circuit Division
2191 Laurelwood Road
Santa Clara, CA 95054-1514
800-456-9229
408-980-4500

Suite 690
16479 Dallas Parkway
Dallas, TX 75248
214-733-4511
Fax: 214-733-4862
(Also optoelectronics)

Manufacturers, Prefixes, Part Number Types, Logo Descriptions & Family Types

SIEMENS COMPONENTS

Suite 206
6525 The Corners Parkway
Norcross, GA 30092
404-449-3981
Fax: 404-449-4522
(Also optoelectronics)

P.O. Box 7931
Mt. Prospect, IL 60056
1-800-77-SIEMENS

Suite 565
1901 N. Roselle Road
Schaumburg, IL 60195
708-884-7009
Fax: 708-884-7599
(Also serves the western USA)

238 Littleton Road
Westford, MA 01886
508-692-0550
Fax: 508-692-2309
(Also optoelectronics)

Suite 435
38701 W. Seven Mile Road
Livonia, MI 48152
313-462-1195
Fax: 313-462-1694
(Also optoelectronics)

Suite 202
307 Fellowship Road
Mt. Laurel, NJ 08054
609-273-6677
Fax: 609-273-8904
(Also optoelectronics)

Suite 606
100 Wood Avenue South
Iselin, NJ 08830-2709
908-603-0600
Fax: 908-603-0669
(Also optoelectronics)

Optoelectronics Division
19000 Homestead Rd.
Cupertino, CA 95014
800-777-4363
408-257-7910
Fax: 408-777-4988

1400 W. 122nd Avenue
Suite 103
Westminster, CO 80234
303-451-5513
Fax: 303-451-6932

522 Belvedere Drive, Suite 112
Kokomo, IN 46901
317-456-1928
Fax: 317-456-3836

Siemens Electromechanical
Components, Inc.
200 S. Richland Creek Dr.
Princeton, IN 47671-0001
812-386-1000
Europe: 49 911 3 00 12 38
49 911 978 3321
http://www.siemens.de/ec

(Relays connectors, fiber optics, hybrids)

Australia:
Siemens, Ltd. Head Office
544 Church Street
Richmond (Melbourne), Vic. 3121
03 4207111
Fax: 03 4207275
TLX: 30425

Siemens AG Osterreich
Postfach 326
1031 Wien
01 71711 5661
Fax: 01 71711 5973
TLX: 1372-01

Belgium:
Siemsne S.A.
Chaussee de Charleroi 116
1060 Bruxelles
02 536 2111
Fax: 02 536 2492
TLX: 21347

Canada:
1180 Courtney Park Drive East
Mississagua, Ont. L5T 1P2
905-564-1995
Fax: 905-670-6563

7300 Transcanada Highway
Point Claire, PQ H9R 5C7
514-426-6103
Fax: 514-426-6125

Denmark:
Siemens A/S
Borupvang 3
2750 Ballerup
44 774477
Fax: 44 774017
TLX: 1258222

Finland:
Siemens Oy
P.O.B. 60
02601 Espoo 0 51 051
Fax: 0 51 05 23 98
TLX: 124 465

France:
Siemens S.A.
39/47 Bd. Ornano
93527 Saint-Denis CEDEX 2
1 49223100
Fax: 1 49223970
TLX: 234077

Germany:
Siemens AG, Salzufern 6-8
10587 Berlin
030 3993 0
Fax: 030 3993 2490
TLX: 17308196 siexnvb

Lahnweg 10
40219 Dusseldorf
0211 399 0
Fax: 0211 399 1481

Rodhelheimer LandstraBe 5-9
60487 Frankfurt
069 797 0
Fax: 069 797 2582
TLX: 4141650

SIEMENS COMPONENTS

Lindenplatz 2
20099 Hamburg
040 2889 0
Fax: 040 2889 3096
TLX: 215584 0

Werner-Von-Siemens Platz 1
30880 Laaten
0511 877 0
Fax: 0511 877 2078
TLX: 922333

Richard-Straub-Strabe 76
81679 Munchen
089 9221 0
Fax: 089 9221 4692
TLX: 529421 19

Von-der-Tann-Strabe 30
90439 Nurnberg
0911 654 0
Fax: 0911 654 6505
TLX: 622251 0

Geschwister-Scholl-Strabe 24
70174 Stuttgart
0711 2076 0
Fax: 0711 2076 2448
TLX: 723941 50

Great Britain:
Siemens PLC
Siemens House
Oldbury, Bracknell
Berkshire RG 12 8FZ
0344 396000
Fax: 0344 396632

Greece:
Siemens AE
Paradissou & Artemidos
P.O.B. 61011
15110 Amaroussio/Athen
01 6864111
Fax: 01 6864299
TLX: 216292

India:
Siemens Ltd.
head Office
134-A Dr. Annie Besant Road, Worli
P.O.B. 6597
Bombay 400018
022 4938786
Fax: 022 4940240
TLX: 1175142

Ireland:
Siemens ltd.
Electronic Components Division
8 Raglan Road, Dublin 4
01 684727
Fax: 01 684633
TLX: 93744

Italy:
Siemens S.p.A.
Semiconductor Sales
Via del Valtorta, 48
21012 Milano
02 66 76 1
Fax: 02 6766 4395
TLX: 330261

Netherlands:
Siemens Netherland N.V.
Postb. 16068
2500 BB Den Haag
070 3333333
Fax: 070 3332790
TLX: 31373

Norway:
Siemens A/S
Ostre Aker Vei 90
Postboks 10, Veitvet
0518 Oslo 5
02 633000
Fax: 02 633805
TLX: 78477

Poland:
Siemens Sp. z.o.o.
ul. Stawki 2, POB 276
00-950 Warszawa
6 35 16 19
Fax: 6355238
TLX: 825 554

Portugal:
Siemens S.A.
Estrada Nacional 117, km 2.6
Alfragide, 2700 Amadora
01 4170011
Fax: 01 4172870
TLX: 62955

Russia:
Siemens AG
1. Donskoj pr., 2
Moskva 117419
095 2 37- 64 76, or -69 11
Fax: 095 237 6614
TLX: 414 385

Singapore:
Siemens Components Pte. Ltd.
Sales and Promotion Office
Industrial products
166 Kallang Way
Singapore 1334
7417418
Fax: 7426239/7422803

South Africa:
Siemens Ltd., Siemens House
P.O.B. 4583
Johannesburg 2000
011 3151950
Fax: 011 3151968
TLX: 450091

South Korea:
Siemens Ltd.
Asia Tower 10th Floor
726 Yeoksam-Dong
Kangnam-Gu
C.P.O. Box 3001, Seoul
02 527 7700
Fax: 02 527 7779

Spain:
Siemens S.A.
Dpto. Componentes
Ronda de Europa, 3
28760 Tres Cantos-Madrid
01 8 03 00 85
Fax: 01 8033926
TLX: 44191

Manufacturers, Prefixes, Part Number Types, Logo Descriptions & Family Types

SIEMENS COMPONENTS — SIGNETICS

Sweden:
Siemens Components
Osterogatan 1
Box 46
S-146 93 Kista
08 7033500
Fax: 08 7033501
TLX: 11672

Switzerland:
Siemens-Albis AG
FreilagerstraBe 28
8047 Zurich
01 495 3111
Fax: 01 495 5050
TLX: 823781-23

Turkey:
Simko Ticaret ve Sarayi A.S.
Meclisi Mebusan Cas.125
Findikli-Istanbul 80040
Turkey
01 251 0900
Fax: 01 249 4953
TLX: 243 33

(Hall effect vane switches, bipolar, unipolar and analog output devices, communications receiver/transmitter ICs, PCI multimedia controllers [jointly w/Zoran], SAW filters and resonators, LEDs [including blue LEDs], LED drivers, intelligent displays, LED numeric displays, light bars, optocouplers [optoisolators], fiber optic emitters, fiber optic photodetectors, photodiodes, phototransistors, photovoltaic cells, laser diodes, microcontrollers, Asynchronous Transfer Mode [ATM] ICs, including SONET and SDH devices, optical encoders, memory switches, power switches, microcontrollers, coprocessors, chime ICs, UARTs, fuzzy logic ICs, EEPROMs, operational amplifiers, picture in picture ICs, tuner ICs, stepper motor ICs, etc.)

Sierra Research & Technology, Inc.
6035 Kerrmoor Dr.
Westlake Village, CA 91362
818-991-1509
Fax: 818-991-1508
E-mail: cores@srti.com

(Development to delivery of silicon microcomputer cores)

Sierra Semiconductor
2075 N. Capitol Ave.
San Jose, CA 95132
408-263-9300
Fax: 408-263-3337

(A fabless company supplying ICs to enhance PC graphics and communications capabilities such as caller ID, mixed signal ICs, DTMF receivers, graphics accelerator and controller ICs, PC audio ICs, and a chip set for speech)

Sierra West Power Systems
300 N. Telshore Ste. 500
Las Cruces, NM 88011
505-522-8828
Fax: 505-522-8766

(DC/DC converters)

Sigma Designs, Inc.
46501 Landing Parkway
Fremont, CA 94538-6421
800-845-8086
510-770-0100

(Video compression ICs used in their products [and by orchid Technology]. Realmagic Chip Sets, MPEG chip integrating both audio and video decoder functions [graphics and video accelerator ICs])

Sigmapower
19060 S. Dominguez Hills Dr.
Rancho Dominguez, CA
90220-6404
310-884-5200

(DC/DC converters)

Sigmax Technology, Inc.
1641 S. Main St.
Milpitas, CA 95035-6262
408-245-2392

(CD-ROM controller ICs)

Signal Processing Technologies (SPT)
Member of the Toko Group
4755 Forge Road
Colorado Springs, CO 80907
800-643-3778, 719-528-2300
Fax: 719-528-2370

6 Misty Ridge Circle
Kinnelon, NJ 07405
201-492-0113
Fax: 201-492-0210

2480 North 1st Street, Suite 260
San Jose, CA 95131
408-432-9210
Fax: 408-432-0545

(A/D and D/A converters, comparators, track and hold amplifiers, filters, voltage regulators, RF amplifiers, power factor controller/pulse width modulator)

Signal Systems Int., Inc.
P.O. Box 470
Lavallette, NJ 08735
908-793-4668
Fax: 908-793-4679

(Mercury switches)

Signetics Co.
See *Philips Semiconductors*

(North American Philips, which has owned Signetics since 1975, restructured and formally dropped the Signetics name on January 1, 1993.)

SII — SILICON STORAGE TECHNOLOGY

SII
See *Seiko Instruments*

SIL-Walker, Inc.
Royal Bldg
5-1-1, Shinjuku-ku
Tokyo 160 Japan
81 3 3341 3651
Fax: 03 3341 3974

(Converter, fax machine, measuring instrument, logic, and image processing ICs)

Silicon Composers, Inc.
470 San Antonio Road
Suite F
Palo Alto, CA 94306
415-843-1135
Fax: N/A
http://www.silcomp.com
info@silcomp.com

(Single board computers, software, mixed signal ASICs, SC32 Stack-chip microprocessor for 32 bit Forth language [device developed at John Hopkins University])

Silicon Designs, Inc.
1445 NW Mall Street
Issaquah, WA 98027
206-391-8329
Fax: 206-391-0446

(Digital integrated accelerometers)

Silicon Detector Corp.
1240 Avenida Acaso
Camarillo, CA 93010-8727
805-424-2884
Fax: 805-484-9935
TLX: 182229

(Optoelectronics; see also Advanced Photonix, Inc.)

Silicon Expertise S.A.
Rue du Bosquet 7
Louvain-la-Nueve 1348
Belgium
32 10 454 904
Fax: 32 10 454 934

(FIFOs)

Silicon General, Inc.
See *Linfinity Microelectronics, Inc.*

Silicon Graphics Computer Systems (SGI)
2011 N. Shoreline Blvd.
P.O. Box 7311
Mountain View, CA 94039
415-960-1980
Fax: 415-961-0595
http://www.sgi.com

(Computer ICs used in video communication equipment)

Silicon Integrated Systems, Corp. (SiS)
2 FL., No. 17, Innovation Road 1
Science Based Industrial Park
Hsin-Chu, Taiwan ROC
886 35 774 922
Fax: 886 35 778 774

240 North Wolfe Road
Sunnyvale, CA 94086
408-730-5600
Fax: 408-730-5639

(PC Chip sets, VGA controllers including multimedia accelerators and Green PC chip sets [devices with power management techniques, graphics controllers and sound ICs])

Silicon Logic, Inc.
550 E. Brokaw Road
San Jose, CA 95161-9048
800-272-8866
408-441-1615
Fax: 408-954-0727

(D/A converters, disk controllers, bus transceivers, Centronics port ICs, serial and parallel port ICs, UARTS, listed as second source for Macronix)

Silicon Magic Corporation
20300 Stevens Creek Blvd.
Cupertino, CA 95014
408-366-8888

(A 1995 start-up company working on embedded memory and multimedia devices [combining graphics, audio, video functions and DRAM], EDO DRAMs, graphics DRAMs)

Silicon Microstructures Inc. (SMI)
46725 Fremont Blvd.
Fremont, CA 94538
510-490-5010
Fax: 510-490-1119

(Bipolar sensors and accelerometers. This company was acquired by Exar Corp. in mid 1995)

Silicon Mountain Design, Inc. (SMD)
5055 Corporate Plaza Drive, Suite 100
Colorado Springs, CO 80919
719-599-7700
Fax: 719-599-7775

(CCDs and programmed FPGAs used in their cameras)

Silicon Sensors, Inc.
Old Highway 18 East
Dodgeville, WI 53533
608-935-2707, Fax: 608-935-2775

(PIN Si photodiodes, standard and custom units available)

Silicon Storage Technology, Inc. (SST)
1171 Sonora Court
Sunnyvale, CA 94086-5308
408-735-9110
Fax: 408-735-9036

Manufacturers, Prefixes, Part Number Types, Logo Descriptions & Family Types

SILICON STORAGE TECHNOLOGY — SILICONIX

(EEPROMs [flash architecture designs])

Silicon Systems, Inc. (SSI)
14351 Myford Road
Tustin, CA 92680
800-624-8999
714-731-7110
714-573-6000
(Call for name of local distributor or representative)
Fax: 714-669-8814,
714-573-6914 or 6906
TWX: 910-595-2809
BBS: 714-544-6525
(14.4k BPS, if trouble call 714-573-6226)
E-mail: spd@ssil.com
http://www.ssi1.com

East Headquarters
(CT, MA, ME, NH, RI, VT):
59 Stiles Road
Salem, NH 03079
603-898-1444
Fax: 603-898-9538

Central Headquarters:
340 W. Butterfield Rd., Suite 4B
Elmhurst, IL 60126
708-832-3111
Fax: 708-832-3172

Administrative Office:
2201 N. Central Expressway,
Suite 132
Richardson, TX 75080
214-669-3381
Fax: 214-669-3495

Automotive Sales Office:
Detroit (F.A.E.)
313-462-2133
Fax: 313-462-2405

Southwest Headquarters:
454 Carson Plaza Drive, Suite 209
Carson, CA 90745
213-532-1524, 3499
Fax: 213-532-4571

Northwest Headquarters:
2840 San Thomas Exp., Suite 140A
Santa Clara, CA 95051
408-980-9771
Fax: 408-748-9488
TLX: 171-200

Northern California:
2001 Gateway Pl., Suite 301 East
San Jose, CA 95110
408-432-7100
Fax: 408-453-5988

Colorado Office:
1860 Lefthand Circle, Suite A
Longmont, CO 80501
303-678-8003
Fax: 303-678-7920

Florida:
2646 S.W. Mapp Rd., Suite 203
Palm City, FL 34990
407-223-1143
Fax: 407-223-8085

Georgia:
6855 Jimmy Carter Blvd.,
Suite 2150
Norcross, GA 30071
404-409-8405
Fax: 404-368-1060

Texas:
690 W. Campbell Road, Suite 150
Richardson, TX 75080
214-479-9170
Fax: 214-479-9172

Europe:
Silicon Systems International
The Business Centre
758-760 Great Cambridge Road
Enfield
Middlesex EN1 3RN
United Kingdom
44 181 443 7061
Fax: 44 181 443 7022

Far East Headquarters:
Silicon Systems, Singapore
3015A Ubi Road 1, #01-01
Kampong UBI Industrial Estate
1440 Singapore
65 744 7700
Fax: 65 748 2431

(Formerly a TDK Group Company, it was acquired by Texas Instruments in mid 1996. The company offers various mixed signal devices including telecommunications devices such as Ethernet, ATM and Modem ICs, and custom ICs, HDD [hard disk drive] devices such as: read/write amplifiers, pulse detectors, filters, head positioning, spindle motor control, controller interface; floppy disk circuits, tape drive circuits and magneto-resistive recording head read/write ICs and automotive IC's such as ignition predrivers, etc.)

Silicon Transistor Corporation (STC)
21 Katrina Road
Chelmsford, MA 01824
508-256-3321
Fax: 508-250-1046
TWX: 710-343-0576
TLX: 200151

(Bipolar and MOSFET devices, diodes, thyristors, transistors, military qualified, custom packaged, etc.)

Siliconix
Siliconix, Inc., Matra MHS
(Formerly Matra Harris, a joint company originally formed by Harris and Matra SA and later Matra sold 50% of the company to Telefunken. Telefunken Electronic GmbH Dialog Semiconductor and Eurosil Electronic are semiconductor operations of Daimler-Benz AG and are sold in the United States under an organization

SILICONIX — SIPEX CORPORATION

named Temic Semiconductor Division [Temic Telefunken Microelectronic GmbH]. Temic Semiconductor Division. Temic is jointly owned by two units of Daimler-Benz)

Deutsche Aerospace and AEG.
2201 Laurelwood Road
Santa Clara, CA 95054-1516
P.O. Box 54951
Santa Clara, CA 95056-0951
1-800-554-5565
408-988-8000
TWX: 910-338-0227
Fax: 408-970-3950, 408-970-3995, 408-727-5414
Faxback (To order datasheets):
408-970-5600
http://www.temic.de/

Matra MHS Electronics Corp.
408-748-9362
Morriston, Swansea SA6 6NE,
United Kingdom
(0792)-310100, and 0344 485757
TLX: 48197
Fax: (0792) 310401
Also (0635)44099

Germany:
Telefunken Electronic GmbH
P.O. B. 3535
D-7100 Heilbronn
(0 71 31) 67-29-45, 67-0
Fax: (0 71 31) 67-23 40
TLX: 728 746 tfk d
Germany (Temic): 0130 857 320

France:
Matra MHS
(1) 30 30.60.70.48 (or .00)
Temic Semiconductor:
1 30 60 71 87

Italy:
(02) 02 331 121
or (02) 489 52258

Scandinavia:
08-733 25 00,
08 733 0090

Singapore: 65-7886668
Hong Kong: 852-724 3377

Japan:
3 3578 0823,
Tomen Electronics 81-3-3506-3696

United Kingdom: 01344 707300

(Matra MHS has SRAMS, microcontrollers and microprocessors and MPEG video ICs, FIFOs, etc. Siliconix has video buffers, MOSFETs, electronic switches, etc. Temic-Telefunken has pulse width modulation ICs, liquid level switch ICs, solid state flasher ICs, RF ICs, IR transceivers, diodes and transistors, etc.)

Silitek Corp.
10Fl, No. 25, Tung Hwa S. Road
P.O. Box 36-110
Taipei, Taiwan
886 2 776-6633,
Fax: 886 2 776 2668, 886 2 711 1382

(Diodes, thyristors)

Silonex, Inc.
331 Cornelia St.
Plattsburgh, NY 12901
518-561-3160, Fax: 518-747-3906

(Optoelectronics)

SIM-XXXX part number
See *ILC Data Device Corporation*

Simco Ticaret ve Sarayi A.S.
(Turkey)
See *Siemens*

Simple Technology, Inc.
3001 Daimler St.
Santa Ana, CA 92705
800-367-7330, 714-476-1180
Fax: 714-476-1209

(DRAM SIMM modules with stacked chips for high memory density, removable solid state hard drives)

Simtek Corp.
1465 Kelly Johnson Blvd.
Colorado Springs, CO 80920
1-800-637-1667
719-531-9444
Fax: 719-531-9481

5696 Peachtree Parkway
Norcross, GA 30092
404-242-2622
Fax: 404-242-2624

2880 Zanker Road #203
San Jose, CA 95134-2122
408-432-7230
Fax: 408-432-7235

Premier House, Union Street
Bristol BS1 2DJ England
0272 49 7762
Fax: 0272 49 7764

(Nonvolatile SRAMs, military qualified parts available)

Sine Wave logo
(With two waves, one slightly out of phase with the other)
See *In-Phase Electronics*

Sino-American Silicon Products, Inc.
8, Industrial East Road
Science Based industrial Park
Hsin-Chu, Taiwan
886 35 772 233
Fax: 886 35 781 706

(Diodes)

Sipex Corporation
Formerly Hybrid Systems, Inc.
(Sipex name or Sx logo)
22 Linnell Circle
Billerica, MA 08121-3985
508-667-8700, Fax: 508-670-9001

Manufacturers, Prefixes, Part Number Types, Logo Descriptions & Family Types

491 Fairview Way
Milpitas, CA 95035
408-945-9080
Fax: 408-946-6191

France:
Sipex S.A.R.L.
30 Rue du Morvan, SILIC 525
94633 Rungix Cedex
33 1 4687 8336
Fax: 33 1 4560 0784

Germany:
Sipex GmbH
Gautinger Strasse 10
82319 Starnberg
49 8151 89810
Fax: 49 8151 29598

Japan:
Nippon Sipex Corporation
Haibara Building
2-3-7 Sotokanda
Chiyoda-Ku, Tokyo 101
81 3 3254 5822
Fax: 81 3 3254 5824

(RS232 line drivers/receivers, AD/DA converters, data acquisition chip sets, analog tile arrays, op-amp macromodels, mixed signal ASICs, EL lamp drivers, etc.)

SIR prefix
See *Irvine Sensors Corp.*

SiS logo
See *Silicon Integrated Systems, Corp.*

SIS logo
See *SIS Microelectronics, Inc.*
(There is a dot over the "i" made up of 3 lines)

SIS Microelectronics, Inc.
P.O. Box 1432
1500 Kansas Avenue, Suite 1D
Longmount, CO 80501
303-776-1667
Fax: 303-776-5947

(AM29030 memory peripheral controller [SMEMC], laser printer peripheral interface controller [SPLC], laser printer data compression IC, etc.)

SITe
Scientific Imaging Technologies, Inc.
P.O. Box 569
Beaverton, OR 97075-0569
503-644-0688
Fax: 503-644-0798
http://www.site-inc.com

(Note: See also ISA Jobin Yvon-Spex [Spex Industries], CCDs and digital electronic imaging products)

SiTEL Sierra b.v.
Netherlands
(This supplier of chips and subsystems for the wireless market was acquired by National Semiconductor in January 1996. It was formerly part of Sierra Semiconductor, San Jose, CA.)

Siward Crystal Technology Co. Ltd.
15, Lane 81
Tan-Fu Road, Sec. 2
Tant Zu Hsing, Tai-Chung,
Taiwan, R.O.C.
886-4-323-4255, or 323-0059
Fax: 886-4-323-8602

One Gateway Drive Plaza Level
Parsippany, NJ 07054
201-898-1234
Fax: 201-898-1235

(Crystals, crystal oscillators, crystal filters [monolithic], TCXOs, VCXOs)

SK prefix
See *Harris Semiconductor* (Formerly RCA), *Level One Technology*

Skan-A-Matic Corporation
See *Clarostat Sensors and Controls*

SIPEX CORPORATION — SLPC

SKM prefix
See *Semikron, Inc.*

Sky Electronics International
166 Tuckahoe Road
Buena, NJ 08310
609-692-0287
Fax: 609-692-0772

(Relays)

SL prefix
See *GEC Plessey Semiconductor,*
Via Technologies,
Windbond Systems

(Former Symphony Laboratories parts. SL is also a prefix for discontinued Fairchild Semiconductor devices.)

SLA prefix
See *S-MOS Systems* (Logic array),
Allegro Microsystems
(Motor driver ICs)

SLD prefix (LED)
See *Sony*

SLI
See *Semiconductor Laser International Corporation*

SLP prefix
See *SenSym, Inc.,*
Beta Transformer Technology
(Pulse transformers)

SLR prefix
See *Siemens,*
ROHM Electronics (LED lamps)

SLO prefix
See *Siemens*

SLPC
(Trademark of SIS Microelectronics, Inc.)

SLR — SMRT

SLR prefix, on LED lamps
See *ROHM Electronics*

SLV prefix
See *Philips*

SLVD prefix
See *Sony*

SLY prefix
See *Siemens*

S+M logo
See *Siemens Components, Inc.*

SM prefix
See *Silicon Magic*,
Smart Modular Technologies
(Memory modules or PCMCIA memory or Fax/modem cards)

SM-A prefix
See *Harris Semiconductor*
(Custom part)

SMx prefix
See *Westcode Semiconductors*

SM-XXX part number
See *ST Olektron Corp.*

Smar Research Corporation
10 Drew Court, Suite 9
Ronkonkoma, NY 11779
516-737-3111, Fax: 516-737-3982

(Communications control ICs for the process control industry including field bus controllers and modem ICs that are pin compatible with NCR devices)

SMART Modular Technologies
45531 Northport Loop, Bldg. 3B
Fremont, CA 94538
1-800-956-7627, 510-623-1231
Fax: 510-623-1434
E-mail: info@smartm.com
http://www.smartm.com

(Memory modules including DRAM, FLASH, SRAM and OTP modules and PCMCIA cards, Fax/modems)

Smart Relay Technologies, Inc.
Commack, NY
516-543-7161
Fax: 516-543-7162

(Solid state relays)

SMC prefix
See *S-MOS Systems*
(Microcomputer, CPU or peripheral),
Standard Microsystems, Corp.,
Advanced Micro Systems
(Motor controller),
Synergy Microwave Corporation
(Filters and couplers, etc.)

SMD
See *Silicon Mountain Design, Inc.*

SMEC, Inc.
12710 Research Blvd
Suite 285
Austn, TX 78759
800-36-7632
512-331-1877
Fax: 512-335-0444

(LC Filters, ferrite beads, NTC thermistors, capacitors, trimmer capacitors, inductors, resistors)

SMEMC
(Trademark of SIS Microelectronics, Inc.)

SMI
See *Silicon Microstructures, Inc.*,
System Microelectronic Innovation GmbH

SMJ prefix
See *Texas Instruments*

SMK Corp.
6-5-5 Togoshi, Shinagawa-Ku
Tokyo 142, Japan
81 3 3785 1111
Fax: 81 3 3785 1122

(Converter ICs, logic ICs, op amps, audio ICs, voltage comparators and voltage regulators, camera, communication and digital video ICs, watch, clock speech synthesis and recognition, television, VCR and telecommunications ICs)

SMM prefix
See *S-MOS Systems*
(Masked programmed memory),
Stanford Microdevices

SMOS
See *S-MOS*

S-MOS Systems, Inc.
A Seiko Epson Affiliate
2460 N. First St.
San Jose, CA 95131
800-228-3964
408-922-0200
Fax: 408-922-0238
E-mail: cards@smos.com
(Cardio (tm) PC motherboards)
WWW: www.smos.com

(Microcomputers, memory ICs, ASICs, graphics ICs [such as VGA controller] and 3D graphics 9accelerator] ICs, disk storage ICs [such as floppy disk controllers], analog switches, real time clock, DC/DC converter, voltage detector, voltage regulator, Hall effect sensors, gate arrays, display controller ICs [including LCD types], and multichip packaging, etc.)

SMP prefix
See *Siliconix, Inc.*

SMRT prefix
See *SenSym, Inc.*

Manufacturers, Prefixes, Part Number Types, Logo Descriptions & Family Types

SMRXX part number
See *Vishay*

(Precision resistors)

SMT prefix
See *Consumer Microcircuits, Ltd.*

SMV prefix
See *Z-Communications*

SN, SNJ prefix
See *Texas Instruments, AMD*

(Note: MMI, which is now AMD, also used an SN prefix. Some of these parts were second sources to TI but some were original MMI devices.)

Societe Generale Semiconductor (SGS)
See *SGS-Thomson*

Sokol Crystal Products, Inc. (SCP)
121 Water Street
Mineral Point, WI 53565
608-987-3363
TLX: 467581

(Quartz crystals and crystal filters)

SoLiCo
Sorenson Lighted Controls, Inc.
75 Locust St.
Hartford, CT 06114
1-800-275-7089
203-527-3092
Fax: 203-527-6047

(LED indicators including blue LEDs and panel mounted units)

Solid Power Corporation
440 Eastern Parkway
Farmingdale, N.Y. 11735
516-694-2883
Fax: 516-694-2849

(Power semiconductors, including military qualified devices)

Solid State Devices, Inc. (SSDI)
14830 Valley View Avenue
La Miranda, CA 90638
213-921-9660
Fax: 213-921-2396

P.O. Box 577
14849 Firestone Blvd
La Mirada, CA 90638
714-670-7734
Fax: 714-522-7424

(Power semiconductors for military, aerospace and industrial applications, including military approved devices, radiation hardened devices, fast recovery diodes and rectifiers, bridge assemblies, etc.)

Solid State (Microtechnology) Electronics Co. Pvt. Ltd.
P.O. Box 7432
J B Nagar Post
Bombay 400059, India
9122 634 5240
Fax: 9122 632 2743
TLX: 11 79325 MTNLIN

(Diodes, transistors, amplifiers, filters, RMS to DC converters, VCOs)

Solid State Electronics Corp.
18646 Parthenia St.
Northridge, CA 91324
818-993-8257
818-993-8259

(Solid state relays, electromechanical choppers)

Solid State, Inc.
46 Farrand St.
Bloomfield, NJ 07003
800-631-2075, 2075
201-429-8700

(Diodes, transistors, linear, digital and interface ICs, etc.)

SMRXX — SOLOMON TECHNOLOGY

Solid State Industries, Inc.
3421 M St. N W, Ste 1752
Washington, DC 20007
202-338-3150
TLX: 64100 SOLIDSTA

(Diodes, thyristors, transistors)

Solid State Scientific, Inc.
(Acquired by Sprague, now Allegro Microsystems)

Solid State Systems
960 Koehl Avenue
Union, NJ 07083
908-688-0222

(Diodes, thyristors, transistors)

Solidas Corporation
100 Century Center Court
San Jose, CA 95112-4512
408-436-8161

(Multilevel [high density] DRAMs, multiple bits stored in a single cell)

Solitron Devices, Inc.
3301 Electronics Way
West Palm beach, FL 33407
407-848-4311
Fax: 407-863-5946
TLX: 51 3435

(Hybrid circuits, voltage regulators, etc.)

SOLO prefix
See *European Silicon Structures*

(Programmable ASICs)

Solomon Technology (USA) Corp.
21038 Commerce Point Drive
Walnut, CA 91789
909-468-3738, Fax: 909-468-3726

SOLOMON — SPACE ELECTRONICS

(Graphics and character LCD modules)

Soltec Corporation
12977 Arroyo Street
San Francisco, CA 91340-1597
800-423-2344, 818-365-0800
Fax: 818-365-7839

(Light emitting diodes and photodiodes)

Sony Corporation
6-7-35, Kita-Shinagawa
Shinagawa-Ku
Tokyo 141 Japan
33 3 5448 2111, 33 3 3448 2111
Fax: 33 3 5447 2244

Sony Corporation of America
Semiconductor Division
10833 Valley View Street
Cypress, CA 90630-0016

Sony Electronics, Inc.
Components Products Company
P.O. Box 6016
Cypress CA 90630-0016
800-288-SONY (i.e., 7669)
714-229-4442, 4270
Fax: 714-229-4333
www.sel.sony.com/semi

Northwest Office:
655 River Oaks Parkway
San Jose, CA 95134
408-944-4314
Fax: 408-433-0834

Central Office:
1200 N. Arlington Heights Road
Itasca, IL 60143
708-733-6072
Fax: 708-773-6068

Southeast Office:
One Copley Parkway, #206
Morrisville, NC 27560
919-380-0786
Fax: 919-467-2963

Northeast Office:
85 Wells Avenue
Newton, MA 02159
617-630-8812
Fax: 617-630-8890

Alaska: 408-432-0190

Kansas, Illinois, Missouri (South):
708-773-6072

Maine, Massachusetts, New Hampshire, Rhode Island, Vermont:
617-630-8812

Puerto Rico: 919-380-0786

Canada:
411 Gordon Baker Rd.
Willowdale, Ontario M2H 2S6
416-499-1414 ext: 2325
Fax: 416-499-8290

Sony Components Products Co. (LCD panels)
1 Sony Drive
Park Ridge, NJ 07656

(Various products including SRAMs, ADCs, computer audio, hybrid and video ICs, including color encoding and decoding, data conversion, synchronization, audio processing, CD-ROM, television, VCR and telecommunications [serial and wireless] ICs, color LCDs, CCDs, GaAs MES FETs, Variable Capacitance diodes, infrared receivers, laser diodes)

Sorep SA
Zone De Courtaboeeuf
7 Avenue Des Andes
Les Ulis, Cedex, 91952
France
33 1 64 46 26 46
Fax: 33 1 69 28 43 96
TLX: 603 665

(Digital, interface, linear and microprocessor support ICs)

Soshin Electric Co. Ltd.
1-18-18 Nakamagome Ohtu-Ku
Tokyo 143
Japan
03 3775 2111; 03 5448 9119
Fax: 03 3775 2071; 03 3775 7092

In the U.S. call distributor
Susco Electronics, Inc.
61A Carolyn Blvd.
Farmingdale, NY 11735
1-800-752-0811
516-249-0811

(Video filters, couplers, mica capacitors, hybrid ICs)

Soundtech Electronics Corporation
P.O. Box 11-73, Peitou
Taipei, Taiwan, R.O.C.
886-2-883-2876
Fax: 886-2-833-2878

(Piezo speakers, buzzers, sirens, transducers)

SP prefix
See *Raytheon* (Transistors),
Sipex Corporation,
Harris Semiconductor
(ESD, electrostatic discharge, protection circuit),
Philips Semiconductors (Signetics),
GEC Plessey Semiconductors
(ECL devices)

Space Electronics, Inc.
4031 Sorrentino Valley Blvd.
San Diego, CA 92121
619-452-4167
Fax: 619-452-5499
E-mail:
102005.1635@compuaserve.com
www:http://www.newspace.com/
industry/spaceelec/home/html

(Magnetic sensors, and qualified parts including SRAMs, bipolar ICs [like LM101, LM111, LM124 and LM139 parts])

Manufacturers, Prefixes, Part Number Types, Logo Descriptions & Family Types

Space Power Electronics, Inc.
305 Jeffrey Lane
Glen Gardner, NJ 08826-9726
908-537-2184, 2185
Fax: 908-537-7522

(Diodes, thyristors, transistors, operational amplifiers)

Space Research Technology, Inc.
31255 Cedar Valley Dr., Suite 309
Westlake Village, CA 91362
818-991-0693
Fax: 818-991-7021

(Error detection and correction ICs, FIFOs, LIFOs, CRC and polynomial error detecting circuits)

Spancom Corp.
528 Weddell Drive, Suite 11
Sunnyvale, CA 94089
408-745-1788
Fax: 408-745-0435

(Microprocessor support ICs)

SPC
See *Solid Power Corporation*

SPC prefix
See *S-MOS Systems*

(3D graphics ICs)

SPC prefix
See *S-MOS Systems*

(Graphics controller)

SPE
See *Space Power Electronics, Inc.*

Spectra Diode Laboratories
See *SDL, Inc.*

Spectrian
550 Ellis Street
Mountain View, CA 94043
415-961-1473, Fax: 415-967-9322

(RF amplifiers)

Spectrum Control, Inc.
6000 West Ridge Road
Erie, PA 16506
814-835-4000
Fax: 814-835-9000
e-mail: 75121.1255@compuserve.com

(EMI/RFI filter components and assemblies including filtered connectors)

Spectrum Devices
13706 Larkway Drive
Sugal Land, TX 77478-2420
713-242-6417
Fax: 713-242-6415

(Blinking LEDs [do not need dc voltage inputs], multichip and clustered LEDs to replace incandescent lamps)

Spellman High Voltage Electronics Corporation
475 Wireless Blvd.
Hauppauge, NY 11788
516-435-1600
Fax: 516-435-1620
http://www.spellmanhv.com
E-mail: sales@spellmanhv.com

(High voltage power supplies)

Spex Industries
See *ISA Jobin Yvon/Spex Division*

Spire Corporation
One Patriots Park
Bedford, MA 01730-2396
617-275-6000
Fax: 617-275-7470
spire.corp@channell.com

(High power diode lasers, wafers, dice and bars and mounted multibar arrays)

SPACE POWER ELECTRONICS — SRM

S.P.K. Electronics Co., Ltd.
2F.-1, No. 312 Jen Ai Road, Sec. 4
Taipei, Taiwan, R.O.C.
886-2-754-2677
Fax: 886-2-708-4124
TLX: 16597 SPKTWN

(Crystals, oscillators, crystal filters, ceramic filters, TCXOs, VCXOs, ceramic resonators)

Sprague-Goodman
1700 Shames Drive
Westburry, NY 11590
516-334-8700
Fax: 516-334-8771

(Surface mount inductors, coils and trimmer capacitors)

Sprague Semiconductor
See *Allegro Microsystems*

SPROC
See *Logic Devices*
(Formerly Star Semiconductor)

SPT (and SPT prefix)
See *Signal Processing Technologies*

SPX prefix
See *Seponix Corp.*

SQ prefix
See *Sierra Semiconductor*

SR prefix
See *GEC Plessy*,
Smart Relay Technology, Inc.
(Solid state relay), *Telbus GmbH*,
Teledyne Relays (Relay)

SRD prefix
See *Siemens*

SRM prefix
See *S-MOS Systems* (RAM),
Telbus GmbH

SRT — STAKTEK

SRT prefix
See *Space Research Technology, Inc.*

SR-XXX part number
See *ILC Data Device Corp.*

SS logo
See *Dimolex Corp.*

SS prefix (Transistors)
See *AVG Semiconductor* (Second source), *Honeywell* (Hall Effect Sensors)

SSC prefix
See *S-MOS Systems* (Standard cell), *TLSI, Inc.* (Sensor to microprocessor interfaces)

SSC-XXXX part number
See *Aydin Vector*

(Super signal conditioner)

SSDI
Solid State Devices, Inc.

SSEC
See *Honeywell Solid State Electronics Center*

SSI
See *Silicon Systems, Inc.*

SSI logo
See *Silicon Systems, Inc.*, *Switching Systems International*

(Power supplies)

SSK prefix
See *Dimolex Corp.*

SSM prefix
See *Saratoga Semiconductor, Corp.*, *Analog Devices*, *Solid State* (Microtechnology) *Electronics Corporation*

SSP-XXXXX part number
See *ILC Data Device Corp.*, *Uwatec AG*
(Sending/measuring IC)

SSQ
Space Station Qualified

SSS
(Formerly Solid State Scientific, then Sprague, now Allegro Microsystems)

SST-1 part number
See *3Dfx*

(3D graphics IC)

SST-XXX (or XXXX) part number
See *Calogic*, *Temic* (Siliconix)

(MOSFET transistor)

SSX prefix
See *SenSym, Inc.*

Stable
See *Dimolex*

ST logo
See *SGS-Thomson Microelectronics*

ST prefix
See *SenSym, Inc.* (Sensors and transducers), *Bourns, Inc.* (Sapphire sensors), *SGS-Thomson Microelectronics*, *Startech Semiconductor Inc.*, *Styra* (PC chip sets), *AVG Semiconductors* (Second source), *Marlow Industries, Inc.* (Thermoelectric coolers), *Bourns, Inc.* (Pressure sensors), *Standard Technologies Corporation* (Math coprocessors).

ST Microsonics Corp.
60 Winter St.
Weymouth, MA 02188
617-337-4200

(Oscillators and hybrid ovenized oscillators)

ST (Signal Technology Corp. subsidiary) Olektron Corp.
28 Tozer Road
Beverly, MA 01915-5579
508-922-0019
Fax: 508-927-9328

(Modulators)

ST Microwave Corporation
340 N. Roosevelt Avenue
Chandler, AZ 85266
602-940-1655
Fax: 602-961-6297

955/975 Benecia Avenue
Sunnyvale, CA 94086
408-730-6300
Fax: 408-733-0254

(DRO products, YIG oscillators and filters, VCOs, DTOs, phase locked oscillators, mixers, preamplifiers, up/down converters, polar frequency discriminators)

Stac Electronics
5993 Avenida Encinas
Carlsbad, CA 92008
619-794-4550
Fax: 619-794-4577

(A supplier of data compression products, such as their LZS [tm] technology and Stacker [R] software.)

Stac XXXX part number
See *Amega Technology*

Stancor
See *White-Rogers*

Staktek Corporation
8900 Shoal Creek Blvd.
Suite 115
Austin, TX 78758
512-454-9531
Fax: 512-454-9409

(3D stacked memory modules)

STANDARD CRYSTAL — STANFORD TELECOM

Standard Crystal Corp.
9940 E. Baldwin place
El Monte, CA 91731
800-423-4578
818-443-2121
Fax: 818-443-9049
TLX: 4998755

(Crystal oscillators)

Standard Microsystems Corp. (SMC)
Components Products Division
80 Arkay Drive
Hauppauge, N.Y. 11788
800-443-SEMI (7364)
(Technical support line)
516-435-6000, 516-273-3100
Fax: 516-231-6004
Faxback: 516-233-4260
BBS: 516-273-4936
FTP: info.smc.com
Internet: http://www.smc.com/
E-mail: chipinfo@smc.com

Ten New England
Business Center Drive, Suite 100
Andover, MA 01810
508-557-5720
Fax: 508-557-9255

Southern Sales Office:
8310 Capitol of
Texas Highway North, Suite 255
Austin, TX 78731
800-644-7364
512-502-0070
Fax: 512-502-0648

Western Sales Office:
2107 North First Street, Suite 660
San Jose, CA 95131
800-404-7364
408-441-0455
Fax: 408-441-0463

European Sales Office:
Englshalkinger StraBe 12
81925 Munich, Germany
49 89 92861170
Fax: 49 89 92861190

U.K Sales Office:
Berkshire Court, Western Road
Bracknell, Berkshire RG12 1RE
United Kingdom
44 13444 18446
Fax: 44 13444 18849

Far East Sales Office:
10F, No. 130, Sec. 4
Nanking, East Road
Taipei, Taiwan R.O.C.
886 2578 7118
Fax: 886 2579 1737

Japan Subsidiary:
Toyo Microsystems, Corp.
Meguro F2 building
8-8 Nakameguro 1 Chome
Meguro-ku, Tokyo 153, Japan
813 5721 2271
Fax: 813 5721 2270

(A fabless company supplying Baud rate generators; display products such as timing controllers and terminal logic controllers, video DACs, etc.; local area network products and data communications products including UARTs, GPIB interfaces; floppy and hard disk drive ICs; keyboard encoders; ethernet controllers, microprocessor products and shift registers.)

Standard Technologies Corporation
P.O. Box 325
Canonsburg, PA 15317
412-746-8696
Fax: 412-745-8596

(Floating point serial math coprocessor)

Standex Electronics
4538 Camberwell Road
Cincinnati, OH 45209
513-871-3777
Fax: 513-871-3779

(Coils, chokes and inductors)

Standish LCD
Division of Standish Industries, Inc.
W7514 Highway V
Lake Mills, WI 53551
414-648-1000
Fax: 414-648-1001

(LCD displays, formerly Hamlin)

Stanford Microdevices
2880 Zanker Road, #203
San Jose, CA 94086
800-764-6642
Fax: 408-730-2621

(MMIC Amplfiers, discrete PHEMTs)

Stanford Research Systems
1290 D. Reamwood Avenue
Sunnyvale, CA 94089
408-744-9040
Fax: 408-744-9049

(Ovenized oscillators, arbitrary waveform generators, function generators, FFT spectrum analyzers, LCR meters)

Stanford Telecom
(Telecommunications)
2421 Mission College Blvd.
(STEL prefix)
Santa Clara, CA 95054-1298
408-748-1010
Fax: 408-980-1066
TELEX: 910-339-9531

1761 Business Center Drive
Reston, VA 22090
703-438-8000

7501 Forbes Boulevard
Seabrook, MD 20706
301-464-8900

2660 Riva Road, #360
Annapolis, MD 21401
301-266-9300

STANFORD TELECOM — STC COMPONENTS

31255 Cedarvalley Drive, #201
Westlake Village, CA 91362
818-707-8700

150F New Boston Street
Woburn, MA 01801
617-935-7020

615 Hope Road, Victoria Plaza
Eatontown, N.J. 07724
908-542-7790

5009 Centennial Boulevard
Colorado Springs, CO 80919
719-594-4475

ASIC Custom Products Division
480 Java Drive
Sunnyvale, CA 94089
408-541-9031
Fax: 408-541-9030

(ASICs and custom products, including wireless burst processors and chirp synthesizers [a synthesizer that creates waveforms with frequencies that change linearly over time])

Stanley Electric Co. Ltd.
2-9-13 Nakameguro
Meguro-Ku Tokyo 153, Japan
81 3 710 2557
Fax: 81 3 792 0007
TLX: 2466623 SECTOK J

Stanley-IDESS S.A.
33 Rue des Peupliers
92000 Nanterre, France
33 1 4781 8585
Fax: 33 1 4786 0916

II Stanley Co. Inc.
2660 Barranca Pkwy.
Irvine, CA 92714
800-533-5331
1-800-LED-LCD1
714-222-0777
Fax: 714-222-0555

4950 West Dickman Road
Battle Creek, Michigan 49015
616-962-3461
Fax: 616-962-3463

Stanley Electric (H.K.) Co., Ltd.
Room 1605A, Silvercord Tower 1, 30 Canton Road
Tsimshatsui, Kowloon, Hong Kong
852 730 1738
Fax: 852 730-1933

Asian Stanley International Co., Ltd.
48/1 Moo 1, Tambol Kukwang, Amphur Ladlumkaew
Pathumthanee, 12140, Thailand
66 2 599 1260
Fax: 66 2 599 1263

(LEDs including 7 segment displays, LCDs and LCD modules, and lamps including cold cathode fluorescent lamps)

Star Micronics America, Inc.
OEM Division
70-D Ethel Road
W. Piscataway, NJ 08854
800-STAR-OEM
908-572-9512
Fax: 908-572-5095

(Audio transducers, printer mechanisms)

Star Semiconductor Corporation
See *Logic Devices, Inc.*

Stars logo
(4 stars arranged in a circular pattern)
See *General Instrument*

Startech Semiconductor, Inc.
1219 Bordeaux Drive
Sunnyvale, CA 94089
800-245-6781
(Evaluation boards containing their products)

408-745-0801
Fax: 408-745-1269
http://www.exar.com/products/star/starmenu.htm

(Frequency generators, drivers/receivers, UARTS, FIFOs. Note: this company is owned by Exar Corporation. For foreign offices, see Exar Corp.)

State of the Art, Inc.
2470 Fox Hill Road
State College, PA 16803-1797
814-355-8004
Fax: 814-355-2714

(Surface mount resistors and resistor networks)

Statek Corporation
A Technicorp Company
512 N. Main Street
Orange, CA 92668
714-639-7810
Fax: 714-997-1256

155 Mildmay Road
Chelmsford, Essex CM2 ODU
England
44 245 352297
Fax: 44 245 252275

(Crystals, crystal oscillators)

STC
See *Silicon Transistor Corporation*

STC prefix
See *SGS-Thomson Microelectronics, Inc.*

STC Components, Inc.
333 W. Higgins Road
Ste 3030
Barrington, IL 60010-9354
800-624-6491
708-490-7150
Fax: 708-490-9707

(Interface ICs)

Manufacturers, Prefixes, Part Number Types, Logo Descriptions & Family Types

STC Components
Quartz Crystal Division
Edinburgh Way
Harlow
Essex CM20 2DE
United Kingdom
0279 626626
Fax: 0279 626626 ext. 2343
TLX: 818746 STC CG G

STC Canon Components (Pty) Ltd.
248 Wikham Road
P.O. Box 62
Moorabbin
Victoria 3189
3 555 1566

STC France SA
4 Allee de l'Astrolabe
Silic 561
94653 Rungis Cedex
France
1 4560 4700
Fax: 1 46875643
TLX: 204978

Ohmstrasse 3, 8044
Unterschleissheim
Germany
89 317 00120
Fax: 89 317 00129

STZ (NZ) Ltd.
P.O. Box 26064
10 Margot St.
Epsom, Auckland 3
New Zealand
649 500019

P.O. Box 286, Baksburg
1460, South Africa
11 528311
TLX: 8814332

STC Components, inc.
636 Remington Road
Schaumburg, IL 60915
312-490-7150
Fax: 312-490-9707
TLX: 910 291 1280

Atrium Executive Center
80 Orville Drive
Bohemia, NY 11716
516-563-0170
Fax: 516-563-0174
TLX: 6973500 HPGEXCTR

325 E. Hillcrest Drive
Suite 101
Thousand oaks, CA 91360
805-495-8667
Fax: 805-495-5803

(Quartz crystals, crystal oscillators, resonators and crystal filters)

STC Defense Systems
155 Mildmay Road
Chelmsford
Essex CH2 004
United Kingdom
44 1245 352 297
Fax: 44 1245 252 275

Brixham Road
Paignton
Devon, TQ4 7BE
United Kingdom
44 0803 550 762

(Diodes, optoelectronics, crystals, crystal controlled oscillators, real time clock modules)

STEL prefix
See *Stanford*

(Telecommunications)

Stetco, Inc.
3344 Schierhorn Ct.
Fraklin Park, IL 60131
800-251-4558
708-671-4208
Fax: 708-671-5270

(Optoelectronics, LEDs)

STG prefix
See *SGS-Thomson Microelectronics, Inc.*

STH prefix
See *Siemens*

STi prefix
See *SGS-Thomson Microelectronics*

STI
See *Semiconductor Technology, Inc.*, *Sigmax Technology, Inc.*

STK prefix
See *Simtek Corporation*

(Nonvolatile SRAMs)

STL prefix
See *Siemens*

STM prefix
See *Siemens*

STMEE prefix
See *SAC-TEC, Inc.*

Stow Laboratories (STOLAB)
7 Kane Industrial Drive
Hudson, MA 01749
508-562-9347
Fax: 508-568-9172

(Linear ICs)

STPSL prefix
See *SGS-Thomson Microelectronics*

Street Electronics
See *Echo Speech Corporation*

Struthers-Dunn/Hi-G Company
(This company closed in June 1994. In February 1995 Magnecraft Electric purchased the Struthers Dunn MIL-R-83536 line [Magnecraft Struthers Dunn]. Also in 1995, Communication

STRUTHERS-DUNN/HI-G — SUPERTEX

Instruments purchased the Hi-G MIL-R-39016 and MIL-R-28776 relay lines.)

Some product may still available from Sherburn Electronics Corporation
175 Commerce Drive
Hauppauge, NY 11788
800-366-3066
Fax: 516-231-1587)

STT prefix
See *3Dfx Interactive, Inc.*

STV prefix
See *SGS-Thompson Microelectronics*

Submicron Technology Ltd.
3600 Peterson Way
Santa Clara, CA 95054
Fax: 408-567-1111

(This company has a privately funded independent wafer foundry to manufacture 8" submicron CMOS wafers in Thailand)

Sumitomo Metal Industries, Ltd.
Development Section Planning Department
Electronic Components Division
1-8, Fusoh-Cho,
Amagasaki, Hyogo, 660 Japan
06 489 5957
Fax: 06 489 5956

J.P.A. Electronics Supply, Inc.
Park 80 West Plaza 1
Saddle Brook, NJ 07662
201-845-0980
Fax: 201-845-5139
TLX: 421611 JPA UI

(LSI chip set ICs for real time image processing including high speed video filter, rank value filter, image data reduction processor, labeling accelerator, projection processor, image data bus controller and a histogram processor)

Summit (tm)
See *United Technologies Microelectronics Center*

Summit Instruments, Inc.
2236 N. Cleveland-Massillon Road
Akron, OH 44333-1255
216-659-3312
Fax: 216-659-3286

(Accelerometers including triaxial types)

Sun Led
See *Sunscreen Company Ltd.*

Sun Microelectronics, Inc.
Sun Microsystems
2550 Garcia Avenue
Mountain View, CA 94043-1100

SPARC Technology Business
1-800-681-8845
415-960-1300, 415-336-0544
Internet: http://www.sun.com/stb/
http://www.sun.com/sparc/
Net.Engine

(SPARC microprocessors and ASIC chip sets for computer workstations)

Sun Opto, Inc.
21221 Commerce Pointe Dr.
Walnut, CA 91789-3056
909-598-8266
Fax: 909-598-4966

(LEDs, LED displays, bargraphs, photodetectors)

Sun Technique Electric Co. Ltd.
451, Cheng Ching Road,
Kaohsiung, Taiwan, R.O.C.
886-7-382-3568, 386-2300,
386-9844
Fax: 886-7-386-2338

(Speakers and miniature speakers)

Sun Wai Electronic Co.
Room 318/319
Chiap Thong Bldg 321
Tokwawan Road
Kowloon, Hong Kong
3 335291 4, 3 654475
Fax: 3 7640090

SunDisk Corp.
See *SanDisk Corp.*

SUNI
Saturn User to Network Interface
See *PMC-Sierra, Inc.*

Sunscreen Company Ltd.
106, Hewlett Center
54 Hoi Yuen Road, Kwun Tong
Kowloon, Hong Kong
852 3450021
Fax: 852 3574178

SuperSPARC
See *Texas Instruments*

Supertex Inc.
1235 Bordaux Dr.
Sunnyvale, CA 94089
1-800-222-9884 (except CA)
408-744-0100
Fax: 408-734-5247
e-Mail: prodinfo@supx.com

Central Sales Office:
1208 Country Club Lane, Suite 108
Fort Worth, TX 76112
817-457-5677
Fax: 817-457-9269

Western Sales Office:
see the central sales office address
800-487-8737

(MOSFETs, high voltage CMOS [driver] ICs 9for flat panel displays, etc., switch mode voltage regulators, consumer and industrial CMOS ICs, DMOS transistors)

Manufacturers, Prefixes, Part Number Types, Logo Descriptions & Family Types

SUR System, Inc.
See *Dimolex Corp.*

Surge Components, Inc.
1016 Grand Blvd.
Deer Park, NY 11729
516-596-1818
Fax: 516-595-1283

(Rectifiers including surface mount devices)

Sussex Semiconductor, Inc.
12551 Towne Lake Dr.
Fort Meyers, FL 33913
813-768-6800
Fax: 813-768-6868

(Diodes, rectifiers, transient suppressors, zener diodes)

Suwa Seikosha Co., Ltd.
See *Epson America, Inc.*

SVD prefix
See *Optek Technology, Inc.*

SVT prefix
See *Optek Technology, Inc.*

SW prefix
See *M/A-COM*

SWD-XXX part number
See *M/A-COM*

Swampscott Electronic Co., Inc.
41 Spinale Road
Swampscott, MA 01907
617-598-4116

(Diodes, thyristors, transistors, linear ICs)

Switching Systems International (SSI)
500 Porter Way
Placentia, CA 92870
714-996-0909
Fax: 714-996-2753
http://www.ssi4power.com

(Power supplies)

SWR prefix
See *Thaler Corporation*

(Sine/cosine oscillator)

Sx
See *Sipex*

SX prefix
See *SenSym, Inc.*
(Sensors and transducers),
American Microsystsms, Inc.

SXO prefix
See *STC Defense Systems*

SXT prefix
See *San Francisco Telecom*

S/XXX part number
See *IBM*

(Microprocessor)

SY prefix
See *Synertek, Inc.*, *Synergy Semiconductor Corporation*

SYC prefix
See *Syvantek Microelectronic Corporation*

SYF prefix
See *Syntaq Technology, Inc.*

SYM prefix
See *Synertek, Inc.*
(M is for military SY devices),
Symbios Logic, Inc.

Symbios Logic, Inc.
1036 Elktron Drive
Colorado Springs, CO 80907
800-856-3093 (Literature division)
http://www.symbios.com
http://www.symbios.com/products/sfc/sftoc.htm (Product catalog)
literature@symbios.com
BBS: 719-573-3562

SUR SYSTEM — SYMBIOS LOGIC

Plant Locations:
2001 Danfield Court
Fort Collins, CO 80525
719-533-7000 (Sales)
719-596-5795
(If get recording press 3016 for technical support)
719-573-3016 (Technical support)
800-636-8022, 316-636-8022
(MetaStor products)

Fax: 719-597-8224, 719-536-3301, 719-533-7240
(Technical support fax)

For Software Drivers/Installation instructions:
BBS: 719-573-3562
FTP.symbios.com; 204,131.200.1
Then to FTP/hub/symchips/scsi/drivers
http://www.symbios.com
E-Mail: symbios@saligent.com

1635 Aeroplaza Drive
Colorado Springs, CO 80916
719-573-3200
3718 N. Roak Road
Wichita, KS 67226-1397

U.S. RAID Hotline:
1-800-440-5606
International RAID Hotline:
316-636-8652

North American Sales Locations:
Eastern Sales Area:
8000 Townline Avenue, Suite 209
Bloomington, MN 55438-1000
612-941-7075

17304 Preston Road, Suite 635
Dallas, TX 75252
214-733-3594

92 Montvale Avenue
Suite 3500
Stoneham, MA 02180-3623
617-438-0381

225

SYMBIOS LOGIC — SYNERTEK

30 Mansell Court, Suite 220
Roswell, GA 30076
404-641-8001

20 Oak Hollow Drive
Southfield, MI 48034
810-746-5056

Western Sales Area:
1731 Technology Drive, Suite 610
San Jose, CA 95110
408-441-1080

3300 Irvine Avenue, Suite 255
Newport Beach, CA 92660
714-474-7095

International Sales Locations:
European Sales Headquarters:
Westendstrasse 193/II
80686 Muenchen
Germany
011 49 89 547470 0
Fax: 49 8231 9644 66
(Technical support)
BBS, use the U.S BBS:
001 719 573 3562

Asia/Pacific Sales Headquarters:
37th Floor, Room 3702,
Lippo Tower
Lippo Centre, 89 Queensway
Hong Kong
011 852 253 00727

(Formerly NCR Microelectronics, this company became a subsidiary of Hyundai Electronics, America, San Jose, CA in November 1994. Gate arrays, ASICs, SCSI ICs, 10base-T Ethernet ICs, RAID controllers, etc. Note: Technical support for AT&T or NCR call NCR Global Support Centers 800-543-9935 or 803-939-6373)

Symbol
(Semiconductors, available from Taitron Components, Inc., 25202 Anza Drive, Santa Clarita, CA 91355-3496, 800-824-8766, 800-247-2232, 805-257-6060; Fax: 800-824-8329, 805-257-6415)

Syntek
1407 116th Avenue, N.E. #117
Bellevue, WA 90004
800-548-8911
Fax: 206-462-7170

(This wafer fab started producing CMOS 4 and 16 bit MCUs [microcontroller units] and domestic appliance ICs in 1994)

Symphony Laboratories
See *Winbond Electronics Corporation*

(Windbond purchased the remaining shares of the company and it is now the research and development group of Winbond. It is now called Winbond Systems Laboratory)

Synergy Microwave Corporation
483 McLean Boulevard
Paterson, NJ 07504
201-881-8800
Fax: 201-881-8361

(Highpass, lowpass and bandpass filters, standard and custom, DC to 2000 MHz range, directional and bidirectional couplers, mixers, power dividers, phase detectors, phase shifters, attenuators, vector modulators, etc.)

Synergy Semiconductor Corp.
3450 Central Expressway
Santa Clara, CA 95051
1-800-788-3297
408-730-1313
Fax: 408-737-0831, 3590

(Corp. HQ) Internet:
info@synergysemi.com
http://www.synergysemi.com
325 Boston Post Road, Unit 1
Sudbury, MA 01776
508-443-1440
Fax: 508-443-1443

(The company also has an investment in a foundry in Frankfurt-on-Oder, System Microelectronic Innovation GmbH [SMI]. ASICs, ECL logic including gates, SRAMs, and logic compatible with standard F100K logic, ClockWorks [tm] Family with standard function ICs such as programmable frequency synthesizers, low skew clock generators, clock distribution and drivers, programmable delay lines, phase locked loops. Functions available as discrete components and telecommunications ICs)

Synertek, Inc.
Subsidy of Honeywell
3001 Stender Way
Santa Clara, CA 95054
408-988-5600
(International sales manager 408-988-5608)
TWX: 910 338-0135

Western Area Sales Office:
150 So. Wolfe Road
Sunnyvale, CA 94086
408-735-0221
TWX: 910-339-9500

Southwest Regional Sales Office:
4401 Atlantic Ave., Suite 101
Long Beach, CA 98087
Tel: 213-428-8776
TWX: 910-341-7705

Central and North Central Sales Office:
1821 Hicks Road
Rolling Meadows, IL 60008
312-991-4620
TWX: 910-687-0244

South Central Regional Sales Office:
1101 E. Arapaho Rd. #149
Richardson, TX 75081
214-235-1105
TWX: 910-867-4723

Eastern and Mid-Atlantic Area Sales Office:
555 Broadhollow Road
Melville, NY 11747
516-752-0900
TWX: 510-224-6247

Northeast Regional Sales Office:
177 Worcester St.
Wellesley Hills, MA 02181
617-431-7630
TWX: 710-383-1582

Southeast Regional Sales Office:
5600 Mariner St.
Suite #219
Tampa, FL 33609
813-870-2222
TWX: 810-876-9148

Northern European Zone Manager:
Honeywell House
Charles Square
Bracknell, Berkshire
United Kingdom
Rg 12 1 EB
Bracknell 24555
TWX: 815 847064

Central European Zone Manager:
Freischuetzstrasse 92
D-8000 Muechen 81
Germany
49 089 9597 0
TWX: 841 5216884

Southern European Zone Manager:
Honeywell SA
Rue Emile Verhaeren 6
92210 St. Cloud
Paris, France
33 01 7711071

Manager Director, European Marketing:
Honeywell Europe SA
Avenue Henri Matisse 16
1140 Brussels, Belgium
32 02 243-1450
HUN BE02
TWX: 24535

Honeywell Pty. Ltd.
Garden Grove Centre
Garden Grove Parade
Adamstown, N.S.W. 2289
049 52 4411
TWX: AA 28338

Units 8/9 Townsville St.
Fyshwick, Australia
062 80 5021
TWX: AA 61583

62 Hopkins St.
Moonah, Tasmania
002 28 0087

150-151 Greenhill Rd.
Parkside, S.A. 5063
08 271-5022
TWX: AA 89553

27 Church St.
Richmond, Victoria
03 429 1933
TWX: AA 31805

389 Stanley Street
South Brisbane
07 246 1255
TWX: AA 42630

Industrial Services Branch
P.O. Box 102
Tamworth N.S.W.
067 67 3172

1 Thorgood St.
Victoria Park, W.A.
09 362 1577
TWX: AA 93063

SYNERTEK — SYNTONIC SYSTEMS

276 Cowper St.
Warrawong, Australia
042 74 0657

Lot 1217 Winnellie Rd.
Winnellie, N.T.
089 84 4492

Honeywell Control Systems. Ltd.
P.O. Box 2196
8 Monohan Rd., Auckland 6
New Zealand
575 479
TWX: NZ 2487

(Custom logic [i.e., FSK modem, DTMF receiver], ROMs, EEPROMs, RAMs, microprocessors, microcomputers and military devices.)

Syntaq Technology, Inc.
380 Stevens Avenue, Suite 214
Solana Beach, CA 92075
619-259-9528
Fax: 619-259-9718

Syntaq Ltd.
Enterprise Court
Northumberland NE 239LZ
United Kingdom
44 1670 731 866
Fax: 44 1670 731 741

(SRAMs, DRAMs, EEPROMs, and memory modules)

Syntar Industries, Inc. (SI)
3400 Brush Hollow Road
Westbury, NY 11590
516-333-2012

(Diodes, thyristors, transistors)

Syntonic Systems, Inc.
17200 NW Corridor Ct.
Beaverton, OR 97006
503-645-5596
Fax: 503-629-8681

S-T SYNTONIC — TAIWAN SEMICONDUCTOR

(Interface ICs)

SYS prefix
See *Syntaq Technology, Inc.*

System Microelectronic Innovation GmbH.
SMI System Microelectric
Wildbahn
Frankfurt-on-Oder,
Markendorf 15236
Germany
335 546 2005
Fax: 335 546 3251

(Note: Owned by Synergy Semiconductor Corporation. FIFOs, ECL parts (10 and 100 prefix), ECL to TTL ICs, Hall Effect ICs, DRAMs, amplifiers, proximity detectors, switches, mixers, LERD drivers)

Systron Donner Microwave
A Thorn EMI Company
13100 Telfair Avenue
Sylmar, CA 91342-9217
818-362-9900
Fax: 818-367-4709

(Frequency converters)

Systron Donner Inertial Division
2700 Systron Drive
Concord, CA 94518-1399
510-682-6161
Fax: 510-671-6590

(Solid state angular rate sensor)

Systronix, Inc.
555 S. 300 E.
Salt Lake City, UT 84111-6398
801-534-1017
Fax: 801-534-1019

(Peripheral controller IC includes UARTs, printer and keyboard control)

Syvantek Microelectronic Corporation
1475 Saratoga Avenue
San Jose, CA 95126
408-252-7988
Fax: 408-752-7996

(SRAMs, ROMs, color look up tables, communications ICs)

SZ prefix
See *SZE Microelectronics GmbH*

SZB prefix
See *Siemens Semiconductors*

SZE Microelectronics GmbH
Hamburger Chaussee 25
Flintbek 24220
Germany
04347 7166 0
Fax: 0 437 7166 12

(EPROMs, microprocessors, DRAMs, coprocessors)

T in a 6 sided polygon
See *Toyocom U.S.A., Inc.*

T logo
See *Teccor*

T prefix
See *Lucent Technologies* (Formerly AT&T Microelectronics)

T3D prefix
See *Trident Microsystems*

T5
See *MIPS Technologies, Inc.*
(Microprocessor)

T9000 series
See *SGS-Thomson Microelectronics*

(INMOS)

TAA prefix
See *SGS-Thomson Microelectronics*

TAC prefix
See *Raytheon Semiconductor*

TACT prefix
See *Texas Instruments*

TAe part number
See *Teledyne Electronic Technologies*, *Microwave Products*
(Formerly Teledyne Microwave Microsystem Products)

TAG Semiconductors, Ltd.
See *SGS-Thomson*

Taiwan Liton Electronic Co., Ltd.
12th Floor 25 Sec. 1
Tunghua S. Road
Taipei, Taiwan
886 2 771 4321
Fax: 886 2 731 0184

(Diodes, LEDs, thyristors, transistors, interface ICs)

Taiwan Semiconductor Manufacturing Co. Ltd. (TSMC)
No. 121
Park Avenue III
Science Based Industrial Park
Hsin-Chu 300, Taiwan, R.O.C.
886 35 780 221
Fax: 886 35 781 545

Building 67, No. 195, Sec. 4
Chung-Hsing Road
Chu-Tung, Hsinchu, Taiwan
886-35-96124
Fax: 886-35-942616

2 FL., No. 8, Alley 16
Lane 235, Pao Chiao Rd.
Hsin Tien City
Taipei hsien, Taiwan R.O.C.
886 2 917 4145
Fax: 886 2 912 2499

Manufacturers, Prefixes, Part Number Types, Logo Descriptions & Family Types

TAIWAN SEMICONDUCTOR — TAT SING

World Trade Center
Strawinskylaan 1145
1077 XX
Amsterdam, The Netherlands
31 20 5753105
Fax: 31 20 5753106

1740 Technology Drive
Suite 660
San Jose, CA 95110
408-437-8762
Fax: 408-441-7713

(This is an independent foundry [the first company in Taiwan to receive ISO 9001 certification] which manufactures parts for a variety of other companies [such as Philips Semiconductors which owns 40% of the company], AMD, Analog Devices, AMS-Austria Mikro Systeme International AG, Music Semiconductor, Oak Technology, Inc., etc. They do not design or sell their own proprietary products. Standard semiconductor devices, such as rectifiers, bridges and transistors, are available from Taitron Components, Inc., 25202 Anza Drive, Santa Clarita, CA 91355-3496, 800-824-8766, 800-247-2232, 805-257-6060; Fax: 800-824-8329, 805-257-6415)

Taiyo Yuden Co. Ltd.
16-20, Ueno 6-Chome, Taito-Ku
Tokyo 110
Japan
81 3 3833 5441

714 W. Algonquin Road
Arlington Heights, IL 60005
708-364-6104
Fax: 708-870-7828
TLX: 910-687-0378

(Capacitors, linear, digital and hybrid ICs)

Takamisawa Electric Co. Ltd.
1134 Tower Lane
Bensenville, IL 60106
708-350-9065
Fax: 708-350-8995
(Relays. Note, this company was acquired by Fujitsu Ltd. in 1995.)

Takenaka Electronic Company, Ltd.
20-1 Narano-Cho, Shironya
Kyoto, Yamashwa-Ku 607
Japan
81 75 581 7111
Fax: 81 75 501 6944

(Optoelectronics, photo ICs for optical sensors)

Talema Electronic, Inc.
3 Industrial Park Drive
St. James, MO 65559
314-265-5500
Fax: 314-265-3350

Talema Nuvotem
Crolley
Donegal, Ireland
353 75 48666

Talema Electronik, GmbH
Riesstrasse 8
W-8034 Germering
Germany
089 84 1000

Talema-Rawmat
1160 #02-05/07
Depot Road
Singapore 0410
65 2719927/8

(Toroidal transformers, coils, chokes)

Tamarack Microelectronics, Inc.
16F-4, No. 1, Fu-Hsing Road
Taipei, Taiwan
886 2 772 7400
Fax: 886 2 776 545

(Ethernet transceivers, ARCnet, Ethernet, ASSP ICs and ASIC design services)

Tampa Microwave Lab, Inc.
12160 Race Track Road
Tampa, FL 33626
813-855-2251
Fax: 813-855-7741

(Phase locked oscillators, loop test translators, converters, synthesized transceivers, satellite simulators)

Tamura Electric Works Ltd
2-2-3, Shimo-Meguro, Meguro-Ku
Tokyo 153, Japan
03 3493 5113
Fax: 03 3493 8319

Tamura Corporation
1-19-43, Higashi-Oizumi
Nerima-Ku
Tokyo 178
03 3978 2111
Fax: 03 3923 0230

Tamura Corporation of America
43352 Business Park
P.O. Box 892230
Temecula, CA 92589
909-699-1270
Fax: 909-676-9482

(Power supplies, hybrid ICs, transformers and inductors)

TAN prefix
See *GHz Technology, Inc.*

Tat Sing Electronics Co., Ltd.
Flat 3 3/F
Kinglet Industrial Building
21-23 Shing Wan Road, Shatin N.T.
Lowloon, Hong Kong
852 605-6607, 852 605-6586
Fax: 852 693-2790

T TAT SING — TECHNITROL

(Transformers and adaptors)

Tatung Company
22, Sec. 3, Chungshan N. Road
Taipei, Taiwan R.O.C.
886 2 592 5252
Fax: 886 2 591 5185

Tatung Company of America, Inc.
Long Beach, CA
310-637-2105
Fax: 310-637-8484

(This company opened a new fab facility for DRAMSs and ASICs, CRTs, monitors)

Tavis Corporation
3636 Highway 49
Marlposa, CA 95338
800-842-6102 (Outside CA)
209-966-2027
Fax: 209-966-4930
(Pressure transducers)

TBB prefix
See *Siemens*

TBC prefix
See *Texas Instruments*

TC prefix
See *Toshiba*, *Ketema* (Temperature IC), *TelCom Semiconductor, Inc.* (Formerly Teledyne Components, Semiconductor), *Texas Instruments* (CCD detectors), *Tamarack Microelectronics, Inc.*, *Hanning Elektro-Werke GmbH*,

TC4S, TC7S series
See *Toshiba*

TCM prefix
See *Texas Instruments*, *Rockwell* (Focal plane arrays), *TelCom Semiconductor, Inc.*

TCS prefix
See *GHz Technology, Inc.*

TDA prefix
See *Philips Semiconductors* (Signetics), *GEC Plessy*, *AVG Semiconductors* (Second source some ICs), *SGS-Thomson Microelectronics*

TDC prefix
See *Raytheon Semiconductor*, *Texas Instruments*

TDE prefix
See *SGS-Thomson Microelectronics*

TDF prefix
See *SGS-Thomson Microelectronics*

TDK Corp.
1-13-1 Nihombashi, Chuo-Ku
Tokyo 103, Japan
81 3 3278 5111
Fax: 81 3 3278 5358

TDK Corporation of America
1600 Feehanville Drive
Mount Prospect, IL 60056
708-803-6100, 708-390-4371
Fax: 708-803-6296

(DC-DC Converters, power transformers, current transformers, common mode choke coils, power ferrite cores, ceramic capacitors)

TD-XXX part number
See *ILC Data Device Corp.*

TE prefix
See *Temex Electronics*

(Crystal filters)

TEA prefix
See *Philips Semiconductor*, *AVG Semiconductors*
(Second source telephone ICs)

TEB prefix
See *Texas Instruments*

TEC prefix
See *Texas Instruments*, *QLogic Corp.*

(Disk drive controllers)

Teccor Electronics, Inc.
1801 Hurd Drive
Irving, TX 75038-4385
214-580-1515
Fax: 214-550-1309
Telex: 79-1600

(Triacs, Quadracs [Teccor trademark, a triac and a diode trigger in the same package or stylized, made up of individual lines; an internally triggered triac] SCRs, rectifiers, Sidacs, Diacs)

Technetics, Inc.
P.O. Box 910, 80306
6287 Arapahoe Ave.
Boulder, CO 80303
303-442-3837
Fax: 303-444-9437

(Power supplies)

Technipower, Inc.
P.O. Box 222
14 Commerce Dr.
Danbury, CT 06813
203-748-7001
Fax: 203-797-9285

(Power supplies, DC/DC converters)

Technitrol, Inc.
1952 East Allegheny Avenue
Philadelphia, PA 19134
215-426-9105
Fax: 215-426-2836
TLX: 834 245

(Delay lines, multivibrators, oscillators, pulse width regulators, Manchester encoders and decoders)

Manufacturers, Prefixes, Part Number Types, Logo Descriptions & Family Types

Technology Products Corporation
2348 Sandridge Drive
Dayton, OH 45439
513-299-9143
Fax: 513-299-8179

314 Harvard Street
Brookline, MA 02146
617-731-0858
Fax: 617-734-1429

105 Willow Avenue
Staten Island, NY 10305
718-442-4900
Fax: 718-442-2124
(A semiconductor business of the IBM Corporation)

Techtrol Cyclonetics
815 market Street
New Cumberland, PA 17070
717-774-2746
Fax: 717-774-6799

(Ovenized crystals)

Tektris Electro Corporation
8130 La Mesa Blvd.
La Mesa, CA 91941
619-589-0444

(DC/DC converters)

Tektronix, Inc.
Hybrid Components Operation
Mail Station 59-420
PO Box 500
Beaverton, OR 97077
503-627-4220, 2515
Fax: 503-627-5560

(Laser diodes, D/A and A/D converters, video amplifiers)

Tektronix Inc.
(Monolithic IC operation, which was part of the Tektronix Components Corporation, was sold to Maxim Integrated Products of Sunnyvale CA in early 1994)

Telbus GmbH
Gunzenhausener Strasse 20
Eching 85386
Germany
89 319 5014
Fax: 89 319 4016
TLX: 5214316

Telcom Devices Corporation
829 Flynn Road
Camarillo, CA 93012
805-445-4500
Fax: 805-445-4502

(InGaAs detectors and arrays)

TelCom Semiconductor, Inc.
(Formerly Teledyne Components Analog Circuits business prior to 12/21/93)
1300 Terra Bella Avenue
P.O. Box 7267
Mountain View, CA 94039-7267
1-800-888-9966 (Literature line)
415-968-9241
415-968-9252 (Voicemail)
Fax: 415-967-1590, 415-940-9633

Central U.S.:
1104 South Mays, Suite 216
Round Rock, TX 78664
512-244-1178
Fax: 512-244-3142

Eastern U.S. and Canada:
800 Turnpike St., Suite 300
North Andover, MA 01845
508-725-3488
Fax: 508-725-3484

52 Guild Street, Suite 16
Norwood, MA 02062
617-769-9420
Fax: 617-769-0469

Latin America:
6 Puddingstone Way
Medway, MA 02053
508-533-2199
Fax: 508-533-2285

European Headquarters and Central Europe:
Telcom Semiconductor GmbH
Lochhamer Strasse 13
D-82152 Martinsried
Germany
49 89 89 56 500
Fax: 49 89 89 56 5022

Southern Europe:
30, Rue de Lowaige
4340 Othee Awans
Belgium
32 41 575418
Fax: 32 41 56584

Northern Europe:
48, Parklands Avenue
Worle; Weston Super Mare
Avon BS 22 OPZ
United Kingdom
44 0 1934 514232
Fax: 44 0 1934 520385

Pacific Rim Headquarters:
10 San Chuk Street
Ground Floor
Kowloon, Hong Kong
852 2324 0122
Fax: 852 22354 9957

(Carious ICs including MOSFET driver power oscillators and motor and PIN drivers, DC/DC converters, power drivers, voltage boosters, ADCs, Voltage to Frequency converters, voltage references [including some LM prefixed devices], voltage detectors, VCOs, temperature controllers and sensors, battery charger ICs, RS-232 transmitter receiver, LCD display drivers. They second some other manufacturer parts, such as Maxim [the TC660 vs the industry 7660])

TECHNOLOGY PRODUCTS — TELCOM

TELE QUARTZ — TELTRON

Tele Quartz GmbH (TQE)
Part of the Tele Quartz Group
Landstrabe
D-6924 Neckarbischofsheim 2
Germany
076268/801-0
Fax: 07268/1435
TLX: 782 359 tqd

Vertriebsburo Nurnberg
D-8500
Nurnberg 70
Euro Quartz GmbH
A-2620 Ternitz
LPE, Laboratoires de
Piezo-Electricite S.A.
F-93110 Rosny Sous-Bois
704-553-0300

Tele Quartz USA, Inc.
3545-H Centre Circle Drive
Ft. Mill, SC 29175

(Quartz crystals, crystal filters)

Telebyte Technology, Inc.
270 Pulaski Rd.
Greenlawn, NY 11740-1601
800-TELEBYT
516-423-3232
(RS422 to RS232 interface converter IC)

Teledyne Components
See *TelCom Semiconductor, Inc.*

Teledyne Crystalonics
See *Crystalonics, Inc.*

Teledyne Microelectronics
Teledyne Electronic Technologies
Electronic Devices Business Unit
12964 Panama Street
Los Angeles, CA 90066
1-800-568-8711 (For data books)
310-577-3825, 310-822-8229
Fax: 310-305-0248, 310-574-2015

(MCMs [multi-chip modules], hybrid microelectronics, facility qualified to MIL-STD-1772, DC/DC converters, power supply bridge modules, stacked memory, fiber optic and photonics products including emitters and detectors)

Teledyne Relays
12525 Daphne Avenue
Hawthorne, CA 90250
213-777-0077
Fax: 213-779-9161

Regional Sales Offices:
Eastern: 201-299-7667,
Fax: 201-299-6998
Southeast: 407-682-9044
North Central: 708-529-1060
Central: 214-348-0898,
708-529-1060
Western: 408-978-8899

Germany: 0611-7636-0,
0611 7636 143,
011 49 611/7636-147
United Kingdom: 081 571-9596
France: 47-61-08-08
Belgium: 02 673-99-88
Japan: 03 3797-6956

(Solid state relays for I/O and power switching)

Teledyne Solid State
A Division of Teledyne Relays
12525 Daphne Avenue
Hawthorne, CA 90250
213-777-0077
Fax: 213-779-9161

Teledyne Electronic Technologies,
Microwave Products
(Formerly Teledyne Microwave
Microsystem Products)
1274 Terra Bella Avenue
Mountain View, CA 94043
415-962-6944
Fax: 415-962-6845

(Microwave filters, T-switches, wireless communications components, GaAs MMIC amplifiers, isolators, delay devices, integrated subassemblies, and merchant MCM capabilities)

Telefunken Electronic GmbH
(See Siliconix Teltonic Berkeley, Inc.)
P.O. Box 277
Laguna Beach, CA 92652
800-854-2436
Fax: 714-497-7331

(Tunable filters, attenuators, VSWR bridges, terminations, RF detectors, coaxial switches, impedance matchers)

Telephonics Corp.
815 Broad Hollow Road
Farmingdale, NY 11735
516-755-7046, 755-7005
Fax: 516-755-7626

(Custom ICs, analog sensor interfaces. See TLSI, Inc.)

Teltone Corporation
22121-20th Avenue SE
Bothell, Washington 98021-4408
1-800-426-3926
206-487-1515
Fax: 206-487-2288

(Telecom components, including DTMF receivers and transceivers, tone detectors and rotary dial pulse counters, etc.)

Teltron GmbH
Bahnhofstrasse 27
Ruhla/Thuringen 99842
Germany
036929 80281
Fax: 036929 62334

(DRAMs, EEPROMs, SRAM modules. See Telbus GmbH)

Manufacturers, Prefixes, Part Number Types, Logo Descriptions & Family Types

TEMEX — TEXAS INSTRUMENTS

Temex Electronics
3030 W. Deer Valley Road
Phoenix, AZ 85027
602-780-1995
Fax: 602-780-2431

(Crystals, crystal filters, LC filters, front end filters and discriminators)

Temic Semiconductors
(A Daimler-Benz company that unites Telefunken Semiconductors, Siliconix, Matra MHS, Dialog Semiconductor and Eurosil Electronic. See Siliconix also, for further contacts).

Temic Telefunken Microelectronic GmbH
Theresienstrasse 2
POB 3535
D-74025
Heilbronn, Germany
47 7131 672423
Fax: 49 7131 672423

(Product line includes blue LEDs)

Tensleep Design Corp.
3809 S. 2nd St.
Austin, TX 78704-7058
512-447-5558

(This company licenses DSP cores and software to other companies.)

Terry Semiconductor, Inc.
6 F., No. 461 Choung-Hwa Road, Sec. 2
Taipei, Taiwan
886 2 307 3922
886 2 301 7598

(Diodes)

Teslaco-Optimum Power Conversion
10 Mauchly
Irvine, CA 92718
714-727-1960

(DC/DC converters)

TESTAS
Turkiye Elektronic Sanayi ve Ticaret A.S.
(Turkish Industries and Trade Corporation)
Kaeanfill Sok, No. 41
006640 Bakanliklar/Ankara, Turkey
33-31-40
Fax: 118-55-06
TLX: 43102 TEST TR

TEW North America
Tew Tokyo Denpa
5903-B Peachtree Industrial Boulevard
Norcross, GA 30092
800-762-0420
Fax: 404-441-3076

(Crystal clock oscillators)

Texas Instruments, Inc. (TI)
Semiconductor Group (SC-91084)
Box 809066
Dallas, TX 75380
800-232-3200 (ASICs)
214-995-6611, 214-997-5469
214-575-6396
(Programmable ASICs)
214-955-2828
(Digital light processing, DMDs-digital micromirror devices)
WWW: http://www.ti.com
(To get customized E-mail and product information, click on "Register Now")
http://www.ti.com/sc/docs
http://www.TI.com/DLT
(Digital light processing, DMDs-digital micromirror devices)

Semiconductor Group Literature Response Center
P.O. Box 172228
Denver, CO 80217
800-477-8924, ext 4500, 2012
(Ext. 3543 is for TI Newsletter subscriptions)

TI Customer Response Center:
1-800-336-5236

(For the nearest TI office or Distributor)
Outside U.S.A. call 214-995-6611
(8am to 5pm CST)
TMS320 Hotline: 713-274-2320,
Fax: 713-274-2324,
BBS: 713-274-2323.
E-mail: 4389750@mcimail.com,
Internet BBS via anonymous ftp
to ftp.ti.com(192.94.94.3)
ftp://ftp.ee.ualberta.ca/pub/dos/TI/

Technical training: 800-336-5236 Ext. 3904 Europe:33 1 30701032, Europe Fax: 49 81 61 80 40 10, Europe BBS: 44 2 34223248

Newsletter: Texas Instruments, Inc.
Attn: Newsletter Staff
DSP Marketing Communications
P.O. Box 1443, MS 737
Houston, TX 77251-1443
DSP Factory Repairs:
713-274-2285
TI Product Information:
214-644-5580

Literature Response Center
P.O. Box 172228
Denver, CO 80217-9270
800-477-8924 ext 3543, 3200

Literature Center
P.O. Box 809066
Dallas, TX 75380-90066
214-995-6611
Customer Response Center
800-336-5236
(Ext. 700 for PALs and FPGAs)
Outside the USA 214-995-6611
(Ext. 700 for PAL's and FPGAs)

Semiconductor Group
Box 809066
Dallas, TX 75380
214-995-6611 Ext. 3990
214-997-2031 (Gate arrays)

TEXAS INSTRUMENTS

Memory, PLD Strategic Marketing
Military Products Division
Semiconductor Group
I-20 & FM 1788
P.O. Box 60448
Midland TX 79711-0448
915-561-6977
Fax: 915-561-6716

Sales Offices:
Alabama
4960 Corporate Drive
Suite N-150
Huntsville, AL 35805-6202
205-837-7530, 205-430-0144

Arizona
8825 N. 23rd Avenue
Suite 100
Phoenix, AZ 85021
602-995-1007, 602-244-7800

California
1920 Main Street
Suite 900
Irvine, CA 92714
714-660-1200
714-660-8140
(Regional Technical Center)

5625 Ruffin Road
Suite 100
San Diego, CA 92123
619-278-9600

5353 Betsy Ross Drive
Santa Clara, CA 95054
408-980-9000
408-748-2222
(Regional Technical Center)

21550 Oxnard Street
Suite 700
Woodland Hills, CA 91367
818-704-8100

San Jose, CA
408-894-9000

Colorado
1400 S. Potomic Street
Suite 101
Aurora, CO 80012
303-368-8000

Connecticut
9 Barnes Industrial Park
So. Wallingford, CT 06492
203-269-0074, 203-265-3807

Florida
2950 N.W. 62nd Street
Suite 100
Fort Lauderdale, FL 33309
305-973-8502, 305-425-7820

4803 George Road
Suite 390
Tampa, FL 33634-6234
813-882-0017

Orlando, FL
407-260-2116

Georgia
5515 Spalding Drive
Norcross, GA 30092
404-662-7900, 404-662-7967
404-662-7945
(Regional Technical Center)

Illinois
515 W. Algonquin
Arlington Heights, IL 60005
708-640-3000, 708-640-2925
708-640-2909
(Regional Technical Center)

Indiana
550 Congressional Drive
Suite 100
Carmel, IN 46032
317-573-6400
(Regional Technical Center)

118 E. Ludwig Road
Suite 102
Fort Wayne, IN 46825
219-482-3311, 4697, 219-489-3860

Indianapolis, IN
317-573-6400

Kansas
7300 College Boulevard
Lighton Plaza
Suite 150
Overland Park, KS 66210
913-451-4511

Maryland
8815 Centre Park Drive
Suite 100
Columbia, MD 21045
301-964-2003

Massachusetts
950 Winter Street
Suite 2800
Waltham, MA 02154
617-895-9100
617-895-9196
(Regional Technical Center)

Michigan
33737 W. 12 Mile Road
Farmington Hills, MI 48331
313-553-1500, 1581

Minnesota
11000 W. 78th Street
Suite 100
Eden Prairie, MN 55344
612-828-9300
(Regional Technical Center)

Missouri
12412 Powerscout Drive
Suite 125
St. Louis, MO 63131
314-821-8400

New Jersey
Parkway Towers
485 E. Route 1
South Islin, NJ 08830
201-750-1050
Edison, NJ 908-906-0033

Manufacturers, Prefixes, Part Number Types, Logo Descriptions & Family Types

TEXAS INSTRUMENTS

New Mexico
2709 Pan American Freeway
N.E. Albuquerque, NM 87107
505-345-2555

New York
6365 Collamer Drive
East Syracuse, NY 13057
315-463-9291

300 Westage Business Center
Suite 140
Fishkill, NY 12524
914-897-2900

1895 Walt Whitman Road
Melville, NY 11747
516-454-6600, 6601

2851 Clover Street
Pittsford, NY 14534
716-385-6770, 6700

North Carolina
2809 Highwoods Boulevard
Suite 100
Raleigh, NC 27625
919-876-2725

Ohio
23775 Commerce Park Road
Beachwood, OH 44122
216-464-6100, 216-765-7258

4200 Colonel Glenn Highway
Suite 600
Beavercreek, OH 45431
513-427-6200

Oregon
6700 S.W. 105th Street
Suite 110
Beaverton, OR 97005
503-643-6758

Pennsylvania
670 Sentry Parkway
Blue Bell, PA 19422
215-825-9500

Puerto Rico
615 Mercantile Plaza Building
Suite 505
Hato Rey, PR 00918
809-753-8700

Texas
12501 Research Boulevard
Austin, TX 78759
512-250-6769

7839 Churchill Way
Dallas, TX 75251
214-917-1264
214-917-3881
(Regional Technical Center)

9301 Southwest Freeway
Suite 360
Houston, TX 77074
713-778-6592

Midland, TX
915-561-7137

8330 LBJ Freeway
Mail Station 8336
Dallas, TX 75265-5303
214-997-3333
Fax: 214-997-6225

Utah
1800 S. West Temple Street
Suite 201
Salt Lake City, UT 84115
801-466-8972

Wisconsin
20825 Swenson Drive
Suite 900
Waukesha, WI 53186
414-798-1001

European Product Information:
Hotline: 33 1 30 70 11 69
Fax: 33 1 30 70 10 32
BBS: 33 1 30 70 11 99
Email: epic@msg.ti.com

Asia Product Information:
852 2 956 7288
Fax: 852 2 956 2200
See Taiwan, Korea,
Hong Kong, Singapore

Australia (and New Zealand):
TI Australia Ltd.
6-10 Talavera Road
North Ryde (Sydney)
New South Wales
Australia 2113
2 878 9000

14 Floor
380 Street
Kilda Road
Melbourne
Victoria, Australia 3004
3 696 1211

171 Philip Highway
Elizabeth, South Australia 5112
8 255 2066

Belgium:
11 Avenue
Jules Bordetlaan 11
1140 Brussels, Belgium
02 242 30 80

Brazil:
Av. Eng. Luiz Carlos Berrini
1461-11o Andar, 04571
San Paulo, SP, Brazil
11 535 5133
Canada

301 Moodle Drive
Mallom Center
Suite 102
Nepean, Ontario, Canada K2H 9C4

613-726-1970
(Regional Technical Center)
Ottawa, Canada 613-726-3201

TEXAS INSTRUMENTS

280 Centre Street East
Richmond Hill,
Ontario, Canada L4C 181
416-884-9181

9460 Trans Canada Highway
St. Laurent, Quebec, Canada
H4S 1R7
514-335-8392

Denmark:
Borupvang 2D, DK-2750
Balleryp, Denmark
45 44687400

Finland:
P.O. Box 86
02321 Espoo, Finland
0 802 6517

France:
8-10 Avenue Morane Saulnier
B.P. 67
78141 Velizy Villacoublay Cedex
France
1 30 70 10 03
33 16 1 30 70 10 05

Germany:
Haggertystrasse 1
8050 Freising
08161 80-00 od. Nbst;
Kurfurstendamm 195-196
1000 Berlin 15
030 8 82 73 65
49 0 81 61 80 4157

Dusseldorfer Strasse 40
6236 Eschborn 1
06196 80 70

Kirchorster Strasse 2
3000 Hanover 51
49 0 511 64 68-0

Maybachstrasse 11
7302 Ostfildern 2 (Nellingen)
0711 3403257

Gildenhofcenter
Hollestrasses 3
4300 Essen 1
201 24 25-0

Holland:
Hogehilweg 19
Postbus 12995
1100 AZ Amsterdam-Zuidoost
Holland
020 5602911

Hong Kong:
8th Floor, World Shipping Center
7 Canton Road
Kowloon, Hong Kong 7351223
DSP Hotline: 852 2956 7268
DSP Fax: 852 2956 1002

Hungary:
Budaorsi u.42
H-1112
Budapest, Hungary
1 1 66 66 17

Ireland:
7/8 Harcourt Street
Dublin 2
Ireland
01 481677

Italy:
Centro Direzionale Colleoni
Pallazio Perseo-Via Paracelso 12,
20041, Agrate Brianza (Mi), Italy
39 039 63221

Via Castello della Magilana 38,
00148 Rome Italy
06 6572651
Via Amendola, 17,
40100 Bologna, Italy
051 554004

Japan:
Texas Instruments Japan Ltd.
Aoyama Fuji Building
3-6-12 Kita-Aoyama Minato-ku
Tokyo, Japan 107
03 3498 2111
Fax: 03 3400 9504

Product Information Center:
0120 81 0026 (within Japan),
03 3457 0972,

International Calls: 813 3457 0972
Fax: 0120 81 0036 (within Japan),
03 3457 1259
International Fax: 813 3457 1259
DSP Hotline: 03 3769 8735,
International Calls: 813 3769 8735
DSP Fax: 03 3457 7071,
International Fax: 813 3457 7071

DSP BBS via NIFTY-Serve:
Type "Go TIASP"

MS Shibaura Building, 9F
4-13-23 Shibaura
Minato-ku
Tokyo, Japan 108
03 3769 8700

Nissho-Iwai Building 5F
2-5-8 Imabashi,
Chuou-Ku, Osaka, Japan 541
06 204 1881

Dai-Ni Toyota Building
Nishi-Kan 7F
4-10-27 Meieki,
Nakamura-Ku, Nagoya, Japan 450,
052 583 8691

Kanazawa Oyama-Cho Dalichi
Seimei Building 6F
3-10 Oyama-Cho,
Kanazawa, Ishikawa, Japan 920
0762 23 5471

Matsumoto Showa Building 6F
1-2-11 Fukashi
Matsumoto, Nagano, Japan 390
0263-33-1060

Daiichi Olympic Tachikawa
Building 6F
1-25-12 Akebono-Cho
Tachikawa, Tokyo, Japan 190
0425-27-6760

TEXAS INSTRUMENTS

Yokohama Business Park East
Tower 10F
134 Gondo-Cho Hodogaya-Ku
Yokohama, Kanagawa 240, Japan
045 338 1220

Nihon Seimi Kyoto Yasaka
Building 5F
843-2, Higashi Shiokohjicho,
Higashi-Iru,
Nishinotoh-In, Shiokohji-Dori,
Shimogyo-Ku, Kyoto, Japan 600
075-341-7713

Sumitomo Seimei Kumagaya
Building 8F
2-44 Yayoi,
Kumagaya, Saitama, Japa, 360
0485-22-2440

2597-1, Aza Harudai,
Oaza Yasaka, Kitsuki, Oita,
Japan 873
09786-3-3211

Korea
Texas Instruments Korea Ltd.
28th Floor Trade Tower
159-1, Samsung-Dong,
Kangnam-Ku Seoul, Korea
2 551 2800
DSP Hotline: 82 2 551 2804
DSP Fax: 82 2 551 2828
DSP BBS: 82 2 551 2914

Malaysia
Texas Instruments Malaysia,
Sdn. Bhd.,
Asia Pacific, Lot 36.1, #Box 93
Menara Maybank,
100 Jalan Tun Perak,
50050 Kuala Llumpur, Malaysia
2306001

Mexico
Texas Instruments de Mexico S.A.,
de C.V. Alfonso Reyes 115, Col.
Hipodromo Condesa Mexico,
D.F., Mexico 06170
5-515-6081

Mexico City
491-70834
(Regional Technical Center)

Norway
Texas Instruments Norge A/s, PB 106,
Refstad (Sinsenveien 53)
0513 Oslo 5, Norway
02 155090

People's Republic of China
Texas Instruments China Inc.
Beijing Representative Office
7-05 CITIC Building
19 Jianguomenwai Dajje
Beijing, China
500-2255 ext. 3750

Philippines
Texas Instruments Asia Ltd.
Philippines Branch, 14th Floor
Ba-Lepanto Building
Paseo de Roxas, Mikati, Metro Manila,
Philippines
2-8176031

Portugal
Texas Instruments Equipamento
Electronico (Portugal) Ltda.,
Eng. Frederico Ulricho,
2650 Moreira Da Maia,
4470 Maia, Portugal
2 948 1003

*Singapore (& India, Indonesia,
Malaysia, Thailand)*
Texas Instruments Singapore
(PTE) Ltd.
Asia Pacific Division
101 Thomson Road #23-01
United Square, Singapore 1130
3508100, Fax: 2536655
DSP Fax: 65 390 7179

Spain
Texas Instruments Espana S.A.,
c/Gobelas 43,
Ctra de la Coruna km 14,
La Florida 28023,
Madrid, Spain
1 372 8051

c/Diputacian,
279-3-5, 08007 Barcelona, Spain
3 317 91 80

Sweden
Texas Instruments International
Trade Corporation (Sverigefilialen)
Box 30, S-164 93 Kista, Sweden
08 752 58 00

Switzerland
Texas Instruments Switzerland AG,
Reidstrasse 6, CH-8953 Dietikon,
Switzerland
01 74 42 811

Taiwan
Texas Instruments Taiwan Ltd.
Taipei Branch,
10th Floor Bank Tower No. 205
Tun Hwa N. Road,
Taipei, Taiwan, Republic of China
2-7139311
DSP Hotline: 886 2 377 1450
DSP Fax: 866 2 377 2718
DSP BBS: 886 2 376 2592
DSP Internet via anonymous ftp to
(140.111.1.10) in directory/vendors/TI/
tms320bbs

Texas Instruments- Acer, Inc.
No. 6, Research New Road II
Science based Industrial park
Hsinchu, Taiwan R.O.C.
886 35 785 112
Fax: 886 35 782 035, 038

(A joint venture to manufacture
DRAMs)

United Kingdom
Texas Instruments, Ltd.
Manton Lane
Bedford, England MK41 7PA
0234 270 111
44 0234 22 3000

TEXAS INSTRUMENTS — THA-XXXXX

TI Die Processors:
Chip Supply: 407-298-7100
Elmo Semiconductor:
818-768-7400
Minco Technology Labs:
512-834-2022

(Linear products including operational amplifiers, comparators, ADCs, voltage comparators and voltage regulators including low-dropout regulators, data acquisition, voltage regulators, data transmission, UARTs, display drivers, intelligent power IC's, memory [including DRAMs], microcontrollers, optoelectronics [including light to frequency converters], telecommunications, speech, MOS memory and a large variety of other products, including military qualified devices, logic, programmable logic devices [PLDs, including gate arrays, mixed signal ASICs], futurebus ICs, Cardbus controllers, digital signal processors [DSPs], SPARC ICs, SCSI terminators, multimedia ICs, Video CD chip sets, video compression chip sets [some with Macrovision patented video copy protection], MPEG, PC audio ICs, Video-CD chip sets, television ICs, Ethernet/Token Ring ICs, Enhanced Transceiver Logic [ETL] for VME-64 bus hot swapping, SONET and ATM ICs, Segmentation and reassembly IC for the SBus also known as the SBus ATM Host Interface [Sahi] IC, speech processors, hybrid, watch and clock ICs, and foundry work, automotive ICs [antibrake systems, theft deterrent systems, airbags]. Note: TI sold their one time programmable antifuse FPGAs to Actel Corp. in 1995.)

Texas Optoelectronics, Inc.
714 Shepherd Dr.
Garland, TX 75042
214-487-0085, Fax: 214-276-8059

(Silicon detectors, PIN diodes, photocells, avalanche photodiodes, custom wavelength LEDs, displays, trans-impedance amplifiers, opto assemblies and hybrids)

TFB prefix
See *Texas Instruments*

TFD prefix
See *Temic* (Telefunken) *Semiconductors*

TFK logo
See *Temic Semiconductors* (Telefunken)

TFL, Time and Frequency Ltd.
P.O. Box 1792
Holon 58117, Israel
972-3-5574107, Fax: 974-3-5574114

(OXCOs, TXCOs and VXCOs)

TGB prefix
See *Texas Instruments*

TGC prefix
See *Texas Instruments*

TGE prefix
See *Texas Instruments*

TGL prefix
See *Amplifonix*

TGU prefix
See *Trident Microsystems*

TGUI prefix
See *Trident Microsystems*

TH prefix
See *Thesys Gesellschaft fur Mikroelektronik GmbH*

TH Electronics
85 Yenping Road South
Taipei, Taiwan
02 371 1101
Fax: 02 381 8050, 02 361 6670

1842 West Grant Road, Suite 102
Tucson, AZ 85745
602-791-7825
(520 area code after 12/31/96)
Fax: 602-791-7938
(520 area code after 12/31/96)
(Solid state relay devision)

97 Gannett Drive
Commack, NY 11725
516-543-7161
Fax: 516-543-7162

(Custom devices including: MCM modules, thick film hybrids, metallized substrates, capacitor and resistor arrays, thermal print heads, pcb assemblies and optically isolated solid state relays)

Thaler Corp.
2015 N. Forbes Blvd.
Tucson, AZ 85745
800-827-6006
602-742-5572
(520 area code after 12/31/96)
Fax: 602-742-9826
,(520 area code after 12/31/96),
520-770- 9222
TLX: 825193

(Sine wave and voltage references and converters including ADCs, sine/cosine oscillator)

THAT logo
See *THAT Corporation*

THAT Corporation
734 Forrest Street
Marlborough, MA 01752
617-229-2500
Fax: 508-229-2590

(RMS level detector, IC Dynamics processor, voltage controlled amplifier ICs.)

THA-XXXXX part number
See *ILC Data Device Corp.*

Manufacturers, Prefixes, Part Number Types, Logo Descriptions & Family Types

THC — THOMSON CONSUMER ELECTRONICS

ThC
See *Thaler Corporation*

Thermik
80 Beamon Road
New Bern, NC 28562
800-624-3292
919-636-5720
Fax: 919-636-5737

(Thermal protectors, temperature limiting switches, thermistor sensors)

Therm-O-Disc
Midwest Components Product Group
1981 Port City Boulevard
P.O. Box 3303
Muskegon, MI 49443
616-777-4100
Fax: 616-773-4214

(Microtemp [trademark] thermal cutoffs, thermistors)

Thermometrics, Inc.
See *Keystone Thermometrics Corp.*

ThermoTrex Corp.
74 West St.
Waltham, MA 02254
617-622-1391
Fax: 617-622-1027

(Thermoelectric modules)

Thesys Gesellschaft Fur Mikroelektronik GmbH
(Thesys Microelectronics)
Haarbergstrasse 61
99097 Erfurt, Germany
49 361 427 8350, 8100
Fax: 49 361 427 6196

Thesys Advanced Electronics
Stefan George Ring 19
Munchen 81929
Germany
89 993 5580
Fax: 89 993 55866

Mittlerer P fad 4
Stuttgart 70499
711 988 9100
Fax: 711 866 1359

Otto-Hahn Strasse 15
Bad Camberg 65520, Germany
6434 5041
Fax: 6434 4277

(Video processor for multimedia applications, scalers, dividers, synthesizers, transceivers, pulse width modulators. This company is a wafer fabrication plant licensed for [and a joint venture of] LSI Logics CMOS and BiCMOS ASICs and standard cell products including microprocessors. This company, formed in late 1992, is also involved in video processing ICs. Thesys was formed in 1992 by privitization of the former East German states [Karl Marx wafer fab]. LSI shares in the company were given back to the German state iof Thuringia and it became 51% owned by Austria Mikro Systeme International AG in 1995. Thesys is run as a separate company.)

Thin Film Technology Corporation (TFT)
1980 Commerce Drive
North Mankato, MN 56003-9923
507-625-8445
sales@thin-film.com

(Chip inductors, chip delay lines, BGA resistor networks, and PECL [positive ECL] delay lines and terminators)

Third Domain, Inc.
P.O. Box 17568
Tucson, AZ 85731-7568
602-290-1820
(520 area code after 12/31/96)
602-885-1189
(520 area code after 12/31/96)

(Video level comparators, video sync strippers, crystal controlled oscillators, video switches, video amplifiers, etc.)

THM prefix
See *Toshiba*

Thomson Consumer Electronics (RCA and GE)
Distributor and Special Products
2000 Clements Bridge Road
Deptford, N.J. 08096-2088
609-853-2417

(A manufacturer of generic replacement semiconductors, the SK series replacement semiconductors [formerly the RCA line] and Japanese JEDEC/ Generic replacement semiconductors.)

Thomson CSF
See *SGS-Thomson*

Thomson-CSF Semiconducteurs Specifiques
Route Departementale 128
BP 46-91401 Orsay Cedex
France
33 1 60 19 70 70
Fax: 33 1 60 19 79 75
TLX: THOM 616780 F

Thomson Military Components
(Thomson Composants Militaires et Spatiaux)
38521 Saint-Egreve
France
50 rue J.P. Timbaud
92402 Courbevoie
Cedex, France
33 1 49 05 39 26
Fax: 33 1 43 34 17 57
TLX: 616 780 F TMS

Thomson Bauelemente GmbH
Perchtinger StraBe 3
D-8000 Munchen 70
Germany
49 89 78 79-0
Fax: 49 89 78 79 145
TLX: 522 916 CSF D

THOMSON CONSUMER ELECTRONICS — TKDD

Thomson Tubes Electroniques
M26 Commercial Complex
Greater Kailash-Part II
New Delhi 110048, India
91 11 644 78 83
Fax: 91 11 645 33 57
TLX: 31.71.443 BCS IN

Thomson Componenti S.p.A.
Via Sergio I, 32
00165 Roma, Italy
39 6 639 02 48 - 638 14 58
Fax: 39 6 639 02 07
TLX: 620683 THOMTE I

Defense and Space Semiconductor Co, Ltd.; (DASCO)
Park Grace Bldg #303
4-32-6 Nishishinjuku
Shinjuku-Ku, Tokyo 160, Japan
81 3 5351 1184
Fax: 81 3 5351 1986

Thomson Electronic Components Ltd.
Unit 4, Cartel Business Centre
Stoudley Road
Basingstoke, Hants RG24 OUG
United Kingdom
44 256 84 33 23
Fax: 44 256 84 29 71
TLX: 858121 TECLUK G

(Linear ARINC 429 Buffer/receiver multichannel ICs, fiber optically coupled CCD arrays, linear array CCD image sensors, area array CCD image sensors, X-Ray Detectors, CCD multiplexer for infrared detectors and special assemblies, etc.)

Thomson Components and Tubes Corporation
Thomson Military and Space Components Division
40 G Commerce Way, P.O. Box 540
Totowa, NJ 07511
201-812-9000, Fax: 201-812-9050
TWX: 710 9877901

(CCDs, microprocessors, etc. This is a sales and marketing office for Thomson in France)

Three
With a vertical line in the lower part of the three (logo)
See *Third Domain*

3 (Three) ATI prefix
See *Optical Imaging Systems, Inc.*

(LCD displays)

(Three) 3Com Corporation
P.O. Box 58145
5400 Bayfront Plaza
Santa Clara, CA 95052-8145
408-764-5000
Fax: 408-764-5001

(ASICs used in their Ethernet and networking products)

(Three) 3Dlabs, Ltd.
Meadlake Pl.
Thorpe Lea Rd.
Egham, Surrey, TW20 8HE
England
44 0 784 470 555
Fax: 44 0 784 470 699
Email: info@3Dlabs.com

3Dlabs Inc.
2010 N. First St.
San Jose, CA 95131
408-436-3455
Fax: 408-436-3458
Email: info@2Dlabs.com

(Formerly DuPont Pixel, silicon [coprocessors] and software drivers for 3-D graphics including 3-D processors known as GLiNT and Permedia).

Three-Five Systems, Inc.
1600 N. Desert Drive
Tempe, AZ 85281
602-389-8600
Fax: 602-389-8801

(High information content LCD and LED displays)

TH-XXXX part number
See *ILC Data Device Corp.*

TI
See *Texas Instruments*

TI prefix
See *Texas Instruments*

Tianma Microelectronics Co. Ltd.
Subsidiary of China Aerotechnology Import-Export Company
Shenzhen, Peoples Republic of China

(LCDs)

TIC Semiconductor
18 West 21st Street
New York, NY 10010
212-675-6722
Fax: 212463-8962

(Diodes, thyristors, transistors)

TIFPLA prefix
See *Texas Instruments*

TIM prefix
See *Toshiba*

TK prefix
See *Toko America*,
Takenaka Electronic Company, Ltd.

TKAD prefix
See *Tektronix, Inc.*

TKDA prefix
See *Tektronix, Inc.*

TKDD prefix
See *Tektronix, Inc.*

Manufacturers, Prefixes, Part Number Types, Logo Descriptions & Family Types

TL — TOKIN CORPORATION

TL prefix
(Usually a TL0XX or TL-XXXXX part number)
See *Texas Instruments*
(Also Astec Semiconductor, Motorola, SGS-Thomson, Second source some parts).

TL Industries, Inc.
2541 Tracy Road
Northwood, OH 43619
419-666-8144
Fax: 419-666-6534

(Memory modules)

TL-SCS part number
See *Texas Instruments*

TLC prefix
See *Texas Instruments*

TLE prefix
See *Texas Instruments*

TLF prefix
See *Silicon Expertise S.A.*

TLSI, Inc.
815 Broadhollow Road
Farmingdale, NY 11735
516-755-7005
Fax: 516-755-7626

(Sensor to microprocessor interfaces, ASICs)

TLC prefix
See *Texas Instruments*

TLCS-XXX part number
See *Toshiba*

(16-bit microcontroller family)

TLV prefix
See *Texas Instruments*

TM prefix
See *Texas Instruments*,
Seeq Technology, Inc.

TMC prefix
See *Raytheon Semiconductor*

(Formerly TRW LSI prefix)

TMG prefix
See *SanRex*

(Isolated Triacs)

TMP prefix
See *Analog Devices*
(Thermostat chip), *Toshiba*
(Networking parts, bus microcontrollers)

TMS prefix
See *Texas Instruments*

TMV prefix
See *Trident Microsystems, Inc.*

TMVP prefix
See *Trident Microsystems, Inc.*

TMX prefix
See *Texas Instruments*

TNT prefix
See *National Instruments*

TODX prefix
See *Toshiba America Electronic Components, Inc.*

TOED prefix
See *Toshiba America Electronic Components, Inc.*

Toei Electronics Co. Ltd.
Kanda Cent Bldg
2-4, Yushima 1-Chome,
Bunkyo-Ku
Tokyo 113
Japan
81 3 3257 1131
Fax: 03 3258 3560

(Linear, analog, and television ICs)

Togai InfraLogic, Inc.
30 Corporate Park
Suite 107
Irvine, CA 92714
714-975-8522
Fax: 714-975-8524

(Fuzzy logic ASICs)

TOIM prefix
See *Temic* (Telefunken)

Tokai Device Co., Ltd.
4-6-33, Meguro, Meguro-ku
Tokyo 153
Japan
03 3791 1181
Fax: 03 3715 1558

(Hybrid ICs)

Tokin Corporation
Head Office:
7-1, Koriyama 6-Chome,
Taihaku-Ku
Sendai, Miyagi-Ken 982, Japan

International Sales Office:
Sumitomo Seimei Sendai Chuo
Bldg, 6-1
Chuo 4-Chome, Aoba-Ku
Sendai 980, Japan
022 211 1281
Fax: 022 211 0975

Hazama Bldg.
5-8 Kita-Aoyama 2-Chrome
Minato-Ku
Tokyo 107, Japan
03-3402-6166
Fax: 03-3497-9756

Tokin America, Inc.
155 Nicholson Lane
San Jose, CA 95134
408-432-8020
Fax: 408-434-0375
http://www.tokin.com

TOKIN CORPORATION — TOSHIBA

9935 Capitol Drive
Wheeling IL 60090
708-215-8802
Fax: 708-215-8804

945 Concord Street
Framingham, MA 01701
508-875-0389
Fax: 508-875-1479

Hong Kong:
Unit 705-707
Chiwan Tower, Park Lane Square
1 Kimberly Road, Tsim Sha Tsui
Kowloon, Hong Kong
730 0028
Fax: 375 2508

Taiwan:
3F-4, No. 57 Fu Shing N. Road
Taipei, Taiwan
02 7728852,
Fax: 02 7114260

Singapore:
140 Cecil Street,
No. 13-01 PIL Bldg.
Singapore
2237076
Fax: 2236093, 2278772

Germany:
Knorrstr. 142,
8000 Munchen 45
Germany
089 311 10 66
Fax: 089 311 35 84

(Chip inductors, coils, chokes, capacitors, EMC filters)

Toko, Inc.
2-1-17, Higashi-Yukigaya
Ohta-Ku
Tokyo 145
Japan
81 3 3727 1161
Fax: 81 3 3728 4690

(Audio and communication ICs)

Toko America, Inc.
1250 Freehanville Dr.
Mt. Prospect, IL 60056-6023
1-800-745-8656
1-708-297-0070

(Voltage regulators, power factor control/PWM ICs, inductors, filters, acoustic wave devices, voltage regulators, hybrid, VCR, and disk drive controller ICs, etc.)

Tokyo Cosmos Electric Co., Ltd.
268, Sobudai 2-Chome
Zama-Shi, Kanagawa 228
Japan
0462 53 2111

(Hybrid ICs)

Tokyo Ko-On Denpa Co. Ltd.
723, Futago, Takatsu-Ku
Kawasaki-Shi, Kanagawa 213
Japan
044 833 0511
Fax: 044 822 0777

(Audio and hybrid ICs)

Tokyo Quartz Company, Ltd.
32-4, Dai-Machi 4-Chome
Hachioji-Shi
Tokyo 193
81 426 24 9191

(HMETS, High Electron Mobility Transistors)

TOLD prefix
See *Toshiba America Electronic Components*

(Laser diodes)

Tomoegawa Paper Co. Ltd.
5-15, Kyobashi 1-Chome,
Chuo-Ku, Tokyo 104, Japan
81 3 3272 4117

(Measuring instrument ICs)

TOP prefix
See *Power Integrations, Inc.*

TOPD prefix
See *Toshiba America Electronic Components*

(Laser diodes)

TORX prefix
See *Toshiba America Electronic Components, Inc.*

Toshiba Corp.
1-1-1 Shibaura, Minatio-Ku
Tokyo, 105-01 Japan
81 03 3457-3423, 81 3 3457 4511
Fax: 81 3 3456 1631/1633

Toshiba America Electronic Components, Inc.
9775 Toledo Way
Irvine, CA 92718
1-800-879-4963, 1-800-457-7777
714-455-2000, 714-453-0224
Fax: 714-859-3963
http://www.toshiba.com

Sunnyvale, CA
408-733-3223
408-43900560
(Programmable ASICs)

Systems IC Division
1060 Rincon Circle
San Jose, CA 95131
800-879-4963

1101-A Lake Cook Road
Deerfield, IL 60015
312-945-1500
Fax: 312-945-1044
(Laser diodes)

USA Regional Sales Offices:
Northwest:
Sunnyvale, CA
408-737-9844
Fax: 408-737-9905

Southwest:
Irvine, CA
714-453-0224
Fax: 714-453-0125

Central:
Dallas, TX:
214-480-0470
Fax: 214-235-4114
Chicago, IL
708-945-1500
Fax: 708-945-1044

Northeast:
Edison, NJ
908-248-8070
Fax: 908-248-8030
Boston, MA
617-224-0074
Fax: 617-224-1096

Southeast:
Atlanta, GA
404-368-0203
Fax: 404-368-0075

Toshiba Electronics Europe GmbH
Hansaallee 181
D-40549
Dusseldorf, Germany
49 211 52960
Fax: 49 211 5296400

Toshiba Electronics (UK) Ltd.
Riverside Way
Camberley, Surrey GU1 53YA
England
44 1726 694600
Fax: 44 1276 682256

(Semiconductors, diodes, IGBTs, FRDs, GTOs, transistors, optoelectronics, photocouplers, CCD sensors, ICs including high power and high temperature units [200-500V and 200 to 300C], hybrid, logic, RISC microprocessors, microcontrollers, masked ROMs, ASICs, FPGAs [licensed from Pilkington Technology] and electronic components for industrial and consumer applications. This company also manufacturers solid state devices, memory cards, DRAMs, and DRAM modules, VRAMs, SRAMs, EPROM, Flash EEPROM, memory cards, CMOS ICs, picture tubes, LCD displays, LEDs, medical tubes, barium ferrite recording media, microwave components, laser diodes, hall effect sensors, microcontrollers, microcomputers, RISC processors, optical transmission devices [fiber optic transmitters and receivers], speech synthesis and recognition, disk drive controller, television, VCR, and telecommunications ICs [including SAR or Segmentation and Reassembly chip sets], and printed circuit boards.)

Total Power International, Inc.
418 Bridge Street
Lowell, MA 01850
508-453-7272
Fax: 508-453-7395

(Power supplies, DC/DC converters)

TOTX prefix
See *Toshiba America Electronic Components, Inc.*

Towa Electron Co., Ltd.
2-44 Muro-Machi
Hadano-Shi, Kanagawa 257
Japan
0463 82 1022

(Hybrid ICs)

Tower Electronics, Inc.
281 South Commerce Circle
Fridley, MN 55432
612-571-3737
Fax: 612-571-5605
E-mail: Towerinc@primenet.com

(DC/DC converters and switching power supplies)

Toyo Communications Equipment Co. Ltd.
2-12-32, Konan, Minato-Ku
Tokyo 108
Japan
03 5462 9600/9610
Fax: 03 5462 9625

(Memory and hybrid ICs)

Toyocom U.S.A. Inc.
Headquarters:
617 E. Golf Rd.
Arlington Heights, IL 60005
(T in a 6 sided polygon)
800-869-6266
708-593-8780
Fax: 708-593-5678

3200 Bristol St., #720
Costa Mesa, CA 92626
800-809-0206
Fax: 714-668-9158

(Surface mount filters, voltage regulators, inc.)

TP prefix
See *Siemens*, *AVG Semiconductors* (Second source telephone ICs), *Supertex* (Discrete semiconductors)

TPC prefix
See *Texas Instruments*

TPIC prefix
See *Texas Instruments*

TPM prefix
See *Datel, Inc.*

TPP or **TPQ** prefix
(Sprague Semiconductor transistor arrays)
See *Allegro Microsystems*

TPR —
TRI-MAG

TPR prefix
See *GHz Technology, Inc.*

TPS prefix
See *Texas Instruments*

TQ prefix
See *TriQuint Semiconductor, Inc.*

TQS logo
See *TriQuint Semiconductor, Inc.*

TR prefix
See *Music Semiconductor*

(Graphics color palette)

Trak Microwave Corporation
4726 Eisenhower Blvd.
Tampa, FL 33634
813-884-1411
Fax: 813-886-2794
TLX: 52-827

Europe:
Dunsinane Avenue
Dundee, Scotland DD2 3PN
44 382 833411, Fax: 44 382 833599

(Microwave and RF products including isolators, circulators, directional couplers, power splitters, mixers, power combiners, variable phase shifters and attenuators, space grade hardware is available)

TRM, Inc.
280 South River Road
Bedford, NH 03110
603-627-6000, Fax: 603-627-6025

(Quadraphase modulators)

Transchem
P.O. Box 667
1521 Pamona Road
Corona, CA 91718-0667
909-371-5751, Fax: 909-371-9602

(DC/DC converters)

Transducer Techniques
43178 Business Park Drive, B-101
Temecula, CA 92950
909-676-3965
Fax: 909-676-1200

(Load cells Force/torque sensors)

Transistor Devices, Inc.
274 South Salem Street
Cedar Knolls, NJ 07969
201-361-6622
Fax: 201-361-7665
http://www.transdev.com
E-mail: info@mailer,transdev.com

(Power supplies, DC/DC converters)

Transistor Specialities, Inc.
See *TSI Microelectronics Corp.*

Trans-Tech, Inc.
Subsidiary of Alpha Industries, Inc.
5520 Adamstown Road
Adamstown, MD 21710
301-695-9400
Fax: 301-695-7065

(Ceramic coaxial transmission elements, dielectric resonators, garnets, ceramic bandpass filters, two and three pole filters)

Transwave Corp.
P.O. Box 489
Vanderbuilt, PA 15486
412-626-0200

(Transistors, optoelectronics)

TranSwitch Corporation
8 Progress Drive
Shelton, CT 06484
203-929-8810
Fax: 203-929-8810
http://www.txc.com
http://www.transwitch.com/

(A fabless company supplying Asynchronous transfer mode [ATM] ICs [SARA IC], SONET and SDH ICs, multiplexer ICs, and Constant Bit Rate ATM Adaption Layer 1 devices [COBRA])

Triad (tm)
See *MagneTek*

(Magnetics)

Triad Semiconductors
See *Music Semiconductors*

Trident Microsystems, Inc.
189 North Bernardo Avenue
Mountain View, CA 94043-5203
415-691-9211, Fax: 415-691-9260

Trident Microsystems Far East Ltd.
Taipei, Taiwan

(A fabless company supplying Windows accelerator ICs, video graphics ICs (NTSC, PAL), 3D graphics ICs, MPEG ICs, graphics accelerator and controller chips and boards, Video Electronic Standards Association United Memory Architecture chip sets, and video capture, overlay and control chip sets.)

Trilithic
9202 East 33rd Street
Indianapolis, IN 46236
1-800-344-2412, 317-895-3600
Fax: 317-895-3613
TLX: 244-344 (RCA)

(Ultraminiature filters, RF components, oscillators, attenuators, fixed, SMD and variable)

TRI-MAG
1601 N. Clancy Ct.
Visalia, CA 93291
209-651-2222
Fax: 209-651-0188
IIT Telex: 4994462
Cable: TRIMAG

TRI-MAG — TSI MICROELECTRONICS

(AC line filters, switching power supplies, DC/DC converters)

Trimax
2747 Rte. 20 East
Cazenovia, NY 13035
315-655-3551

(Miniature thermal circuit protectors)

Tri Source, Inc.
415 Howe Avenue
Shelton, CT 06484
203-924-7030, Fax: 203-924-7045

(Plug in VME power supplies)

TriQuint Semiconductor, Inc.
Computing and Networking Division
2300 Owen Street
Santa Clara, CA 95054
408-982-0900
http://www.triquint.com/
E Mail: cust.serv@tqs.com

3625A S.W. Murry Blvd.
Beaverton, OR 97005
503-644-3535
Fax: 503-644-3198

European Headquarters:
GiGA A/S
Mileparken 22
DK-2740 Skovlunde, Denmark
45 44 92 61 00
Fax: 45 44 92 59 00

(ASICs, GaAs, fiber optic, GaAs power amplifiers, data communication ICs, memory ICs, MMIC amplifiers, clock buffers, ASTM ICs, etc.)

TriTech Microelectronics Pte. Ltd.
(Semiconductor)
No. 2 Science Park Drive
Singapore Science Park
Republic of Singapore 0511
65-777-2566
Fax: 65-777-7482

(RTCs - single chip voice storage controller, real time clocks, pen based [input processor] computer chip sets, DSPs, caller ID chip with integrated DTMF tone transmitter, customer specific ASICs [such as the audio codecs and I/O for Creative Technologies Sound-Blaster audio add on cards. This company is a spin off from Chartered Semiconductor and has had agreements with Simtek Corp in the past. It is the design arm of Singapore Technologies Semiconductors [who also owns Chartered Semiconductor Pte. Ltd.])

Triton Chip Set
See *Intel Corp.*

Tri-Tronics Company, Inc.
P.O. Box 25135
Tampa, FL 33622-5135
813-886-4000
Fax: 813-884-8818

(Miniature optical sensors, rugged and waterproof versions available)

TRM prefix
See *Hitachi Semiconductor* (SONET devices), *Tekram Technology*

TRT in a circle logo, Tohritsu Co.
See *RO Associates*

TRU prefix
See *Lucent Technologies* (Formerly AT&T Microelectronics)

TRV prefix
See *Hitachi Semiconductor*

(SONET devices)

TRW LSI Products Div.
See *Raytheon Semiconductor*
(In La Jolla)

*TriTronics
1306 Continental Dr.
Abingdon, MD 21009
"Phone" - 1-800-638-3328*

(A/D, D/A converters, linear, signal synthesis, imaging, transform, correlators, vector arithmetic/filters, fixed and point arithmetic, memory and storage products)

TS prefix
See *Thomson Components and Tubes Corp.*, *TSI Microelectronics Corp.*

TSB prefix
See *Texas Instruments*, *Thomson Components*

TSC prefix
See *Telcom* (formerly Teledyne Components), *Texas Instruments* (Programmable ASICs)

TSC America
22048 Sherman Way, Suite 109
Canoga Park, CA 91303
818-346-3467
Fax: 818-346-3385

Technical Information:
516-361-9499, Fax: 516-265-6241

(Very high speed motion control IC for servo and stepper motors)

Tseng Laboratories, Inc.
6 Terry Drive
Newtown, PA 18940
215-968-0502
Fax: 215-860-7713

(Graphics ICs, including Windows accelerators and Super VGA devices)

TSI Microelectronics Corporation
(Formerly Transistor Specialities, Inc.)
5 Southside Road
Danvers, MA 01923
508-774-8722, Fax: 508-774-0939
TWX: 710-347-0309

T - U TSI — UCM

(Custom hybrids and special products)

TSL prefix
See *Texas Instruments*

TSP prefix
See *Texas Instruments*

TSR prefix
See *Thompson Components*

TSS prefix
See *Texas Instruments*

TST-XXXX part number
See *Beta Transformer Technology, Corp.*

TTE, Inc.
2251 Barry Avenue
Los Angeles, CA
800-776-7614
310-478-2791
Fax: 800-445-2791, 310-445-2791

(Filters including Gaussian, Butterworth and Chebyshev designs)

TTLP prefix
See *Technitrol, Inc.*

Tudor Technology Ltd.
5145 Campus Dr.
Plymouth Meeting, PA 19462
800-777-0778
Fax: 610-828-0635

(Thermocouples, RTDs and other temperature sensing equipment)

Tundra Semiconductor Corp.
See *Newbridge Microsystems*

Turbo IC
1031 E. Duane Avenue
Sunnyvale, CA 94086-2625
408-739-1818

(EEPROMs)

Turck Inc.
3000 Campus Drive
Minneapolis, MN 55441
612-553-7300
Fax: 612-553-0708
Applications: 1-800-544-7769

(ICs used in their proximity sensors)

TVP prefix
See *Texas Instruments*, *Trident Microsystems, Inc.*

TVS prefix
See *FCI Semiconductor*

(Transient voltage suppressors)

TWD prefix
See *Amplifonix*

Two in a circle logo,
See *Allegro Microsystems*

(ICs and transistors. For capacitors contact United Chemicon and Vishay Sprague).

TWR prefix
See *Datel, Inc.*

TXC-XXXXX part number
See *TranSwitch Corporation*

TXO prefix
See *STC Defense Systems*

TXT logo
See *Texet Corp.*

U prefix
See *InterFet Corporation* (FET devices), *Temic Semiconductors* (Telefunken), *Mikroelektronik Dresden GmbH*

U underlined
See *Unitrode*

U over a T in a circle logo
See *Uni-Tran Semiconductor Corp.*

uA prefix
(A term first used to denote a micro amplifier, by Fairchild Semiconductor in 1964, which was acquired by National Semiconductor in 1987. Second sourced material available from Texas Instruments, Inc.)

UBA prefix
See *Philips Semiconductors*

UC prefix
See *Unitrode*
(Some devices second sourced by Astec Semiconductor)

UCB prefix
See *Philips Semiconductors*

UCC prefix
See *Unitrode Integrated Circuits Corporation*

UCE, Inc.
35 Rockland Road
Norwalk, CT 06854
203-838-7500
Fax: 203-838-2566

(Optoelectronics)

Uchihashi Estec Co. Ltd.
International Trade Division
9-14, Imazu-Kita 2-Chome
Tsurumi-Ku
Osaka 538, Japan
06 962 6661
Fax: 06 962 6669
TLX: J64377 ELCUT

Chatham Components, Inc.
33 River Road
Chatham, NJ 07928
201-635-8075
Fax: 201-635-6345

(Thermal fuses [cutoffs])

UCM prefix
See *Datel, Inc.*

UDC prefix
See *Staktek Corporation*

(Memory stack modules)

UDR prefix
See *GHz Technology, Inc.*

UDT Sensors, Inc.
An OSI (Opto Sencors) Company
12525 Chadron Avenue
Hawthorne, CA 90250
310-978-0516
Fax: 310-644-1727

(Photodetectors, trans-impedance amplifiers, microelectronics and hybrids, opto-assemblies, injection and transfer molded packaging)

uE logo
(Actually a Greek mu symbol followed by an E)
See *Micro Electronics, Ltd.*

uE over an m logo
(Actually a Greek mu symbol followed by an E over the m)
See *EM Microelectronic-Marin SA*

UEC prefix
See *Provector AB*

U-FO, Top-Tech Co., Ltd.
Fl. 5-3, No. 186, Sec. 1
Keelung Road
Taipei, Taiwan
886 2 753 0753
Fax: 886 2 753 0754

(Diodes)

UHP prefix
(Sprague Semiconductor)
See *Allegro Microsystems*

ULA prefix
See *Ferranti*

Ultra Analog, Inc.
47747 Warm Springs Blvd
P.O. Box 14164
Fremont, CA 94539
510-657-2227
Fax: 510-657-4225

(Digital and interface ICs)

UltraLogic
(Trademark of Cypress Semiconductor)

Ultra Technology
2512 10th St.
Berkeley, CA 94710
510-848-2149
http://www.dnai.com/~jfox/

(Forth microprocessor [a MISC processor, an improved version of a RISC processor])

Ultra Volt, Inc.
CS9002
Ronkonkoma, NY 11779-9002
800-9HV Power
516-471-4444
Fax: 516-471-4696

(Regulated high voltage DC to DC converters)

UM prefix
See *Unicorn Microelectronics, Corp., AVG Semiconductors*
(Second source devices)

UM-X part number
See *Mercury United Electronics*

(Quartz crystal)

UMC
See *United Microelectronics Corp.*

UMEC
See *Universal Microelectronics Co., Ltd.*

UMF prefix
See *Philips Semiconductor*
(Formerly Signetics)

UMIL prefix
See *GHz Technology, Inc.*

UNIAX Corporation
6780 Cortona Drive
Goleta, CA 93117-3022
805-562-9293

(Full color, plastic LED designs)

Unicorn Microelectronics Corporation
3350 Scott Blvd, Bldg 48 & 49
Santa Clara, CA 95054
408-727-9239
Fax: 408-492-1720
TLX: 172730 NMC SNTA

(Counters, LCD drivers, Fan controllers, DTMF ICs, Pulse and Tone dialers, character generators, FIFOs, LIFOs, RAMs, SRAMs, DMA controllers, clock generators, bus controllers, timers, priority interrupt controllers, UARTS)

Unipac Optoelectronics Corp.
No. 3, Industry E. Rd. III
Science-Based Industrial park
Hsinchu, Taiwan R.O.C.
886 35 772 700
Fax: 886 35 772 730

(Small size active matrix [TFT] LCDs)

Uniphase
163 Baypointe Parkway
San Jose, CA 95134
4080434-1800
Fax: 408-433-3838
United Kingdom: 44 0438 745055
Germany: 49 089 3196026
Japan: 3 3226 6321

UNIPHASE — UNITRODE INTEGRATED CIRCUITS

(Laser diodes)

Uniphase Telecommunications Products (UTP)
Subsidiary Uniphase Corporation
1289 Blue Hills Avenue
Bloomfield, CT 06002
203-769-3000
Fax: 203-769-3001

(Fiber optic lasers, fiber optic gyros, CATV/RF products)

Unipower Corporation
3900 Coral Ridge Dr.
Coral Springs, FL 33065
954-346-2442
Fax: 954-340-7901
E-mail: sales@unipower-corp.com

(Power supplies)

United Chequers (Hong Kong) Ltd.
RM. 1101-2 Mongkok Commercial Center
16 Argyle Street
Mongkok, KLN, Hong Kong
852 391-7306, 852 391-6725
Fax: 852 789-3205
TLX: 52688 CKEPS
Cable: Chequershk

(Ceramic filters and resonators, quartz crystals)

United Electronics Corp. Ltd.
See *Fer Rite Electronics Ind. Co., Ltd.*

United Microelectronics Corp. (UMC)
No. 13, Innovation Road I
Science Based Industrial park
Hsinchi, Taiwan, R.O.C.
886 35 782 258
Fax: 886 35 774 767

8F, No. 233-1 Bao Chiao Rd
Hsin Tien, Taipei County
Taiwan, ROC
886-2-918-1589, ext. 6305
(PC Chip sets), 035 782 258
Fax: 886-2-918-0188, 035 774 767
No. 3 Industry East Road

Science-Based Industrial Park
Hsinchu, Taiwan
035-773131
Fax: 035 774 767

Sales Office:
9th Floor, No. 201-26, Tunhua North road
Taipei, Taiwan, R.O.C.
886 2 7152455
Fax: 886 2 7166291
http://www.umc.com.tw/ (Chinese)

United Microelectronics (Europe) B.V.
Hoekenrode 2, 1102 BR
Amsterdam Zuidoost
The Netherlands
020-970-766
Fax: 020-977-826
TLX: 11677 UMC NL

UMC K.K.
3F, 20-29, 2-Chome Takanawa,
Minato-Ku
Tokyo, Japan
03-280-3661
Fax: 03-280-3663

United Microelectronics Company, Ltd.
RM 1003, Tower B,
Hundhom Comm. Centre
37-39, MA TAU WAI Rd,
Hunghom, Kowloon, Hong Kong
852-765-7122
Fax: 852-765-7483

In the U.S. contact:
Integrated Technology Express, Inc.
Milpitas, CA
408-934-7330
Fax: 408-934-7337

(Various computer ICs including CRT controller, memory controllers, real time clocks [RTCs], hard/floppy disk controllers, communications, microprocessor, microcontrollers, SRAMs, ROMs, display, I/O, "green" PC chip sets, Pentium class chip sets, and peripheral ICs, LCD displays, flash memory, DRAMs, CD-ROM ICs, Video CD voice ICs, etc.)

United-Page, Inc.
481 Getty Avenue
Patterson, NJ 07503
201-279-7500

(Diodes, thyristors, transistors)

United Technologies Microelectronics Center
1575 Garden of the Gods Road
Colorado Springs, Colorado 80907-3486
719-594-8000, 8014

Marketing:
800-722-1575
719-594-8166

Technical Information:
719-594-8252

Literature Requests:
800-645-UTMC
http://www.utmc.com

(A fabless company supplying SRAMs, ROMs, gate arrays, radiation hardened parts, MIL-STD-1553 protocol devices including their trademarked SuMMIT device)

Unitrode Integrated Circuits Corporation
7 Continental Blvd.
Merrimak, NH 03054-0399
603-424-2410, 429-8610
Fax: 603-424-3460
http://www.unitrode.com

Manufacturers, Prefixes, Part Number Types, Logo Descriptions & Family Types

(A/D flash converters, step-up voltage regulators, switched capacitor voltage converter with regulator, power amplifiers, power driver and interface controllers, power factor correction ICs, linear regulators, PWM [pulse width modulator] ICs for precision motor control, electronic circuit breakers, line drivers, SCSI terminators [including plug and play types], IR receiver ICs, motion control ICs etc.)

Universal Microelectronics Co., Ltd. (UMEC)
No. 3, 27th Road
Taichung Industrial Park
Taichung 40813, Taiwan
886 4 359 0096, 886 4 359 0100
Fax: 886 4359 0129, 3590063
TLX: 57477 UMEC

UMEC International Corporation
2535 W. 237th St., Suite 113
Torrance, CA 90505
310-326-7072
Fax: 310-326-7058

UMEC Elektxronische
Komponenten GmbH
Krcuzenstrasse 80
74076 Heilbronn, Germany
7131 76170
Fax: 7131 761720

Rm. 504
41-5
Nihonbashi Hamacho 3-Chome
Chuo-Ku, Tokyo 103
Japan
81 3 3808 1467
Fax: 81 3 3808 1477

(Magnetic components such as ISDN chokes, transformers, pulse transformers, integrated filter modules, power transformers, inductors and sensors, telecom transformers and chokes, PCMCIA cards [including memory and fax/modem cards with internal DAA and Ethernet LAN cards])

Universal Semiconductor, Inc.
1925 Zanker Road
San Jose, CA 95112
408-279-2830, 408-436-1906
Fax: 408-436-1125

(Optoelectronics, thyristors, transistors, digital, interface and microprocessor support ICs)

Unizon
See *Komatsu Group*
(Available from Taitron Components, Inc., 25202 Anza Drive, Santa Clarita, CA 91355-3496, 800-824-8766, 800-247-2232, 805-257-6060; Fax: 800-824-8329, 805-257-6415)

Unisys Corp.
Communications Systems Division
Government Systems Group
(DSP Components)
640 North 2200 West
Salt Lake City, UT 84116-2988
801-594-2001 (Main), 6044
(Technical information), 4440
(Product information and orders), 3737
Fax: 801-594-5908 or 4127

(Spread spectrum demodulator for wireless communications applications)

UNR prefix
See *Datel, Inc.*

UPA prefix
See *Datel, Inc.*, *NEC*
(California Eastern Labs for dual transistor arrays)

uPB prefix
See *NEC, Inc.*

uPC prefix
See *NEC, Inc.*

uPD prefix
See *NEC, Inc.*,
AVG Semiconductors
(Second source some ICs)

UNITRODE INTEGRATED CIRCUITS — USAR

uPF prefix
See *NEC, Inc.*

UPM prefix
See *Datel, Inc.*

UR prefix
See *USAR Systems, Inc.*

US prefix
On vacuum fluorescent displays (VFDs)
See *Futaba*

U.S. Crystal Corporation
3605 McCart Street
Ft. Worth, TX 76110
800-433-7140
817-921-3014
Fax: 817-923-0424

(Crystals)

U.S. Elco, Inc.
See *Cosel U.S.A., Inc.*

U.S. Sensor
1832 West Collins Avenue
Orange, CA 92867
714-639-1000
Fax: 714-639-1220

(Thermistors, probes and assemblies)

U-Shin Ltd.
7-2, Nishi-Shimbashi 1-Chome,
Minato-Ku
Tokyo 105
Japan
81 3 3502 5241

(Car and hybrid ICs)

USAR Systems, Inc.
568 Broadway
New York, NY 10012
212-226-2042
Fax: 212-236-3215

U - V

USAR — VALTRONIC

(Keyboard encoders, user programmable keyboards, keyboard protocol converters [i.e., PC to RS232, PC to Parallel, PC to HP-HIL], DEC keyboard encoder IC, access bus devices, embedded pointing devices)

USC prefix
See *Datel, Inc.*

UT prefix
See *United Technologies Microelectronics Center*,
Universal Microelectronics Co., Ltd.

(Magnetic components such as transformers, chokes, inductors, filters)

UTMC
See *United Technologies Microelectronics Center*

UTP
See *Uniphase Telecommunications Products (UTP)*

UTV prefix
See *GHz Technology, Inc.*

UVC Corporation
11 Suzanne Lane
Pleasantville, NY 10570
914-769-7334

(Image processing, video processing and timing ICs)

Uwatec AG
Engenbuhl 130
Hallwil 5705
Switzerland
41 64 542 940
Fax: 41 64 542 280

(Sensing and measuring ICs)

UWR prefix
See *Datel, Inc.*

UZF-XXX part number
See *Aromat Corporation*

(Fiber-optic photoelectric sensors)

V prefix (DMOS product)
See *Supertex, Inc.*
and *Calogic Corp.*

(If a memory device [such as a DRAM] see *Mosel-Vitelec Corp.*, *NEC* [Microprocessor], *EZ Communications* [VCOs], *V3 Corporation*, *EM Microelectronic* [Voltage surveillance ICs, level shifter ICs])

V3 Corporation
2348 Walsh Avenue, Suite G
Santa Clara, CA 95051
800-488-8410
408-988-1050
Fax: 408-988-2601
v3info@terraport.net
http://www.vcubed.com/

759 Warden Avenue
Toronto, Ontario
Canada M1L 4B5
416-285-9188

(PCI bus bridge ICs [single and dual], burst DRAM controllers, system controllers, plug-in evaluation boards, motherboards and stand alone boards.)

VA prefix
See *Intel Corporation, VTC, Inc.*

VAC prefix
See *Cypress Semiconductor*

Vadem
1960 Zanker Rd.
San Jose, CA 95112-4126
408-943-9301, 408-467-2141
Fax: 408-943-9735, 408-467-2199

(A fabless company supplying PC Chip sets, PCMCIA interface controller and LCD/CRT controller ICs, PCMCIA host adaptors)

Vaisala Sensor Systems
100 Commerce Way
Woburn, MA 01801-9922
617-933-4500
Fax: 617-933-8029

(Humidity sensors)

Valor Electronics, Inc.
A GTI Company
9715 Business Park Avenue
San Diego, CA 92131-1642
619-537-2600
Fax: 619-537-2525
E-mail: valormkt@valorsd,
mhs.compuserve.com

Flat B
7/F., K.K. Industrial Building 5 Mok Cheong Street Tokwawan, Kowloon, Hong Kong
852 2333 0127
Fax: 852 2363 6206

SteinstraBe 68
81667 Munchen Germany
49 89 48022823
Fax: 49 89 484743

(Pulse transformers, filters, modules, DC/DC converters, delay lines)

Valpey-Fisher Corporation
Subsidiary of Matec
75 South Street
Hopkinton, MA 01748
1-800-982-5737
508-435-6831
Fax: 508-497-6377

(Crystal oscillators)

Valtronic USA, Inc.
6168 Cochran Rd.
Solon, OH 44139
216-349-1239
Fax: 216-349-1040

Valtronic AG
Les Charbonnieres 1343
Switzerland
41 2184 11041
Fax: 41 2184 11360
TLX: 459 402

(This company is a turnkey manufacturer of small DIP, SIP, and surface mount COB [chip on board] and COC [chip on chip] modules including CMOS SRAM modules and UV EPROM modules)

Value The Customer (tm)
See *VTC, Inc.*

VAM prefix
See *GHz Technology, Inc.*

Vanguard Electronics Co., Inc.
1480 W. 178th St.
Gardena, CA 90248-3277
213-323-4100

(Pulse transformers, magnetics)

Vanguard International Semiconductor Corporation (VISC)
123, Park Avenue, 3,
Science Based Industrial Park
Hinchi, Taiwan R.O.C.
886 35 770 355, Fax: 886 35 788 575

(Formerly the Sub-Micron Lab of the Taiwan government, it was acquired by a group of investors led by TSMC in 1995 to manufacture DRAMs and SRAMs. It introduced a 64M DRAM in March 1996.)

Vari-L Company, Inc. (VARIL)
11101 E. 51st Avenue
Denver, CO 80239
303-371-1560, Fax: 303-371-0845

(Mixers, transformers, voltage controlled oscillators, phase-locked loop synthesizers, frequency doublers, switches)

Varitronix, Ltd.
4/F Liven House
61-63 King UIP St.
Kwun Tong, Kowloon
Hong Kong
3 894317
TLX: 36643 V TRAX HX

VL Electronics, Inc.
3250 Wilshire Blvd., #1301
Los Angeles, CA 90010
213-738-8700
Fax: 213-738-5340

(Optoelectronics, LCD displays)

Varo, Inc.
2800 W. Kingsley Road
Garland, TX 75046
P.O. Box 40676
Garland, TX 75040
214-840-5531
Fax: 214-840-5151

(DC/DC converters, diode bridges)

Vatell Corporation
P.O. Box 66
Christiansburg, VA 24073
703-961-2001
Fax: 703-953-3010

(Sensors to measure heat flux and temperature concurrently)

VB prefix
See *SGS-Thomson Microelectronics*

VCA prefix
See *Burr-Brown Corp.*

VCA Associates
7131 Owensmouth
Suite B87
Canoga Park, CA 91303
818-704-9202
Fax: 818-704-9310

(Linear ICs)

VCD prefix
See *Mosel-Vitelic Corporation*

VCP
See *Integrated Information Technology*

(RISC controller)

VCR prefix
See *InterFET Corporation*

(FET devices)

VDAC prefix
See *Lambda Advanced Analog*

VE logo
See *Vanguard Electronics Co., Inc.*

Vector Technology Ltd.
5/6 Roseheyworth Business Park
Abertillery, Gwent, U.K. NP3 1SP
0495 320222
Fax: 0495 320484

(Laser integrated detectors)

Vectron Laboratories, Inc.
166 Glover Ave.
Norwalk, CT 06856-5160
203-853-4433
Fax: 203-849-1423

(Crystal oscillators)

Vectron Technologies, Inc.
267 Lowell Road
Hudson, NH 03051
1-800-NEED-FCP
603-598-0070
Fax: 603-598-0075

(Clock and voltage controlled oscillators, SAW filters, clock recovery and data retiming units for telecommunications and data communications)

VENTRONICS — VICTOR COMPANY

Ventronics, Inc.
346 Monroe Avenue
P.O. Box 142
Kenilworth, NJ 07033
908-272-9762
Fax: 908-272-7630

(This company has joint agreements to manufacture ceramic disc capacitors, wall mounted power supplies, magnetic component, and varistors. They also distribute various other components including delay lines, fans, relays, LCDs, diodes, rectifiers, piezo devices, potentiometers, resistors and PC boards.)

Verdure Industries, Inc.
Theft-Block
P.O. box 24886
Omaha, NE 68124
402-691-8300
Fax: 402-691-8400

(This company used to market ICs, including the C-MART-1000, a standalone asynchronous digital receiver transmitter which could function as a data acquisition system, but now produces parts for internal use.)

Veritech Microwave
111-B Corporate Blvd.
South Plainfield, NJ 07080
908-769-0300
Fax: 908-769-0330

(Microwave linear amplifiers)

VersaMod (tm),
See *Power Systems, Inc.*

Vertex Semiconductor Corp.
See *Toshiba*

VES prefix
See *VLSI Technology*

VF prefix
See *Valpey-Fisher*

VFAC prefix
See *Valpey-Fisher*

VGI/VIKAY
(LCDs and LCD modules)

VGLC prefix
See *Vitesse Semiconductor Corp.*

VGT prefix
See *VLSI Technology, Inc.*,
Philips Semiconductor

(Programmable ASICs)

VGX prefix
See *VLSI Technology, Inc.*

VG-XXX part number
See *Vadem*

VH prefix
See *Raltron Electronics Corp.*

(Oscillators)

VIA logo
See *Via Technologies, Inc.*

Via Technologies, Inc.
5020 Brandin Ct.
Fremont, CA 94538
510-683-3300
Fax: 510-683-3301

Taipei, Taiwan
(Part of Taiwan's First International Computer, Inc, [FIC] and owned by Formosa Plastics Group [of Taiwan] in 1992)
http://www.fic.com.tw
(First International home page)

(PC chip sets including EISA and VESA bus controllers, ISA bus controllers, PCI chip sets, Video Electronic Standards Association United Memory Architecture chip sets, power management, multimedia ICs, and Pentium chip sets. Their chip sets include the Diana, Pluto, Apollo, Lyra, Lynx, and Pegasus lines.)

VIC prefix
See *Cypress Semiconductor*

Vicor Corporation (Express)
23 Frontage Road
Andover, MA 01810
508-470-2900, Fax: 508-475-6715
http://www.vicr.com

Westcor
Division of Vicor
560 Oakmead Parkway
Sunnyvale, CA 94086
408-522-5280
Fax: 408-774-5555

Europe:
Carl-von-Linde Strasse 15 D-85748
Garching-Hochbruck, Germany
49 0 89/3 29 27 63-66
Fax: 49 0 89/3 29 27 67

Asia-Pacific:
10F. No. 7, Ho Ping E. Road, Sec. 3
Taipei, Taiwan
886 2 708 8020, 886-2-736 4349,
886-2-736 4284
Fax: 886 2 755 5578,
886-2-733 7341

(Power supplies, DC/DC convertors, military versions available. Note: JT PowerCraft, Inc., part of JT Electronics Corporation [Tokyo, Japan] is licensed to manufacture and sell DC/DC converters and power supplies based upon Vicor Corporation power supply technology).

Victor Company of Japan, Ltd.
3-12, Moriya-Cho
Kangawa-Ku
Yokohama-Shi
Kanagawa 221, Japan
045 450 1641
Fax: 045 450 1672

VICTOR COMPANY — VITESSE

(CD player ICs)

VID prefix
See *Advanced Risc Machines*

Vidar - SMS Co. Ltd.
No. 145, Section 2
Chung Shan N. Rd.
Taipei, Taiwan R.O.C.
886 2 521 1100
Fax: 886 2 521 3300

Sun Moon Star, Inc.
San Jose, CA
408-452-7811
Fax: 408-452-1411

(Power supplies)

VIDC prefix
See *Advanced RISC Machines*

Video International Development Corporation
65-16 Brook Avenue
Deer Park, NY 11729
516-243-5414
Fax: 516-243-4314

(TV multi-standard converter IC for PAL, SECAM, NTSC, PAL-M and PAL-N)

Videologic, Inc.
1001 Bayhill Dr.
Suite 310
San Bruno, CA 94066
1-800-578-5644
415-875-0606

(PowerPlay64 coprocessor, video ICs, display cards)

VideoMail, Inc.
568-4 Weddell Drive
Sunnyvale, CA 94089
408-747-0223
Fax: 408-747-0225

(Image processing, video processing and timing ICs)

Viper family of GaAs gate arrays
See *Vitesse Semiconductor Corp.*

Virtual Chips
2107 N. First St., Suite 100
San Jose, CA 95131
408-452-1600, Fax: 408-452-0952
http://www.vchips.com
Email: sales@vchips.com

(PCI cores for IC designs including synthesizible Universal Serial Bus [USB] Cores. This company was acquired by Phoenix Technologies, a supplier of PC BIOS software, in 1996 and is now part of their Special Products Division.

VISC
See *Vanguard International Semiconductor Corporation*

Vishay Resistors
63 Lincoln Highway
Malvern, PA 19355-2120
610-644-1300
Fax: 610-296-0657,
Sales: 610-640-9081

(Various fixed and variable resistors. Vishay, started in 1962, was named after a no longer existing Polish town where President Felix Zandman was born. Note: Vishay owns Dale, Sfernice, part of Sprague, Roederstein and Vitrimon)

Visic, Inc.
1109 Mckay Drive
San Jose, CA 95131
408-434-3100
Fax: 408-263-2511
TLX: 278807

(Memory modules)

Visicom Laboratories, Inc.
Emerging Technologies Division
17301 West Colfax, Suite 150
Golden, CO 80401
303-277-0271, Fax: 303-277-1505

(This company emulates obsolete modules using new [VHDL] technologies. The generic information on the functionality of the circuitry enables the emulation of the circuit using currently available gate arrays, and can also be used to emulate the circuit in any future available technology. This technique was used to develop Universal Standard Electronic Modules [UniSEM] to replace obsolete modules in the U.S. Navy standard computer the AN/UYK-44 MRP. [Note: MicroLithics Corporation is another division of this company])

Vitarel, Inc.
3572 Corporate Court
San Diego, CA 92123
619-292-8353
TLX: 69-7855

(Life support medical electronics, custom designed hybrid parts manufactured to meet a variety of military specifications)

Vitec Electronics Corp.
8530 W. Roosevelt Avenue
Visalia, CA 93291
209-651-1535
Fax: 209-651-1538

(Common mode chokes and EMI suppression inductors)

Vitelic
See *Mosel-Vitelic*

Vitesse Semiconductor Corp.
741 Calle Plano
Camarillo, CA 93012
1-800-VITESSE
805-388-3700
805-388-7408 (Gate arrays)

V VITESSE — VORNE INDUSTRIES

(Digital gallium arsenide [GaAs] VLSI devices, gate arrays, high speed clock drivers, fiber channel chip sets, PC Chip sets, optoelectronic ICs, SONET/SDH ICs. Note: Vitesse is French for "Speed")

Vivid Semiconductor, Inc.
7402 W. Detroit St. #120
Chandler, AZ 85226
602-961-3200
602-951-6760
(Fast response line for urgent calls)
Fax: 602-961-0200

(A fabless company founded in 1993. Mixed signal ICs including LCD column drivers for TFT LCDs or passive LCDs with active addressing)

Vixel Corporation
325 Interlocken Parkway
Broomfield, CO 80021-3484
303-460-0700

(Vertical cavity laser emitting products)

VL prefix
See *VLSI Technology, Inc.*,
VLSI Technology Inquiries, *Opti, Inc.*

V-L
See *Varitronix, Ltd.*

VL Electronics, Inc.
See *Varitronix, Ltd.*

VLC prefix
see *Third Domain, Inc.*

VLSI - VLSI Technology Inc.
8375 South River Parkway
Tempe, AZ 85284
1-800-872-6753
(Advertising response line)
602-752-8574, 6302
Fax: 602-752-6000
http://www.vlsi.com

200 Parkside Drive
San Fernando, CA 91340
602-752-6202
408-434-7956 (Gate arrays)
408-434-7520
(Field programmable ASICs)

1109 McKay Dr.
San Jose, CA 95131
619-625-4620
408-434-3000, 3100, 3218, 7673, 7969
Fax: 408-263-2511

(PC Chip sets, Fuzzy Logic ICs, field programmable ASICs, gate arrays, T1/E1 line-interface units, data encryption processors, CDPD [cellular digital packet data] ICs, modem/decoder ICs for digital video broadcasting, ATM and WAN IC, Video Electronic Standards Association United Memory Architecture chip sets, etc.)

VLSI Technology Inquiries
200 Parkside Dr.
San Francisco, CA 91340
602-752-6202

(ISA bus controllers)

VLSI Vision, Ltd.
See *VVL*

VM prefix
See *VLSI Technology, Inc.*,
VTC, Inc.

VMA prefix
See *Veritech Microwave*

VN prefix
See *Zetex*, *Motorola*,
SGS-Thomson Microelectronics, Inc.

(Solid state relays)

VNA-XX part number
see *Mini-Circuits*

(Amplifier)

VNH prefix
See *SGS-Thomson Microelectronics, Inc.*

VNS-XXXXXX part number
See *VLSI Technology, Inc.*

V/O Electronorgtechnica
32/34 Smolenskaya-Sennaya Square
Moscow, 121200
Russia
205 00 33
TLX: 411385, 411386

(Diodes, transistors, and various ICs)

Volgen America, Inc.
(A Tokyo Keidenki Company)
32980 Alvaredo-Niles Road
Union City, CA 94587
510-429-7170
Fax: 510-429-7173

(DC/DC converters, metal cased units available)

Voltage Multipliers, Inc.
8711 W. Roosevelt Avenue
Visalia, CA 93291
209-651-1402
Fax: 209-651-0740
TLX: 530484

(Fast recovery rectifier chips, transient voltage suppressors, hermetically sealed chips available)

Voltronics Corporation
100-10 Ford Road
Denville, NJ 07834
201-586-8585, Fax: 201-586-3404

(Thermistors)

Vorne Industries, Inc.
1445 Industrial Drive
Itasca, IL 60143-9804
708-875-3600, Fax: 708-875-3609

(Large character displays)

VP prefix
See *Eletech Electronics, Inc.*,
GEC Plessy, VLSI Technology, Inc.

VPS prefix
See *VLSI Technology, Inc.*

VR prefix
See *GEC Plessy*

VR prefix
See *NEC Corp.*

(Microprocessors)

VRE prefix
See *Thaler Corporation*

(Voltage references)

VS prefix
See *Vitesse Semiconductor, NCR,
EG&G Reticon* (CCD devices),
Vivid Semiconductor, Inc.
(LCD column drivers), *VTC, Inc.,
Intel Corporation*

VSC prefix
See *Vitesse Semiconductor,
Third Domain, Inc.,
Philips Semiconductor*

(Programmable ASICs)

VSIS, Inc.
(A 1996 spin-off from Mitsubishi Electronics [Sunnyvale, CA], to build systems on chips [software in silicon]. It is working on parts for set-top box applications, graphics and networking)

VSL prefix
See *Vitesse Semiconductor*

VSM prefix
See *Third Domain, Inc.*

VSP prefix
See *Vitesse Semiconductor Corp.*

VT prefix (such as VT8XCXXXX)
See *Via Technologies, Inc.*

VTB prefix
See *EG&G*

VTC Inc.
2800 East Old Shakoppe Road
Bloomington, MN 55425
612-853-5100
Fax: 612-853-3355

Burkenweg 6
Bachnehring 8091
Germany
8071 2722
Fax: 8071 5858

(Logic families including AC, ACT, FCT, buffers, operational amplifiers, A/D converters, comparators, encoder/decoders, frequency synthesizers, NTDS ICs, line drivers/receivers, read/write preamplifiers for thin film heads)

VTI logo
See *Vectron Technologies, Inc.*

VTV prefix
See *GHz Technology, Inc.*

VVL (VLSI Vision Ltd.)
Aviation House
31 Pinkhill
Edinburgh, Scotland EH12 8BF
44 131 539 7111
Fax: 44 131 539 7141
E-mail: info@VVL.CO.UK

18805 Cox Avenue, #260
Saratoga, CA 95071
408-374-4722
E-mail: dl@vvl.co.uk

In the U.S. you can also contact the distributor:
Marshall Electronics, Inc.
P.O. Box 2027
Culver, CA 90231
800-800-6608
Fax: 310-391-8926

(Video imaging ICs such as a digital camera on a chip, combined CCD and electronics on a chip)

VXI Electronics, Inc.
4607 S.E. International Way
Milwaukie, OR 97222
503-652-7300

(Voltage regulator modules for microprocessors)

VXO prefix
See *STC Defense Systems*

VxP prefix
See *AuraVision Corp.*

VXS prefix
See *TEW North America*

VY prefix
See *VLSI Technology, Inc.*

W prefix
See *Lucent Technologies*
(Formerly AT&T Microelectronics),
*Tseng Labs, IC Works, Inc.,
Wacom Technology, Corp.,
Winbond Electronics Corp.*

W underlined in a circle logo
See *Westinghouse Electric Corporation*

W under a dot with two lightning bolts (logo),
See *Wickman*

(Fuses)

W32 prefix
See *Tseng Laboratories, Inc.*

WAC prefix
See *Integrated Telecom Technology*

WACOM — WAVETEK

Wacom Technology Corp.
501 Southeast
Columbia Shores Blvd
Suite 300
Vancouver, WA 98661
360-750-8882
Fax: 360-737-8061

(Graphics tablets, chip sets for battery-less pen)

WAH III Technology Corporation
16 Digital Drive
Novato, Ca 94949-5760
8 Digital Drive
Novato, CA 94949-5759
415-883-1693, ext. 283

(Miniature AM LCD displays for projection displays, head mounted displays and virtual reality goggles)

Waferscale Integration, Inc. (WSI, Inc.)
47280 Kato Road
Freemont, CA 94538
800-877-6220
800-832-6974 (800 TEAM WSI)
800-562-6363 (in CA)
510-656-5400
Fax: 510-657-5916, 510-657-8495
TLX: 289255
Canada: Intelatech, Inc.
416-629-0082

Northeast Regional Sales:
239 Littleton Road, Suite 3C
Westford, MA 01886
617-692-4989
Fax: 617-692-4083

Midwest Regional Sales:
Barrington Pointe, Suite 400
2300 N. Barrington Road
Hoffman Estates, IL 60195
312-490-5318
Fax: 312-884-9423

South West Regional Sales:
17011 Beach Blvd., Suite 900
Huntington Beach, CA 92647
714-843-9407
Fax: 714-841-9083

Mid-Atlantic Regional Sales:
3 Neshaminy Interplex, Suite 301
Trevose, PA 19047
215-638-9617
Fax: 215-638-7326

South East Regional Sales:
101 Washington Street, Suite 14
Huntsville, AL 35801
205-539-7406
Fax: 205-539-7449

North West Regional Sales:
Call Fremont, CA

WSI France:
33 1 69320120
Fax: 33 1 69320219

(Low voltage field programmable microcontroller peripheral ICs, EPROMs)

Wall Industries, Inc.
5 Watson Brook Road
Exeter, NH 03833
603-778-2300
Fax: 603-778-9797

(Power supplies, DC/DC converters)

Walmsley Microsystems, Ltd.
Aston Science Park
Love Lane
Birmingham B7 4BJ
United Kingdom
44 21 6281800, 44 21 359 0981
(International calls)
Fax: 44 21 6281801

(Optoelectronics, linear and digital ICs)

Warren G-V
One Apollo Drive
Whippany, N.J. 07981
201-386-1200
Fax: 201-386-9331

(Solid state airflow switches in TO-3 and TO-5 packages)

Watkins-Johnson Company
3333 Hillview Avenue
Palo Alto, CA 94304-1223
1-800-WJ1-4401
415-493-4141, 415-813-2972
Fax: 415-813-2402, 415-813-2515
TLX: 34 8415

Watkins Johnson International
Windsor, England

(Amplifiers and mixers, items for spacecraft use available)

Wavelength Electronics, Inc.
P.O. Box 865
Bozeman, MT 59771
406-587-4910
Fax: 406-587-4911
Email: wavelength@imt.net

(Laser diode drivers units and temperature controllers)

Waveline Solid State, Inc.
P.O. Box 718
West Caldwell, NJ 07006
201-226-9100
Fax: 201-226-1565
TWX: 710-734-4324

(PIN diode control devices including attenuators, switches and phase shifters)

Wavetek Corporation
Communications Division
5808 Churchman Bypass
Indianapolis, IN 46203
1-800-851-1202
317-788-9351
Fax: 317-782-4607

WAVETEK — WHITE-ROGERS

(RF and microwave components including filters and fixed or programmable attenuators - stripline/microstrip)

Way-Tech Enterprise Co. Ltd.
10th Floor, No. 245
Ba Der Road, Sec. 2, Taipei, Taiwan
886 2 721 0826, 886 2 777 2725
Fax: 886 2 777 2726

(Diodes)

WD prefix
See *Western Digital*

WDC prefix
See *Western Digital*

WDC within a circle logo
See *(The) Western Design Center, Inc.*

WE prefix
See *White Microelectronics*

WEDSP prefix
See *Lucent Technologies*
(Formerly AT&T Microelectronics)

Weitek Corporation
1060 East Arques Avenue
Sunnyvale, CA 94086
1-800-758-7000 (UNIX products)
408-738-8400 (Main number)
408-738-5765 (DOS products)
Fax: 408-522-7504
Faxback: 800-827-8708

(A fabless company supplying workstation microprocessors [such as the Sparc Power uP], graphics ICs, GUI ICs, windows/graphics accelerators, graphics and Video Electronic Standards Association United Memory Architecture chip sets, and core logic chip sets, etc.)

Wenzel Associates
1005 La Posada Drive
Austin, TX 78752
512-450-1400
Fax: 512-450-1490
http://www.wenzel.com

(Crystal oscillators)

Westcode Semiconductors, Inc.
3270 Cherry Avenue
Long Beach, CA 90807
310-595-6971
Fax: 310-595-8182

Westcode Semiconductors Ltd.
P.O. Box 57
Chippenham
Wiltshire, England SN15 1JL
0249 444524
Fax: 0249 659448
TLX: 44751

(Standard and fast recovery diodes, phase controlled SCRs, regenerative gate inverter SCRs, distributed gate inverter SCRs)

The Western Design Center, Inc.
2166 East Brown Road
Mesa, AZ 85213
602-962-4545
Fax: 602-835-6442
wdesign@wdesignc.com
http://www.wdesignc.com/

(A fabless company supplying microcontrollers and microprocessors and an industrial control computer called the Mensch)

Western Digital Corporation
8105 Irvine Center Dr.
Irvine, CA 92718
714-932-5000, 4900
Fax: 714-932-6294
http://www.wdc.com/
ftp://ftp.wdc.com/

(A fabless company supplying mixed-voltage, video graphics ICs including a single chip LCD VGA controller with integrated Windows acceleration and 32 bit local bus support, MPEG ICs. Note: Western Digital sold its multimedia graphics IC business, including paradise graphics accelerator cards and RocketChip ICs, to Philips Semiconductors in 1995.)

Westinghouse Electric Corporation
Epitaxial Power Transistor Line
See *Solid State Devices, Inc.*
Westinghouse Electric Corporation
P.O. Box 746
Baltimore, MD 21203
410-765-1000

(Digital signal processing ICs, military specification parts available)

WF prefix
See *White Microelectronics*

WFM prefix
See *White Microelectronics*

WG prefix
See *Westcode Semiconductors*

White Microelectronics
White Technology, Inc.
4246 E. Wood Street
Phoenix, AZ 85040
602-437-1520
Fax: 602-437-9120

(Military qualified/tested microprocessor modules, memory ICs, SRAM, EEPROM, and FLASH devices and modules)

White-Rogers
Emerson Electric Co.
9797 Reavis Road
St. Louis, MO 63123
314-865-8799
Fax: 314-638-2400

(Stancor transformers, relays and contactors)

WICKMAN USA — WOLFSON

Wickmann USA, Inc.
4100 Shirley Drive
Atlanta, GA 30336
404-699-7820, Fax: 404-699-9176
http://www.wickmannusa.com

Wickman-Werke GmbH
Abtl. VGT, Annenstr. 113
D-58453, Witten, Germany

(Circuit protection devices including fuses and fuseholders)

Wilco Corporation
6451 Saguaro Court
Indianapolis, IN 46268
317-293-9300, Fax: 317-293-9462

(Inductors including surface mount units)

Wilcoxon Research
21 Firstfield Road
Gaithersburg, MD 20878
1-800-WILCOXON
301-330-8811, Fax: 301-330-8873

(Accelerometers, output sensors, junction boxes, splashproof cable assemblies)

Wilorco, Inc.
729 W. Anaheim St.
Long Beach, CA 90813
213-775-6592

(DC/DC converters)

Winbond Electronics Corp.
No. 2, R&D Road VI
Science-Based Industrial Park
Hsinchu, Taiwan R.O.C.
No. 4, Creation Road III
Science Based Industrial Park
Hsinchu, Taiwan R.O.C.
886 035-770 066
Fax: 886 035 774527,
886 35 789 467

123 Hoi Bun Road
World Trade Center
Kowloon, Hong Kong
852 7516 0237
Fax: 852 7552 2064

US Sales Office:
Winbond Electronics North America Corp.
2730 Orchard Parkway
San Jose, CA 95134
408-943-6666
Fax: 408-943-6668
http://www.winbond.com.tw

Winbond Systems Laboratory
4000 Burton Drive
Santa Clara, CA 95054
408-986-1701
Fax: 408-986-1771

3350 Scott Blvd
Santa Clara, CA 95054
408-982-0381
Fax: 408-982-9231

One authorized distributor is:
UrChoice Electronics
408-987-7888
Fax: 408-987-7898

(This company second sources some SGS-Thomson ICs, computer ICs and has also licensed the flash memory architecture designs of Silicon Storage Technology [SST], Sunnyvale, CA, CD ROM ICs, MPEG and JPEG chip sets, SRAMs, DRAMs, PA-RISC architecture embedded controllers [licensed from Hewlett-Packard], Ethernet ICs, Cache Memory, MPEG ICs, consumer electronics ICs. Note: Symphony Laboratories, a supplier of PC Chip sets including PCMCIA controllers-second source with Omega Micro, Inc.; became Winbond Systems Laboratory in late 1995)

Windsor Corporation
Ceramic Lighting Division
410 South 96th Street, #7
Seattle, WA 98108
206-767-4070

(Flat panel fluorescent lamps for LCD backlighting, hot or cold cathode designs)

WinSystem
715 Stadium Drive
Arlington, TX 76011
817-274-7553
Fax: 817-548-1358

(Single board computers with custom ICs)

Wireless Logic Inc. (WLI)
2021 North Capitol Avenue, Unit B
San Jose, CA 95132
408-262-1876
Fax: 408-262-2903

(Spread spectrum IC, I/O stage IC)

WJ logo
See *Watkins-Johnson*

WK-XXX part number
See *Westcode Semiconductors*

WLT-XXXX part number
See *Wireless Logic Inc.*

WM prefix
See *Wolfson Microelectronics*

WMS prefix
See *White Microelectronics*

WMT prefix
See *Gentron Corp.*

Wolfson Microelectronics
Lutton Court, 20 Bernard
Edinborough, EH8 9NX, Scotland
44 131 667 9386
Fax: 44 131 667 5176
TLX: 727659

Manufacturers, Prefixes, Part Number Types, Logo Descriptions & Family Types

WOLFSON — XICOR

(A fabless semiconductor company offering cellular radio ICs, D/A converters, PCI interface ICs, color scanning ICs)

World Magnetics
810 Hastings Street
Traverse City, MI 49686
616-946-3800
Fax: 616-946-0274

(Pressure switches and sensors)

World Products, Inc.
19654 8th St. E., P.O. Box 517
Sonoma, CA 95476
707-996-5201
Fax: 707-996-3380
TLX: 171715 WORLD PRO SNON

(Diodes)

World Wide Component Distributors
18 Stern Avenue
Springfield, N.J. 07081
1-800-222-6268
201-467-6264
Fax: 201-467-8519

(Specializes in Japanese Semiconductors and generic replacement parts.)

WRLS prefix
See *RF Products, Inc.*

WS prefix
See *Waferscale Integration, Inc.*, *White Microelectronics*

WSI
See *Waferscale Integration, Inc.*

WS-XXXXX-XXXX part number,
See *White Technology, Inc.*

Wustlich Opto-Electronic Components
P.O. Box 8765
Philadelphia, PA 19101-8765
800-235-7880, Fax: 215-365-7245

(LEDs, manufactured in Germany)

Wuxi Huajing Electronics Group
Semiconductor Device Factory
165 Qingshi Road
Wuxi Jiangsu Province
Peoples Republic of China, 226171

(One of the Peoples Republic of China's largest IC vendors in 1993)

WX-XXXX part number,
(Transistors sold by Westinghouse Electric in 1953)

Wyse Technology, Inc.
3471 N. First Street
San Jose, CA 95134-1803
800-438-9973, 408-473-1200
Fax: 408-473-1222

(Chip sets, board level components and symmetrical multiprocessor systems)

X prefix
See *Xicor*, *Philips Semiconductor* (Custom part), *TSC America* (Motor motion control IC), *Harris Semiconductor* (Custom part), *Kyopal C. Ltd.*

XC prefix
See *Xlinx, Inc.*
(Motorola if Digital Signal Processor)

XDA to **XXXDA** prefix
See *GEC Plessy*

XE prefix
See *XECOM*

XECOM Inc.
374 Turquoise Street
Milpitas, CA 95035
408-945-6640, Fax: 408-942-1346
E-mail: info@xecom.com

(Telephone interface hybrid modules)

Xentek, Inc.
A Taiyo Yuden Company
1770 La Costa Meadows Drive
San Marcos, CA 92069
619-471-4001
Fax: 619-471-4021

(CCFT [cold cathode fluorescent tubes] inverters, DC/DC converters [for LCD modules], SIP units available; switching power supplies)

Xerox
Headquarters:
800 Long Ridge Road
P.O. Box 1600
Stanford, CT 06904
203-968-3000
Fax: 203-325-6078

(ICs for their own products)

Xicor, Inc.
1511 Buckeye Drive
Milpitas, CA 95035
408-432-8888
Fax: 408-432-0640
http://www.xicor.com
http://www.xicor.com/xicor/menulink/linkpdf.htm
E-mail: info@smtgate.xicor.com
BBS: 800-258-8864
(Download source code, etc., setting 8N1)

Northeast Area:
83 Cambridge St., Unit 1D
Burlington, MA 01803
617-273-2110
Fax: 617-273-3116

Southeast Area:
201 Park Place, Suite 201
Altamonte Springs, FL 32701
407-767-8010
Fax: 407-767-8912

XICOR — XYCOM

Mid-Atlantic Area:
50 North Street
Danbury, CT 06810
203-743-1701
Fax: 203-794-9501

North Central Area:
953 North Plum Grove Road
Suite D
Schaumburg, IL 60173
708-605-1310
Fax: 708-605-1316

South Central Area:
9330 Amberton Parkway
Suite 137
Dallas, TX 75243
214-669-2022
Fax: 214-644-5835

Southwest Area:
4141 MacArthur Boulevard, Suite 205
Newport Beach, CA 92660
714-752-8700
Fax: 714-752-8634

Northwest Area:
2700 Augustine Drive, Suite 219
Santa Clara, CA 95054
408-292-2011
Fax: 408-980-9478

Northern Europe Area:
Hawkins House
14 Black Burton Road, Carterton
Oxford 0X8 3QA
United Kingdom
44 993/844 435, Fax: 44 993/841 029
Telex: 851 838029

Central Europe Area:
Xicor GmbH
Technopark Neukeferloh
Bretonischer Ring 15
W-8011 Grasbrunn bei Muenchen
Germany
49 89/461 0080, Fax: 49 89/460 5472
Telex: 841 5213883

Southern Europe Area:
Xicor Sarl
27 Avenue de Fontainebleau
94270 Le Kremlin Bicetre
France
33 1/46 71 49 00
Fax: 33 1/49 60 03 32
Telex: 842 632160

Japan:
Xicor Japan K.K.
Suzuki Building, 4th Floor
1-6-8 Shinjuku, Shinjuku-ku
Tokyo 160, Japan
81 33/225 2004
Fax: 81 33/225 2319

Asia/Pacific Area:
Call the main headquarters in Milpitas, CA.

(EEPROMs, NOV RAMs (non-volatile RAMs), digital potentiometers, EEPOTs, security ICs)

XL prefix
See *Exel Microelectronics*

Xlinx, Inc.
2100 Logic Drive
San Jose, CA 95124-9920
1-800-231-3386
408-559-7778
Fax: 408-559-7114
E-mail: @xlinx.com
World Wide Web: http://www.xlinx.com
Europe: 44-932-349401
Japan: 81-3-297-9191
Asia: 852-3-720-0900

(A fabless company supplying reprogrammable FPGAs [including antifuse sea of gates devices], EPLDs, military screened devices available)

XPX prefix
See *Data Instruments, Inc.*

(Pressure transducers)

XR prefix
See *Exar Corp.*

XSIS Electronics, Inc.
12620 W. 63 St.
Shawnee, KS 66216
913-631-0448
Fax: 913-631-1170
TLX: 437227

(Oscillators, including parts qualified to MIL-O-55310)

XTAL Technologies, Ltd.
28 Millrace Drive
Lynchburg, VA 24502
804-385-8300
Fax: 804-385-8100

(Monolithic crystal filters, package filters, crystals)

XSTPSL prefix
See *SGS-Thomson Microelectronics*

XTR prefix
See *Burr-Brown*

(Transmitter)

XX part number
(On a SOT package)
See *Hewlett-Packard*

(MMIC amplifiers)

XXX number (AXX number)
(If voltage, low voltage dropout regulators)
See *Toko America, Inc.*

XXXXX/BEAJC, or **/BEBJC**
or similar part number suffix
See *Motorola*

Xycom, Inc.
750 N. Maple Road
Saline, MI 48176
800-367-7300
313-429-4971
Fax: 313-429-3543

Manufacturers, Prefixes, Part Number Types, Logo Descriptions & Family Types

(Digital ICs)

Yamaha Systems Technology, Inc.
100 Century Center Court
San Jose, CA 95112
1-800-543-7457, 408-467-2300
Fax: 408-437-8791
http://www.yamaha.com/

Head Office:
Yamaha Corporation
10-1, Nakazawa-Cho, Hamamatsu-Shi
Shizuoka 430
053 460 1111, Fax: 053 465 1105

(Logic, Vodem, FAX modem, speech digitizer, CD player ICs, PC audio and television ICs, FM and wavetable sound synthesizer, ISDN ICs, graphics controllers, etc.)

Yaskawa Electric Corporation
2-1, Kurosaki Shiroishi,
Kita-Kyushu-Shi
Fukuoka 806, Japan
093 645 8800, Fax: 093 631 8837

(Hybrid ICs)

YBG prefix
See *Siemens*

YGV prefix
See *Yamaha Systems Technology, Inc.*

YIM prefix
See *Optotek Ltd.*

YL prefix
See *Siemens*

YM prefix
See *Yamaha Systems Technology*

Yokogawa Electric Corporation
9-32, Naka-Cho 2-Chome
Musashino-Shi
Tokyo 180, Japan
0422 52 5555

(Hybrid ICs)

YSI, Inc.
P.O. Box 1476
Dayton, OH 45482-0264
800-765-4974
Fax: 513-767-9353
E-mail: Info@YSI.com

(Temperature sensors)

YTM prefix
See *Yamaha Systems Technology*

Yugengaisha Sakugi Densi Seisakusho
472, Shimo-Sakugi, Sakugi-Mura
Futami-Gun, Hiroshima 728-01
Japan
082 455 2311

(Car ICs)

Yutaka Electric Manufacturing Co., Ltd.
228, Kariyado, Nakahara-Ku,
Kawasaki-Shi,
Kangawa 211
Japan
044 411 2171
Fax: 044 435 4555

(Hybrid ICs)

Z prefix
See *Zilog, Inc.*, *Synertek* (Second source some products), *Advanced Micro Devices*, *Zyrel, Inc.* (SRAMS); *Omron Electronics, Inc.* (Displacement sensors)

Zahir
A trademark of Cariger, Inc.
(Tricolor programmable bar graph displays)

ZB prefix
See *MSI Electronics, Inc.*

XYCOM — ZETEX **X - Z**

ZC prefix
See *Zetex*

Z-Comm logo
See *Z-Communications*

Z-Communications
9939 Via Pasar
San Diego, CA 92196
619-621-2700
Fax: 619-621-2722

(Miniature voltage controlled oscillators)

ZDT prefix
See *Zetex*

ZDX prefix
See *Zetex*

Zentrum Mikroelektronik Dresden GmbH
Grenzstrasse 28
D-01009 Dresden, Germany
49 351 88 22 306
Fax: 49 351 88 22 337

(Nonvolatile memory, in partnership with Simtek Corp. on nonvolatile SRAM devices. Also a wafer foundry. See also Pacific Silicon Technologies)

Zetex, plc
Fields New Road
Chadderton, Oldham, OL9 8NP
United Kingdom
44 1161-627-5105 (Sales)
44 1161-627-4963
(General inquiries)
Fax: 44 1161-627-5467
TLX: 668038

Drosselweg 30
8000 Munchen 82
089 430 90 29
Fax: 089 439 37 64

Z

ZETEX — ZYREL

87 Modular Avenue
Commack, NY 11725
516-543-7100
Fax: 516-864-7630

47 Seabee Lane
Discovery Bay
Hong Kong
987 9555
Fax: 987 9595

(Transistors, diodes, MOSFETs, RF transistors, optoelectronic devices)

Zevex, Inc.
5175 Greenpine Drive
Salt Lake City, UT 84123
801-264-1001
Fax: 801-264-1051

(Ultrasonic [non-invasive] air bubble and fluid detectors)

ZIA
(A computer drive chip alliance of Zilog, Inc.; International Microelectronic Products, Inc.; and Allegro Microsystems, Inc., Located in San Jose, CA 800-722-CHIP)

Ziatech Corporation
1050 Southwood Drive
San Luis Obispo, CA 93401
805-541-0488
Fax: 805-541-5088

(Single board computers with some custom ICs)

Zilog, Inc.
210 East Hacienda Ave.
Campbell, CA 95008-6600
800-662-9826 (within the US)
408-370-8000
Fax: 408-370-8056
http://www.zilog.com

(Digital signal processors [DSP ICs], microcontrollers, peripheral circuits, multimedia, TV remote control ICs, TV captioning signal decoders, TV controllers for multilanguage displays, and PC audio products. Note: various discontinued military NMOS parts are still available from Zeus Electronics [an Arrow Company] 1-800-52-HI-REL)

ZM prefix
See *Zetex*

ZNA prefix
See *GEC Plessy*

Zoran Corporation
2041 Mission Blvd.
Santa Clara, CA 95054
408-986-1314
Fax: 408-986-1240
200 Reservoir Street
Suite 300
Needham Heights, MA 02194
614-449-5990

1821 Walden Office Square
Suite 426
Schaumburg, IL 60173
312-397-0710

Zoran Microelectronics, Ltd.
Advanced Technology Center
P.O. Box 2495
Haifa, 31024 Israel
011-972-4-533-175

(Digital signal processors [DSP ICs], digital filter processors, MPEG and JPEG chip sets, PCI multimedia controllers [jointly w/Siemens], Dolby surround sound digital decoding IC, etc. for medical, commercial and military applications. DVD add in cards)

ZPSD prefix
See *WSI, Inc.*

ZR prefix
See *Zoran Corp., Zetex Corp.*

ZS prefix
See *Zetex*

ZT prefix
See *Zetex*

ZTS prefix
See *Integrity Technology Corp.*

(Ceramic resonators)

ZTX prefix
See *Zetex*

ZVC prefix
See *Zetex*

ZVN prefix
See *Zetex, AVG Semiconductors*
(Second source devices)

ZVNL prefix
See *Zetex*

ZVP prefix
See *Zetex*

ZY prefix
See *Zyrel, Inc.*

Zycad Corporation
GateField Division
47100 Bayside Parkway
Fremont, CA 94538
510-623-4400
Fax: 510-623-4550
http://www.zycad.com (Zycad)

(Programmable gate arrays)

ZyMOS Corporation
See *Appian Technology, Inc.*

Zyrel, Inc.
Pruneyard Tower One, #680
Campbell, CA 95008
408-879-0199
Fax: 408-879-0442

(Successive approximation registers, SRAMs

ZYREL—Additional Notes

NOTES:

1. "X" in the listings represents any number.

2. Some numerical part numbers, when second sourced by other vendors, often have letters before the numerical prefix. For example, Intel has a 29C prefix, but various other vendors make this programmable device series, including Texas Instruments, with a TCM29C prefix and AMD with an AM29C prefix. Not all alternate vendors are shown.

LOGOS

All logos are the property of their respective owners and are reproduced herein for identification purposes only.

ABB HAFO — AMERICAN ELECTRONIC COMPONENTS, INC.

ABB Hafo

ALI
Acer, Inc.

ACON, INC.

Actel Corporation

Adaptec, Inc.

ADAPTIVE NETWORKS

Advanced Analog (Lambda)

Advanced Linear Devices

Advanced Hardware Architectures, Inc.

AMS
Advanced Memory Systems

Advanced Micro Devices, Inc. (AMD)

Advanced Micro Systems, Inc.

Advanced Power Technology (APT)

Advanced Semiconductor, Inc.

AEG
AEG Corporation

Aeroflex Laboratories, Inc.

ASi
Alden Scientific, Inc.

ALEPH INTERNATIONAL

Allegro Microsystems, Inc.

ALLEN-BRADLEY
A ROCKWELL INTERNATIONAL COMPANY

Y Alliance
Alliance Semiconductor Corporation

Alpha Industries, Inc.

ALPS Electric, Inc.

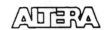
Altera Corporation

American Electronic Components, Inc.

Logos

AMERICAN MICRO SEMICONDUCTOR — ASTEC

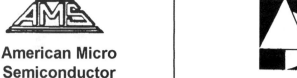
American Micro Semiconductor

AMI.

American Microsystems, Inc. (AMI)

American Power Devices, Inc.

American Zettler, Inc.

Amperex®
Amperex Electronic Corporation
(See Philips Components)

Amplifonix

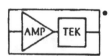
Amptek, Inc.

Anadigics, Inc.

Analog Devices

Analog Systems

Analogic Corporation

AND

Anderson Electronics, Inc.

Antel Optronics, Inc.

ANZAC

APC Ltd.

Apex Microtechnology Corporation

ARK Logic, Inc.

Aromat Corporation

Array Microsystems

◆ ASAHI GLASS CO.,LTD.

AKM
Asahi Kasei Microsystems Co., Ltd.

Astec Semiconductor

267

A - C AT&T — CHIPS AND TECHNOLOGIES

The Component Identifier & Source Book

AT&T Microelectronics

Atmel Corporation

Atmos Technology, Inc.

AuraVision Corporation

Avance Logic, Inc.

Avantek

AVX Corporation

Aydin Vector

BENCHMARQ

BETA TRANSFORMER TECHNOLOGY CORPORATION

BethelTronix, Inc.

BI Technologies Corporation
(Beckman Industrial Corp.)

BiT — Bipolar Integrated Technology, Inc.

BKC International Electronics, Inc.

Brooktree Corporation

BT&D TECHNOLOGIES

BusLogic

CALIFORNIA MICRO DEVICES

CAMBRIDGE ACCUSENSE, INC.

Capacitec

CARDINAL COMPONENTS INC.

CATALYST SEMICONDUCTOR

Centon Electronics

CENTRONIC
Centronic, Inc.

CHERRY SEMICONDUCTOR

CHINFA / Hotaihsing Powertech

Chips and Technologies, Inc.

CMD Technology, Inc.

Circuit Technology, Inc.
(GEC Plessy Semiconductors)

CIRRUS LOGIC

CKE

Clairex Electronics
(Clarostat Sensors and Controls)

CODI Corporation
(Discontinued)

Coherent, Inc.

Comlinear Corporation

Communications Instruments, Inc.

COMMUNICATION TECHNIQUES, INC.

Compensated Devices, Inc.

COMPONENT TECHNOLOGY

COMPUTER PRODUCTS

Concurrent Logic, Inc.

CONNOR WINFIELD

CONVERSION DEVICES, INC.

COUGAR COMPONENTS

CREE RESEARCH, INC.

Crosspoint Solutions, Inc.

CMD TECHNOLOGY — DALSA INC.

CRYSTAL Semiconductor Corporation

CRYSTAL

CTS Microelectronics Corp.

Cybernetic Micro Systems

Cypress Semiconductor

Daden Associates, Inc.

DAEWOO

DAICO INDUSTRIES, INC.

DALLAS SEMICONDUCTOR

DALSA INC.

The Component Identifier & Source Book

DATA DELAY DEVICES — EVERLIGHT

Data General

Datel

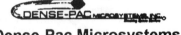
DATATRONICS

Dense-Pac Microsystems, Inc.

Densitron Corporation

DIODES INCORPORATED

Display International

DOLPHIN INTEGRATION

DowKey Microwave Corporation

DSP Communications, Inc.

DYMEC

EG&G

EIC SEMICONDUCTOR

élantec

Electronic Designs, Inc.

edi
Electronic Devices, Inc.

ELECTRONIC MEASUREMENTS, INC.

Electronic Technology Corporation

Elmwood Sensors

EM Microelectronic — Marin Sa

Emulex Micro Devices

Endicott Research Group, Inc.

EC^2 engineered components company

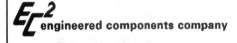
Epitaxx, Inc.

Epitex Electronics Ltd.

ERICSSON
Ericsson Components, Inc.

EUROM FlashWare Solutions

EVERLIGHT
Everlight Electronics Co.

Logos

EXAR — GENTRON E - G

Exar Corporation

EXEL Microelectronics

Exponential Technology, Inc.

Fagor Electronic Components, Inc.

F

F

Fairchild Semiconductor

Fermionics *Opto-Technology*

Ferranti Semiconductors
(GEC Plessy)

Film Microelectronics, Inc.

Fis, Inc.

Fox Electronics

Frequency Electronics, Inc.

Fuji Electric Co., Inc.
(Collmer Semiconductor)

FUJITSU

Fujitsu Microelectronics, Inc.

Galileo Technology, Inc.

GATEFIELD

GATEWAY PHOTONICS CORPORATION

General Electric
(Harris Corporation, Thomson Consumer Electronics)

GENERAL INSTRUMENT GI

GENERAL MICROWAVE

General Semiconductor Industries, Inc.
(Protek Devices, General Instrument Power Semiconductor Div.)

Genesis Microchip, Inc.

GENNUM CORPORATION

Gentron Corporation

G - I GERMANIUM — INDUSTRIAL DEVICES

Germanium Power Devices Corporation

Gigabit Logic
(Triquint Semiconductor, Inc.)

Gilway Technical Lamp

GlobTek Inc.

GoldStar Gold Star Co., Ltd.

GoldStar Semiconductor, Ltd.

Greenwich Instruments Ltd

GTE Corporation

Guardian Electric Manufacturing Co.

Hammond Manufacturing

Harris Semiconductor

Hayashi Denkoh Co., Ltd.

HEI inc.

Herotek, Inc.

HEWLETT PACKARD

Hexawave, Inc.

HI-SINCERITY MICROELECTRONICS

Hitachi

Hittite Microwave Corporation

Hokuriku Electric Industry Co., Ltd.

Holmate Technology Corporation

HUGHES
Hughes Semiconductor Products Center

Hybrid Systems

Hybrid Systems Corporation
(Sipex Corporation)

HY-CAL Engineering

HyComp, Inc.

I-Cube

IC DESIGNS

Idea, Inc.

ISI IDEAL SEMICONDUCTOR, INC.

DDC
ILC DDC
(Data Device Corporation)

Industrial Devices, Inc.

Logos

Industrial Electronic Engineers, Inc.

INFINITE TECHNOLOGY CORPORATION

Information Storage Devices, Inc.

inmos

Intech Microcircuits
(Lambda Advanced Analog)

Integrated Circuit Systems, Inc.

Integrated Circuit (IC) Works

Integrated Circuits Incorporated

Integrated Device Technology, Inc.

Integrated Power Semiconductor, Ltd.

Integrated Silicon Solution Inc.

intel

Intelligent Motion Systems, Inc.

INTERNATIONAL CMOS TECHNOLOGY, INC

International Microelectronic Products

International Rectifier

Intersil
(Harris Corporation)

INDUSTRIAL ELECTRONIC ENGINEERS — ITI FERRO TEC

Intronics, Inc.

IQ Systems, Inc.

Isocom PLC

ISO LINK Inc.

Isotek Corporation
(Alpha Resistors)

ISOTEMP RESEARCH, INC.

ITI Ferro Tec

I - L

IXYS — LOGITEK, INC.

IXYS Semiconductor Corporation

jbm ELECTRONICS, INC.

JKL Components Corporation

JRC New Japan Radio Co., Ltd.
NJR Corporation (A Subsidiary of New Japan Radio Co., Ltd.)

Kawasaki LSI

KDS America
a Daishinku Corporation

KEC Korea Electronics

Keystone Thermometrics

KOA Speer Electronics, Inc.
KR Electronics, Inc.

Krypton Isolation, Inc.

Kyocera Corporation

LakeShore Cryotronics, Inc.

Lambda Electronics Inc.

Lansdale Semiconductor, Inc.

Lansdale Semiconductor, Inc.

Laser Diode Incorporated
Laser Diode Products, Inc.
(Laser Diode, Inc.)

LDP Laser Diode Products Inc.

Lattice Semiconductor Corporation

Leecraft
Leecraft Manufacturing Co., Inc.
(Lighting Components & Designs)

Level One Communications, Inc.

Linear Technology

LinFinity Microelectronics

Lite-On, Inc.

Logic Devices Incorporated

Logitek, Inc.

Logos

LSI — MITEQ   L-M

LSI COMPUTER SYSTEMS INC.

LTI

Lucas Deeco Corporation

Lumex Opto/Components, Inc.

MX MACRONIX, INC

Marconi Electronic Devices, Ltd.
(GEC Plessy)

marktech international, corp.

Matra Harris

Matsushita

MAXIM

MEDIA VISION
Media Vision Technology, Inc.
(Aureal Semiconductor)

MELCHER Ⓜ
Melcher AG

Mercury United Electronics, Inc.

Merrimac Industries, Inc.

MICREL
Micrel Semiconductor, Inc.

MICROCHIP
Microchip Technology, Inc.

Micro Linear

MicroModule Systems

MicroNetworks

Mii
Micropac Industries, Inc.

Micro Power Systems

M/A-COM

Microwave Associates
(M/A-Com)

MSC
Microwave Semiconductor Corporation
(SGS-Thomson)

MicroWave Technology, Inc.

Mitel Semiconductor

MITEQ

M – N MITSUBISHI — NEWPORT

The Component Identifier & Source Book

Mitsubishi

Mitsumi Electric Co., Ltd.

Modular Devices, Inc.

Monolithic Memories, Inc.
(Advanced Micro Devices [AMD])

Monolithic Sensors, Inc.

Morgan Matroc, Inc.

Mos Technology
(Commodore Semiconductor)

Mosel Vitelic

Motorola

MSK
M.S. Kennedy Corporation

M-Systems
Flash Disk Pioneers

M-tron
M-Tron Industries, Inc.

MURATA ERIE NORTH AMERICA

MUSIC SEMICONDUCTORS

NATIONAL INSTRUMENTS

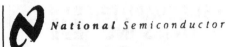
National Semiconductor

National Semiconductor

NS
National Semiconductor

NCR
NCR Microelectronics Products
(Symbios Logic, Inc.)

nCube
nCube Corporation

NEC
NEC Electronics Inc.

NES
NEW ENGLAND SEMICONDUCTOR

Neuralogix, Inc.
(Adaptive Logic, Inc.)

NEWBRIDGE
Newbridge Microsystems
(Tundra Semiconductor Corp.)

newport
Newport Technology

Logos

NEXCOM — PANJIT N - P

Nexgen Microproducts, Inc.

NIC Components Corp.

Nippon Steel Semiconductor Corporation

NMB Semiconductor Co., Ltd.
(Nippon Steel Semiconductor)

NTE (New Tone) Electronics, Inc.

Okaya Electric Industries Co., Ltd.

Oki Semiconductor

Omnirel Corporation

OMNIYIG, INC.

OPTEK

Optek Technology, Inc.
(Optron)

Opto Technology, Inc.

ORBIT SEMICONDUCTOR, INC.

Origin

Origin
Origin Electric Co., Inc.

Omega Micro, Inc.

OXLEY
Oxley, Inc.

PMC
PACIFIC MICROELECTRONICS CORPORATION

PACIFIC MONOLITHICS

Panametrics

PANJIT SEMICONDUCTOR

OAK TECHNOLOGY, INC.

Ohmtek
(A company of Vishay)

277

PARADIGM TECHNOLOGY — PULSE MICROWAVE

Paradigm Technology, Inc.

Parametric Industries

Performance Motion Devices

Performance Semiconductor Corporation

Pericom Semiconductor Corporation

Phihong Enterprise Co., Ltd.

Philips Semiconductor

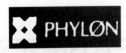
Photron Semiconductor

Phylon, Inc.

Piezo Technology, Inc.

Pilkington Microelectronics, Ltd.

Pixel Semiconductor
A Cirrus Logic Company

Plessy Semiconductor Corp.
(GEC Plessy Semiconductors)

PLUS LOGIC
Plus Logic
(Xilinx)

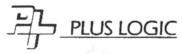
Polara Engineering, Inc.

Powerex, Inc.

Power Integrations, Inc.

POWER TRENDS

PPC Products Corporation

Precision Monolithics Inc.
(Analog Devices)

Premier Magnetics Inc.

PROTEK DEVICES

Pulse

pulse Microwave

Logos

Quality Semiconductor, Inc.

QuickLogic Corporation

REALTEK
Realtek Semiconductor Corporation

Reliability, Inc.

RF Products, Inc.

RHOMBUS INDUSTRIES INC.

RLC ELECTRONICS, INC.

Rockwell International

Rohm Electronics Division

RSM Sensitron Semiconductor

SAMSUNG Semiconductor

SanKen
Sanken Electric Co., Ltd.

SANYO
Sanyo Semiconductor Corporation

QT — SEIKO-EPSON Q - S

Saratoga Semiconductor

Saratoga Semiconductor Corporation

SAWTEK INCORPORATED

SCHURTER

Sciteq Electronics, Inc.

SDL

Seagate Microelectronics Limited

SEASTAR OPTICS INC

Seiko-Epson, Epson America, Suwa Seikosha Co.

SEIKO INSTRUMENTS — SILICON GENERAL

Seiko Instruments, Inc.

Semicoa Optoelectronics

Semicon, Inc.

Semiconductor Technology, Inc.

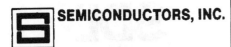
SEMICONDUCTORS, INC.

Semitronics Corp.

Sensitron Semiconductor
(RSM Sensitron Semiconductor)

Sensory Circuits, Inc.

SEPONIX CORPORATION

SGS-Thomson
(Mostek)

SGS-Thomsom Microelectronics

Shindengen Electric Manufacturing Co., Ltd.

Siemens

SIERRA SEMICONDUCTOR

Signal Processing Technologies

signetics

Silicon General, Inc.
(Linfinity Microelectronics, Inc.)

Logos

SILICON STORAGE TECH. — STANDARD MICROSYSTEMS

Silicon Storage Technology, Inc.

Silicon Sensors, Inc.

Solitron Devices, Inc.

SILICON SYSTEM INCORPORATED

Skan-A-Matic Corporation
(Clarostat Sensors and Controls)

 SOLOMON
Solomon Technology Corporation (USA)

SILICON TRANSISTOR CORP.

S-MOS SYSTEMS

S·MOS SYSTEMS
SMOS Systems
(Seiko Epson)

Space Electronics, Inc.

Siliconix, Inc.

 Sokol
Crystal Products, Inc.

SPECTRIAN

S
Sipex Corporation

Solid State, Inc.

SPRAGUE
THE MARK OF RELIABILITY
Sprague Semiconductor
(Allegro Microsystems)

Silicon Integrated Systems

Solid State Devices, Inc.

ST Olektron Corporation

SIS Microelectronics, Inc.

Solid State Scientific
(Allegro Microsystems)

S-T STANFORD — TEXAS INSTRUMENTS

Stanford Microdevices

Stanford Telecommunications
(Stanford Telecomm)

Stanley Electric Co.

Star Semiconductor Corporation
(Logic Devices, Inc.)

State of the Art, Inc.

Statek Corporation

STC Components

Summit Instruments, Inc.

Supertex Inc.

Symbol

SYNERGY SEMICONDUCTOR

SYNERTEK
Synertek, Inc.

Syntar Industries, Inc.

Taiwan Semiconductor Manufacturing Co., Ltd.

Taiyo Yuden Co., Ltd.

TDK Corporation

TECCOR ELECTRONICS, INC.

TelCom Semiconductor, Inc.

TELE QUARZ GROUP

Teledyne

TELEFUNKEN electronic

TELTONE
Teltone Corporation

TEMEX
Temex Electronics

TEXAS INSTRUMENTS

Logos

Thaler Corporation

Third Domain, Inc.

Thomson CSF

Three-Five Systems, Inc.

Toko, Inc.

TOSHIBA

Toshiba Corporation

TOWER ELECTRONICS INC.

Trak Microwave Corporation

Transducer Techniques

Transistor Specialties Inc.
(TSI Microelectronics)

Transwitch Corporation

Trident Microsystems, Inc.

TRILITHIC

TriQuint SEMICONDUCTOR

TRW
TRW LSI Products
(Raytheon Semiconductor)

Tundra Semiconductor Corporation
(Newbridge Microsystems, Inc.)

Uchihashi Estec Co., Ltd.
(Elcut)

THALER — V3 CORPORATION T - V

UDT SENSORS, INC.

uniphase

UMC
UNITED MICROELECTRONICS CORP

UNITED TECHNOLOGIES MICROELECTRONICS CENTER

UNITRODE

UNIVERSAL MICROELECTRONICS
Universal Microelectronics Co., Inc.

unizon
KOMATSU GROUP

V^3 Corporation
a wholly owned subsidiary of V3 Inc

V - Z VANGUARD — ZORAN

Vanguard Electronics Co., Inc.

VARO INC

Ventronics, Inc.

Verdure Industries, Inc.

VLSI Technology, Inc.

VLSI Technology, Inc.

VTC Incorporated

Waferscale Integration, Inc.

Weitek Corporation

Westcode Semiconductors, Inc.

The Western Design Center, Inc.

Western Digital Corporation

Westinghouse Electric Corporation

White Technology, Inc.
(White Microelectronics)

Wickman USA, Inc.
(Wickman-Werke)

Wilcoxon Research

xecom

Xicor, Inc.

Xilinx, Inc.

ZETEX

ZIA

Zilog, Inc.

ZORAN
Zoran Corporation

PRODUCT LISTINGS

While we have strived to make the following list complete, it may not be all inclusive of the devices made by the manufacturers contained in this publication.

In addition, for parts that have many sources of interchangeable products (such as capacitors, thermistors, resistors, etc.), not all vendors are listed in this publication.

With product lines constantly being added or discontinued it is impossible to track all the devices offered by any manufacturer. The Author and Publisher disclaim any liability or responsibility for errors in the listing.

ACCELEROMETERS — ASICS

Accelerometers:

Analog Devices (Silicon types)
Columbia Research Laboratories, Inc.
Dytran Instruments, Inc.
EG&G IC Sensors, EG&G Reticon
Endevco
Entran Devices, Inc.
Jewell Electrical Instruments
Kistler Instrument Corporation
Lucas NovaSensor
PCB Piezotronics, Inc.
Robert Bosch Corporation
Sensotek, Inc.
Setra Systems, inc.
Silicon Designs, Inc.
Silicon Microstructures Inc. (SMI) (Exar Corp.)
Summit Instruments, Inc.
Wilcoxon Research

Amplifiers (and Op Amps):

Analog Devices
Advanced Laser Devices
Advanced Milliwave Laboratories
Advanced Photonics, Inc.
Alpha Industries
Amplica, Inc.
Amplifonix
Amtek, Inc.
Anadigics, Inc.
Analog Systems
Anzac
Apex Microtechnology
BT&D Technologies (Optical)
Penstock, Inc.
Burr-Brown Corp.
Calogic Corporation
Celeritek
Columbia Research Laboratories, Inc. (Strain gauge)
Comlinear Corp.
Cougar Components
Daewood Electronic Components Company Ltd.
Datel, Inc.
DBS Microwave, Inc.
Dexter Research Center, Inc.
EMF Systems, Inc.
Elantec, Inc.
Electro-Optical Systems, Inc.
EM Research Engineering, Inc.
FEI Communications, Inc.
Film Microelectronics, Inc.
Graseby Optronics (Graseby Infrared)
Harris Semiconductor (Including video)
Herotek, Inc.
Hitachi America, Ltd.
Hittite Microwave Corporation
InterFET-ITAC
ITS Electronics, Inc.
New Japan Radio Corporation
JCA Technology, Inc.
KDI/triangle Electronics, Inc.
Kulite Semiconductor Products (Transducer)
LG Semicon
Linear Technology Corporation
Linfinity Microelectronics, Inc.
Litton Solid State
LTI
Maxim Integrated Products
Melles Griot
Memtech Technology Corporation (Sense)
MICRA Corp.
Micro Networks
Microwave Solutions, Inc.
Motorola, Inc.
M.S. Kennedy Corporation (MSK)
MX-Com, Inc.
National Hybrid, Inc.
Omnirel Corp.
On Chip Systems
Opamp Labs, Inc.
Optical Communication Products, Inc.
Optical Electronics, Inc. (OEi)
Petrond Microwave
Philips Semiconductors - Signetics
Phoenix Microwave Corp.
Plessy Tellumat
Q-bit Corporation
Raytheon Company
RF Micro-Devices, Inc.
Samsung Microwave Semiconductor, Inc.
Sanken Electric Co. Ltd.
Siemens Components, Inc.
Signal Processing Technologies (SPT)
Silicon Systems, Inc. (SSI)
Solid State (Microtechnology) Electronics Co. Pvt. Ltd.
Space Power Electronics, Inc.
Spectrian
System Microelectronic Innovation GmbH.
Tektronix, Inc. (Video)
Teledyne Solid State
Texas Instruments, Inc. (TI)
Texas Optoelectronics, Inc.
Third Domain, Inc. (Video)
TriQuint Semiconductor, Inc.
UDT Sensors, Inc.
Unitrode Integrated Circuits Corporation
Veritech Microwave
VTC Inc.
Watkins-Johnson Company

ASICS:

Asahi Kasei Microsystems
AVG Semiconductors
BethelTronix, Inc. (BTI)
BMR Labs, Inc.
Chia Hsin Livestock Company
Chip Express Corporation

Cypress Semiconductor
Elmos Electronik GmbH
EM Microelectronic-Marin S.A.
Fujitsu Microelectronics, Inc.
Future Technology Devices
    International. Ltd. (FTD)
GENNUM Corporation
Hitachi
Honeywell Solid State Electronics
    Center (SSEC)
Integrated Circuit Systems, Inc.
    (ICS)
KAOS Semiconductor
KMOS Semi-Custom Designs
LSI Logic Corporation
Lucent Technologies
Maxtek Components Corporation
Microchip Technology, Inc.
Motorola
NEC Electronics, Inc.
OnSpec Electronic, Inc.
Orbit Semiconductor, Inc.
Philips Semiconductors - Signetics
Qualcomm, Inc.
RTG, Inc.
San Francisco Telecom, Inc.
Semefab, Ltd.
SICOM, Inc.
Silicon Composers, Inc.
Sipex Corporation
Symbios Logic
Tatung Company
TI
TLSI, Inc.
Toshiba Corp.
TriTech Microelectronics Pte. Ltd.
VLSI Technology

Attenuators:

Alpha Industries, Amplifonix
Barry Industries, Inc.
Component General, Inc. (CGI)
Daico Industries, Inc.

General Microwave
JFW Industries Inc.
Kay Elemetrics, Corp.
KDI/triangle Electronics, Inc.
Litton Solid State
Loral Microwave-Narda
Lucas Weinschel
M/A (Microwave Associates)
Com Semiconductor (Silicon)
Products, Inc.
M/A-COM Gallium Arsenide
Products, Inc.
Maxtek Components Corporation
Merrimac Industries, Inc.
Mini-Circuits
Miteq
Samsung Microwave
Semiconductor, Inc.
Synergy Microwave Corporation
Telonic Berkeley, Inc.
Trak Microwave Corporation
Trilithic
Waveline Solid State, Inc.
Wavetek Corporation

Audio ICs:

AMD
Analog Devices
Asahi Glass Co., Ltd.
Asahi Kasei Microsystems, Inc.
Alanta Signal Processors, Inc.
Audio Digital Imaging, Inc.
Aureal Semiconductor
Avance Logic, Inc.
Burr-Brown
C&S Electronics
Chrontel, Inc.
Consumer Microcircuits, Ltd.
    (Filters)
Crystal Semiconductor
Echo Speech Corporation
E-mu Systems, Inc.
ESS Technology, Inc.
Exar Corporation
Fujitsu
Hamai Electric Lamp Co. Ltd.

ASICS — AUTOMOBILE ICs

Hitachi
Integrated Circuit Systems, Inc.
(ICS)
ITT Semiconductors
Matsushita
Mitsubishi
Mitsumi
NEC
New Japan Radio
Newbridge Microsystems, Inc.
Nikkoshi Co. Ltd.
Nippon Precision Circuits Ltd.
Oak Technology, Inc. (OTI)
Oai Electric
Opti Inc.
Philips Semiconductors - Signetics
Ricoh Company Ltd.
ROHM Electronics
Sakata USA Corporation
Sanken Electric
Sanyo Electric
SGS-Thomson Microelectronics
Sierra Semiconductor
Sigma Designs, Inc.
Silicon Magic Corporation
SMK Corp.
Sony Corporation
Texas Instruments, Inc. (TI)
Toko, Inc.
Tokyo Ko-On Denpa Co. Ltd.
Toshiba Corp.
TriTech Microelectronics Pte. Ltd.
    (Semiconductor)
Yamaha Systems Technology, Inc.
Zilog, Inc.

Automobile ICs:

Analog Devices
Daimler Benz (Temic)
Delco Electronics
Fuji Electric
Fujitsu Ltd.
Hamai Electric lamp. Co., Inc.

A - C AUTOMOBILE ICs — CCDs

Hitachi
Intel
ITT Semiconductors
Matsushita
Mitsubishi
NEC
New Japan Radio Corp.
Motorola
Nippon Precision Circuits, Inc.
Oki Electric
Omron Corp
Sakugi Densi Seisakusho
Sanken Electric
Sanyo Electric
TI
Toki, Inc.
Toshiba Corp

Avalanche Photodiodes:

Advanced Photonics
Electron Tubes, Inc.
Janos Technology, Inc.
Texas Optoelectronics, Inc.

Bar graphs:

Cariger, Inc.
IDEA, Inc.
Ledtronics, Inc.
Lumex Opto/Components, Inc.
Marktech International Corp.

Battery Management/ Charger ICs:

Avasem Corporation
Benchmarq Microelectronics, Inc.
Dallas Semiconductor (Battery charger ICs)
Enstore R&D GmbH (Battery charger/monitor)
IDV Solutions O2Micro
Operating Technical Electronics

Sanyo Energy (USA) Corp.
Seiko-Epson
SGS-Thomson Microelectronics
Shoreline Electronics, Inc.
TelCom Semiconductor, Inc.
Bubble Memory:
Memtech Technology Corporation

BUS Controllers:

ACC Microelectronics Corp.
Aeroflex labs (1553)
Appian Technology
Logic Devices, Inc.
National Hybrid, Inc.
Smar Research Corporation
Unicorn Microelectronics Corporation
Via Technologies, Inc.
VLSI Technology Inquiries

Camera ICs:

Fujitsu Microelectronics, Inc.
Hitachi
Minolta Camera Co. Ltd.
Mitsumi Electric Co. Ltd.
Motorola
NEC
New Japan Radio Corp.
Rhythm Watch Company, Ltd.
SMK Corp.
TI
Toshiba
VVL (VLSI Vision Ltd.)

Capacitors:
(Note: This reference contains only a few sources for capacitors as many vendors have parts that are interchangeable and identification of a specific vendor is often not necessary)

Alpha Industries
AVX Corporation
California Micro Devices (CMD)

Cornell Dubilier
Dale Electronics, Inc.
Diablo Industries
Dielectric Laboratories, Inc.
GTE Corporation
Interpoint Corp.
ITT Components
KOA Corp.
Mallory
Mini-Systems, Inc. (MSI)
Murata Mfg. Company (Headquarters)
MuRata/Erie North America
NTE (New Tone) Electronics, Inc.
NIC Components Corp.
Nichicon Corporation
Nissei Denki Pvt., ltd.
Philips Components, Discrete Products Division
RF Electronics, Inc.
Robert G. Allen Co.
ROHM Electronics
Semi Dice, Inc.
Semtech
SMEC, Inc.
Soshin Electric Co. Ltd.
Sprague-Goodman (Trimmers)
Taiyo Yuden Co. Ltd.
TDK Corp.
Tokin Corporation
Ventronics, Inc.
Voltronics Corporation (Trimmers)

CATV ICs:

Anadigics, Inc.
BethelTronix, Inc. (BTI)
Motorola, Inc.
Uniphase Telecommunications Products (UTP)

CCDs:

Dalsa Inc.
Eastman Kodak Co.
EG&G Solid State Products

Group
Hualon Microelectronics Corp. (HMC)
Kinseki, Ltd. (CCD delay lines)
Lockheed Martin Fairchild
Thomson Components and Tubes Corporation

CD ICs:

Acer, Inc.
Axis Communications
Burr-Brown
Chrontel, Inc. (Clock Synthezider for Video-CD ICs)
Echo Speech Corporation
Fujitsu
Hitachi
Matsushita Electric Industrial Co. Ltd.
Mitsubishi
Mitsumi Electric Co. Ltd.
Motorola
New Japan Radio Corp.
Nikkoshi Co. Ltd.
Nippon Precision Circuits Ltd.
Oak Technology, Inc. (OTI)
Oaki Electric
On Chip Systems
Rancho Technology, Inc.
Rohm
Sanyo Electric
Sigmax Technology, Inc.
Sony Corporation
Texas Instruments, Inc. (TI)
United Microelectronics Corp.
Victor Company of Japan, Ltd.
Winbond Electronics Corp.
Yamaha Systems Technology, Inc.

Cellular ICs:

American Microsystems, Inc.
Analog Devices Asaha Kasei Microsystems Co. Ltd.,
BethelTronix, Inc. (BTI)
Celeritek
CommQuest Technologies, Inc.
Display Technologies, Inc. (DTI)
DSP Communications, Inc.
GEC Plessy Semiconductors
Hitachi America, Ltd.
Intellon Corporation
LTI
M/A (Microwave Associates) Com Semiconductor (Silicon) Products, Inc.
M/A-COM Gallium Arsenide Products, Inc.
Matsushita Electric Industrial Co. Ltd.
Micronetics
MX-Com Inc.
National Semiconductor
Oki Semiconductor
Pacific Communications Sciences, Inc. (PCSI)
Penny Technologies, Inc.
Peregrine Semiconductor Corporation
Philips Semiconductors - Signetics
Plessy Tellumat
Pulse
Punjab Semiconductor Devices
Punjab Wireless Systems, Ltd.
San Francisco Telecom, Inc. (SFT)
SiTEL Sierra B.V.
Sony Corporation
Stanford Telecom (Telecommunications)
Teledyne Solid State
Unisys Corp.
VLSI - VLSI Technology Inc.
Wireless Logic Inc. (WLI)
Wolfson Microelectronics

Circuit Protection Devices/ Transient Voltage Suppressors:
(See also MOVs)

AVX Corporation
Diodes, Inc.
Electronics Industry (USA) Co. Ltd. (EIC)
Ericsson Components AB
FCI Components
Frederick Components International Ltd.
General Instrument
Harris Semiconductor
International Semiconductor, Inc.
Marcon Electronics Co., Ltd.
Okaya Electric Industries Co. Ltd.
Paccom Electronics
Phoenix Contact, Inc.
Protek Devices
RF Electronics, Inc.
Semicon Components, Inc.
Semtech
Sussex Semiconductor, Inc.
Voltage Multipliers, Inc.
Wickmann USA, Inc.

Circulators/Isolators:

Aerotek Co. Ltd.
Dorado International Corporation
FDK America, Inc.
Loral Microwave-Narda
M/A (Microwave Associates) Com Semiconductor (Silicon) Products, Inc.
M/A-COM Gallium Arsenide Products, Inc.
Trak Microwave Corporation

Clock Circuits:
(See also RTCs, Real Time Clocks)

Applied Microcircuits Corp.
Asti Pacific Corp.
AVX
Chrontel, Inc.
Cypress Semiconductor
EM Microelectronic-Marin S.A.

CLOCK CIRCUITS — COMPARATORS

Fujitsu Microelectronics, Inc.
IC Designs
IC Works (ICW)
Integrated Circuit Systems, Inc. (ICS)
Lansdale Semiconductor, Inc.
Matsushita Electric Industrial Co. Ltd.
MicroClock
Micro Networks
MSK Corporation
New Japan Radio Company, Ltd.
Nippon Precision Circuits Ltd.
Oki Semiconductor
Pericom Semiconductor Corp.
PLX Technology
Premier Magnetics, Inc.
Ricoh Company Ltd.
Seiko-Epson
SMK Corp.
Synergy Semiconductor Corp.
Texas Instruments, Inc. (TI)
TriQuint Semiconductor, Inc.
Unicorn Microelectronics Corporation
Vitesse Semiconductor Corp.

CODECS:

AMD
Analog Devices
Asahi Kasei Microsystems Co., Ltd.
Audio Digital Imaging, Inc.
Chrontel, Inc. (Multimedia audio)
Consumer Microcircuits, Ltd.
Digital Equipment Company (DEC) (Video)
Digital Voice Systems, Inc. (DVSI) (Voice)
ESS Technology, Inc.
Fincitec OY
GEC-Plessy (A-Law)
Harris Semiconductor (A-Law)
LSI Logic Corporation (Reed-Solomon)
MX-Com Inc. (CVSD, voice)
Oki Semiconductor (Telecommunications)
Samsung (Telecommunications)
Siemens (Telecommunications)
TriTech Microelectronics Pte. Ltd. (Semiconductor)

Chokes/Inductors/Transformers:

Allen Avionics
American Precision Industries
APC Ltd.
Associated Components Technology
AVX Corporation (Thin film inductors)
Beta Transformer Technology
Coilcraft
Dale Electronics, Inc.
Datatronics
Delta Electronics, Inc.
Ecliptek Corp.
Electronic Precision Components
Electronic Techniques (Anglia), Ltd. (ETAL)
Elytone Electronics Co., Ltd.
Engineered Components Co. (EC2)
Excel Cell Electronic Co., Ltd. (ECE)
Fer Rite Electronics Ind. Co., Ltd.
Group West International
Hammond Manufacturing
Integrity Technology Corp. (Fax/modem couplers)
James Electronics (Equipment and Supplies), Inc.
JBM Electronics, Inc.
Jerome Industries Corporation
J.W. Miller Magnetics
Kappa Technologies, Inc. (Kappa Networks, Inc.)
KOA Corp.
Lik Hang Electronic Co. Ltd. (LH)
MagneTek
Magnetic Circuit Elements, Inc.
Magnetico, Inc.
Manutech, Inc.
Midcom
Mini-Circuits
Nano-Pulse Industries
Newport Technology, Inc.
NIC Components Corp.
OPT Industries, Inc.
PCA Electronics, Inc.
PICO Electronics, Inc.
PREM Magnetics, Inc.
Premier Magnetics, Inc.
Pulse
RCD Components, Inc.
RF Prime, Inc.
Rhombus Industries, Inc.
Rubycon America, Inc.
Schaffner EMC, Inc. (for EMI suppression)
Shogyo International Corp.
SMEC, Inc.
Sprague-Goodman
Standex Electronics
Talema Electronic, Inc.
Tamura Electric Works Ltd
Tat Sing Electronics Co., Ltd.
TDK Corp.
Tokin Corporation
Universal Microelectronics Co., Ltd. (UMEC) (for ISDN)
Valor Electronics, Inc.
Vanguard Electronics Co., Inc.
Vari-L Company, Inc. (VARIL)
Vitec Electronics Corp. (EMI suppression)
White-Rogers

Comparators:

Advanced Linear
AMD
Allegro Microsystems
Analog Devices

Product Listings

Analog Systems
Brooktree
Burr-Brown
GEC Plessy
LG Semicon
Harris Semiconductor
Holt
LeCroy
Linear Technology
Maxim
Mitsubishi
Motorola
National Semiconductor
New Japan Radio Corp.
Performance Semiconductor
Philips Semiconductor
Quality Semiconductor
Raytheon
Rohm
Samsung
Sanyo
SGS-Thomson
TI
TLSI
Toshiba
Unitrode
VTC

Converters, V/F-F/V:
(Voltage-Frequency, Frequency-Voltage)

Advanced Analog
Analog Devices
Burr-Brown
Datel
Exar
SGS-Thomson
Sharp
Micro Networks
National Semiconductor
Raytheon
TelCom

Crystals (and Crystal Oscillators):

Alpha Components
Andersen Labs
AVX
Amplifonix
Anderson Electronics, Austron, Inc.
Bliley Electric Company
Bomar Crystal Company
Cal Crystal Lab., Inc. Comclok, Inc.
Cardinal Components
Connor-Winfield Corporation
Crystek Crystal Corp.
CTS Knights, Inc.
Dale Electronics, Inc.
Datum, Inc.
Ecliptek Corp.
ECS Inc.
EMF Systems, Inc.
Electro Dynamics Crystal Corporation (EDC)
FOX Electronics
Frequency Electronics, Inc.
Harris Semiconductor
Hong Kong Crystals Company (HKC)
Hybrids International, Ltd.
Hy-Q International (USA), Inc.
International Crystal Mfg. Co., Inc. (ICM)
Isotemp Research Inc.
ITT Components
KDS America
Kinseki, Ltd.
KTS Electronics
KVG GmbH
Kyocera Corp.
Lap-Tech, Inc.
M-tron Industries, Inc.
Mercury United Electronics, Inc.
MF Electronics Corp.
Micro Crystal
Milliren Technologies, Inc. (MTI)
Monitor Products Co. Inc.
Motorola

COMPARATORS — DC/DC

Murata Mfg. Company (Headquarters)
MuRata/Erie North America
MuRata Electronics
NEL Frequency Controls, Inc.
Oak Frequency Control Group (OFC)
Piezo Crystal Co.
Piezo Technology, Inc. (PTI)
Precision Devices, Inc. (PDI)
Q-Tech Corporation
Raltron Electronics Corporation
SaRonix
Siward Crystal Technology Co. Ltd.
Sokol Crystal Products, Inc. (SCP)
S.P.K. Electronics Co., Ltd.
Statek Corporation
STC Components
Techtrol Cyclonetics
Tele Quartz GmbH (TQE)
Temex Electronics
United Chequers (Hong Kong) Ltd.
U.S. Crystal Corporation
Xtal Technologies, Ltd.

DC/DC:

Abbott Electronics, Inc.
Acon, Inc.
Acopian
Lambda Advanced Analog
Anadigics
Analog Devices
Analog Modules
Apex Microtechnology
ASTEC America
AT&T Power Systems
Avionic Instruments
Burr-Brown Corp.
Calex Manufacturing Co. Inc.
Chinfa/Hotaihsing Powertech

DC/DC — D/A and A/D CONVERTERS

Computer Products, Power Conversion
Conversion Devices, Inc. (CDI)
Cosel U.S.A. Inc.
D1 International, Inc.
Datel, Inc.
Delta Electronics, Inc.
EG&G Power Systems, Inc.
Elantec, Inc.
ELDEC Corporation
Ericsson Components AB
ETRI, Inc.
FCI Components
Forton/Source
Gamma High Voltage Research, Inc.
HDL Research Lab, Inc.
IBM Corp.
ILC-DDC Data Device Corporation
International Power Devices, Inc. (IPD)
International Power Sources, Inc.
Intronics, Inc.
JBM Electronics, Inc.
JT PowerCraft, Inc.
Kappa Technologies, Inc. (Kappa Networks, Inc.)
Kepco Inc.
Lambda Electronics, Inc.
Lambda Advanced Analog
Linear Technology Corporation
Lockheed Martin Federal Systems
LZR Electronics, Inc.
Maxim Integrated Products
Melcher Inc.
Micro Linear Corp.
Micropac Industries, Inc.
Modular Devices, Inc. (MDI)
Murata Mfg. Company (Headquarters)
MuRata/Erie North America
Nano-Pulse Industries
Natel Engineering Co.

National Semiconductor Corporation
Newbridge Microsystems, Inc.
Newport Technology, Inc.
Nidec/Power Gerneral
Omega Power Systems, Inc.
Omega Research Limited
Omnirel Corp.
Orion Industries, Inc.
Pentastar Electronics, Inc.
Polytron Devices, Inc.
Powercube Corp.
Power Convertibles
Power General
Power Sensors Corporation
Power Solutions, Inc.
Power Systems, Inc.
Premier Magnetics, Inc.
Raytheon Company
Reliability, Inc.
RO Associates
Semiconductor Circuits, Inc.
Semtech
Sierra West Power Systems
Sigmapower
S-MOS Systems, Inc.
Technipower, Inc.
Tektris Electro Corporation
TelCom Semiconductor, Inc.
Teledyme Microelectronics
Teslaco-Optimum Power Conversion
Total Power International, Inc.
Tower Electronics, Inc.
Transchem
Transistor Devices, Inc.
TRI-MAG
Varo, Inc.
Vicor Corporation (Express)
Volgen America, Inc.
Wall Industries, Inc.
Wilorco, Inc.
Xentek, Inc.

D/A and A/D Converters:
(Digital-Analog and Analog-Digital)

Advanced Analog
Analog Devices
Advanced Linear Devices
Analog Systems
Analogic
Asaki Kasei Microsystems, Co., Ltd.
Brooktree
Burr-Brown
Datel
Fujitsu
GEC Plessy
Harris Semiconductor
IDT
Maxim
Micro Linear
Micro Networks
Micro Power
Mitsubishi
Motorola
National Semiconductor
NEC
Philips Semiconductor
Raytheon
Sanyo
SGS-Thomson
Sharp
Siemens
Signal Processing Technology
Sipex
Sony
TI
TelCom Semiconductor

Cross Point Switches:
American Microsystems, Inc.
Analog Devices
Dallas Semiconductor
Harris Semiconductor
Maxim
Mitel
Mitsubishi
Motorola

D/A and A/D CONVERTERS — DIODES

SGS-Thomson
Silicon Systems

DAT Drive ICs:

Analog Devices
Asahi Kasei Microsystems, Ltd.
Burr-Brown
Fujitsu
Hitachi
Matsushita
Mitsumi
New Japan Radio
Motorola
Nippon Precision Circuits, Inc.
Oki Electric
Sony
TI
Toshiba

Data Acquisition (and A/D converters for data acquisition):

Advantech
Analog Corp.
Analog Devices
Aydin Vector
BethelTronix, Inc. (BTI)
Cybernetic Micro Systems
Datel, Inc.
Edge Technology
Harris Semiconductor
ITC Microcomponents, Inc.
Linear Technology Corporation
Micro Networks
National Semiconductor Corporation
OnSpec Electronic, Inc.
Sipex Corporation
Texas Instruments, Inc. (TI)
Verdure Industries, Inc.

Delay Lines:

Allen Avionics
Andersen labs,
Bel Fuse, Inc.
Brooktree Corp.
Coilcraft
Control Electronics Co. Inc.
Dale Electronics, Inc.
Dallas Semiconductor
Data Delay Devices, Inc.
Datatronics
Delta Electronics, Inc.
Electronic Precision Components
Elmec Technology of America
Engineered Components Co. (EC2)
ESC Electronics Corporation
Hytek Microsystems, Inc.
JBM Electronics, Inc.
Kappa Technologies, Inc. (Kappa Networks, Inc.)
Kinseki, Ltd.
Matthey Electronics
Murata Mfg. Company (Headquarters)
MuRata/Erie North America
Newport Technology, Inc.
PCA Electronics, Inc.
Penny Technologies, Inc.
Polara Engineering, Inc.
Premier Magnetics, Inc.
Pulse
RCD Components, Inc.
RF Monolithics
Rhombus Industries, Inc.
Synergy Semiconductor Corp.
Technitrol, Inc.
Thin Film Technology Corporation (TFT)
Vaisala Sensor Systems
Ventronics, Inc.

Diodes/Transistors:

Advanced Semiconductor
American Microsemiconductor
AVG Semiconductors
Central Semiconductor Corp.
Collins Electronics Corp.
Com Semiconductor (Silicon) Products, Inc.
Compensated Devices, Inc.
Comset Semiconductors SprL
Continental Device, India Ltd. (CDIL)
Custom Components, Inc.
Digitron Electronic Corporation
Diodes, Inc.
Dionics, Inc.
Diotec Electronics Corp.
Directed Energy, Inc.
Displays, Inc.
Eastron Corporation
Edal Industries, inc.
Electronics Industry (USA) Co. Ltd. (EIC)
Elite Semiconductor Products, Inc.
Elm State Electronics, Inc.
English Electric Valve, Ltd.
Eupec
Fema Electronics Corp.
Frontier Electronics Co., Ltd.
Fuji Electric
Fuji Semiconductor
Fullywell Semiconductor Co., Ltd.
GD Rectifiers Ltd.
GEC Plessy
GEM Asia Enterprise Co. Ltd.
General Transistor Corp. (GTC)
Germanium Power Devices
Giddings & Lewis Advanced Circuitry Systems
Goldentech Discrete Semiconductor, Inc.
Herrmann KG
Hi-Sincerity Microelectronics Corp.
High Voltage Component Associates (HVCA)
Hind Rectifiers, Ltd.
Hi-Tron Semiconductor Corp.
Hughes Semiconductor Products

DIODES — DISK DRIVE ICs

Hutson Industries
HV Component Associates, Inc.
Hybrid Semiconductors & Electronics, Inc. (HSE)
International Electronic Products of America (IEPC)
International Diode Corporation
International Rectifier
International Semiconductor, Inc.
IXYS Corp.
Joelmaster Systems, Ltd.
KIC Corporation
Knox Semiconductor, Inc.
Koep Precision Standards, Inc.
Kung Dar Electronics Co. Ltd.
Leadtron Enterprise Corp.
Lion Enterprises Corp.
Litton Solid State
Lucas Stability Electronics, Ltd.
M-Pulse Microwave
M/A (Microwave Associates)
Maida Development Company
Master Instrument Company
Max-Lion Corp.
Metelics Corp.
Micro Electronics, Ltd.
Microwave Diode Corp.
Mistral SpA
Mospec Semiconductor Corp.
Motorola
MSI Electronics, Inc.
New England Photoconductor (NES)
New England Semiconductor (NES)
New Japan Radio Corp.
New Jersey Semiconductor Products Co., Inc.
NIC Components Corp.
Nihon Inter Electronics Corp.
Nihon Semiconductor, Inc.
Oneida Electric Mfg. Inc.
Origin Electric Co. Ltd.
Parametric Industries, Inc.
PD&E, Inc.
Philips Components
Photron Semiconductor Corp.
Power Products International, Ltd.
Power Semiconductors, Inc.
Powerex
Q Source, Inc.
Qua-tron Electronics Industrial Co., Ltd.
Rectron Electronic Enterprises, Inc.
Renard Manufacturing, Co., Inc.
RF Electronics
Sakata USA Corporation
SanRex Corporation
Sansha Electric Mfg. Co. Ltd.
Secowest Italia SpA
Semelab Plc.
Semicon Components, Inc.
Semicon, Inc.
ST Semiconductors of Indiana
Semiconductor Technology, Inc. (STI)
Semiconductors, Inc.
Semitron Industries, Ltd.
Semitronics Corporation
Semitron Semiconductors
Semtech
SEP Electronic Corporation
Shindengen Electric Mfg. Co. Ltd.
Silicon Transistor Corporation (STC)
Silitek Corp.
Sino-American Silicon Products, Inc.
Solid State Devices, Inc.
Solid State (Microtechnology) Electronics Co. Pvt. Ltd.
Solid State, Inc.
Solid State Industries, Inc.
Solid State Systems
Space Power Electronics, Inc.
STC Defense Systems
Sussex Semiconductor, Inc.
Swampscott Electronic Co., Inc.
Syntar Industries, Inc. (SI)
Taiwan Liton Electronic Co., ltd.
Terry Semiconductor, Inc.
TIC Semiconductor
Toshiba Corp.
U-FO, Top-Tech Co., Ltd.
United-Page, Inc.
Ventronics
V/O Electronorgtechnica
Way-Tech Enterprise Co. Ltd.
Westcode Semiconductors, Inc.
World Products, Inc.
Zetex Plc.

Disk Drive ICs:

Adstar (IBM)
AMI Microsystems
Exar
Fujitsu
Hitachi
Hualon Microelectronics Corp.
IMP, Inc.
LG Semicon
Matsushita
Micro Linear
Mitac
Oki
NEC
National Semiconductor
Mitsubishi
Ricoh
Rohm
Sakata
SGS-Thomson
Silicon Logic
S-MOS
Silicon Systems
Standard Microsystems Corporation
Toko America
Toshiba
United Microelectronics Corp.
Western Digital Corporation

Displays:

American led-gible (LED)
American Bright Optoelectronics Corp. (LED)
Anders Electronics (LED)
Babcock Display Products (Plasma)
Crystaloid (LCD)
Dale Electronics, Inc. (Plasma)
Data International Co. Ltd. (LCD)
Densitron Corporation (Electroluminescent, plasma and LCD)
Displaytech, Inc. (Ferroelectric LCD on silicon)
Display Technologies, Inc. (LCDs)
Fujitsu Microelectronics, Inc. (Plasma)
Futaba (Vacuum fluorescent)
Hunter Components (LCD)
IDEA, Inc. (LED)
IEE, Industrial Electronic Engineers, Inc. (Vacuum fluorescent)
Info-Lite (large)
Kingbright Electronic Co. Ltd. (LED)
Kyocera Industrial Ceramics (LCD)
Ledtech Electronics Corp. (LED)
LXD Inc. (LCD)
Marktech International Corp. (LED)
Matsushita Electric Industrial Co. Ltd. (Plasma)
Nan Ya Technology Corporation (LCD)
NEC (LCD)
Noritake Co. Ltd. (Vacuum fluorescent displays)
NTC/American Bright Optoelectronics Corporation (LED)
Optotek Ltd. (LED)
Optrex Corp. (LCD)
Photonics Imaging (AC gas)
PixTech, Inc. (FED (Field emission))
Planar Systems, Inc. (EL)
Plasmaco, Inc. (Plasma)
Robert G. Allen (LED)
ROHM Electronics (LCD)
S.I. Diamond Technology, Inc. (Field-emmission displays)
Samsung Display Devices (LCD, AM-LCD, VFD)
Seiko Instruments, USA Inc. Liquid Crystal Display Dept. (LCD displays)
Sharp Corporation Japan (LCDs, AM-LCDs, electroluminescent displays)
Shelly Associates, Inc. (Graphic LCD modules-passive matrix color and monochrome, discrete LEDs)
Siemens Components, Inc. (Intelligent digits)
Standish LCD (LCD displays)
Stanley Electric Co. Ltd. (LED, LCD modules)
Sunscreen Company Ltd. (LED)
Texas Optoelectronics, Inc. (LED)
Three-Five Systems, Inc. (LCD and LED displays)
Toshiba Corp. (LCD)
United Microelectronics Corp. (LCD)
Varitronix, Ltd. (LCD)
Vorne Industries, Inc.
WAH III Technology Corporation (Miniature AM LCD displays for projection, head mounts, virtual reality goggles)

DRAMs:

Acer, Inc.
Advanced Electronic Packaging
Alliance Semiconductor
Advance Data Technology
Cypress Semiconductor
EDI
Hitachi
Hyundai
LG Semiconductor
Micron
Mitsubishi
Mosaic
Mosel-Vitelic
Motorola
MSIS
NEC
OKI
Panasonic
Samsung
SGS-Thomson
TI
Toshiba
IBM

DSPs, Digital Signal Processors:

Analog Devices
Lucent Technologies
Cirrus Logic
DSP Group
Fujitsu
GEC Plessy
Harris
Hitachi
IBM Microelectronics
Infinite Solutions
Intel
ITT
Motorola
National Semiconductor
NEC
Philips
Star Semiconductor
Texas Instruments
VLSI

D - E DSPs — ETHERNET ICs

Zilog
Zoran

DUARTs:

Exar
Macronix Inc.
Motorola
Philips Semiconductors
VLSI Technology

ECL:

(Note: See manufacturer listing for other logic families)

Advance Data Technology, Inc.
Applied Microcircuits Corporation
Engineered Components Co.
    (Delay lines)
Fujitsu Microelectronics
GEC Plessy Semiconductors
Loral-Microwave FSI
National Semiconductor
Motorola Semiconductor
Synergy Semiconductor
System Microelectronic
    Innovation, GmbH

EEPROMS:

Advance Data Technology
Asahi Kasei Microsystems
Atmel
Catalyst Semiconductor
Dense-Pac
EM Microelectronics
Exel
Hitachi
Hughes
Microchip Technology
Mitsubishi
Mosaic
Motorola
National Semiconductor

NEC
Oki Semiconductor
Rohm
S-MOS
Samsung
SEEQ
Seiko
Seimens
SGS-Thomson
Toshiba
White Technology
Xicor

EISA Bus ICs:

Brooktree
Dallas Semiconductor
Datel, Inc.
PLX Technology

Electronic Ballast ICs:

Holt Integrated Circuits, Inc.
Micro Linear Corp.
Philips
Siemens

Electronic Potentiometers:

Analog Devices
Dallas Semiconductor
Xicor

Encryption ICs:

Intel
Lintel NV/SA
National Semiconductor
Newbridge Microsystems
Mykotronx, Inc.
Rainbow Technologies, Inc.
VLSI - VLSI Technology Inc.

Engine Control ICs:

Delco Electronics Corporation
Dinan (Performance Engineering)

Intel Corp.
Onset Computer

EPLDs:

AMD
American Micro Semiconductor
Atmel
Cypress Semiconductor
ICT
National Semiconductor
SGS-Thomson

EPROMs:

Advance Data Technology
AMD
Asahi Kasei Microsystems Atmel
Catalyst Semiconductor
Cypress Semiconductor
Dense-Pac
Greenwich Instruments
Hitachi
ISSI
Microchip Technology
Mosaic
MSIS
National Semiconductor
Oki
Philips
Ricoh
Seeq
TI
Toshiba
Waferscale

Ethernet ICs:

AMD
Fujitsu
Micro Linear
Philips
SEEK
Silicon Systems
Symbios Logic
Western Digital

FAX/Modem ICs:

AMD
AMI
Analog Devices
Asahi Kasei Microsystems, Ltd.
Clarion Co., Ltd.
Fuji Electric
Fujitsu
Hamai Electric lamp Co., Ltd.
Hitachi
Intel
Matsushita
Mitsubishi
NEC
New Japan Radio Corp.
Nippon PRecision Circuits, Inc.
Oki Electric
Ricoh
Rohm
Sanyo Electric
Seiko Epson
Sharp
Sil-Walker, Inc.
TI
Toko, Inc.
Toshiba
Yahama

FDDI ICs:

AMD
Analog Devices
Fujitsu
Hitachi
Matsushita
Mitsubishi
NEC
New Japan Radio, Corp.
Motorola
Oki Electric
Ricoh
Rohm
Sanken Electric
TI
Toko
Toshiba

Ferroelectric RAMS (FRAMS):

Enhanced Memory Sytems, Inc.
Hitachi America, Ltd.
Matsushita Electric Industrial Co. Ltd. Frams
Ramtron

Ferrite Beads/Devices:

Aerotek Co.
Alpha Industries
Dorado International Corporation
Fer Rite Electronics Ind. Co., Ltd.
MMG-North America
NIC Components Corp.
SMEC, Inc.
TDK Corp.

FIFOs:

AMD
Aptos Seiconductor
Advance Data Technology
Cypress Semiconductor
Dallas Semiconductor
GEC Plessy
Harris Semiconductor
IDT
Mosel-Vitelic
Music Semiconductor
Paradigm
Philips
Quality Semiconductor
National Semiconductor
Oki
Raytheon
Sharp
TI

Filters:

AAK Corp.
Aerovox Corp.
Allen Avionics (video)
Alpha Components

Andersen Labs (SAW),
Asahi Kasei Microsystems, ltd.
Burr-Brown
Cermetek
Datel
EG&G-Reticon
Exar
Harris Semiconductor
IMP
Linear Technology
Maxim
MicroLinear
National Semiconductor
SGS-Thomson

Flash Memory:

Added Value Electronics Distribution, Inc.
Alliance Semiconductor
AMD
Atmel
Hitachi
Intel
Micron
Mitsubishi
Samsung
SGS-Thomson
Toshiba
SanDisk

FPGAs/Custom ICs:

ABB Hafo
Added Value Electronics Distribution, Inc.
Advanced Linear
Alden Scientific
Aptos Semiconductor
Asahi Kasei Microsystems
Actel
American Microsemiconductor
Applied Micro Circuits
Aspec Technology

F - G FPGAs — GRAPHICS IMAGING

Atmel
Cherry Semiconductor
Crosspoint Solutions
DSP Group
EM Microelectronic
Exar
Fujitsu
Gennum
Linfinity Microelectronics
GEC-Plessy
Harris Semiconductor
Holtek
Honeywell
IMP
Integrated Circuit Systems
Lucent Technology
Matra MHS
MitsubishiMotorola
National Semiconductor
NEC
OKI
Panasonic
Performance
Raytheon
Ricoh
S-MOS
Samsung
Sanyo
SGS-Thomson
Sharp
Sierra Semiconductor
Sipex
Sony
Space Electronics
Synergy Semiconductor
Symbios Logic
TI
Toshiba
UTMC
Universal Semiconductor
Vertex
Vitesse
VLSI Technology

Xlinx
ZyMOS

Fuses:

Altech
Bussman
Littlefuse
San-o
Schurter
Wickmann

Futurebus ICs:

Cable and Computer Technology, Inc.
National Semiconductor Corporation
Philips Semiconductor
Texas Instruments, Inc. (TI)

Fuzzy Logic ICs:

Adaptive Logic
Fujitsu Microelectronics, Inc.
Hitachi America, Ltd.
Omron Corporation
Siemens Components, Inc.
Togai InfraLogic, Inc.
VLSI - VLSI Technology Inc.

GaAs:

Alpha Industries
Amplifonix
Anadigics
Avantek
Celeritek
Daico Industries, Inc.
FEI Communications, Inc.
Harris Semiconductor
Herotek, Inc.
Hexawave, Inc.
Hittite Microwave Corporation
M/A (Microwave Associates) Com Semiconductor
(Silicon) Products, Inc.
M/A-COM Gallium Arsenide Products, Inc.
Matsushita Electric Industrial Co. Ltd. Frams
Motorola, Inc.
Murata Mfg. Company (Headquarters)
MuRata/Erie North America
National Hybrid, Inc.
Northrup-Grumman Corporation
Northrop Electronics Systems Division
Oki Semiconductor
Phase IV Systems, Inc.
Rockwell International Semiconductor Systems
Samsung Microwave Semiconductor, Inc.
Sanyo Semiconductor Corp.
Sony Corporation
Teledyne Solid State
TriQuint Semiconductor, Inc.
Vitesse Semiconductor Corp.

Graphics/Imaging:
(See also Image Processors)

Alliance Semiconductor
Appian Technology
Arcobel Graphics
Ark Logic, Inc.
ATI Technologies, Inc.
Avance Logic, Inc.
Chips and Technologies, Inc.
Cirrus Logic Inc.
DY 4 Systems, Ltd.
8x8 Inc.
Genesis Microchip, Inc.
Integrated Circuit Systems, Inc. (ICS)
Integrated Device Technology, Inc. (IDT)
Intergraph Corp.
LSI Logic Corporation
Macronix Inc.

Product Listings

GRAPHICS — IMAGE PROCESSING

Matrox Electronic Systems, Ltd.
Music Semiconductors
NEC Electronics, Inc.
NeoMagic Corporation
Number Nine (Computer) Visual Technology Corporation
NVidia Corporation
Oak Technology, Inc. (OTI)
Opti Inc.
Paradise Systems, Inc.
Pericom Semiconductor Corp.
S3, Inc.
Samsung Semiconductor, Inc.
Sierra Semiconductor
Sigma Designs, Inc.
Silicon Integrated Systems, Corp. (SiS)
Silicon Magic Corporation
S-MOS Systems, Inc.
3Dlabs, Ltd.
Trident Microsystems, Inc.
Tseng Laboratories, Inc.
VSIS, Inc.
Weitek Corporation
Western Digital Corporation
Yamaha Systems Technology, Inc.

Hall Effect Devices:

American Electronic Components
Analog Devices
Cherry Semiconductor Corp.
F.W. Bell
Hitachi America, Ltd.
Honeywell Micro-Switch Division
ITT Semiconductors
ITT-Intermetall
Matsushita Electric Industrial Co. Ltd.
Micro Switch
Murata Mfg. Company (Headquarters)
MuRata/Erie North America
New Japan Radio Company, Ltd.
Ohio Semitronics, Inc.
Optek Technology Inc.
Sakata USA Corporation

Sanyo Semiconductor Corp.
Seiko-Epson
Sharp Corporation Japan
Siemens Components, Inc.
S-MOS Systems, Inc.
System Microelectronic Innovation GmbH.
Toshiba Corp.

HIPPI (High Performance Parallel Interface) ICs:

Applied Micro Circuits Corp.
GTE Microelectronics

Hybrid ICs:

AAD & Co.
AEL Industries
Alden Scientific
Ambit Microsystems
Analog Devices
Aptek Williams
Asahi Glass
Astec Pacific Corp.
Beam Electronics Industrial Co., Ltd.
Burr-Brown
Clarion, Co., Ltd.
Elbex Video, Ltd.
FDK Corp.
Fuji Electric
Fujitsu
Goyo Electronics Co., Ltd.
High Reliability Components Corp.
Hitachi
Hokuriku Electric Industry Co., Ltd.
IAM Electronics Co., Ltd.
Iwata Electric Co., Ltd.
Iwatsu Precision Co., Ltd.
Japan Resistor Mfg. Co., Ltd.
KOA Corp.
Kokusai Electric Co., Ltd.
Marcon Electronics Co., Ltd.
Matsushita

Mitsubishi
Mitsumi
Murata Manufacturing Co., Ltd.
NEC
New Japan Radio Co.
NGK Insulators, Ltd.
Nichicon Corp.
Nihon Inter Electronics Corp.
Nikkohm Co., Ltd.
Nippon Avionics Co., Ltd.
Nippon Chemi-Con Corp.
Nippon Seiki, Co. Ltd.
Nitsuko Corp.
Ono Sokki Co., Ltd.
Origin Electric Co., Ltd.
Risho Kogyo Co., Ltd.
Rohm
Saitama Parts Industrial Co., Ltd.
Sanken Electric
Sankyo Seiki Mfg. Co., Ltd.
Sanshin Electric Co., Ltd.
Sanyo
Sharp
Shindengen Electric Mfg. Co., Ltd.
Sony
Soshin Electric Co., Ltd.
Stanley Electric Co., Ltd.
Tamura Corp.
Tokai Device Co., Ltd.
Toko, Inc.
Tokyo Ko-On Dempa Co., Ltd.
Toshiba
Toyo Communication Equipment Co., Ltd.
Yaskawa Electric Corp.
Yukata Electric Mfg. Co., Ltd.

Image Processing ICs:

Analog Devices
Burr-Brown
Fujitsu
Hitachi

I - L IMAGE PROCESSING — LEDs

ITT Semiconductors
Matsushita
Mitsubishi
Mitsumi
NEC
Nippon Precision Circuits, Inc.
Oki Electric
Ricoh
Rohm
Sharp
Sil-Walker, Inc.
Toko, Inc.
Toshiba
Internet ICs:
LSI Logic Corporation
JTAG boundary ICs:
National Semiconductor Corporation
Quality Semiconductor, Inc. (QSI)

Lamps (Bulbs):

(Note: This reference contains only a few sources for lamps as many vendors have parts that are interchangeable and identification of a specific vendor is often not necessary)

Aborn Electronics (IR)
Altair Corp.
Chicago Miniature Lamp, Inc.
Dialight Corp. (solid state)
GBC, Inc.
Gilway Technical Lamp
JKL Components Corporation (Miniature fluorescent)
LCD Lighting, Inc. (Miniature fluorescent)
Lighting Components & Designs
Lumex Opto/Components, Inc.
NTC/American Bright Optoelectronics Corporation
Quality Technologies Corporation (Solid state)
Shogyo International Corp.
SoLiCo (Solid state)
Spectrum Devices (Solid state)
Stanley Electric Co. Ltd. (Miniature fluorescent)

JPEG/MPEG ICs:

Acer, Inc.
Atlantic Signal Processors, Inc.
AuraVision Corporation
ATI Technologies, Inc.
C-Cube Microsystems
8x8 Inc.
Hyundai Electronics America (Industries Co., Ltd.)
LSI Logic Corporation
Lucent Technologies
Matra MHS
Mitsubishi Electronics America, Inc.
Odeum Microsystems, Inc.
Philips Semiconductors - Signetics
S3, Inc.
SGS-Thomson Microelectronics
Sigma Designs, Inc.
Texas Instruments, Inc. (TI)
Trident Microsystems, Inc.
Western Digital Corporation
Winbond Electronics Corp.
Zoran Corporation

LAN ICs:

AMD
Crystal Semiconductor
Intel Corp.
National Semiconductor
Standard Microsystems Corp.

Laser Diodes:

Blue Sky Research
BNR
Coherent, Inc.
Ensign-Bickford Company
Epitaxx, Inc.
Fermionics Opto-Technology
Gateway Photonics Corporation
High Power Devices, Inc.
Hitachi America, Ltd.
Laser Diode Products, Inc.
Lucent Technologies
Mitsubishi Electronics America, Inc.
Optical Communication Products, Inc.
Philips Key Modules
ROHM Electronics
SDL, Inc.
Seastar Optics, Inc.
Siemens Components, Inc.
Sony Corporation
Tektronix, Inc.
Toshiba Corp.
Uniphase

LEDs:

Aborn Electronics
Altair Corp.
American Zettler
American Bright Optoelectronics
Bright LED Electronics Corporation
Chicago Miniature Lamp, Inc.
Cree Research, Inc.
Industrial Devices, Inc.
Data Display Products (DDP)
Dialight Corp.
EG&G Solid State Products Group
Electrodynamics, Inc.
ELMA Electronic, Inc.
Elna America, Inc.
E-O Communications, Inc.
Everlight Electronics Co., Ltd.
Gilway Technical Lamp
Hewlett-Packard Company
Huayue Electronic Devices HY Electronics Corporation Industry Co.
Industrial Devices, Inc. IDI
Kingbright Electronic Co. Ltd.

LEDs — MICROPROCESSORS L - M

Lamp Technology, Inc.
LED Technology, Ltd.
Ledtech Electronics Corp.
Ledtronics, Inc.
Lite-On, Inc.
Lumex Opto/Components, Inc.
Marktech International Corp.
Micro Electronics, Ltd.
Micropac Industries, Inc.
Microsemi Corporation
Mitsubishi Cable America, Inc.
NTC/American Bright
Optoelectronics Corporation
Oxley, Inc.
Philips Semiconductors
Shelly Associates, Inc.
Siemens Components, Inc.
SoLiCo
Texas Optoelectronics, Inc.
UNIAX Corporation

Macrovision ICs:

Texas Instruments, Inc. (TI)

Manchester Encoders:

GEC-Plessy
Harris Semiconductor
Technitrol, Inc.

Memory Controllers:

ACC Microelectronics Corp.
Benchmarq Technology
California Micro Devices (CMD)
Dallas Semiconductor
United Microelectronics Corp. (UMC)

Microcontrollers:

4 Bit Microcontrollers: Fujitsu, Hitachi, Lucent Technologies, Mitsubishi, National Semiconductor, NEC
Oki Semiconductor

Microcontrollers, 8 Bit:

Adaptive Logic
Atmel
Dallas Semiconductor
Fujitsu
Harris Semiconductor
Hitachi
Intel
Luceht Technologies
Microchip Technology
Motorola
National Semiconductor
NEC
Oki Semiconductor
Philips
Rohm
SGS- Thomson
Siemens
Silicon Systems
Texas Instruments
Toshiba
Western Design Center
Zilog

Microcontrollers, 16 and 32 Bit:

Hitachi
Lucent Technologies
Matsushita
Matra MHS
Mitsubishi Electric
Motorola
National Semiconductor
NEC
Samsung
WSI

Microprocessors:

Actel
Advance Data Technology, Inc.
Advanced RISC Machines, Ltd.
Allied Signal Aerospace
AMD
Cyrix
GEC Plessy
Hitachi
IBM
IDT
Intel
LSI Logic
Motorola
NEC
TI

Microprocessors, CISC:

Acer
Advanced Micro Devices
Cypress Semiconductor
Cyrix
Digital Equipment Corporation
IBM Microelectronics
Integrated Information Technologies
Intel
Motorola
National Semiconductor
NEC
NexGen
SGS- Thomson
Texas Instruments
United Microelectronics Corporation
VLSI Technology
Weitek

Microprocessors, RISC:

Acer
Advanced Micro Devices
Advanced RISC Machines
Digital Equipment Corporation
Fujitsu
GEC Plessy
Lucent Technology (AT&T)
Hewlett Packard
Hitachi

MICROPROCESSORS — MOTION-MOTOR CONTROL ICs

Hyundai
IBM Microelectronics
Integrated Device Technology
Intel
Macronix
Matsushita Electric Corp.
Mitsubishi Electric Corp.
MIPS Technologies
Motorola
NEC
NKK Steel
Oki Electric
Pericom Semiconductor (Pioneer Electronics)
Philips
Renaissance Microsystems
Ross Technology
SGS- Thomson (Inmos)
Samsung
Sharp Electronics
Siemens
Sun Microsystems
Texas Instruments
Toshiba
United Microelectronics Corporation
VLSI Technology
Winbond Electric Corp.

MIL-STD-1553 Bus ICs:

Aeroflex Labs
ILC-DDC

Military ICs:

Actel Corp.
Advanced Analog
Aeroflex Circuit Technology Corp.
Allied Signal Aerospace
Altera Corporation
American Microsystems, Inc.
Analog Devices
Apex Microtechnology Corp.
Atmel
Lucent Technology (AT&T)
Austin Semiconductor
C-MAC Microelectronics
Comlinear Corp.
Crystal Semiconductor
CTS Microelectronics
Cypress Semiconductor
Datel, Inc.
ILC- DDC
Elantec (selected contracts only)
Electronic Design, Inc. (EDI)
Exar Corporation
GEC Plessy
Harris Semiconductor
Hewlett-Packard
Holt, Inc.
Honeywell SSEC
Hycomp, Inc.
Integrated Device Technology (IDT), Inc.
Intel Corp.
Interpoint Corp.
Lansdale Semiconductor
Lattice Semiconductor
Linear Technology Corp.
Linfinity Microelectronics, Inc.
Logic Devices, Inc.
Loral Corporation
LSI Logic Corp.
Marconi Circuit Technology
Matra Harris (MHS)
Maxim Integrated Products
Micrel, Inc.
Micropac Industries, Inc.
Micro Power Systems
Micro Networks Company
M.S. Kennedy Corp.
Natel Engineering
National Hybrid, Inc.
National Semiconductor
Omnirel Corp.
Performance Semiconductor
Philips Semiconductors
Raytheon Semiconductor
David Sarnoff Research Center
Semelab PLC
SGS-Thomson
Signal Processing Technologies, Inc. (SPT)
Siliconix, Inc.
Simtek Corp.
Sipex Corp.
Solitron Devices
Texas Instruments
Thomson Components and Tubes, Inc.
United Technologies, Microelectronics Center
Unitrode Integrated Circuits Corp.
VLSI Technology, Inc.
Waferscale Technology, Inc.
White Technology, Inc.
Xicor Corp.
Xlinx Inc.
Zilog, Inc.

Motion/Motor Control ICs:

Allegro
Cybernetic Micro Systems
Ericcson
GEC Plessy
ILC-DDC
LG Semicon
Linfinity
LSI Components
Harris Semiconductor
Mitsubishi
Motorola
National Semiconductor
NEC
New Japan Radio
Phillips Semiconductor
Rohm
Sanyo
SGS-Thomson
Silicon Systems
Solitron
Texas Instruments
Toko
Unitrode

Multimedia ICs:
(See also MPEG, and JPEG)

Altech International
Alliance Semiconductors
Ark Logic
AuraVision Corporation
Media Vision
Aztech Systems, Ltd.
BethelTronix, Inc. (BTI)
Crystal Semiconductor
Digital Research in Electronic Acoustics & Music SA (Dream)
E-mu Systems, Inc.
Hitachi America, Ltd.
Integrated Circuit Systems, Inc. (ICS)
Matrox Electronic Systems, Ltd.
NVidia Corporation
OnSpec Electronic, Inc.
Raytheon Company
Siemens Components, Inc.
Silicon Integrated Systems, Corp. (SiS)
Silicon Magic Corporation
Texas Instruments, Inc. (TI)
Thesys Gesellschaft Fur Mikroelektronik GmbH
Via Technologies, Inc.
Zilog, Inc.
Zoran Corporation

Neural Network ICs:

Adaptive Solutions, Inc.
Echelon Corporation
Oxford Computer, Inc.
Noise Generators:
Noise/Com

NVRAMs (Nonvolatile RAM):

Advance Data Technology
AMD
Atmel
Hitachi
Intel
Macronix
NEC
Samsung
SGS-Thomson
Sharp
Toshiba

Optocouplers:

AEG (Telefunken),
CP Clare Corp. (CPC)
Hewlett-Packard Company
Interpoint Corp.
Isocom PLC
Siemens Components, Inc.

Optoelectronics:

Aborn Electronics
Advanced Optoelectronics
Advanced Semiconductor, Inc.
Advani Oerlikon, Ltd.
Alpha Products, Inc.
American Microsemiconductor
American Bright Optoelectronics
A.P.I. Electronics
Bedford Opto Technology, Ltd.
Big-Sun Electronics Co. Ltd.
China Semiconductor Corporation
Citizen Electronics Co. Ltd.
Clarostat Sensors and Controls Group
Collins Electronics Corp.
Computer Management & Development Service
Comset Semiconductors SprL
Denyo Europa GmbH
Devar, Inc.
Dionics, Inc.
Displays, Inc.
DO Industries, Inc.
EG&G Solid State Products Group
EG&G Judson
EG&G Reticon
EG&G Optoelectronics
Elec-Trol, Inc.
Eltech Instruments, Inc.
English Electric Valve Co. Ltd.
Excel Technology International Corp.
Fasco Industries, Inc.
Finlux Inc.
Fujitsu Kiden Ltd.
Germanium Power Devices Corp.
HEI, Inc.
Hewlett-Packard Company
High Voltage Semiconductor Specialists, Inc.
Hitachi America, Ltd.
HI-WIT Electronics Co. Ltd.
Honeywell Optoelectronics
ICO-Rally Corporation
Instrument Design Engineering Associates
Interpoint Corp.
Jiann WA Electronics Co. Ltd.
Kanematsu USA, Inc.
Kinematics & Controls Corp.
Kingbright Electronic Co. Ltd.
Kodenshi Corp.
LED Technology, Ltd.
Lohuis International
Siemens Components
Marl International Ltd.
Master Instrument Corp.
Max-Lion Corp.
Mentor GmbH & Co.
Micropride Ltd.
Microwave Diode Corp.
Morrihan International Corp.
Motorola, Inc.
NTC/American Bright Optoelectronics Corporation
Takenaka Electronic Company, Ltd.
Opto Technology, Inc.
Oriel Corp.
Ortel Corp.
PCI Inc.

O - P OPTOELECT. — POWER SUPPLIES

PED Ltd.
Philips Semiconductors
Powertech, Inc.
Proxitronic
Raytheon Company
Refac Electronics Corp.
Robert G. Allen
Segor-Optoelectronique
Semetex Corporation
Semicoa Optoelectronics
ST Semiconductors of Indiana
SEP Electronic Corporation
Sharlight Electonics Co., Ltd.
Shian Yih Electronic Industry Co.
Shokai Far East Ltd.
Silicon Detector Corp.
Silonex, Inc.
STC Defense Systems
Stetco, Inc.
Takamisawa Electric Co. Ltd.
Texas Instruments
Texas Optoelectronics, Inc.
Toshiba Corp.
Transwave Corp.
UCE, Inc.
Universal Semiconductor, Inc.
Varitronix, Ltd.
Walmsley Microsystems, Ltd.

PALs:

AMD
Honeywell Solid State Electronics
Microchip Technology, Inc.
Philips Semiconductors

Parity ICs:

Applied Microcircuits Corp.

PCI Bus/Bridge ICs:

Adaptec
Appian Technology
Applied Microcircuits Corp.,
California Micro Devices (CMD)
Digital Equipment Company (DEC)
Initio Corporation
Newbridge Microsystems, Inc.
PCTech
PLX Technology
Sand Microelectronics (PCI cores)
Siemens Components, Inc.
V3 Corporation
Via Technologies, Inc.
Virtual Chips (PCI Cores)
Wolfson Microelectronics
Zoran

PECL:

Arizona Microtek, Inc.
Motorola
Synergy Semiconductor
Thin-Film Technology Corporation

Photo Diodes/Detectors/Sensors:

ABB Hafo
Advanced Photonix
AEG (Telefunken)
Antel Optronics,
Advanced Optoelectronics
Aleph International
Centronic, Inc.
Everlight Electronics Co., Ltd.
Hamamatsu Corporation
Interpoint Corp.
Takenaka Electronic Company, Ltd.

PIN Diodes:

Advanced Photonix
Alpha Industries
Amplifonix
Antel Optronics,
BT&D Technologies
Com Semiconductor (Silicon) Products, Inc.
Epitaxx, Inc.
Film Microelectronics, Inc.
M/A (Microwave Associates)
M/A-COM Gallium Arsenide Products, Inc.
Maxtek Components Corporation
Optical Communication Products, Inc.
Silicon Sensors, Inc.
TelCom Semiconductor, Inc.
Texas Optoelectronics, Inc.
Waveline Solid State, Inc.

PLDs (Programmable Logic Devices):

Actel
Atmel
Altera
AMD
Cypress
Lattice
Lucent Technologies
Philips Semiconductor
TI
Xlinx

PLL (Phase Locked Loops):

ERSO
Exar Corporation
Harris Semiconductor
Micrel
Mitel Semiconductor
Motorola
On Chip Systems
Philips Semiconductor
SGS-Thomson
Synergy Semiconductor Corp.

Power Supplies:

Abbott Electronics
Acopian
Advanced High Voltage

POWER SUPPLIES — PRESSURE SENSORS

Advanced Power Solutions
American High Voltage
Astec Amperica
Astrodyne
ATC Power Systems, Inc.
Ault, Inc. Autek Power Systems
Babcock Display Products
Bertan High Voltage Corporation
Computer Products, Power Conversion
Condor DC Power Supplies, Inc.
Conversion Equipment Corporation (CEC)
Cosel U.S.A. Inc.
Cui Stack, Inc.
Delta Electronics, Inc.
EG&G Power Systems, Inc.
Electron Tubes, Inc.
Electronic Measurements, Inc.
Elpac Power Systems
EMCO High Voltage Company
Farnell Advance Power, Inc.
Forton/Source
Gamma High Voltage Research, Inc.
GlobTek, Inc.
Golden Pacific Electronics, Inc. (GPE)
Group West International
HC Power, Inc.
International Power Devices, Inc. (IPD)
Jerome Industries Corporation
Jetta Power Systems, Inc.
JT PowerCraft, Inc.
Keltron Power Systems, Inc.
Kepco Inc.
Lambda Electronics, Inc.
Lambda Novatronics, Inc.
Logitek, Inc.
Lorain Products
Lucent Technologies
Lutze, Inc.
LZR Electronics, Inc.
Matsusada Precision Devices, Inc.
Mean Well Enterprises Co. Ltd.
Megapower
Melcher Inc.
Nidec/Power Gerneral
Omega Power Systems, Inc.
Operating Technical Electronics, Inc. (OTE)
OPT Industries, Inc.
Oryx Power Products
Phihong Enterprise Company, ltd.
Phoenix Contact, Inc.
Polytron Devices, Inc.
Powercube Corp.
Power-One
Power Sensors Corporation
Power Solutions, Inc.
Power Switch Corporation
Raytheon Company
Power Technology Inc.
RO Associates
Semiconductor Circuits, Inc.
Shindengen Electric Mfg. Co. Ltd.
Spellman High Voltage Electronics Corporation
Switching Systems International (SSI)
Tamura Electric Works Ltd
Technetics, Inc.
Technipower, Inc.
Total Power International, Inc.
Tower Electronics, Inc.
Transistor Devices, Inc.
TRI-MAG
Tri Source, Inc.
Unipower Corporation
Ventronics, Inc.
Vicor Corporation (Express)
Vidar - SMS Co. Ltd.
Wall Industries, Inc.
Xentek, Inc.

Power Supply ICs:
(Including Pulse Width Modulator ICs)

Analog Systems
Aptex Microtechnology
Astec Semiconductor
Cherry Semiconductor
Fujitsu
Linear Technology
Linfinity
Micrel
Micro Linear
Mitsubishi
Motorola
National Semiconductor
NEC
Samsung
SGS-Thomson
Supertex
TelCom
Texas Instruments
Unitrode

Pressure Sensors/Transducers:

Bourns, Inc.
Data Instruments, Inc.
Delco Electronics Corporation
Dytran Instruments, Inc
Endevco
Entran Devices, Inc.
Foxoboro/ICT, Inc.
Fujikura America, Inc.
Hobbs Corporation (Pressure and vacuum switches)
Honeywell Solid State Electronics Center (SSEC)
IMO TransInstruments
Lucas NovaSensor
Lucas Schaevitz
Monolithic Sensors, Inc.
Motorola, Inc.
NeXt Sensors
Sensotek, Inc.
Sen-Sym Inc.
Servoflo Corporation
Setra Systems, inc.
Tavis Corporation
World Magnetics

P - R PROMs — REFERENCES

PROMs:

AMD
Aptos Semiconductor
Atmel
Cypress Semiconductor
Greenwich Instruments
Harris Semiconductor
Lansdale
Microchip Technology
National Semiconductor
Philips
Raytheon
SEEQ
Texas instruments
UTMC

Proximity Detectors/Sensors/Switches:

Ascom Microelectronics
Altech
Baumer Electric, Ltd.
Hermetic Switch, Inc.
Pepperl + Fuchs, Inc.
Quality Technologies Corporation
System Microelectronic Innovation GmbH.
Turck Inc.

Radiation:

ABB Hafo
APP Optics (UV Radiation detector)
Dexter Research Center, Inc. (Sensing thermopiles)
Electro-Optical Systems, Inc. (IR radiation detection)
Graseby Optronics (Graseby Infrared) (Blackbody radiation sources)
Harris Semiconductor (Radiation hardened ICs)
Honeywell Solid State Electronics Center (SSEC) (Radiation hardened ICs)
ILC-DDC Data Device Corporation (Radiation shielded)
Integrated Device Technology, Inc. (IDT) (Radiation hardened, enhanced ICs)
LSI Logic Corporation (Radiation hardened ICs)
National Semiconductor Corporation (Radiation hardened ICs)
SICOM, Inc. (Radiation hardened ICs)
Solid State Devices, Inc. (SSDI) (Radiation hardened parts)
United Technologies Microelectronics Center (Radiation hardened parts)

Reed Solomon (and Error Correction):

Advanced Hardware Architectures
Cirrus Logic Inc.
Eastman Kodak Company
ERSO
Fujitsu Microelectronics, Inc.
Integrated Device Technology, Inc. (IDT)
LSI Logic Corporation
Space Research Technology, Inc.

Reed Relays/Switches:

CP Clare Corp. (CPC)
Excel Cell Electronic Co., Ltd. (ECE)
Gem Electronics, Inc.
Hamlin
Hasco Components International Corp.
Hermetic Switch, Inc.
Meder Electronic GmbH
Okita Works Co. Ltd.

References/Regulators (Voltage):

Alpha Semiconductor
AMD
Analog Devices
Analog Systems
Astec Semiconductor
AVG Semiconductors
Burr-Brown Corp.
Cherry Semiconductor Corp.
EM Microelectronic-Marin S.A.
Fujitsu Microelectronics, Inc.
GBC, Inc.
Harris Semiconductor
Hitachi America, Ltd.
Holtek Microelectronics, Inc.
Hycomp, Inc.
iC Haus GmbH
LG Semicon
Linear Technology Corporation
Linear Technology Corporation
Linfinity Microelectronics, Inc.
Maxim Integrated Products
Micrel Semiconductor, Inc.
Micro Linear Corp.
Mitsumi Electric Co. Ltd.
Motorola, Inc.
National Semiconductor Corporation
New Japan Radio Company, Ltd.
Omnirel Corp.
Power Integrations, Inc.
Power Trends, Inc.
ProSemi GmbH
Ricoh Company Ltd.
ROHM Electronics
Sanken Electric Co. Ltd.
Seiko-Epson
SemeLab Plc.
Semtech
Shindengen Electric Mfg. Co. Ltd.
Signal Processing Technologies (SPT)

Product Listings

SMK Corp.
Solitron Devices, Inc.
Supertex Inc.
TelCom Semiconductor, Inc.
Texas Instruments, Inc. (TI)
Thaler Corp.
Toko, Inc.
Toyocom U.S.A. Inc.
Unitrode Integrated Circuits Corporation

Relays:

Aleph International
Altech Corp.
American Zettler
Anritsu America
Aromat Corp.
Astralux Dynamics, Ltd.,
Babcock Display Products
Communications Instruments, Inc. (CII) (TO-5 styles)
Cornell Dubilier
Coto Wabash
Crydom Company
Dionics, Inc.
Dow-Key Microwave Corp.
Electrodyne
Electromatic Controls Corp.
Farnell Relay Products
Film Microelectronics, Inc.
Fujitsu Microelectronics, Inc.
Gentron Corporation
Guardian Electric Manufacturing Co.
Hasco Components International Corp.
Jewell Electrical instruments (Meter relays)
Kilovac Corporation
Magnecraft/Struthers-Dunn (MSD) (TO-5 relays)
Micropac Industries, Inc.
NEC Electronics, Inc.
NTE (New Tone) Electronics, Inc.
Okita Works Co. Ltd.
Omron Corporation

Opto 22
SGS-Thomson Microelectronics
Sky Electronics International
Smart Relay Technologies, Inc.
Solid State Electronics Corp.
Teledyne Relays (TO-5 relays)
TH Electronics
Ventronics, Inc.
White-Rogers

Resistors:

(Note: This reference contains only a few sources for resistors as many vendors have parts that are interchangeable and identification of a specific vendor is often not necessary)

Barry Industries, Inc.
Bourns, Inc.
Caddock Electronics, Inc.
California Micro Devices (CMD)
Component General, Inc. (CGI)
Dale Electronics, Inc.
Diablo Industries
ETF Technology
General Resistance
Interpoint Corp.
KDI/triangle Electronics, Inc.
KOA Corp.
Lumex Opto/Components, Inc. (Photo)
Mini-Systems, Inc. (MSI)
NTE (New Tone) Electronics, Inc.
NIC Components Corp.
Ohmite Manufacturing Co.
Ohmtek
Philips Components, Discrete Products Division
Rawmat Electronics (S) Pte Ltd.
RCD Components, Inc.
RF Electronics, Inc.
RF Power Components, Inc.
SMEC, Inc.
State of the Art, Inc.
Ventronics, Inc.
Vishay Resistors

REFERENCES/ REGULATORS — RS-232 ICs

Resonators:

AVX Corporation
H.K. Crystals Co.
Hong Kong Crystals Company (HKC)
Integrity Technology Corp.
Murata Mfg. Company (Headquarters)
MuRata/Erie North America
RF Monolithics
ROHM Electronics
Siemens Components, Inc.
S.P.K. Electronics Co., Ltd.
STC Components
Trans-Tech, Inc.
United Chequers (Hong Kong) Ltd.

RS-232 ICs:

Elantec
Exar
Mitsubishi
Motorola
National Semiconductor
Philips
Rochester Electronics
SGS-Thomson
Texas Instruments
RTCs (Real Time Clocks) (See also clock circuits):
Benchmarq Microelectronics, Inc.
EM Microelectronic-Marin S.A.
ERSO
FOX Electronics
Ricoh Company Ltd.
S-MOS Systems, Inc.
STC Defense Systems
TriTech Microelectronics Pte. Ltd. (Semiconductor)
United Microelectronics Corp. (UMC)

S-T SATELLITE ICs — TELEVISION ICs

Satellite ICs:

BethelTronix, Inc. (BTI)
CommQuest Technologies, Inc.
GEC Plessy Semiconductors
LSI Logic Corporation
Odeum Microsystems, Inc.
Tampa Microwave Lab, Inc.

SAW (Surface Acoustic Wave) Devices:

Andersen labs
AVX Corporation
Fujitsu Microelectronics, Inc.
GEC Plessy Semiconductors
Oki Semiconductor
RF Monolithics
Sanyo Video Components
Sawtek Inc.
Siemens Components, Inc.
Vectron Technologies, Inc.

Sbus ICs:

Cypress Semiconductor
LSI Logic

SCSI Controllers/Terminators:

Adaptec
AdvanSys
Aeronics
AMD
Cirrus Logic
Dallas Semiconductor
Fujitsu
Logic Devices
Silicon Systems
Sony
Symbios Logic
Unitrode

SONET:

Applied Microcircuits Corp.
BethelTronix, Inc. (BTI)
Integrated Circuit Systems, Inc. (ICS)
Integrated Telecom Technology, Inc. (IgT)
Laser Diode Products, Inc.
Lucent Technologies
PMC-Sierra
SGS-Thomson Microelectronics
Siemens Components, Inc.
Texas Instruments, Inc. (TI)
TranSwitch Corporation
Vitesse Semiconductor Corp.

SRAMs:

Alliance Semiconductor
Aptos Semiconductor
Advance Data Technology
Cypress
Hitachi
IDT
Micron
Mitsubishi
Motorola
NEC
Samsung
Sony
Toshiba
Winbond

Stacked Memory Modules:

Cray Research, Inc.
Cubic Memory, Inc.
Dense-Pac Microsystems, Inc.
Simple Technology, Inc.
Staktek Corporation
Teledyne Microelectronics
TRW

Telecommunication (and Telephone) ICs:

Acapella Ltd.
Applied Microcircuits Corp.
AMD
Asahi Kasei Microsystems, Co., Ltd.
Fujitsu
Hewlett-Packard
Hitachi
ITT Semiconductors
Kern Co., Ltd.
Matsushita
Mitsumi
NEC
New Japan Radio, Corp.
Motorola
Nippon Precision Circuits, Inc.
Oki
Rohm
Sanken Electric
Sanyo Electric
Seiko Instruments
Sharp
Sony
Teltone
TI
Toko, Inc.
Toshiba

Television ICs:

AI Tech International
Asahi Glass
Fujitsu
Hamai Electric lamp, Co., Inc.
Hitachi
ITT Semiconductors
Matsushita (Panasonic)
Mitsubishi
Mitsumi
NEC
New Japan Radio Co, Inc.
Motorola
Oki Electric
Rohm

Sanken Electric
Sanyo
Sharp
Sony
TI
Toei Electronics, Co., Ltd.
Toko, Inc.
Toshiba
Yamaha

Thermistors:

(Note: This reference contains only a few sources for thermistors as many vendors have parts that are interchangeable and identification of a specific vendor is often not necessary)

Alpha Thermistor & Assembly, Inc.,
BetaTherm Corporation
Dale Electronics, Inc. Interpoint Corp.
Keystone Thermometrics Corporation
KOA Corp.
Maida Development Company
Murata Mfg. Company (Headquarters)
MuRata/Erie North America
NIC Components Corp.
Philips Semiconductors
Quality Thermistor, Inc.
SensorTec, Incorporated
SMEC, Inc.
Therm-O-Disc
U.S. Sensor
Voltronics Corporation

Thermocouples:

(Note: This reference contains only a few sources for thermocouples as many vendors have parts that are interchangeable and identification of a specific vendor is often not necessary)

ACK Technology
Astra Net, Inc.
Exergen Corporation (IR types)
Marshall Thermocuoples
Nanmac Corporation
Pyromation, Inc.
Raytek, Inc. (non-contact)
SensorTec, Incorporated
Tudor Technology Ltd.

Thermoelectric Coolers:

E-TEK Dynamics, Inc.
Hytek Microsystems, Inc.
ITI FerroTec
Marlow Industries, Inc.
Melcor
ThermoTrex Corp.

Thyristors/Triacs/SCRs:

Advanced Semiconductor
Advani Oerkikon Ltd.
Alpha Industries
American Microsemiconductor
American Power Devices
A.P.I. Electronics
Astro-Craft,
Comset Semiconductors SprL
Dionics, Inc.
Douglas Randall, Inc.
FR Industries, Inc.
GEC Plessy Semiconductors
ICM International Controls & Measurement
International Power Semiconductor
Isocom PLC
Loras Industries, Inc.
Microwave Diode Corp.
Microwave Technology, Inc.
Motorola, Inc.
Powertech, Inc.
SanRex Corporation
Teccor Electronics, Inc.
SemeLab Plc.

TELEVISION ICs — UARTs   T - U

SID Microeletronica
Universal Semiconductor, Inc.

Transceivers:

American Microsystems
BethelTronix, Inc. (BTI)
Crystal Semiconductor
Hewlett-Packard Company
Level One Communications, Inc.
Linear Technology Corporation
Linfinity Microelectronics, Inc.
Micro Linear Corp.
Motorola, Inc.
National Hybrid, Inc.
Pericom Semiconductor Corp.
Pulse
Robert Bosch Corporation
Silicon Logic, Inc.
Siliconix
Tamarack Microelectronics, Inc.
Tampa Microwave Lab, Inc.
Thesys Gesellschaft Fur Mikroelektronik GmbH

Transient Suppressors:

(See Circuit Protection Devices Transistors)

Advanced Microelectronic Products
Advanced Semiconductor
Advani Oerlikon, Ltd.
Alpha Industries
American Microsemiconductor
A.P.I. Electronics
Applied Reasoning Corp.
Aristo-Craft

UARTs:

California Micro Devices (CMD)
Exar Corporation

U - V UARTs — VRAMs

Harris Semiconductor
LG Semicon
Mitsubishi
Philips Semiconductors
Siemens Components, Inc.
Silicon Logic, Inc.
Silicon Systems
Standard Microsystems Corp.
Startech Semiconductor, Inc.
Systronix, Inc.
Texas Instruments, Inc. (TI)
Toshiba
Unicorn Microelectronics Corporation
Western Digital

Varistors:

CKE Inc.
Harris Semiconductor
Maida Development Company
NIC Components Corp.
Panasonic
Ventronics, Inc.

VCR ICs:

Analog Devices
Asahi Glass Co., Ltd.
Fuji Electric
Fujitsu
Matsushita
Mitsubishi
Hitachi
NEC
New Japan Radio, Corp.
Motorola
Nippon Precision Circuits, Inc.
Oki Electric
Rohm
Sanken Electric
Sanyo
Sony
TI

Toko, Inc.
Toshiba

Video ICs (Pixel Controllers):

Acer, Inc.
Advanced RISC Machines, Ltd.
AI Tech International
Alliance Semiconductor
Analog Devices
ARK Logic, Inc.
Array Microsystems
Arcobel Graphics B.V.

Voice/Speech Processors/Synthesis:

Analog Devices
Asahi Chemical Industry Co., Ltd.
Fujitsu
Hitachi
ITT Semiconductors
ISD
Matsushita
Mitsubishi
NEC
New Japan Radio Co.
Motorola
Nippon Precision Circuits, Inc.
Oki Electric
Ricoh
Sanyo Electric
Seiko Epson
Sharp
Toshiba

Voltage Regulators:

Analog Devices
Burr-Brown
Calogic Corp.
Cherry Semiconductor
Elantec Semiconductor, Inc.
Harris Semiconductor
Linear Technology Corp.
M.S. Kennedy Corp.
Maxim Integrated Products

Micrel, Inc.
Micro Linear Corp.
Mitsubishi Electronics
Motorola Semiconductor
NTE Electronics
National Semiconductor Corp.
Power Integrations Inc.
Power Trends, Inc.
Raytheon Semiconductor
Rohm Electronics
SGS-Thomson Microelectronics, Inc.
Samsung Semiconductor
Semtech Corp.
Siemens Components, Inc.
Solitron Devices, Inc.
Supertex, Inc.
Temic Semiconductor
TelCom Semiconductor, Inc.
Texas Instruments, Inc.
Thomson Components
Toko America
Unitrode Corporation.

VRAMs:

Advance Data Technology
Brooktree Corp.
Dense-Pac Microsystems, Inc.
Hitachi
IBM Corp.
LG Semicon
Miteq
Mitsubishi
Mosaic
Mosel-Vitelic, Inc.
NEC Electronics, Inc.
OKI
Panasonic
Ramtron
Samsung Semiconductor, Inc.
Sharp
Texas Instruments
Toshiba Corp.

Watch ICs:

Asti Pacific Corp.
Fujitsu
Hamai Electric Lamp Co. Ltd.
Hitachi
ITT Semiconductors
Matsushita
NEC
New Japan Radio Co.
Nippon Precision Circuits, Inc.
Oki Electric
Ricoh
Seiko Epson
Seiwa Techno Systems Corp.
TI
Toko, Inc.

Wireless (see Cellular)

SOURCES OF HARD-TO-FIND OR OBSOLETE COMPONENTS

SOURCES OF HARD-TO-FIND OR OBSOLETE COMPONENTS

The problem of obsolete parts is growing as the rapid pace of technology development obsoletes even the most advanced products. The new product innovation cycle is only 3 to 4 years. The life cycle of parts is about five to six years (down from a 10-year availability), with some parts only lasting 2 years.

The reasons behind component obsolescence are:

1. The technology used to produce the part is obsolete (a new technology is used to produce most of the product line).

2. The devices are only selling in low volume (thus the fabrication facility could be better utilized to produce a part that is more profitable for the company.

3. The cost to transfer from one wafer fabrication process to another is too costly.

4. There are technical incompatibilities in transferring the process to other wafer fabrication lines.

5. Corporate mergers cause product lines to be consolidated and redundant fabrication facilities are closed. (Equipment upgrades, and common testers for components may also cause components to become obsolete if it is considered too costly to manufacture the device with the new equipment, or to write test software for the new test systems.)

Usually, if a part is going to be discontinued, manufacturers provide a 6 to 12 month time frame for final orders of the devices. Some sources of obsolete parts are listed below. Aftermarket manufacturers plan for support of obsolete product for at least 10 years after the device is discontinued by the original manufacturer. (The first place to contact for obsolete material is the original manufacturer. They can tell you whether any remaining stock and dies were transferred to an aftermarket manufacturer, or whether they still have stock available. This stock can be either in the U.S., or in an overseas location.)

Organizations that Track Part Obsolescence

The Navy manages a Microcircuit Obsolescence Management Program (MOM) that identifies devices (by types and package styles) that are being discontinued by integrated circuit manufacturers, and provides alternate sources of the devices where possible. Information is available from the Government and Industry Data Exchange Program (GIDEP).

Component Obsolescence notices are distributed through the GIDEP DMSMS (Diminishing Manufacturing Sources and Material Shortages) Notices. Many manufacturers supply discontinued notices to GIDEP on their products. (Texas Instruments uses this as their main way of notifying users that parts are going to be discontinued.) Information on the GIDEP program is available from:

Government and
Industry Data Exchange Program
GIDEP OPERATIONS CENTER
P.O. Box 8000
Corona, CA 91718-8000
909-273-4677
DSN: 933-4677
Fax: 909-273-5200

The MOM program became part of the Naval Surface Warfare Center, Crane Division and the Naval Air Warfare Center, and is now known as the DTC, Diminishing Manufacturing Sources (DMS) Technology Center. They can be reached at 1-800-DMS-4886. They now help in finding solutions to obsolete part problems to keep military systems/ platforms operational. Their bulletin board is accessed by calling 317-306-4992 or DSN 369-4992. Detailed information regarding the bulletin board is provided during the initial log-in.

Obsolete parts, and new sources for components, are also tracked by Electronic Buyers' News (in the "Last Runs" column):

Electronic Buyers' News
CMP Publications, Inc.
600 Community Drive
Manhasset, N.Y. 11030-3875

Sources of Hard-To-Find or Obsolete Components

Subscriptions (Address changes):
Electronic Buyers' News
P.O. Box 2020
Manhasset, N.Y. 11030-3875

(See the listings for magazines and technical publications, below.)

Another organization that tracks obsolete parts is:

TacTech
Transition Analysis of Component Technology
22700 Savi Ranch Parkway
Yorba Linda, CA 92686
714-974-7676
FAX: 714-921-2715

This company offers the Defense/Aerospace industry an electronic military microcircuit information service. The working elements for the networking service consists of: 1) comprehensive microcircuit library (which includes alternate sources for parts and whether military qualified versions of a device are available); 2) weapon system usage library; 3) comprehensive discontinuance notification; 4) customer program impact; 5) a microcircuit life cycle projectory system; and 6) a parts list risk analysis. The service can be installed on a mainframe computer or accessed via a PC and a modem. A newsletter is also issued by the company.

Sources of Obsolete Components

Companies that manufacture devices discontinued by the original manufacturer:

1. American Power Devices
   7 Andover Street
   Andover, MA 01810
   and:
   69 Benett Street
   Lynn, MA 01905
   508-475-4074
   Fax: 508-475-8997

This manufacturer, in business for over 24 years, produces industrial and military semiconductor devices. Included in their product line are Stabistors and multichip devices that are direct replacements for discontinued General Electric, Unitrode MPD series and Motorola MZ 2360 and 2361 series.

2. Calogic Corporation
   237 Whitney Place
   Freemont, CA 95439
   510-656-2900
   FAX: 510-651-1076, 3025

This company, which has been offering IC foundry service to various manufacturers for over eight years, has for the last two years been purchasing some obsolete product lines from different manufacturers. They manufacture some of the parts discontinued by Topaz, Intersil and Siliconix (small signal discretes including n and p JFETs, single and duals; n and p channel MOSFETs, singles and duals, enhancement and depletion mode; Monolithic Dielectrically Isolated dual transistors; and monolithic Junction Isolated dual transistors). They also make equivalents to National Semiconductor devices. Standard data book product, special items and military screening are available. (Note: the remaining inventory of Siliconix Power MOSFET series 2N6758JANTX thru 2N6800JANTX was acquired by Hamilton Hallmark in May 1996. Contact them at 10950 Washington Boulevard, Culver City, CA, 310-558-2000 Fax: 310-558-2809).

3. Caton Connector Corporation
   20 Wapping Road
   Kingston, MA 02364
   617-585-4315
   Fax: 617-585-2973

This company manufactures connectors from Burndy Corporation (such as the EC, MMC, PD, ME, Sealock and YIC/YOC series) and from U.S. Components, Inc. (including their MI, 625, MIG, MH, MIWC, UTP, U980/U900 and UPCR/UP2CR series).

4. Central Semiconductor Corp.
   145 Adams Ave.
   Hauppauge, NY 11788
   516-435-1110
   Fax: 516-435-1824

This company manufactures surface mount Schottky rectifiers that can be used to replace the MBRS120 series that is no longer available from Motorola Semiconductor.

5. David Sarnoff Research Center
   Subsidiary of SRI International
   CN 5300
   Princeton, NJ 08543-5300
   609-734-2437, 2000
   Fax: 609-734-2075, 2992, 2443

   (SRI International)
   333 Ravenswood Ave.
   Menlo Park, CA 94025-3493
   415-859-3285
   Fax: 415-859-2844

The GEM program (Generalized Emulation Microcircuits) is a result of an R&D initiative by the Defense Logistics Agency and the Defense Electronics Supply Center with the guidance and support from the Weapons System Improvement Group within the Office of the Secretary of Defense. The GEM system has the capability to produce IC devices that are form, fit and function equivalent to original devices at a quality level that satisfies testing in accordance with MIL-STD-883C. GEM system IC devices have been successfully circuit board and system level tested by government laboratories such as NWSC Crane, IN; AFLC Ogden, UT; AFLC, Warner-Robbins, GA; ET&DL, Ft. Monmouth NJ. (one such device was the Raytheon Semiconductor RM2505 digital multiplier used in the AN/SQS-56 surface ship sonar receiver and the BQQ-5 submarine sonar system). In late 1993 they produced 40 to 50 new ICs, costing between $25K to $60K. Costs vary due to complexity, availability of documentation and samples and whether similar parts were already emulated.

GEM is a flexible integrated manufacturing system capable of producing tested ICs within ten weeks from order, and provides a source for otherwise non-available replacement ICs.

The GEM system uses developed technologies, including the BiCMOS process, to produce ICs by characterization testing to produce a part that emulates the original device. Wafers are fabricated using the David Sarnoff Research Center IC Foundry. The current cost (as of January 1992) to produce an emulation replacement IC with the prototype system is approximately $50K.

Other information on the "GEM" Program is available from:

John Christensen, GEM Program Manager,
    DLA-PRM 703-274-6445
Rick Easter, NSWC, Crane 812-854-1528
Harvey Hanson, NOSC 619-553-2674
Raymond Grillmeier, DESC 513-296-6064
Lloyd S. Peters, SRI International 415-859-3650,
    FAX 415- 859-2844
James S. Crabbe, David Sarnoff Research Center
    609-734-3299

6. DPA Labs, Inc.
   2251 Ward Avenue
   Simi Valley, CA 93065
   805-581-9200
   Fax: 805-581-9790
   Internet: DPALABS@AOL.COM

This lab which does component screening and failure analysis can also provide form, fit and function military obsoleted components.

7. General Transistor Corporation
   216 W. Florence Ave.
   Inglewood, CA 90301
   310-673-8422
   Fax: 310-672-2905

This company manufactures transistors discontinued by such manufacturers as RCA and Motorola, as well as second sourcing other available devices.

8. GTE Microelectronics
   77 "A" Street
   Needham Heights, MA 02194-2982
   1-800-544-0052
   Fax: 617-455-2088

Obsolete parts are recreated using FPGA, Gate-Array, Standard Cell technology or by the design of a plug-in module. Whole obsolete PC cards can also be recreated. Devices manufactured in obsolete technologies can also be supported.

9. InterFET-ITAC (Formerly ITAC Hybrid Technology)
   322 Gold Street
   Garland, TX 75042
   214-487-1287
   Fax: 214-276-3375

Sources of Hard-To-Find or Obsolete Components

This company manufactures high temperature (200°C) operational amplifiers that can replace similar products discontinued by Burr-Brown Corporation.

10. ISI-Ideal Semiconductor Inc.
    46721 Freemont Blvd.
    Fremont, CA 94538
    510-226-7000
    Fax: 510-226-1564

This company, established in 1987, manufactures obsolete parts using wafers or tooling supplied by the original manufacturer. Devices can also be reverse engineered and emulated using standard cell devices. Devices can be supplied qualified to MIL-STD-38510, MIL-STD-883 and MIL-STD-19500, MIL-STD-750 and MIL-STD-202. Microcircuits and semiconductors from a variety of original manufacturers, including AMD, Harris, National Semiconductor, IDT, Signetics, Quality, Samsung and Zytrex can be supplied.

11. Lansdale Semiconductor
    2929 S. 48th St., Suite #2
    Tempe, AZ 85282
    602-438-0123
    Fax: 602-438-0138
    http://ssi.syspac.com/~lansdale/

This company, which has been in business over 27 years, manufactures older technology products such as RTL, DTL, TTL and memory devices. Their product line includes devices formerly manufactured at the Signetics company closed bipolar wafer fabrication line in Orem, Utah, and Intel M82XX type bus controller and clock generator drivers. They also have Motorola Semiconductor discontinued military CMOS, MECL and Schottky logic lines. They have the ability to process parts to MIL-STD-883. They have a minimum order policy of $500. (Note: Motorola Semiconductor Products Sector's Commercial Plus Technologies Operation [CPTO] is supplying die for North American military and aerospace customers to: Chip Supply, Inc., Elmo Semiconductor Corp., Minco Technology Labs, Inc., Semi Dice, Inc. [Addresses are provided below.] Motorola announced it was dropping all military products effective June 30, 1996. [They have also ceased making any product in ceramic + DIP packages.].)

12. Linear Systems
    4042 Clipper Court
    Fremont, CA 94538
    510-490-9160
    Fax: 510-353-0281

This company manufactures linear ICs and replacements for discontinued Motorola small signal transistors.

13. Micrel Inc.
    1849 Fortune Drive San Jose, CA 95131
    408-944-0800
    Fax: 408-944-0970

This company has been processing wafers for mature and obsolete MOS technologies since 1978. The available devices include many CMOS and Metal Gate devices formerly manufactured by RCA, National Semiconductor and Rockwell. Micrel uses the same tooling and test programs as the original manufacturer. They can also reverse engineer parts, if necessary from die photographs. Microcircuits can be processed to MIL-STD-883 Class B or S and also radiation hardened ICs.

14. Micro Networks
    324 Clark Street
    Worcester, MA 01606
    508-852-5400
    Fax: 508-853-8296
    508-852-8456

This manufacturer has pin-for-pin compatible parts (their MN3290 DACs) for Burr Brown Corporation's military DAC700.

15. Microsemi Corporation
    2830 S. Fairview Street
    Santa Ana, CA 92704
    714-979-8220
    Telex: 4720306

    8700 East Thomas Road
    Scottsdale, AZ 85252
    602-941-6300
    TWX: 910-950-1320

    2830 South Fairview St.
    Santa Anna, CA
    714-979-8220

800 Hoyt St.
Broomfield, CO 80020
303-469-2161
Fax: 303-469-2161

Raytheon Semiconductor transistor products are now manufactured by Microsemi, Inc.

16. Omnirel Corporation
    205 Crawford Street
    Leominster, MA 01453
    508-534-5776
    Fax: 508-537-4246

This company acquired Motorola's Military Bi-Polar transistor business in 1995 from the Motorola Commercial Plus Technologies Operation (CPTO).

17. R&E International, Inc.
    210 Goddard Blvd. Suite 100
    King of Prussia, PA 19406
    1-800-253-7007
    215-992-0727
    Fax: 215-992-0734

This company, founded in 1987, manufactures, and has stock of, the CMOS SCL4000 series parts formerly manufactured by Solid State Scientific (S Cubed) and Sprague Semiconductor (now Allegro Microsystems, Inc.). Military screened parts (JM38510), DESC and MIL-STD-883 parts are available.

18. Reliable Electronic Manufacturing (REM)
    Division of Wyvern Technologies
    1205 E. Warner Avenue
    Santa Ana, CA 92705
    1-800-962-1085
    714-966-0710
    Fax: 714-556-7014

This company can design and manufacture replacements for obsolete components as well as designing a variety of arrays, networks, piggyback boards and SIP and DIP packaged parts. Their specially-designed parts include RC networks, field programmable modules and active modules. Devices can be designed to meet military specifications.

19. Rochester Electronics, Inc.
    10 Malcolm Hoyt Drive
    Newburyport MA 01950-4018
    508-462-9332
    Fax: 508-462-9512
    Suite 2D, Britannia House
    Leagrave Road
    Luton, Bedfordshire, England LU3 1RJ
    011 44 01 582 488680
    Fax: 011 44 01 582 488681

Discontinued and Custom Packaged military and commercial semiconductors. They also may have in stock some hard to find items that have not been discontinued. This vendor, which has been in business over 9 years, has the facilities to custom package semiconductor dies and also manufactures discontinued parts from die masters. This aftermarket manufacturer is the authorized distributor for obsolete products from various original manufacturers including AMD, Allegro MicroSystems, Altera, Harris Semiconductor, Intel, Linear Technology, Texas Instruments, Microchip Technologies, Inc., Standard Microsystems (SMC), and National (and Fairchild) Semiconductor.

20. Semiconductors, Inc.
    3680 Investment Lane
    Riviera Beach, FL 33404
    800-327-6183
    407-842-0305
    Fax: 407-845-7813
    TLX: 62014516

This company manufactures germanium transistors which have been discontinued by other manufacturers. They also manufacture diodes, SCRs, triacs and voltage regulators (second source to National Semiconductor).

21. MSIS Semiconductor
    (Formerly MSI/Scorpion Semiconductor)
    1999 Concourse Drive
    San Jose, CA 95131
    408-944-6270, 6271
    Fax: 408-944-6272

This company produces the full line of P-Channel Silicon gate MOS technology products formerly supplied by AMD (Advanced Micro Devices). They offer products and design services in N-Channel and

CMOS process technology. There is a minimum order policy of $500 for commercial parts and $1000 for military parts.

22. New England Semiconductor
    6 Lake Street
    Lawrence, MA 01841
    508-794-1666
    Fax: 508-689-0803

A manufacturer of military qualified semiconductors including bipolar (EPI-Base, MESA and Planar process) devices and ultrafast recovery rectifiers.

23. Reliance Merchandising Company
    1-800-796-9629

This company has hard to find potentiometers.

24. Solid State Electronics Corporation
    18646 Parthenia Street
    Northridge, CA 91324
    818-993-8257 (Also used as fax number)

This company sources electromechanical choppers, used in precision dc amplifiers, voltmeters and servo motors. Devices available include stock from companies that have discontinued the parts (i.e., Airpax, Bristol, Stevens Arnold, Brown Converters, etc.) or are manufactured by companies under private labeling agreements.

25. Sunset Silicon Products
    402A Ridgefield Circle
    Clinton, MA 01510
    508-365-6108 (Phone and Fax)

This company recreates the obsolete part functionally using either the original design process, or by new design tools such as gate arrays (the parts are thus either emulated or recreated). If the part is emulated, circuit equivalency is targeted through simulation using model parameters and design rules from existing processes. The part is then manufactured using a gate array or a standard cell with CMOS or Bipolar processing. Using this approach the basic design specifications are realized but the identical performance range of the original part may not be achieved. If the part is recreated, the device is reverse engineered with the process and the circuit is made as close to the original as possible. Components and processes available include: Analog and Digital, Integrated Circuits, Transistors, Diodes, Unijunctions, Hybrids with the following processes; High frequency, Small Signal, PMOS, NMOS, CMOS, TTL, ECL, Linear, High Voltage, Low Noise, Resistor Matching, etc. A one-time NRE fee which varies from about $25K to $45K is charged for the design process.

26. TSI Microelectronics
    (Formerly Transistor Specialities, Inc.
    Hybrid Division)
    5 Southside Road
    Danvers, MA 01923
    508-774-8722
    Fax: 508-774-0939
    TWX: 710 347-0309

This company manufactures custom hybrids and custom packaged parts. They also manufacture obsolete high reliability IC's and transistors that are used in military programs.

27. Visicom Laboratories, Inc.
    Emerging Technologies Division
    17301 West Colfax, Suite 150
    Golden, CO 80401
    303-277-0271
    Fax: 303-277-1505

This company emulates obsolete modules using new (VHDL) technologies. The generic information on the functionality of the circuitry enables the emulation of the circuit using currently available gate arrays, and can also be used to emulate the circuit in any future available technology. This technique was used to develop Universal Standard Electronic Modules (UniSEM) to replace obsolete modules in the U.S. Navy standard computer the AN/UYK-44 MRP.

(Note: MicroLithics Corporation is another division of this company).

Companies that stock obsolete parts, or die, or locate sources of stocked obsolete material:

When discussing part requirements with part brokers be prepared to provide a target price, or the highest amount you would be willing to pay for the device.

(Note: Commercial components from different manufacturers may not be equivalent. These differences may range from some parameter variances to parts with the same part number being totally different.)

1. A.C.P. (Advanced Computer Products), Inc.
   ACP Components
   1317 East Edinger
   Santa Ana, CA 92705
   714-558-8822
   800-347-3423
   Fax: 714-558-1603

This company can supply current and hard to find/obsolete material (including IC's, semiconductors, capacitors, crystals, and diodes).

2. Act Electronics
   Parts Department
   2345 E. Anaheim Street
   Long Beach, CA 90804
   214-433-0475

Service manuals and repair parts for Grundig stereo equipment.

3. All Electronics Corp.
   P.O. Box 567
   Van Nuys, CA 91408
   1-800-826-5432
   818-904-0524
   FAX: 818-781-2653

Various surplus parts, including obsolete items.

4. America II Electronics
   2600 118th Avenue N
   St. Petersburg, FL 33716
   (Also known as A-1 Electronics)
   A-1 Electronics: 800-736-4397
   813-572-9933
   Fax: 813-572-9944
   America II Electronics: 800-767-2637
   813-573-0900
   Fax: 813-572-9696
   http://www.america2.com

(Note: The purchasing division of this company is known as The IC Exchange, 2620 118th Avenue North, St. Petersburg FL 33716, 813-573-0900, Fax: 813-572-9944)

Established in 1989, this company (in memory products) has an inventory of over 10 million ICs and concentrates on second source inventories and obsolete parts. They do deal with all types of parts, electrical, electromechanical, etc.

5. American Design IC Components
   400 County Avenue
   Secaucus, N.J. 07094
   201-601-8999
   Fax: 201-601-8991

This company has a stock of many discontinued parts.

6. American IC Exchange
   (Formerly Bally Micro)
   27 Journey
   Aliso Viejo, CA 92656
   800-821-8858
   800-229-7690
   800-634-6936
   714-362-6555
   Fax: 714-362-5333
   E-mail: ictrader@aice.com
   http://www.aice.com

This worldwide distributor specializes in locating obsolete parts, and keeps a large inventory in stock, including parts from American (from Analog Devices to Zilog), Japanese and Korean suppliers (such as Fujitsu, Goldstar, Hitachi, Mitsubishi, Samsung and Toshiba). They also have the capability to manufacture SRAM and DRAM modules.

7. American Microsemiconductor, Inc.
   133 Kings Rd.
   Madison, N.J. 07940
   201-377-9566
   FAX 201-377-3078

Specializes in obsolete and hard to find Japanese and U.S. parts. Has a network of suppliers to help in locating material.

8. Amps Abundant
   1891 N. Gaffey St.
   Units K&L
   San Pedro, CA 90731
   1-800-233-0559
   Fax: 310-833-9154

Sources of Hard-To-Find or Obsolete Components

This company stocks replacement power semiconductors, including obsolete devices.

9. Audio Parts Company
   1070 South Orange Drive
   Los Angeles, CA 90019
   800-999-5559
   213-933-8141

This specialty parts distributor has replacement parts for some items no longer available in the U.S. including Bohsei (TV sets), Garrard (turntables) and Wollensak (tape recorders).

10. Baldwin Components
    2 Trade Zone Drive
    Ronkonkoma, NY 11779-9709
    1-800-645-5608
    516-588-5700
    Fax: 516-588-5441

This distributor, which directly imports components, has over 500 million surface mount components in stock (including resistors, trimmer potentiometers, capacitors and crystals).

11. Belmont Trading Company
    8141 N. Austin Avenue
    Morton Grove, IL 60053
    1-800-REFURBS
    847-581-9201
    Fax: 847-581-9206
    http://www.belmont~trading.com

This company buys and sells recycled ICs.

12. Best Innovations
    178-8 Ohguchi-Nakamichi, Kanagawa-Ku
    Yokohama-Shi 221
    Japan
    045 402 0068
    045 402 1161
    International Mobile Phone: 81 30 2283889

This company has secondhand (used and surplus) ICs including memory ICs, VRAMs, SIMMs, SRAMs, EPROMs, flash memory, MPUs and FM sound generators. They also sell used computer disk drives, and equipment.

13. Blue Fin Technologies
    55 Green Street
    Portsmith, NH 03801
    603-433-2223
    Fax: 603-433-6437
    http://www.bluefin.com

This supplier has material in stock including hard-to-find and obsolete components.

14. Bruin Electronics
    1100 Seminary Street
    Rockford, IL 61108
    800-573-8934
    815-987-2775
    Fax: 815-987-2776
    http://www.BruinEl.com
    E-mail: BruinEl@aol.com

This company has refurbished hard to find and obsolete parts.

15. Burlington Microelectronics
    (Division of Solid State Testing Laboratory)
    56 Middlesex Turnpike
    Burlington, MA 01803
    617-273-5657
    Fax: 617-273-4896

    This company may have obsolete part die in stock.

16. Chaffin Electronics, Inc.
    Route 1
    330 Lambro Lane
    Franklin Furnace, OH 45629
    1-800-821-7208
    614-574-4456, 6906
    Fax: 614-574-2124

This company, which has been in business for 10 years, specializes in a variety of parts including programmable parts.

17. Channel-Tek International Corporation
    814 South Military Trail
    Deerfield Beach, FL 33442
    800-229-4751
    305-421-2227
    Fax: 305-421-1865

International Toll-Free Numbers:
   Denmark Fax: 800 1 7535
   England Fax: 0800 89 5099
   France Fax: 0590 2025
   Germany Fax: 0130 82 09 75
   Norway Fax: 800 12583
   Sweden Fax: 020 79 7578

This company can supply a variety of components (amplifiers, circuit breakers, semiconductors, ICs, hardware, relays, resistors, switches, terminals, terminal blocks, varistors) and can locate obsolete or hard to find parts.

18. Chip Tech, Ltd.
   585 Merrick Road
   Lynbrook, NY 11563
   800-762-4536
   516-593-3333
   Fax: 516-593-7515

This company specializes in hard-to-find and obsolete components, including passive parts.

19. Chip Supply, Inc.
   7725 N. Orange Blossom Trail
   Orlando, FL 32810-2696
   407-298-7100
   Fax: 407-290-0164
   75664-3074@COMPUSERVE.com

A supplier of tested "known good die" from Micron Semiconductor, Inc., Allegro, AMD, Analog Devices, Atmel, Brooktree, Harris Semiconductor, IDT, Linear Technology, Micro Power Systems, National Semiconductor, Temic, Teledyne, Thomson Components, Texas Instruments and Xilinx. They also have die to support discontinued Motorola military products.

20. Classic Components Corp.
   Corporate Headquarters
   23605 Telo Avenue
   Torrance, CA 90505
   310-539-5500
   Fax: 310-539-4500
   http://206.14.133.66
   http://www.class-ic.com

3252 S. Fair Lane
Tempe, AZ 85282
602-414-1400
Fax: 602-414-0500

2393 Teller Road, #106
Newbury Park, CA 91320
805-499-7499
Fax: 805-498-7775

2121 Old Oakland Road
San Jose, CA 95131
408-434-1600
Fax: 408-434-0999

2005 Cypress Creek Road
Ft. Lauderdale, FL 33309
305-771-1411
Fax: 305-771-2311

33 Cokac Loop
Ronkonkoma, NY 11779
516-588-4445
Fax: 516-558-1116

1826 B. Kramer Lane
Austin, TX 78758
512-832-1222
Fax: 512-832-8444

Europe (Headquarters):
Kimpler St r. 286
4150 Krefel
Krefeld, Germany
49 02151 399993
Fax: 49 02151 311181
Munich Office:
089 4623660
Fax: 089 46236655

Ramat-Gan, Israel
03613 1426
Fax: 03525-4601

This distributor has over $40 million dollars in inventory which includes obsolete parts.

Sources of Hard-To-Find or Obsolete Components

21. Commodity Components International (CCI)
    100 Summit Street
    Peabody, MA 01960
    800-688-7980
    508-538-0020
    Fax: 508-538-3633
    Email: cc1@cci-inc.com
    http://www.cci-inc.com

This company, with its SEMI Search network can locate sources of discontinued or hard to find IC's and semiconductors all over the world.

22. Dataronics
    237350 Blueberry Hill #12
    Conroe, TX 77385
    713-367-0562
    FAX: 713-292-4914

This liquidator has a large quantity of parts, circuit boards and peripherals and can locate anything from microcircuits to platen knobs.

23. Dependable Component Supply
    3620 Park Central Blvd North
    Pompano Beach, FL 33064
    800-336-7100
    954-974-7100
    Fax: 954-974-7101

    2740 West Magnolia Avenue
    Burbank, CA 91505
    800-654-8900
    818-558-7100
    Fax: 818-558-7108

This company, which was started in 1986, with their "Global mainframe parts Finding System," can find discontinued and hard to find parts and also does part recycling.

24. DERF Electronics Corporation
    1 Biehn St.
    New Rochelle, N.Y. 10801
    1-800-431-2912 (Outside NY)
    914-235-4600
    Fax: 914-235-2138

    HC 3 Box 932-1
    Scott Thomson Road
    Old Town, FL 32680
    1-800-874-3373 (outside FL)
    352-542-8002
    Fax: 352-542-3370

In business since 1946, this company buys surplus material and may have obsolete parts in their inventory.

25. Digitron Electronics
    144 Market Street
    Kenilworth, NJ 07033
    1-800-526-4928
    908-245-4928
    Fax: 908-245-0555

This independent stocking distributor finds discontinued and hard to get American, European and Japanese semiconductors and ICs.

26. Dodd Electronics
    P.O. Box 112
    New York, N.Y.
    914-739-5700
    FAX: 914-739-5854

Stocking distributor of obsolete and discontinued integrated circuits.

27. Dynasty Electronic Supply
    A Division of Pace Electronics, Inc.
    200 South Semoran Blvd
    Orlando, FL 32807
    800-447-8589
    407-381-5908
    Fax: 407-381-0785

This distributor specializes in locating passive parts including crystals (and oscillators), capacitors and inductors/transformers. They can locate parts with their World Search (tm) system. They also sell various other electronic components such as semiconductors, connectors, fuses, potentiometers, relays, tools, test equipment and computers and accessories.

28. EDLIE Electronics
    2700 Hemstead Turnpike
    Levittown, LI N.Y. 11756-1443
    516-735-3330
    800-645-4722

Various surplus parts including tubes and IC's.

29. ELECTROnet
    Electronic Parts inventory Database
    Division of Stocknet Corporation
    P.O. Box 550483
    Dallas, TX 75355
    214-503-9844
    Fax: 214-503-9897
    E-mail: electro@stknet.com
    http://www.stknet.com

This company has an Internet electronic parts database to locate stock electronic parts from the current inventories of most major franchised distributors of first sale merchandise. A free trial period is available to potential users.

30. Electronic Expediters International
    Division of Electronic Expediters, Inc.
    14828 Calvert Street
    Van Nuys, CA 91411-2774
    P.O. Box 9
    Van Nuys, CA 91408-0009

    818-781-1910
    Fax: 818-782-2488
    Telex: 910 495 1751 Elec Expd Van

This company has supplies of hard to find and obsolete parts, including semiconductors, ICs (some surface mounted parts), military and industrial components (resistors, shock mounts, tubes, switches, circuit breakers, fans, filters, coils, connectors, cable, knobs, lamps, etc.). A catalog is available.

31. Electronic Material Industries
    10153 1/2 Riverside Drive #108
    Toluca Lake, CA 91602
    818-842-5953, (after 5 pm 818-846-7676)
    Fax: 818-845-3613

This company will locate obsolete and hard to find military capacitors, resistors, transistors and integrated circuits.

32. Electronic Salvage Parts
    2706 Middle Country Road
    Centereach, N.Y. 11720

Various surplus parts.

33. Electronics Unlimited
    20525 Manhattan Place
    Torrance, CA 90501
    310-328-3664
    Fax: 310-328-8367
    E-mail: ELCUN@IX.NETCOM.COM

This distributor may have hard-to-find parts in inventory or can locate parts through its network.

34. Electrospec
    24 East Clinton Street
    Dover, N.J. 07801
    1-800-631-9616
    201-361-6300
    Fax: 201-361-7868

This company locates obsolete or hard to find wire, cable, tubing and electrical connectors.

35. Elmo Semiconductor Corp.
    7590 North Glenoaks Boulevard
    Burbank, CA 91504-1052
    818-768-7400
    Fax: 818-767-7038
    TWX: 910-321-2943
    TLX: 69-8181

This company packages and screens IC die. They also have die to support discontinued Motorola military products.

36. Florida Semiconductor
    Division of Florida Reclamation
    1933 West Copans Road
    Pompano Beach, FL 33064
    954-969-9796
    Fax: 954-969-8823 (sales), 8804 (purchasing)

This company may have obsolete parts in their refurbished parts inventory.

Sources of Hard-To-Find or Obsolete Components

37. Fox Electronics
    309 East Brokaw Road
    San Jose, CA 95112-4208
    408-437-1577
    Fax: 408-437-9299
    http://www.foxelectronics.com

This company recycles parts from printed circuit cards and handles company excess inventories. They have obsolete and hard-to-find parts (parts are identified as New or Pulls).

38. GALCO Indistrial Electronics
    26010 Pinehurst Drive
    Madison Heights, MI 48071
    1-800-521-1615
    810-542-9090
    Fax: 810-542-8031

In addition to providing field and depot repair services for electronic controls, they have a large inventory of parts (such as power semiconductors, fuses, and motor controls), which contains obsolete items.

39. General Components
    927 Calle Negocio
    San Clemente, CA 92672
    1-800-944-3463
    714-361-8800
    FAX: 714-361-0062

In business since 1983, this company has a computerized parts search system (GEN-COM) that includes surplus inventory from OEM's worldwide, as well as distributor material.

40. Graveyard Electronics
    2600 118th Avenue N
    St. Petersburg, FL 33716
    800-833-6276
    813-571-2000
    Fax: 813-571-2018

This stocking distributor, a sister company of America II, specializes in locating obsolete parts.

41. GTC
    216 West Florence
    Inglewood, CA 90301
    310-673-8422
    Fax: 310-672-2905

This company, in business since 1976, has many current devices in stock and specializes in obsolete semiconductors.

42. Gulf Components, Inc.
    5100 North Federal Highway
    Ft. Lauderdale, FL 33308
    1-800-345-2680
    954-492-5383
    Fax: 954-491-8725
    http://www.gulfcomponents.com
    E-mail: gulf@ix.netcom.com

This independent distributor has sources capability to locate military specification and obsolete parts.

43. H&R Enterprises
    21521 Blythe Street
    Canoga Park, CA 91307
    818-703-8892
    Fax: 818-703-5920

This company specializes in hard-to-find/obsolete ICs, transistors and diodes, both military and commercial parts.

44. Hamilton Hallmark
    An Avnet Company
    1-800-668-0852

This distributor bought several million dollars worth of the remaining Motorola military products inventory in the fall of 1995, prior to the deadline for military parts orders of September 29, 1995 imposed by Motorola's Commercial Plus Technologies Operation (CPTO).

45. Hi-Tech Component Distributors, Inc.
    320 North Nopal Street
    Santa Barbara, CA 93103
    805-966-5454
    Fax: 805-966-2354

This distributor finds obsolete and hard to find parts, and also sells hard and floppy drives and computer peripherals.

46. HLK & Associates, Inc.
    1305 SOM Center Road
    Cleveland, Ohio 44124
    1-800-222-3855
    1-800-351-1521 (OEM Hotline)
    216-442-1444
    Fax: 216-442-1412

    1-800-351-1522 (OEM Hotline)
    http://www.industry.net/hlk

This company specializes in finding hard-to-find or discontinued military and commercial parts.

47. Impact Components
    (Formerly JPE-Jarrah Pacific Electronics)
    2300 Boswell Road
    Suite 120
    Chula Vista, CA 91914
    800-326-5139 (1-800-IC Hotline)
    619-421-4808
    Fax: 619-421-5704

    Japan Office:
    4-7-24-307 Heiwadai
    Nerima-Ku, Tokyo
    Japan
    03 3559 2733
    Fax: 03 3559 2838

This company specializes in hard to find semiconductors and IC's. They often sell to distributors and have sources in Southeast Asia and Europe and have a sister company in Australia.

48. Innovative Technology, Inc.
    3821 Moana Way
    Santa Cruz, CA 95062
    408-462-6547
    Fax: 408-479-4818

This company has stock and buys excess OEM inventory and sells and locates hard to find parts such as DRAMs, SRAMs, TTL, Linear and Analog ICs and capacitors.

49. I.T.I. - Imminent Technologies, Inc.
    22529 39th Avenue South East
    Bothell, Washington 98021
    In California:
    619-384-5001
    Fax: 619-384-5003
    In Washington:
    206-485-8030
    Fax: 206-485-7258

This distributor specializes in locating hard to find parts including semiconductors, IC's and passive devices. They have been in business since 1990.

50. International Circuit Sales Corporation (ICS)
    1702 East Highland Avenue
    Suite 403
    Phoenix, AZ 85016
    800-427-7862
    602-224-5322
    Fax: 602-224-5014

    1275 Kennestone Circle
    Marietta, GA 30066
    800-842-0471
    404-427-9906
    Fax: 404-427-9580

This company locates hard-to-find or obsolete parts through its worldwide supplier network.

51. Jacques Ebert Associates, Inc.
    44 School Street
    Glen Cove, N.Y. 11542
    800-645-2666
    516-671-6123

This company stocks and locates hard-to-find capacitors.

52. Jameco Electronics
    1355 Shoreway Road
    Belmont, CA 94002
    415-592-8097
    FAX: 415-592-2503
    415-595-2664
    Telex: 176043

May have some obsolete or hard to find parts in inventory, but occasionally has a flyer sale and disposes of the old material at reduced prices.

53. Jerome Industries, Inc.
    74 West Cochran Street, #B
    Simi Valley, CA 93065
    805-527-5893
    Fax: 805-527-6684

Specializes in Western Digital, Paradise, Faraday, and Chips and Technology IC's but can locate components from a variety of other manufacturers (from Analog Devices to Zilog) and also locate passive components from such companies as AMP, Kemet, Molex, etc.

54. JTM
    539 Valley Gate Road
    Simi Valley, CA 93065
    805-527-9228
    Fax: 805-527-2710
    URL: http://www.jtment.com
    E-mail: jtm@jtment.com, jtm@earthlink.net, and jtm@west.net

Specializes in current, obsolete and hard to find devices from Western Digital, Faraday (Division of Western Digital), Chips and Technology, VLSI, Brooktree, Xlinx, Sierra, Opti, National Semiconductor, and Paradise.

55. Keen Electronics Corporation
    31225 Bainbridge Road
    Cleveland, OH 44139
    800-367-5336
    216-498-3460
    Fax: 216-498-3470

This company (a sister company of Targetronix) locates allocated, hard to find and obsolete ICs, transistors, diodes, rectifiers, passive components (including resistors) and electromechanical parts.

56. Krueger Company
    1544 West Mineral Road
    Tempe, AZ 85283
    800-245-2235
    602-820-5330
    Fax: 602-820-1707

This company has "refurbished" (reconditioned) parts that have been pulled from circuit boards, and excess OEM inventory stock. They specialize in DRAM's, SRAM's, EPROM's, EEPROM's, EPLD/FPGA erasable logic, microprocessors and microcontrollers. They also may have interface, telecommunications, linear, logic, diodes, transistors, capacitors, resistors, connectors, sockets and oscillators.

57. Lantek Corporation
    46 Main Street
    Sparta, NJ 07871
    201-729-9010
    Fax: 201-729-4953

This distributor is part of an international group that can source obsolete or hard to find components internationally with its on-line computer search capability.

58. LECTRO Components, Inc.
    154 Easy Street
    Carol Stream, IL 60188
    708-690-0520
    Fax: 708-690-0563

This distributor stocks various Kulka terminal blocks and can customize blocks for different applications.

59. LJ Enterprises
    68 Railroad Avenue
    Valley Stream, NY 11580
    516-872-5000
    Fax: 516-872-5081, 5082
    (Later in 1995, moving to Newark, DE)

With over 9 years of experience in the aerospace, import and export fields, this company specializes in current and obsolete Japanese semiconductors and passive components (for audio and video equipment). Besides their own inventory they have access to overseas components inventories.

60. Luke Systems, International
    27827 Via Amistosa, Suite 101
    Agoura Hills, CA 91301
    818-991-9373
    Fax: 818-991-4654

This distributor specializes in locating obsolete parts and can search over 500 worldwide companies for hard to find components, including AMD devices.

61. Micro-C Corporation
    11085 Sorrento Valley Court
    San Diego, CA 92121
    619-552-1213
    1-800-723-1357
    FAX: 619-552-1219

    Micro-C/I.I. Ltd.
    1 Whittle Place
    South Newmoor Industrial Estate
    Irvine, Scotland KA11 4HR
    011-44-294-221836
    011-44-294-2211837
    Fax: 011-294-221838

This company specializes in recycling or providing "Reconditioned" ICs, parts pulled off of circuit assemblies. They also sell, with their logo, flash EPROMs, static RAMs and some processor ICs.

62. MicroRam Electronics, Inc.
    222 Dunbar Court
    Oldsmar, FL 34677
    1-800-MICRO-71
    813-854-5500
    Fax: 813-855-3844

This independent distributor stocks hard to find and allocated parts.

63. Micro Wholesale International
    196 Technology, Suite H
    Irvine, CA 92718
    714-753-0360
    Fax: 714-753-0363

This company sells IBM spare parts and may have obsolete ICs in stock.

64. Milex Electronix Corporation
    85 Engineers Road
    Hauppauge, NY 11788
    800-645-5033
    516-231-1500
    Fax: 516-434-1333

This company buys surplus inventory and specializes in finding discontinued or hard-to-find items.

65. Minco Technology Labs, Inc.
    1805 Rutherford Lane
    Austin, TX 78754
    512-834-2022
    Fax: 512-837-6285

This company packages and screens IC die, they also have die to support discontinued Motorola military products.

66. Mission
    7 Bendix
    Irvine, CA 92718
    714-859-1300
    Fax: 714-859-4700

This company specializes in finding memory ICs including DRAMs, SRAMs and TTL and CPU modules.

67. MIT Distributors
    4125 Keller Springs Road
    Suite 160
    Dallas, TX 75244-2035
    1-800-MIT-FIND (1-800-648-3463)
    214-733-3322
    Fax: 214-733-0048

This company, which was founded in 1991, specializes in locating hard-to-find and obsolete parts (including VRAMs and DRAMs). They can search the inventories of over five hundred distributors in the U.S., Canada and Europe.

68. Monarchy
    380 Swift Avenue
    Unit 21
    South San Francisco, CA 94080
    415-873-3055
    800-922-7755

This company specialized in finding Memories, Static RAMs and EPROMs.

69. Nettix Electronix, Inc.
    1111 Alderman Drive
    Suite 280
    Apharetta, GA 30202
    404-751-1911
    Fax: 404-751-7188

Sources of Hard-To-Find or Obsolete Components

This company specializes in finding obsolete ICs (including microprocessors and SRAMs) and semiconductors.

70. The Network Group
    31364 Via Colinas
    Suite 103
    Westlake Village, CA 91362
    818-889-1400
    Fax: 818-889-4987
    E-mail: sales@netgroup.com

This 12-year-old company has a global network to search over 3,200 stocking locations for all types of electronic components including obsolete parts.

71. Network International Component Trading (NICT)
    Box 787
    18 East Main St.
    Malone, NY 12953-0787
    518-483-8181
    Fax: 518-483-8273

This company buys and sells surplus ICs and semiconductors and specializes in obsolete and hard-to-find components.

72. New England Circuit Sales, Inc. (NECX)
    292 Cabot Street
    Beverly MA 01915
    508-927-8250
    Fax: 508-922-1341

This parts broker (independent distributor), established over 10 years ago, has a trademarked "Part/Find" system, a computerized database with over 3000 worldwide part inventories (including Japan, Taiwan and China). The database contains listings of parts from other part brokers, obsolete parts, company excess inventories, etc. They also have an international network called "The Exchange (sm)".

73. NICR
    2640 118th Avenue North
    St. Petersburg, FL 33716
    800-950-3344
    813-523-2000
    Fax: 813-573-1505

This company has over 5 million refurbished (reused) ICs in stock, including EPROMs, SRAMs, DRAMs and Flash Memory ICs.

74. NOW Electronics
    50 Gerald Street
    P.O. Box 829
    Huntington, N.Y. 11743
    516-351-8300
    Fax: 516-351-8354

This company makes lifetime buys and stocks obsolete parts. If devices are not in stock they use their computerized "Semi-Search" system to locate material from a worldwide network of part brokers, distributors and OEMs. This company also stocks die and can custom manufacture devices to meet MIL-STD-883C screening and processing requirements (they are certified to MIL-I-45208).

75. #1 Components, Inc.
    95 Horseblock Road
    Yaphank, NY 11980
    800-790-0013
    516-345-2400
    Fax: 516-345-2406

This distributor, which specializes in surface mount parts and value added services, may have in stock (or can find) obsolete and hard to find parts.

76. OEM (Optical, Electronic and Mechanical) Parts, Inc.
    3029 N. Hancock Ave.
    Colorado Springs, CO 80907
    719-635-0771
    Fax: 719-475-2249

This company specializes in finding various components including ICs and tubes.

77. Orange Coast Components
    26685 Madison Ave.
    Murrieta, CA 92562
    800-736-3726
    909-698-5555
    Fax: 909-698-7661

This vendor may be able to supply obsolete parts from inventory, or find the parts through IC Link (R) Electronic Component Marketing Network.

78. Pacific Diversified Components
    608 South Venice Boulevard
    Venice, CA 90291
    800-669-7329
    310-822-5022
    Fax: 310-822-6062

This company has a global search network to find any type of circuit board component.

79. Peacock Industrial PTE LTD
    53 Jalan Besar
    Singapore 0820
    65 299 9552
    Fax: 65 299 9138

This company specializes in audio, video and TV components. See also Solingen Ltd.

80. Performance Memory Products
    (Also known as Performance Electronics)
    1565 Creek St., Suite 101
    San Marcos, CA 92069
    800-255-8607
    619-471-5383
    Fax: 619-471-9691
    http://www.memorywld.com/~memory/chips/index/html

    In Arizona: Memory World
    1438 Scottsdale Road
    Tempe, AZ 85281
    800-424-1968
    Fax: 602-994-0776

(Also in Australia, 118 Willoughby Rd, Suite 9, Crowsnest New Australia 2065, 011-62-9064533, Fax: 906-1871; and in the United Kingdom: 86-90 High Street, Yiewsley, West Drayton, Middx, UK. 011-44-895-420100, Fax: 895-442238)

This company has manufacturers' replacements for and has discontinued and obsolete memory modules from, various original manufacturers including TI, Toshiba, Hitachi, Mitsubishi, Fujitsu, OKI and NEC.

81. Prime Tech, Inc.
    1210 Warsaw Road
    Suite 900
    Rosewell, GA 30076
    404-594-8608
    Fax: 404-594-8631

This company can supply European parts, ranging from electrical, electronic, electromechanical and mechanical items for different machines and equipment. They can also locate electronic controls/circuit boards for a variety of foreign made machines. In addition they are able to locate hard to find Japanese items through European distributors. This company also provides consultation services in translating technical information and redesigning/modifying plant equipment and systems.

82. Qualified Parts Laboratory (QPL), Inc.
    333 Soquel Way
    Sunnyvale, CA 94086
    408-737-0992
    Fax: 408-736-8708

This company locates obsolete parts or die. They can package the die and perform various screening tests to qualify the packaged part for military Class B or S use.

83. RH Electronics, Inc.
    4083 Oceanside Blvd., Suite G
    Oceanside, CA 92056
    619-724-2800
    Fax: 619-724-3133

This company will go back to the original manufacturer and try to get devices manufactured (if the quantity warrants it), or will obtain enough information to get another company to manufacture the obsolete device. They can source commercial or military qualified parts, including capacitors, resistors, connectors, switches, ICs, transistors, fuses, disk drives, transformers, circuit breakers, relays, filters and lamps.

84. R.W. Electronics, Inc.
    206 Andover Street
    Andover, MA 01810
    508-475-1303
    Fax: 508-475-1461

This company, in business since 1981, specializes in global distribution and sourcing of such material as ICs, semiconductors, disk drives and computer peripherals,

85. **Semi Dice, Inc.**
    Semi Dice SoCal
    High Rel Division
    10961 Bloomfield St.
    P.O. Box 3002
    Los Alamitos, CA 90720
    310-594-4631
    714-952-2216
    Fax: 310-430-5942

    Semi Dice North
    1600 Wyatt Dr. #14
    Santa Clara, CA 95054
    800-325-6030 (Outside CA)
    408-980-9777
    Fax: 408-980-9499

    Semi Dice East
    24 Norfolk Ave.
    So. Easton, MA 02375
    508-238-8344
    Fax: 508-238-9204

    Semi Dice Long Island
    12-5 Technology Dr.
    East Setauket, NY 11733
    516-941-3190
    Fax: 516-941-3185

Note: Call 1-800-345-6633 outside CA and MA.

This company packages and screens IC die, they also have die to support discontinued Motorola military products. The company is a franchised supplier of bare die from various companies including Allegro Microsystems, Inc. (bare die and MOS capacitors); Analog Devices (PMI); BKC Semiconductors; Calogic; Compensated Devices, Inc.; International Rectifier; Philips; Sipex; Semtech; Sussex Semiconductor, Inc. They also have passive components from Amitron; Califronia Micro Devices; Johanson Dielectrics; Novacap; Semi Films Division; State of the Art, Inc.; Tansitor Electronics, Inc., and Vishay (Sprague).

86. Semitech Inc.
    Cooper Run Executive Park
    334 Cooper Road, Bldg A
    Berlin, NJ 08009
    609-768-0030
    Fax: 609-768-5690

This company can locate obsolete parts including ICs, diodes and transistors. They can provide a copy of their computerized inventory run.

87. SG Industries, Inc.
    Manchester, MA
    Barcelona
    Spain
    Sao Paolo, Brazil

A semiconductor trading company formed in 1995 that focuses on finding obsolete parts.

88. Solingen Limited
    14/F, Flat B, Good Year Fty. Bldg.
    119-121 How Ming St., Kwun Tong
    Kowloon, Hong Kong
    852 790 6613
    Fax: 852-341-3587

This company specializes in audio, video and TV components. See also Peacock Industrial PTE LTD.

89. Stock Source (tm)
    57, boulevard Anatole France
    BP 172
    93304 Aubervilliers Cedex
    France
    33 1 48 34 59 00
    Fax: 33 1 48 34 32 49
    sns@calvanet.calvacom.fr

This company specializes in hard-to-find electronic components including IC, optoelectronics, telecommunication circuits, quartz crystals and crystal oscillators)

90. Sun Light Electronics, Inc.
    16011 Foothill Blvd.
    Irwindale, CA 91706
    818-969-0165
    Fax: 818-969-0965
    http://www.SunLight USA.com
    E-mail: EICSUN@aol.com

This distributor has an inventory of semiconductors and ICs and can find discontinued and hard to find items including Japanese parts.

91. Stack Electronics
    200 W. Main Street
    Babylon, NY 11702
    516-321-6086
    Fax: 516-321-5662

    565 Turnpike Street
    N. Andover, MA 01845
    508-681-9977
    Fax: 508-681-9976

This company, using its "Stack Track" (tm), can locate hard to find parts, specializing in surface mount components.

92. Sun Light Electronics, Inc.
    16011 Foothill Blvd
    Irwindale, CA 91706
    818-969-0165
    Fax: 818-969-0965

This distributor has stock of discontinued and hard to find transistors, ICs, SCRs, Triacs and diodes as well as passive and memory parts.

93. Targetronix
    Division Target Electronic Corporation
    6551 Cochran Road
    Solon, OH 44139
    1-800-321-3807
    216-248-7930
    Fax: 216-248-4571

This 17 year old ISO 9002 distributor locates hard to find and obsolete military and commercial parts (ICs, discrete semiconductors, connectors, passive parts and electromechanical parts) from a variety of international sources.

94. TELTECH Research Corporation
    Technical Knowledge Service
    2850 Metro Drive
    Minneapolis, MN 55425-1566
    612-829-9000
    800-833-8330

This company offers a vendor locator service to source specialized parts, materials, equipment and services. They find emergency, or secondary sources of material and determine plant locations, order capacity, stock status and lead time. They also track technologies, patents and offer access to technical experts to answer questions.

95. Trans-World Electronics
    15304 E. Valley Blvd.
    City of Industry, CA 91748
    800-822-1236

This company has parts and servicing information for Multitech, Dyna Tech, Spectrum, and HiTech brands of equipment.

96. Universal Integrated Circuits
    361 Randy Road
    Carol Stream, IL 60188
    1-800-8070-ICS
    708-665-2073
    Fax: 708-665-3803

This company has over 2 million refurbished (reused) ICs in stock.

97. Universal Systems and Components
    1575 Westwood Blvd., Sta 204
    Los Angeles, CA 90024
    310-477-0322
    Fax: 310-479-1447

This distributor, which does searches for obsolete components, handles ICs, diodes, transistor, capacitors and resistors.

98. Unlimited Industries, Inc.
    516-321-9490
    Fax: 516-321-9140

This distributor has ICs in stock and has a global tracking system to search for electronic components.

99. Yuga Enterprise
    705 Sims Drive #03-06
    Shun Li Industrial Complex
    Singapore 387 384
    65 741 0300
    Fax: 65 749 1048

Sources of Hard-To-Find or Obsolete Components

A supplier of Japanese ICs and transistors.

100. Zeus Electronics
(An Arrow Company)
1-800-52-HI-REL
6 Cromwell Street, Suite 100
Irvine, CA 92718
714-581-4622
Fax: 714-454-4355

6276 San Ignacio Avenue, Suite E
San Jose, CA 95119
408-629-4789
Fax: 408-629-4792

37 Skyline Drive
Building D, Suite 3101
Lake Mary, FL 32746
407-333-3055
Fax: 407-333-9681

1140 West Thorndale Avenue
Itasca, IL 60143
708-595-9730
Fax: 708-595-9896

25 Upton Drive
Wilmington, MA 01887
508-658-4776
Fax: 508-694-2199

100 Midland Avenue
Port Chester, NY 10573
914-937-7400
Fax: 914-937-2553

3220 Commander Drive
Carrolton, TX 75006
214-380-4330
Fax: 214-447-2222

Zilog has made Zeus Electronics the sole agent for the sale of their military manufactured products in April 1995. Zeus Electronics can supply product discontinued by Zilog. Zeus, with its affliation to TACTEC can also provide information on substitute or alternate manufacturers of discontinued semiconductors.

Sources of Vacuum Tubes

Vacuum tube manufacturers:

1. California Tube Laboratory, Inc.
   1305 17th Avenue
   Santa Cruz, CA 95062-3096
   800-824-3197
   408-475-2939

This company remanufactures triodes, klystrons, magnetrons, electron guns, ion pumps and other tubes for various applications. They also sell new tubes.

2. Richardson Electronics Ltd.
   40W267 Keslinger Road
   LaFox, IL 60147
   708-208-2401
   708-208-2200
   Fax: 708-208-2550
   TLX: 283461
   1-800-RF Power (1-800-348-5580)
   http://www.rell.com
   Canada: 800-348-5580
   Latin America: 708-208-2200

   Richardson Electronics (Europe) Ltd.
   Inspring House
   Searby Road
   Lincoln LN2 4DT
   England
   0522 542631
   Fax: 0522 545453
   TLX: 56175REEL UK

   Richardson Electronique S.A.
   Direction Commerciale
   1/9, Avenue du Marais
   Batiment Sophocle
   Parc des Algorithmes
   95108 ARGENTEUIL Cedex
   France
   1 34 26 4000
   Fax: 1 34 26 4020
   TLX: 606550 F RICH TUB

Richardson Electronics Italy S.r.L.
Viale L. Ariosto, 492/G
50019 Sesto Fiorentiono (FI)
055 420 10 30
Fax: 055 421 07 26
TLX: 571403 GEBTD

Distretto Nord Italia
P. le Biancamano n. 1
20154 Milano
Italy: 2 331 04 220
Fax: 055 421 07 26

Richardson Electronics Iberica S.A.
Calle Hierro, 9
1a Pianta Nave 10 Edificio Legazpi
28045 Madrid
Spain
1 528 37 00
Fax: 1 467 54 68

Camps Y Fabres 3-11
08006 Barcelona
Spain
3 415 83 03
Fax: 3 415 53 79

Richardson Electronics GmbH
Benzstrasse 28, Bau B
W-8039 Puchheim
Germany
089 80 02 13 1
Fax: 089 80 02 13 8

Richardson Electronics Japan Co., Ltd.
Tachibana Building
1-22, 3-Chome, Negishi
Taito-Ku
Tokyo 110, Japan
3 3874 9933
Fax: 3 3874 9944

Richardson Electronics Pte. Ltd.
Block 25 Bendemeer Road
#01-587
Singapore 1233
65 298 4974
Fax: 65 297 2459
TLX: 33471

A manufacturer of various types of tubes and power semiconductors and a distributor for various electronic components.

3. Svetlana Electron Devices, Inc.
   3000 Alpine Road
   Portola Valley, CA 94028
   415-233-0429
   Fax: 415-233-0439

This is a Russian manufacturer of power grid and modulator electron tubes.

Companies that stock, or locate, vacuum tubes:

1. Alltronics
   2300 Zanker Road
   San Jose, CA 95131
   408-943-9773
   Fax: 408-943-9776

This company has vintage radio vacuum tubes in stock.

2. Antique Electronic Supply
   6221 South Maple Ave.
   Tempe, AZ 85283
   602-820-5411
   FAX 602-820-4643

A source of supply for vacuum tubes.

3. Daily Elexs, Div. E
   Box 5029
   Compton, CA 90224
   213-774-1255
   FAX 213-603-1348

A source of vacuum tubes.

4. Daily Electronics
   10914 NE 39th St.
   Suite B-6
   Vancouver, WA 98682
   800-346-6667

This company has over 4 million tubes in stock.

Sources of Hard-To-Find or Obsolete Components

5. Diers
   4276-SC2 North 50th Street
   Milwaukee, WI 53216-1313

Old radio and TV tubes.

6. EDLIE Electronics
   2700 Hemstead Turnpike
   Levittown, LI N.Y. 11756-1443
   516-735-3330
   800-645-4722

Various surplus parts including tubes and ICs.

7. Electronic Expediters International
   See above for address.

8. Fair Radio Sales Co.
   P.O. Box 1105
   1016 E. Eureka Street
   Lima, Ohio 45802
   419-223-2196
   419-227-6573

Surplus new and used equipment, some ICs, and vacuum and oscilloscope CRT tubes.

9. International Components Corp.
   105 Maxess Rd.
   Melville, N.Y. 11747
   516-293-1500
   FAX: 516-293-4983

A source of supply for vacuum tubes.

10. New Sensor Corporation
    133 Fifth Avenue
    New York, N.Y. 10011
    212-529-0466
    1-800-633-5477
    Fax: 212-529-0486

This company has tubes from worldwide sources (Russia, China, Yugoslavia, Germany, Czeckslovakia and the U.S.) and can burn-in and match tubes.

11. OEM (Optical, Electronic and Mechanical) Parts, Inc.
    See preceding address in sources for ICs

12. Penta Laboratories, Inc.
    21113 Superior Street
    Chatsworth, CA 91311
    800-421-4219
    818-882-3872
    Fax: 818-882-3968

A supplier of various types of tubes including data storage, flash, geiger-mueller, discharge, image orthicon, miniature, plumbicon, power, , receiving, strobe, switching, transmitting, etc.

13. RF Parts
    435 South pacific Street
    San Marcos, CA 92069
    800-RF-Parts
    619-744-0500
    Fax: 619-744-1943

This distributor stocks transmitting tubes (such as EIMAC) and RF power transistors, power modules, power FETs and Japanese RF parts (Mitsubishi, Toshiba).

14. Steinmetz
    7519 Maplewood Avenue
    R.E., Hammond, IN 46324

This organization can supply lists of old and new tubes that can be supplied.

15. Thor Electronics
    P.O. Box 707
    321 Pennsylvannia Ave.
    Linden, NJ 07036
    1-800-666-8467
    908-486-3300
    Fax: 908-486-0923

This company directly imports Japanese and Taiwanese semiconductors, and also sells electron tubes, connectors, wire/cable, and so on. They also sell military qualified parts and locate hard to find obsolete devices. A database of the parts the company has is available on diskette.

16. Tucker Electronics
    1717 Reserve Street
    Garland, TX 75042
    800-527-4642
    Texas: 800-749-4642
    214-340-0631

Has new and "pulls" (tubes removed from equipment) and surplus electronic test equipment.

Companies that Manufacture Obsolete Computer Boards

1. Atlantean Microsystems, Ltd.
   Robjohns House
   Navigation Road
   Chelmsford, Essex
   CM2 6HE
   United Kingdom
   (0245) 494292
   Fax: (0245) 494184

This company manufactures circuit cards for the National Semiconductor Starplex development system, Multibus 1 and Cimbus boards; Motorola's Exorcisors, Exorcsets, Exormacs, VME/10, Micromodules, Versamodules, and VME I/O bus boards; and Intels Multibus 1 boards.

2. Micro Industries
   8399 Green Meadows Dr. N
   Westerville, OH 43081
   800-369-1086

This company manufactures discontinued circuit boards for Intel and National Semiconductor systems, including the National Semiconductor Starplex system.

Organizations that Offer On-Line Parts Locator Services

1. Data-Connection, Inc.
   15231 Alton Parkway #200
   Irvine Spectrum, CA 92718
   1-800-LINK-ME-1
   714-753-8000
   Fax: 714-753-9999

This 14-year-old networking company has offered BrokerLink, an on-line database service, since July 1993. This network lists the ICs of independent distributors in one common database. The service can be searched via a part number or the database can be downloaded to your computer. This service is billable monthly.

2. Fastparts, Inc.
   2026 West Iowa Street
   Chicago, IL 60622
   312-862-4553
   708-933-5346
   Fax: 312-486-0547
   seat@fastparts.com

This organization has an interactive on-line parts trading network. Network members anonymously list "bid" and "ask" prices for parts. Searches are made by a manufacturer and manufacturer part number. Details on quantity available, date code, part marking and part (shipping) packaging are provided. There is a yearly fee for the service.

2. ILS-Inventory Locator Service, Inc.,
   A Ryder System Company
   3965 Mendenhall Road
   Memphis TN 38115
   800-233-3414
   901-794-4784
   Fax: 901-794-1760

A database service is provided where worldwide suppliers list their inventories (electrical, electronic, and mechanical) and capabilities with this independent organization. Quantity and condition (new, used, or overhauled) of the part are included. There is also a listing by part number of companies that overhaul parts/equipment.

3. Netlogic Incorporated
   16 Technology, Ste. 107
   Irvine, CA 92718
   1-800-638-5644 or 714-453-0600
   Fax: 714-453-9210

This software networking company offers "Parts Locator," a buying and selling network that lists sources of materials for sale (the seller negotiates a price). The entire database can be downloaded or you can do searches for the manufacturer's part number, quantity, description and date code. Membership is either quarterly or annually.

SOURCES OF FURTHER INFORMATION ON PARTS

SOURCES OF FURTHER INFORMATION ON PARTS

This book does not include details on specific IC functions or parameters or on distributors or manufacturer representatives. This data can be found in the following publications.

1. D.A.T.A. Business Publishing
   P.O. Box 6510,
   Englewood, CO 80155-9832,
   A division of Information Handling Services.
   303-799-0381
   1-800-447-4666

Reference books on digital, interface, linear, memory and microprocessor integrated circuits and transistors. The books summarize device parameters and provide listings of identical/equivalent parts from different manufacturers.

2. Berkley Design Technology, Inc.
   39355 California St., Suite 206
   Fremont, CA 94538
   510-791-9100
   Fax: 510-791-9127
   E-mail: bdti@bdti.com

This company has a report, Buyer's Guide to DSP Processors that compares and analyzes 23 competitive digital signal processor chip families.

3. EEM, Electronic Engineers Master Catalog
   Hearst Business Communications, Inc.,
   645 Stewart Avenue,
   Garden City, N.Y. 11530,
   516-227-1300
   Fax: 516-227-1901
   http://eemonline.com

EEM is a multi-volume set of books that contain product information from leading electronics manufacturers (including catalog pages), and distributor listings. There is also the EEM Local Sources which is available in nine regional editions listing local distributors in that region for components.

4. Electronics Representatives Association (ERA)
   20 E. Huron
   Chicago. IL 60611
   312-649-1333

This manufacturers' representative organization has available a publication called the Locator, an international directory of manufacturers' representatives in the electronics industry. In late 1995 Harris Infosource International (Twinsburg, OH) sold the Electronic Representatives Directory (ERD), launched in 1948, to this organization. This ERD directory lists over 5,300 electronics representatives.

5. Harris Publishing Co.
   2057 Aurora Road
   Twinsburg, OH 44087-1999
   216-425-9000
   800-888-5900
   Fax: 800-643-5997
   216-425-7150

This publisher issues the European Electronics Directories and other international directories such as Scott's for Canada as well as directories for Italy, Mexico, Puerto Rico and former Soviet Republics. They also issue Industrial Manufacturers Directories, separate issues for different states of the U.S.. Directories are also available on CD ROM and for LAN use.

6. IC Master
   Hearst Business Communications
   645 Stewart Ave.
   Garden City, N.Y. 11539
   516-227-1300
   Fax: 516-227-1901

This reference lists over 80,000 ICs by function. Alternate sources, IC manufacturers, distributors and local sales offices are listed. Specification sheets for some of the ICs are also included in this set of 3 books. The IC Master is available on CD-ROM and the IC Master Directory of Manufacturers' Data Pages and the Alternate Source Directory (ASD) are available on computer disk.

7. ICE Integrated Circuit
Engineering Corporation
"Worldwide Survey of IC Manufacturers
and Suppliers Profile"
15022 N. 75th Street
Scottsdale, AZ 85260-2476
602-998-9780
Fax: 602-948-1925
Internet: 73512.304@compuserve.com

Detailed profiles of semiconductor manufacturers including location of facilities including fabrication facility capabilities, company overview and market strategies, key strategic alliances, financial performance history, and key management personnel.

8. Modern IC Databook
WEKA Publishing Inc.,
97 Indian Field Road,
Greenwich, CT 06830.
203-622-4177

This databook contains data on various types of ICs including: linear, digital, consumer, microcontroller, microprocessor, coprocessor, digital signal processor, data communication, interface, A/D, D/A, synchro-resolver, VCR, optoelectronic, and transducer IC's. Periodic supplements expand the information contained in the book, which includes a functional index, detailed component data sheets and manufacturer address and alternate source list.

9. Nova Electronik GmbH
Donatusstr. 158
D-50259 Pulheim
Postfach 2103
D-50250 Pulheim Germany
02234 984 17-0
Fax: 02234 984 17-19
E-mail: NOVA-GmbH@t-online.de

Nova Electronik, Inc.
Radix International, Inc.
3741 Venture Drive, Ste. 335
Duluth, GA 30136
770-497-9717
Fax: 770-497-0784

This company has IC-Scout (c), a CD-ROM that lists worldwide IC manufacturers, distributors and component part numbers. Various types of data searches can be done. They also have ALABEL (c) (All About Electronics) an electronic component finder on CD-ROM that lists sources of active, passive and electromechanical components.

In addition to these companies, there is a component information service called PartNet, that allows search by characteristics for mechanical and electronic components from numerous vendors simultaneously. The Web address is: http://part.net

APPENDIX

Information Deleted from Listings Because of Insufficient, Incorrect, or Unavailable Data

AB MIKROELEKTRONIK — ASP SOLUTIONS

AB Mikroelektronik GmbH
I-Brandstatter-Str-2/Postfach
(Multichip modules per ICE)

Adobe
(PixelBurst coprocessor for acceleration of high resolution images)

Advanced Semiconductor Manufacturing Company of Shanghai (ASMC)
(Formerly Shanghai Radio Factory Number 7, this is a company jointly owned by the Chinese Government, Philips Semiconductors and Northern Telecom. It can manufacture bipolar, CMOS and nonvolatile memory devices)

Advanced Semiconductors, Inc.

Advanced Telecommunications Modules, Ltd.
UK
ATM, Inc., U.S.
(Asynchronous transfer mode IC chip sets)

Advancel Logic
Cupertino, CA
(A supplier of ATM cores)

Aerospace Display Systems
Manchester, NH
(High contrast dichroic LCDs for aerospace applications)

Alcatel Business Systems
France
ASIC Video-CODEC

Alcatel Espace
(A licensee of the 3-D component packaging scheme known as Tripod [or generically known as MCM-V, V for vertical] from Thomson-CSF)

Alcatel Mietec
Oudenaarde, Belgium
(Developing chip set to support Asynchronous Data Subscriber Loop [ADSL] for data communications over twisted pair copper wire)

Alcatel of North America
7635 Plantation Drive
Roanoke, VA 24019
540-265-0600
Fax: 540-265-0618

Algorex Corporation
145 Marcus blvd.
Hauppauge, NY 11788
(Linear DC servo amplifiers, parts that are screened to military specifications available)

Alpha Microcircuits
Im Technologiepark 1
Frankfurt Oder 15236
Germany
335 557 1752
Fax: 335 557 1759
What do they make?

Alphatec Electronics
Thailand
(IC manufacturing and test)

Alpha-TI Semiconductor Co. Ltd.
Alpha Technopolis
Chaschernsao Province

(A DRAM fab partnership formed by businessman Charn Uswachoke and TI in Thailand in late 1995.)

American Korean International (AKI)
Originally part of American Microsystems, Inc.
Still in business?

American Semiconductor Europe (Germany)
A manufacturer of high temperature semiconductors.

Amex Electronics

Anam Semiconductor & Technology
Seoul, South Korea
(Licensee of TSCI DSP core)

Anshan Semiconductor Factory (ASMC)
Liaoning, China
(A state-owned foundry.)

Angstrem
Russia

Apta
5341 Derry Avenue
Agoura Hills, CA 91301-4510
818-865-1237
Is this the correct company address?
Address from an Internet search.
(Custom hybrids and MCMs)

Artist Graphics
(3D graphics products)

Asahai Microsystems, Inc.
Japan
(Still in business?)
Originally part of American Microsystems, Inc.

Ashtech
San Mateo, CA
(ASICs for their Global Positioning Systems [GPS], products jointly developed with Philips Semiconductors)

ASIC Technology Santa Clara, CA
(An affiliate of UMC and ASIC and a library vendor founded in 1995)

ASP Solutions, Ltd.
Las Vegas, NV ?
(Xium IC for massively parallel computing)

Appendix

ASPEX MICROSYSTEMS — CTS A - C

Aspex Microsystems, Ltd.,
Uxbridge, Middlesex, England
(VLSI Asociate String processor [Vasp])

Atrament, Inc.
Farmingdale, NY
(Thermoelectric [peltier] coolers)

Aucera Technology Corp.
Taipei, Taiwan
(Chip resistors, capacitors and hybrid ICs)

Aydin Microwave
(GaAs fet and bipolar solid state amplifiers, militray, commercial and satellite communications applications)

AZ Displays, Inc.
Aliso Viejo, CA 92656
Fax: 714-643-1423 (Human resources)
(LCDs)

Batron
(LCDs)

Beijing Electron Tube Factory
Beijing, China

(Established in 1956, this state run factory manufacturers vacuum tubes and discrete semiconductors.)

Beyschlag
(Resistor networks)

Bharat Electronics, Ltd. (BEL)
Jalahalli
Bandalore 560 013
Fax: 91 812 343119
TLX: 0845 2244 BEL
(Semiconductors including diodes, transistors, and some ICs.)

BH Electronics
(Magnetics)

Binford Electronics
(On a NET listing)

Bookham Technology, Ltd.
Chilton, England
(Integrated optical circuits. Newbridge Networks, Ltd. is an investor.)

British Telecommunications plc
(Video compression chip sets)

Broadcom Corporation
Los Angeles
Irvine, CA
714-829-5555
(Amplitude modulation and digital transmission ICs for digital cable television, fast ethernet ICs.)

Cabletron
(ASICs used in their hub modules.)

Calvert Electronics International, Inc. (CEI)
Still in business?

Cambridge Display Polymers, Ltd. (CDT)
Cambridge, England
(Light-emitting polymer displays)

Cast, Inc.
Pomona, NY
(ASIC cores including devices from AMD and IDT)

Centre Suisse D'Electronique et de Microtechnique S.A.
(ICE rept)

Centro de Technologia de Semiconductores
(ICE rept)

Chartered Semiconductor Technologies Pte Ltd.
Singapore
(This semiconductor foundry produces nonvolatile SRAMs, some devices sold by SIMTEK Corporation in the US. Chartered has also licensed the flash architecture designs of Silicon Storage Technology [SST], Sunnyvale, CA)

China Electronics Corporation
Wuxi, China
(This firm has an agreement with Atmel Corporation to manufacture Atmel parts)

Chromatic Research, Inc.
Sunnyvale, CA
Mountain View, CA
415-254-1600
Phone discontinued 7/16/96
(Video/graphics controller ICs, VLIW multi media processor (Mpact))

Chunghua Picture Tubes Co, Ltd.
Taiwan
(LCDs)

Coloray Display Corporation
Fremont, CA
Division of Scriptel Holdings, Inc., of Columbs, OH
(Flat panel displays)

Conditioning Semiconductor Device Corporation (CSdc)
Still in business?

Crystalonics, Inc.
Is this company still in business?
147 Sherman Street
Cambridge, MA 02140
415-967-1590
(Formerly Teledyne Crystalonics, this company is a supplier of replacements for military qualified Motorola transistors.)

CTS Knights, Inc.
Subsidary of CTS Corporation
Also known as CTS Frequency Control Division
1201 Cumberland Avenue
West Lafayette, IN 47906
317-463-2565
Control devices
(MAC)

D - F DOLPHIN — FORSCHUNG.

Dolphin SCI Technology A.S.
Oslo, Norway
(Scalable coherent interface (SCI) ICs. Note: CMOS versons are available from LSI Logic Corporation)

Dominion Semiconductor LLC
Manassas, VA
(A joint venture partnership established between IBM and Toshiba Corporation scheduled to start production of DRAMs in 1998. Production output will be to IBM and Toshiba.).

DpiX
Palo Alto, CA
(LCDs. Note: A Xerox PARC spin-off company)

DSP Semiconductors, Inc.
(DSP core and multimedia ICs)

DSP Semiconductors, Ltd.
Givat Scmuel, Israel

DSP Telecommunications, Inc.
(Mobile communications and automotive application ICs including chip sets for cellular phones and mobile radios)

Edge Semiconductor
San Diego, CA
(A designer and supplier of ICs for ATE equipment)

EE Tech, Inc.
(MC68000 microprocessor, 1980 book)

Electronic Designs, Inc. (EDI)
42 South Street
Hopkinton, MA 01748
508-435-2341
Fax: 508-435-6302
Los Angeles, CA
Do they have an office here?

Electronic Transistors Corp. (ETC)

(NPO) Electronica
Voronezh, Russia
(DSPs, microprocessors, power MOSFETs, microwave devices)

Eletec Instruments, Inc.

Elex
Alexandrov, Russia
(A semiconductor operation 70 miles northeast of Moscow)

FET Electronics Ltd.
Newbury
FET Electronics, Inc.
Houston, TX

(Darlington transistor arrays, transistors, diodes, power ICs, op amps, voltage regulators)

EMA Microwave
(Mixers, pin diodes, MMIC assemblies, commercial and military devices, etc.)

Emcore Corporation
Somerset, NJ
908-271-9090
(LEDs, including gallium Nitride GaN blue LEDs)

Emerging Display Technologies
(LCD modules, standard and custom)

EMM
(Electronic Memory modules? Microprocessors in 1980 book)

Ensoniq
(In a NET listing)

Environmental Research Institute of Michigan
313-994-1200
(Programmable pipeline element [PPE] VLSI IC, to accelerate digital images.)

EOS
See *Eos Corporation*

Eos Corporation
805-484-9998
(Miniature power supplies)

Essex International
(SX-200 microprocessor, 1980 book)

ESTI Telecommunicacion-UPM
Spain
(Real-time, harmonic matching Pitch to MIDI converter, multimedia codecs)

Eurodia GesmbH Comp. Co.
Still in business?
Waldgasse 37
A-1101
Wien, 132150
Austria

Eurotech
(Programmable parts such as EPROMs)

FED Corporation
Microelectronics Center of North Carolina (MCNC)

FED Corporation
Hopewell Junction, NY
(FED, field emission displays)

Fiberoptic Alignment Solutions
(Visible laser diodes)

Flat Panel Display Co. Ltd. (FPD)
Eindhoven, Netherlands
(Active matrix LCDs; a company owned by Philips, Thomson, Sagem and Merck)

Forschungsgesellschaft fur Informationstechnik
(F.I.T., Research Institute for Information Technology)
Germany

Appendix

(In 1992 started selling magnetic sensors from ceramic superconductors, superconducting quantum interference devices, SQUIDs)

FQL
?
(FQL 0008A 6E, in Matsushita Floppy Drive)

Fujitsu-Towa
(Formerly Pal-Tech Electronics, a manufacturer of multichip modules, thick film hybrids, flex circuit assemblies, wireless communications modules and SAW filters)

G2 logo

Galvantech
(Listed ICE 5/96)

Gems Sensors Division
Plainville, CT
800-321-6070
(Solid-state relays, zener barriers)

General Microchip, Inc.
Markham, Ontario
Canada
(Image/video resizing IC)

GMT Microelectronics
Valley Forge, PA
610-666-7950
(A foundry)

Goal Electronics, Inc.
185 Frobisher Dr.
Waterloo, Ontario, Canada
519-725-0932
(IC Master. Make?)

G&S Engineering
Moreno Valley, CA
(Hall Effect speed sensors)

G.T. Microwave, Inc.
P.O. Box 369
Denville, NJ 07834
(Reflective [RF, microwave] switches)

Halbleiterwerk Frankfurt On-der-Oder
Germany
(Bipolar semiconductors. Note: This company is in a joint venture with Synergy Semiconductor Corporation)

Hall Technologies, Inc.
4042 Clipper Ct.
Fremont, CA 94538-6540
Old address?
Milipitas, CA

Handock Co., Ltd.
Taipei, Taiwan
886 02 772 3618
Fax: 886 02 772 3619
(Optoelectronics)

Headland Products, Inc.
Milipitas, CA
Toronto
Part of LSI Logic Corporation
(A/D converters. ADC16M1)

Hosiden (Hoshiden?) Corporation
Japan
(LCDs)

Huajing Microelectronics Group Co.
China Huajing Electronics Group
Wuxi, China
(A discrete, bipolar and CMOS foundry in a venture with Siemens to manufacture DRAMs)

Iljin Group
South Korea
(In 1995 allied with IDT to use their SRAM technology)

Image Quest Technologies, Inc.
Fremont, CA
(TFT active matrix LCDs, this company started in 1992 funded by and 55% owned by Hyundai Electronics Industries Co., Ltd., of Korea)

FORSCHUNG.— INTERCONNECT

Imtech, Inc
702 Avenue R
Grand Prarie, TX 75050

InfoChip Systems, Inc. (ICE rept)
2840 San Thomas Expressway
Santa Clara, CA 95051
408-727-0514
Fax: 408-727-4190
What do they make?

In-Phase Electronics

Instant Circuit Corporation (ICE rept)
Instant Circuit Holdings, Inc.

Institute of Microelectronics Circuits and Systems (IMS)
Duisberg Germany
(Part of Germany's national Fraunhofer research network. Microcontroller cores and microcontrollers.)

Integral
Minst, Belarus
(CMOS logic, analog circuits for TV, video and audio equipment)

Integrated Systems
http://www.isi.com

Intelligent Resources Integrated Systems
Arlington Heights. IL
(Interactive television chip sets.)

Interconics
Power Supplies?
225 E. Vista Industria
Compton, CA 90220

Interconnect Technology
Sarawak, East Malaysia
Contact; Catherine Figueria: 408-467-9940
J. Shelton Associates: 214-239-5119

I - L INTERCONNECT — LEUZE

(A semicounductor foundry that is a joint venture between National Semiconductor and the Malaysian government.)

Integraphics Systems, Inc.
(A 1993 startup)
408-982-8588

(The Cyber2000 allows computer viewing on a conventional television set [a multimedia graphics accelerator that incorporates a TV encoder])

Intermettal Semiconductor
Germany

International
Newbury Park, CA
(GaAs ICs)

International Computers Ltd.
(Partly owned by Fujitsu/Nokia)
http://www.icl.co.uk/
ICs?
(Flat screen displays, passbook printers and systems for retailers and petrol [gas] stations. This division in Finland.)

International Devices, Inc.

Interphase Corporation?
13800 Senlac Drive
Dallas, TX 75234
214-919-9000
Fax: 214-919-9200

Interuniversities MicroElectronics Center (IMEC)
Leuven, Belgium
(Direct sequence spread-spectrum ASICs, developed with Sait Systems and the European Space Agency, silicon retina)

IRT Corporation
Electronic Systems Div.
3030 Callan Road
San Diego, CA 92121
619-450-4343 (Disconnected)
TLX: 69-5412
(Nuclear event detectors)

Isotronics, Inc. ?
Industrial park
New Bedford, MA 02746

ITT Semiconductors
A good address in the U.S.?
500 Broadway
Lawrence MA 01841

iTV Corporation
San Mateo, CA
(MISC [Minimal Instruction Computer chip])

Keltron Power Devices, Ltd. ?
Shoranur Road
Malangunnathukavu Trichur District
Kerala 680581
India
(Transistors)

Khandelwal Electronics and Finance Ltd. (KEFL)
Bombay, India
(Diodes)

KK-Chip
P.O. Box 3215
2601 DE Delft
The Netherlands
(Chip directory, producing silicon. A foundry?)

Kobe Steel
Kobe Kyongo Corp.
Hyogo-Ku
Kobe Japan
(A foundry company)

Korea Electronics Co., Ltd.
9F Chamber Bldg 45
4-KA, Namdaemum-Ro Chung-Ku
C.P.O. 4896
Seoul, Korea
Fax: 756-5800
TLX: KECOS K27424
(Diodes, optoelectronics, transistors, thyristors, linear ICs)

Koyo
(Crystal oscillators)

KSC Semiconductor
MA?
(Zeners)

Lanstar Semiconductor, Inc.
Arlington, TX
(A fabless DRAM supplier using technology licensed from Texas Instruments)

Laserpath
1986 listing
Still in business?
160 Sobrante Way
Sunnyvale, CA 94086
408-773-8484
Fax: 408-736-2697
TLX: 510-601-0583
(Quick turnaround [one day] gate arrays)

Laser Components GmbH
Am Weidegrund 10
D-8038 Groebenzell, Germany
(Diode laser systems)

Laser Display Technologie GmbH (LDT)
Gera Germany
(Diode pumped solid state blue laser which can be used in a projection laser display)

Leuze Electronic
Owen, Germany
(Optoelectronic sensors)

Lien Hsing Integrated Circuits
Taiwan
(A new contract fab, opened in late 1996. This company is a partnership between UMC [United Microelectronics Corporation] Alliance Semiconductor and S3, Inc.)

Linear Technology Corporation
Do they have a Hong Kong office?

Linfinity Microelectronics
Is this still a valid address?
7382 Bolsa Avenue
Westminster, CA 92683

Maxim Integrated Products
http://www.maxim-ic.com
(Various ICs including linear devices, op amps, comparators, ADCs, references, interface ICs, microprocessor supervisors, DC/DC converters, wideband and current feedback amplifiers, switching regulators, voltage regulators, voltage monitors, power management ICs, analog multiplexers, switched capacitance filters, etc. Note: Samples of up to 2 pieces are available from Sunnyvale CA at Ext 6215. The company will also take small orders of 100 pieces, or less if there is a long delivery schedule, and the need is urgent.)

Magellan
(ASICs for their Global Positioning Systems [GPS])

Malaysian Institute of Microelectronic Systems
Kuala Lumpur, Malaysia,
(Started operation in 1993 as a 1.5 micron CMOS ASIC foundry)

Marvell Semiconductor, Inc.
San Mateo, CA
(Founded in February 1995, this company sells PMRL [partial response, maximum likelihood response] ICs for disk drives.)

Massachusetts Component, Inc.
617-246-1784
(This company packages and screens IC die)

MECL
Tiajin, China
(Motorola China Electronics, Ltd.; a Motorola fab subsidiary incorporated in 1992)

MED prefix
See *MedianiX*

MedianiX
Mountain View, CA
(Founded in September 1994 and 50% owned by New Japan Radio Corporation. Audio multimedia ICs including ICs for Dolby prologic and karoake machines.)

Mercury Semiconductors
Portion purchased by Cypress Semiconductor?

Messerschmitt-Boelkow-Blohm
Germany
(Associated with Temic Telefunkin Microelectronic, a licensee of the 3-D component packaging scheme known as Tripod [or generically known as MCM-V, V for vertical] from Thomson-CSF)

Microelectronics and Computer Technology Corporation (MCC)
Austin, TX
(This company is working on advanced MCM packaging)

Microcosm Communications, Ltd.
Portsmouth, England
(A fabless semiconductor start-up [in 1996] providing chip sets for fiber optic communications, such as 10 Base FL devices)

Micron Display Technology, Inc.
Boise, ID
(Field-emmission displays)

MicroUnity Systems Engineering, Inc.
Sunnyvale, CA
(San Jose???)
(A 1996 startup company. Communications [media] processor IC, cable modems, mixed signal codecs)

Midas Microelectronics Corp.
Taipei, Taiwan
(Sold in the U.S. by Skywell Electronics Corp., in southern and northern CA.)
(Hybrid ICs)

Mid-West Microelectronics
New York (Offices)
Missouri (Foundry)
816-251-5228
(A foundry for semiconductor designers)

Mikron
Zelenograd, Russia

Morion
St. Petersburgh, Russia
(This crystal and oscillator company is majority owned by KVG and Staudte Engineering [Cedar City, UT])

MOS Technology
Still in business?
3330 Scott Blvd.
Santa Clara, CA 95051
(Became Commodore Semiconductor, microprocessors, MCS prefix?)

Mosaic Microsystems, Ltd.
Kent, England
(This supplier of RF/IF ICs for communication circuits, including cordless telephones and direct broascast satellite systems, was acquired by Analog Devices in 1996. Mosaic, Inc., in the U.S., was also acquired.).

M - O
MOSYS — OREN

MoSys (Monolithic Systems)
San Jose, CA
Santa Ana, CA
(DRAMs and speciality memory devices)

M.S. Transistor ?
From RH Electronics Brochure

National Semiconductor
Check address
Deleted for Sams Delivery
Singapore:
National Semiconductor Asia Pacific Pte. Ltd.
Southpoint 200
200 Cantonment Road #13-01
Singapore 0208
Singapore
65 225-2226
Fax: 65 225-7080
Telex: NATSEMI RS 50808

NCI ?
5900 Voss Road
West Palm Beach FL, 33407

NCI Systems
Division of Communications Techniques
(DTOs, Digitally Tuned Oscillators)

NCM Corporation
20393 Kent Way
Los Gatos, CA
408-926-0290
Discontinued 7/30/96
From IC Master
Who are they?

Nekko
(In Radio Shack surplus board; had other NEC parts)

Nestor, Inc.
One Richmond Square
Providence, RI 02906
(Neural network ICs, designed by Nestor, Inc., and Intel for intelligent character recognition systems)

Nichia Chemical Industries, Ltd.
Anan, Tokushima
Japan
(Pure green LED with a brightness of 6 candela, 525 nm wavelength, at 20ma current)

Niec/Qmi, Inc.
Caliente/Twin oaks, CA
800-996-3088
(MOSFETs and FET switching transistors)

NKK Corporation
Korea
(Semiconductors including masked ROMs, SRAMs, RISC microprocessors, and mass storage optical products)

Noise/COM
(Thought they were moving to Ramsey, NJ, in 1995?)
E. 49 Midland Avenue
Paramus, NJ 07652
201-261-8797
Fax: 201-261-8339
(Noise soure instruments and references)

Nordic VLSI
Trondheim, Norway
(ASICs under joint agreement with Harris Corp.)

North American Semiconductor, Ltd.

North Light Displays
(A subsidiary of S.I. Diamond Technology)

Novalog, Inc.
Costa Mesa, CA
714-429-1122
(Infrared transceivers. This company is owned by Irvine Sensors.)

NS Electronics
Bangkok, Thailand
(An IC manufacturing and test plant)

Nucleonic Products Company
Still in business?
(Division of Thomson)
P.O. Box 1454
Canoga Park, CA 91304
213-887-1010
TWX: 910-494-1954
TLX: 651-479

6660 Variel Avenue
Canoga Park, CA 91303
(Various semiconductors including Ge, Si, voltage regulators, zener diodes, band switch, and variable capacitance diodes, transistors, etc.)

Omega Micro, Inc.
155 North Wolf Road
Sunnyvale, CA 94086
408-522-8895
Fax: 408-774-9330

440 Oakmead Parkway
Sunnyvale, CA 94086
408-992-1100
Fax: 408-774-9330
(PCMCIA controller ICs and related items.)

Issue of 12/14/92 has this address:
44160 Plymouth Oaks Drive Plymouth
MI 48170
313-416-8500
Fax: 313-416-8520

Oren Semiconductor Ltd.
Yoqne'am, Israel
Santa Clara, CA
(A company formed by Zoran Corporation and The Goldtron Group of Singapore in 1995, to develop ICs to correct flawed television images, including ghost canceller ICs)

OZ Optics, Ltd.
244 Westbrook Road
Unit #2
West Carleton Industrial park
Carp, Ontario K0A 1L0
Canada
Phone?
(Pigtailed laser diodes)

Pacific Microelectronics Center (PMC)
Vancouver British Columbia
(Multichip modules)

Is this the same company as:
Pacific Microelectronics Corporation (PMC)
10575 Southwest Cascade Blvd.
Portland, OR 97223
1-800-622-5574
503-684-5657
Fax: 503-620-8051
http://www.pmcnet.com/~pmc
Email: pmc@pmc@pmcnet.com ?
(Design services for design, assembly, testing of multichip modules, electronic circuit boards and IC packages including BGA packages)

Pacific Microsonics
CA
(High- Defination Compatible Digital ICs for audio CD playback)

Pacific Technologies Group
Santa Clara, CA
408-764-0644
Fax: 408-496-6142
(Computer core logic ICs)

Panafacom
(MN prefix microprocessors, 1980 book)

PC Tech
(A unit of Micron Electronics)
(Controller ICs)

Philips Photonics
Israel: 0 3 6450315 ?
 Fax: 0 3 493272 ?
Are these phone numbers correct?

Pico Systems, Inc.
(At the Enviromnental Research of Michigan, ERIM)
313-994-1200
(MCMs with antifuse technology)

PictureTel Corporation
USA
(Audio CODECs for video conferencing)

Picvue Electronics Ltd.
Taiwan
(LCDs; sister company is Polytronix, Inc., in TX)

PIMC
Boeblingen Hulb, Germany
(A Philips Electronics, NV and IBM company, formed in October 1994, to share memory capacity at an IBM memory fab facility)

Pioneer
NJ
(Per Pioneer [distributor] representative, no affliation. This company buys parts and re-marks them)

Pioneer Electronics
Japan
(Licensed the organic electroluminescent [EL] flat panel display technology from Kodak)

Pioneer New Media Technologies, Inc.
600 E. Crescent Avenue
Upper Saddle River, NJ 07458-1846
(Digital imaging chip sets for JPEG, MPEG pr H.261 digital images.)

Plessy Tellumat
South Africa
Address of the HQ?

In the U.S. contact:
LNY Sales, Inc.
548 Sunrise Highway
West Babylon, NY 11704-6003
516-661-8900
(Low noise wireless communications amplifiers)

PNY Electronics, Inc.
200 Anderson Ave.
Moonachie, NJ 07074
201-438-6300
Fax: 201-438-9144
Do they have an office in CA?

Santa Clara, CA
Point Nine Technologies
(DMOS RF units)

PolyVision
NY, NY ?
Subsidiary of the Alpine Group
(Flat panel displays)

PowerTV
A subsidiary of Scientific Atlanta
Cupertino, CA
(Eagle IC, a multimedia IC, for set-top TV boxes)

President Internation Corp.

Princeton Microsystems, Inc.
37 Station Drive
Princeton Junction, NJ 08550
(VCOs)

Qudos Technology Ltd. (ICE rept)
Didcot, England
(ASIC prototypes, MCM modules)

Racom Systems, Inc.
Englewood, CO
(Ferroelectric memories)

R - S RAMBUS — SIRF TECHNOLOGY

Rambus Inc.
Mountain View, CA
(DRAMs)

Rendition, Inc.
Mountain View, CA
(3-D rendering [graphics processor] ICs, including the Verite IC)

Resotech, Inc.
(Dielectric resonator oscillators; phase-locked, voltage-controlled and free-running)

Rhopoint
(Thermistors)

Roland ?
(Photo in Pacific Microelectronics Corp brochure on a Digital Voice Module, part of Rogers Instrument Corp? 714-727-2100? Manufacturer of monitors, plotters, multisync [1983 catalog])

RWG AG
Germany
(In partnership with American Semiconductor, Europe [Germany] to design high-temperatue semiconductors)

Sarif, Inc.
(Polysilicon LCDs)

Semefab, Ltd.
What is the UK headquarters address?
United Kingdom

Ste 800
1901 N. Roselle Road
Schaumburg, IL 60195
708-490-6475
(ASICs and various ICs)

Semi Processes, Inc.
See *SPI* ?

Semiconductors, Ltd.

Setron
(MOVs, metal oxide varistors)

Shanghai Belling Microelectronics Manufacturing Co., Ltd.
Shanghai, Peoples Republic of China
(A joint venture of Alcatel Bell of Belgium and state-owned companies, Shanghai Radio Factory No. 14 and Shanghai Bell Telephone Equipment Manufacturing Co. Ltd.)
(Microprocessors [Intel 8088 and 8086 compatible], ASICs and ICs for digital telephone switching systems, phone sets, remote controls, clocks, and consumer and communications products)

Shanghai Gulf Semiconductor Factory
Shanghai, Peoples Republic of China
(Rectifiers and schottky diodes)

Shanghai Hua Xu Microelectronics Co.
(VCR ICs in a joint venture with Matsushita Electric Industrial Co. and Matsushita Electronics. The Shanghai assembly operation is called Shanghai Matsushita Semiconductor, Ltd.)

Shenzhen Electronics Group (SEG)
Peoples Republic of China (State owned)
Shougang NEC
Beijing, China
(A joint venture between Shougang Corp. [40%] and NEC [60%], this wafer fab started manufacturing CMOS and bipolar TV ICs, 64K DRAMS and telecom ICs in 1993)

Shuttle Technology
Fremont, CA
(Designs and manufactures ASICs for OEM printed circuit board configuration or cable solutions)

Signum Systems
171 E. Thousand Oaks Blve, Suite 202
Thousand Oaks, CA
805-371-4608
(Listed IC Master. What do they make?)

Silicon and Software Systems
Ireland
(Image processing, video processing and timing ICs)

Silicon Dynamic
Sunnyvale, CA
(Formed by Oki America, Inc., in early 1996, this company develops technologies for multimedia and telecommunications, including codecs, MPEG, DSP and RISC microprocessor products.)

Silicon Sensors, GmbH
Berlin, Germany
(Silicon photodetectors. See Electron Tubes, Inc., Rockaway, NJ)

Silicon Video Corporation
Cupertino, CA
(Plant, San Jose, CA)
(Thin CRT version of field-emmission displays)

Single Chip System Corporation
16885 West Bernardo Drive
Suite 295
San Diego, CA 92127
(A fabless company supplying radio frequency ID labels containing a programmed IC)

Sintek
Hong Kong
(This wafer fab started manufacturing SRAMs and ASICs in 1993)

Sirf Technology, Inc.
Santa Clara, CA
(Founded in 1995 this company is involved in consumer GPS [Global Positioning Systems] and GPS ICs)

Skynet
408-945-6355
Fax: 408-945-6357

Southern California:
714-252-1239
Fax: 714-252-1241
Fax Express: 516-227-1435
http://www.skynetusa.com
(Power supplies)

South African Micro-Electronic Systems Ltd. (Sames)
Pretoria, South Africa
(51% owned by Austria Mikro Systems International AG [AMS])
(A BiCMOS and CMOS wafer fab that manufactures ICs for the South African telecommunications industry)

Spase BV
Nijmegen, Netherlands
(MPEG ICs and CD-I [Compact Disk Interactive] ICs)

SPI - Semi Processes, Inc.?
1971 N. Capitol Avenue
San Jose, CA 95132
What do they make?

Styra
(PC chip sets, including secound sources for Chips and Technology parts)

Suntac, ST62C006 Japan, computer part
(PC chip sets, including secound sources for Chips and Technology parts)

Superex Electronics Corporation
?
Still in business?
What do they make?
151 Ludlow Street
Yonkers, N.Y.10705

Switch-Power
San Jose, CA
(Power supply regulators for the Pentium processor)

Symetrix Corporation
Colorado Springs, CO
(Ferromagnetic RAMs or FRAMs)

Syntek
1407 116th Avenue, N.E. #117
Bellevue, WA 90004
800-548-8911
Fax: 206-462-7170
Do they have a plant in Taiwan?
(This wafer fab started producing CMOS 4, 16 bit MCUs [microcontroller units] and domestic appliance ICs in 1994)

Symbol
Address ?
(Semiconductors, available from Taitron Components, Inc., 25202 Anza Drive, Santa Clarita, CA 91355-3496, 800-824-8766, 800-247-2232, 805-257-6060; Fax: 800-824-8329, 805-257-6415)

Tadpole Technology
(SPARCBook MCM IC)

Taisil Electronic Materials Corporation
Hsinchu Science Based Industrial Park
Hsinchu, Taiwan
(A silicon wafer foundry that is a joint venture between MEMC [Monsanto Electronic Materials Corporation] Electronic Materials, Inc., China Steel Corporation of Taiwan Taiwan State Bank, Chiao Tung Bank, and China Development Corporation, a Taiwan investment Bank.)

Taiwan Research Service Organization
Hsinchu, Taiwan
(Multichip modules using facilities of PMC, Pacific Microelectronics Center, Vancouver, British Columbia)

Tatung Company
22, Sec. 3, Chungshan N. Road
Taipei, Taiwan R.O.C.
886 2 592 5252
Fax: 886 2 591 5185

Tatung Company of America, Inc.
Long Beach, CA
310-637-2105
Fax: 310-637-8484
(This company opened a new fab facility for DRAMSs and ASICs. Has this been postponed [see EBN 12/11/95] on the Hsinchu plant? Otherwise, in the CRT and Monitor business)

TCSI Corp.
(Formerly Teknekron Communications Systems)
(A DSP software vendor that licenses its custom DSP core)

TCTM acronym ?
GIDEP ALERT ON MIL-S-19500/343
JAN2N2857
GIDEP ALERT 3/8/93

Teac Corporation
Do they make ICs also?
Musashino Center Bldg.
1-19-18 Naka-Cho
Musashino, Tokyo, 180
Japan
0422 52 5041

Teac America, Inc.
Data Storage Products Division
7733 Telegraph Road
Montebello, CA 90640
213-726-0303, 727-7609
Fax: 213-727-7621
TLX: 677014 TEAC MBLO
San Jose, CA: 408-437-9055
Chicago, IL: 708-490-5311
Austin, TX: 512-329-1037
Boston, MA: 508-683-8322

T - U   TEAC — UNITED MEMORIES

Teac Deutschland GmbH
Arbergerstr. 10
D-8036
Herrsching, F.R. Germany
08152-37080
(Computer drives)

Shinjuku-Ku
Is this facility for ICs?
Tokyo, Japan

Tech Semiconductor Singapore Pte., Ltd.
Singapore
(A joint venture of TI, HP, Canon Inc. and the Singapore government [Economic Development Board]. This wafer fab started in 1993)
(Captive market CMOS DRAMs, the first product 16M DRAMs, for the owner companies)

Technical Data Freeway
Concord, MA
http://www.tdf.com
(RISC processor, telecomm, multimedia and microcontroller cores. This company has an exclusive agreement with Actel for their FPGAs. The Argonaut RISC processor core comes from Argonaut Technologies, Ltd., United Kingdom)

Technical Development Company
Bethpage, NJ
(ISA bus chip sets)

Temwell Corporation
Fax: 886-2-5515250
(This is the ad fax in RF Design magazine)
(Dual-balanced mixers, power splitters)

Texas Instruments
Are these locations still valid?

1 Sierra Gate Plaza
Suite 255B
Roseville, CA 95678
916-786-9206

370 S. North Lake Boulevard
Suite 1008
Altamonte Springs, FL 32701
407-260-2116

Iowa
373 Collins Road N.E. Suite 201
Cedar Rapids, IA 52402
319-395-9550

8 Woodlawn Green
Suite 100
Charlotte, NC 28217
714-527-0930, 704-522-5487
Correct area code?

Washington
5010 148th Avenue N.E.
Building B, Suite 107
Redmond, WA 98052
206-881-3080

Texas Memory Systems
Houston, TX
(DSP ICs including the swiFFT IC)

Texet Corp.

(Three-D) **3Dfx Interactive, Inc.**
Mountain View, CA
(3-D rendering ICs)

TIP prefix
(Transistors)
See *Texas Instruments*, *AVG Semiconductor* (Second source)
Is this also true for Motorola?

T&O
(In ad for Kingston Technology
800-337-3812
http://www.kingston.com/vbnw.htm)
(Crystals)

Tohoku Semiconductor (ICE rept)
Tokyo Japan
(A joint venture of Toshiba Corporation and Motorola, Inc., to manufacture microprocessors and ASICs)

Tower Semiconductor Ltd.
Migdal Ha'Emek, Israel
(A foundry formed by the partial buyout of National Semiconductor im March 1993)

Transitron Electronic Corporation
Still in business?
168-182 Albion Street
Wakefield, MA 01880

100 Unicorn Park Drive
Woburn, MA 01801
617-933-9640
(Diodes, transistors, thyristors, various ICs)

TwinStar Semiconductor, Inc.
Richardson, TX 75083-4104
Fax: 214-578-5914 (Human resources)
(A DRAM joint venture formed in January 1995, by TI and Hitachi Ltd.. Product will be sold under the TI or Hitachi labels, not as TwinStar)

TXC logo
?
(Oscillators)

ULSI Systems, Inc.
(Coprocessors)

UMax
Taiwan
(A DRAM foundry in partnership with Mitsubishi)

United Memories, Inc.
(Originally formed by Ramtron International Corporation and Nippon Steel Semiconductor)

U - Z
UNI-TRAN — ZF MICROSYSTEMS

Uni-Tran Semiconductor Corp.
Still in business?
Rt 2
Lake Ariel, PA 18436
717-698-5617

Universal Semiconductor, Inc.
1925 Zanker Road
San Jose, CA 95112
408-279-2830, 408-436-1906
Fax: 408-436-1125
(Optoelectronics, thyristors, transistors, digital, interface and microprocessor support ICs)

Unizon
Komatsu Group
Address?
(Available from Taitron Components, Inc., 25202 Anza Drive, Santa Clarita, CA, 91355-3496, 800-824-8766, 800-247-2232, 805-257-6060; Fax: 800-824-8329, 805-257-6415)

UPI Semiconductor

UTRON Technology, Inc.
Taiwan
(Foundry)

Valence Semiconductor (HK) Ltd. (ICE rept)
Flat 7A, Tai Po Commercial Centre
152-172 Kwong Fuk Road
Toi Po New Territories
Hong Kong
852 656 1151
Fax: 852 652 2301
What do they make?

Valgo GmbH
Still in business?

Videologic, Inc.
1001 Bayhill Dr., Suite 310
San Bruno, CA 94066
1-800-578-5644
415-875-0606
Related to UK company VideoLogic?
(PowerPlay64 coprocessor, video ICs, display cards)

VisionTek
Gurnee, IL
(Memory modules)

VTT (ICE rept)
?

WaferDrive Corporation
Scotts Valley, CA?
Not here per phone co. 8/8/95
(Stacked memory used in PCMCIA cards)

Walburn Devices, Inc.

Watkins-Johnson Company
3333 Hillview Avenue
Palo Alto, CA 94304-1223
1-800-WJ1-4401
415-493-4141, 415-813-2972
Fax: 415-813-2402, 415-813-2515
TLX: 34 8415

Full address for England?
Watkins Johnson International
Windsor, England
(Amplifiers and mixers; items for spacecraft use available)

Xtechnology, Inc. (Xtec)
Santa Clara, CA
A subsidiary of Integrated Information Technology, Inc. (IIT)
(Video processor ICs)

YE Data
(Computer IC in floppy drive?)

Zenamic
(MOVs [metal oxide varistors])

ZF MicroSystems, Inc.
Palo Alto, CA
(PC computers in ultraminiature modules)

OBSOLETE PARTS DISTRIBUTORS

Apricot International
Address?
813-321-3008
Fax: 813-327-0200
This company locates allocated or obsolete components.

Greenpeace Electronics
Part of ZD Integrated Circuits
Largo, FL
An independent distributor that specializes in the resale of obsolete products and excess inventory purchases.

M.S. Hitech
Address?

PROMPT® Publications

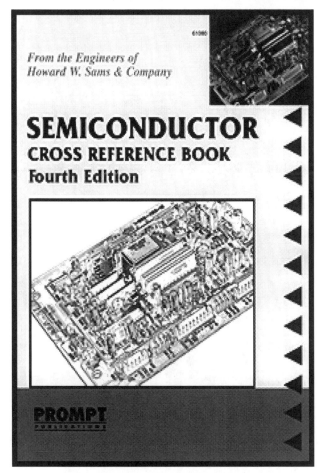

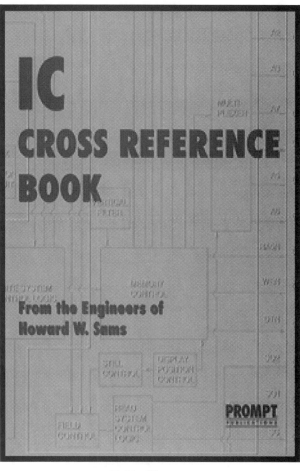

Semiconductor Cross Reference Book
Fourth Edition
by Howard W. Sams & Company

From the makers of PHOTOFACT® service documentation, the *Semiconductor Cross Reference Book* is the most comprehensive guide to semiconductor replacement data. The volume contains over 475,000 part numbers, type numbers, and other identifying numbers. All major types of semiconductors are covered: bipolar transistors, FETs, diodes, rectifiers, ICs, SCRs, LEDs, modules, and thermal devices.

$24.95
Paper/668 pp./8-1/2 x 11"
ISBN#: 0-7906-1080-9
Pub. Date 10/96

IC Cross Reference Book
Second Edition
Howard W. Sams & Company

The engineering staff of Howard W. Sams & Company assembled the *IC Cross Reference Book* to help readers find replacements or substitutions for more than 35,000 ICs and modules. It is an easy-to-use cross reference guide and includes part numbers for the United States, Europe, and the Far East.

$19.95
Paper/192 pp/8-1/2 x 11"
ISBN: 0-7906-1096-5
Pub. Date 11/96

Call us today for the name of your nearest distributor. **800-428-7267**

PROMPT® Publications

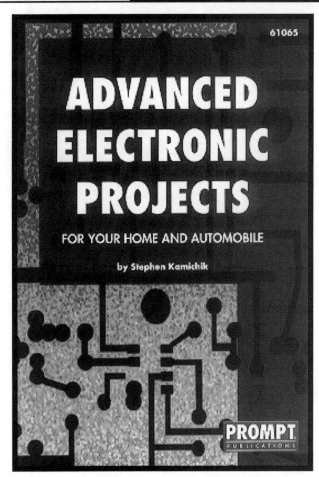

Advanced Electronic Projects for Your Home and Automobile
by Stephen Kamichik

You will gain valuable experience in the field of advanced electronics by learning how to build the interesting and useful projects featured in *Advanced Electronic Projects*. The projects in this book can be accomplished whether you are an experienced electronic hobbyist or an electronic engineer, and are certain to bring years of enjoyment and reliable service.

$18.95
Paper/160 pp./6 x 9"/Illustrated
ISBN#: 0-7906-1065-5
Pub. Date 5/95

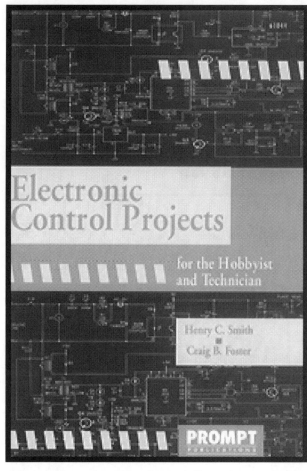

Electronic Control Projects for the Hobbyist and Technician
by Henry C. Smith and Craig B. Foster

Would you like to know how and why an electronic circuit works, and then apply that knowledge to building practical and dependable projects that solve real, everyday problems? Each project in *Electronic Control Projects* involves the reader in the actual synthesis of a circuit. A complete schematic is provided for each circuit, along with a detailed description of how it works, component functions, and troubleshooting guidelines.

$16.95
Paper/168 pp./6 x 9"/Illustrated
ISBN#: 0-7906-1044-2
Pub. Date 11/93

800-428-7267   Call us today for the name of your nearest distributor.

PROMPT® Publications

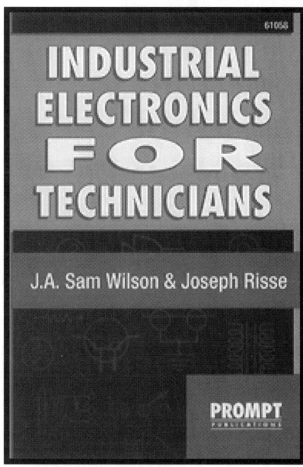

Industrial Electronics for Technicians
by J. A. Sam Wilson and Joseph Risse

Industrial Electronics for Technicians provides an effective overview of the topics covered in the Industrial Electronics CET test, and is also a valuable reference on industrial electronics in general. This workbench companion book covers the theory and application of industrial hardware from the technician's perspective, giving you the explanations you need to understand all of the areas required to qualify for CET accreditation.

$16.95
Paper/352 pp./6 x 9"/Illustrated
ISBN#: 0-7906-1058-2
Pub. Date 8/94

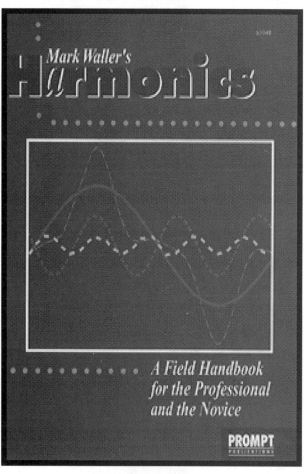

Mark Waller's Harmonics
A Field Handbook for the Professional and the Novice

Many operational problems can be solved through an understanding of power system harmonics. As life/safety issues become more and more important in the world of electrical power systems, the need for harmonic analysis becomes ever greater. This book is the essential guide to understanding all of the issues and areas of concern surrounding harmonics and the recognized methods for dealing with them.

$24.95
Paper/132 pp./7-3/8 x 9-1/4"
ISBN#: 0-7906-1048-5
Pub. Date 3/94

Call us today for the name of your nearest distributor. **800-428-7267**

PROMPT® Publications

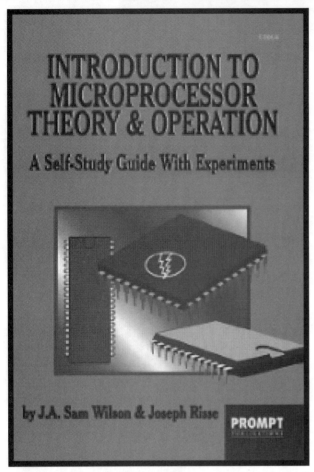

Introduction to Microprocessor Theory & Operation
A Self-Study Guide With Experiments
by J.A. Sam Wilson and Joseph Risse

Introduction to Microprocessor Theory & Operation takes you into the heart of computerized equipment and reveals how microprocessors work. By covering digital circuits in addition to microprocessors and providing self-tests and experiments, this book makes it easy for you to learn microprocessor systems.

$16.95
Paper/212 pp./6 x 9"
ISBN#: 0-7906-1064-7
Pub. Date 2/95

Schematic Diagrams
The Basics of Interpretation and Use
by J. Richard Johnson

Step-by-step, *Schematic Diagrams* shows you how to recognize schematic symbols and their uses and functions in diagrams. You will also learn how to interpret diagrams so you can design, maintain, and repair electronics equipment. Subjects covered include component symbols and diagram formation, functional sequence and block diagrams, power supplies, audio system diagrams, computer diagrams, and more.

$16.95
Paper/208 pp./6 x 9"/Illustrated
ISBN#: 0-7906-1059-0
Pub. Date 9/94

800-428-7267   Call us today for the name of your nearest distributor.

PROMPT® Publications

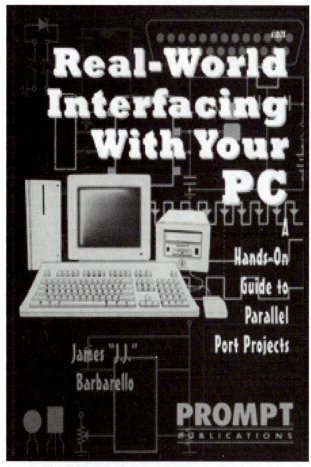

Real-World Interfacing with Your PC
James "J.J." Barbarello

Modern software allows users to do everything from balance a checkbook to create a family tree. Interfacing, however, is truly the wave of the future. *Real-World Interfacing With Your PC* provides all the information necessary to use a PC's parallel port as a gateway to electronic interfacing. Includes a chapter on project design and construction techniques, a checklist for easy reference, and a recommended inventory of starter electronic parts to which readers at every level can relate.

$16.95
Paper/119 pp/7-3/8 x 9-1/4"/Illustrated
ISBN#: 0-7906-1078-7
Pub. Date 3/96

Simplifying Power Supply Technology
by Rajesh J. Shah

Simplifying Power Supply Technology is an entry point into the field of power supplies. It simplifies the concepts of power supply technology and gives the reader the background and knowledge to confidently enter the power supply field.

$16.95
Paper/160 pp./6 x 9"/Illustrated
ISBN#: 0-7906-1062-0
Pub. Date 3/95

Call us today for the name of your nearest distributor. **800-428-7267**

PROMPT® Publications

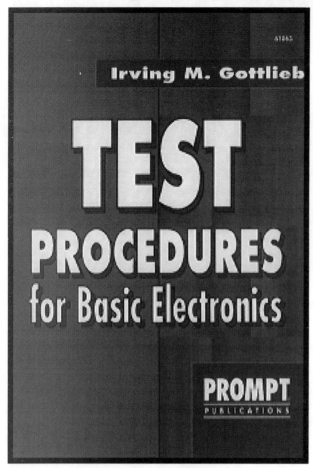

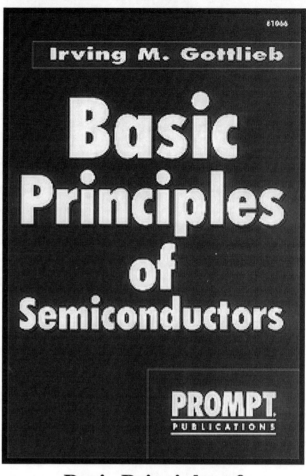

Test Procedures for Basic Electronics
by Irving M. Gottlieb

Basic Principles of Semiconductors
by Irving M. Gottlieb

Test Procedures for Basic Electronics covers many useful electronic tests and measurement techniques, with emphasis on the use of commonly available instruments. Students, hobbyists, and professionals are provided with the whats and whys of obtaining useful results, whether they are repairing a modern CD player or restoring an antique radio.

$16.95
Paper/356 pp./7-3/8 x 9-1/4"
ISBN#: 0-7906-1063-9
Pub. Date 12/94

Few books offer the kind of concise and straightforward discussion of electrical concepts that is found in *Basic Principles of Semiconductors*. From an exploration of atomic physics right through a detailed summary of semiconductor structure and theory, the reader goes on a step-by-step journey through the world of semiconductors. Some of the subjects covered include electrical conduction, transistor structure, power MOSFETs, and Gunn diodes.

$14.95
Paper/161 pp./6 x 9"/Illustrated
ISBN#: 0-7906-1066-3
Pub. Date 4/95

800-428-7267 Call us today for the name of your nearest distributor.

PROMPT® Publications

Basic Electricity
Revised Edition, Complete Course
by Van Valkenburgh, Nooger & Neville, Inc.

From a simplified explanation of the electron to AC/DC machinery, alternators, and other advanced topics, *Basic Electricity* is the complete course for mastering the fundamentals of electricity. The book provides a clear understanding of how electricity is produced, measured, controlled, and used. A minimum of mathematics is used in the direct explanation of primary cells, magnetism, Ohm's law, capacitance, transformers, DC generators, AC motors, and other essential topics.

$24.95
Paper/736 pp./6 x 9"/Illustrated
ISBN#: 0-7906-1041-8
Pub. Date 2/93

Basic Solid-State Electronics
Revised Edition, Complete Course
by Van Valkenburgh, Nooger & Neville, Inc.

Modern electronics technology manages all aspects of information—generation, transmission, reception, storage, retrieval, manipulation, display, and control. A continuation of the instruction provided in *Basic Electricity*, *Basic Solid-State Electronics* provides the reader with a progressive understanding of the elements that form various electronic systems. Electronic fundamentals covered in the illustrated, easy-to-understand text include semiconductors, power supplies, audio and video amplifiers, transmitters, receivers, and more.

$24.95
Paper/944 pp./6 x 9"/Illustrated
ISBN#: 0-7906-1042-6
Pub. Date 2/93

Call us today for the name of your nearest distributor.   **800-428-7267**

PROMPT® Publications

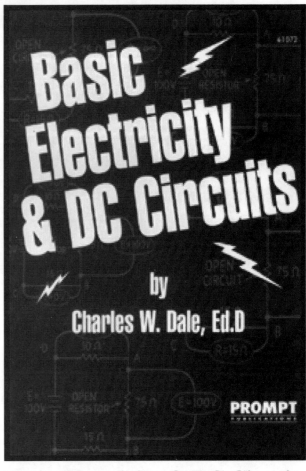

Tube Substitution Handbook
Complete Guide to Replacements
for Vacuum Tubes and Picture Tubes
by William Smith and Barry Buchanan

The most accurate, up-to-date guide available, the *Tube Substitution Handbook* is useful to antique radio buffs, old car enthusiasts, ham operators, and collectors of vintage ham radio equipment. In addition, marine operators, microwave repair technicians, and TV and radio technicians will find the *Handbook* to be an invaluable reference tool. Diagrams are included as a handy reference to pin numbers for the tubes listed in the *Handbook*.

$16.95
Paper/ 149 pp./6 x 9"/Illustrated
ISBN#: 0-7906-1036-1
Pub. Date 12/92

Basic Electricity & DC Circuits
by Charles W. Dale, Ed.D

Electricity is constantly at work around your home and community, lighting rooms, running manufacturing facilities, cooling stores and offices, playing radios and stereos, and computing bank accounts. Now you can learn the basic concepts and fundamentals behind electricity and how it is used and controlled. *Basic Electricity and DC Circuits* shows you how to predict and control the behavior of complex DC circuits. The text is arranged to let you progress at your own pace, and the concepts and terms are introduced as you need them, with many detailed examples and illustrations.

$34.95
Paper/928 pp./6 x 9"/Illustrated
ISBN#: 0-7906-1072-8
Pub. Date 8/95

800-428-7267 Call us today for the name of your nearest distributor.

PROMPT® Publications

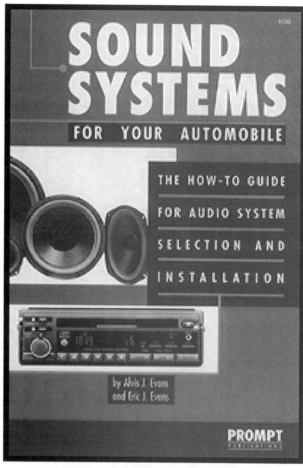

Speakers for Your Home and Automobile

How to Build a Quality Audio System
by Gordon McComb, Alvis J. Evans, and Eric J. Evans

The cleanest CD sound, the quietest turntable, and the clearest FM signal are useless without a fine speaker system. With easy-to-understand instructions and illustrated examples, this book shows how to construct quality home speaker systems and how to install automotive speakers.

$14.95
Paper/164 pp./6 x 9"/Illustrated
ISBN#: 0-7906-1025-6
Pub. Date 10/92

Sound Systems for Your Automobile

The How-To Guide for Audio System Selection and Installation
by Alvis J. Evans and Eric J. Evans

Whether you're starting from scratch or upgrading, this book will show you how to plan your car stereo system, choose components and speakers, and install and interconnect them to achieve the best sound quality possible. Easy-to-follow steps, parts lists, wiring diagrams, and fully illustrated examples make planning and installing a new system easy.

$16.95
Paper/124 pp./6 x 9"/Illustrated
ISBN#: 0-7906-1046-9
Pub. Date 1/94

Call us today for the name of your nearest distributor.   **800-428-7267**

PROMPT® Publications

Advanced Speaker Designs for the Hobbyist and Technician
by Ray Alden

This book shows the electronics hobbyist and the experienced technician how to create high-quality speaker systems for the home, office, or auditorium. Every part of the system is covered in detail, from the driver and crossover network to the enclosure itself. You can build speaker systems from the parts lists and instructions provided, or you can actually learn to calculate design parameters, system responses, and component values with scientific calculators or PC software.

$16.95
Paper/136 pp./6 x 9"/Illustrated
ISBN#: 0-7906-1070-1
Pub. Date 7/94

Theory & Design of Loudspeaker Enclosures
Dr. J. Ernest Benson

Considered one of the world's leading experts on loudspeaker enclosures, Dr. J. Ernest Benson originally published the contents of this book as a three-part journal series in his native Australia. Now released for the first time in the United States, his book is filled with illustrations, helpful charts, and important design concepts. The information in the *Theory & Design of Loudspeaker Enclosures* will not only explain the theory of loudspeaker design, but also provide practical applications for every speaker enthusiast.

$19.95
Paper/256 pp/6 x 9"/Illustrated
ISBN: 0-7906-1093-0
Pub. Date 9/96

800-428-7267 Call us today for the name of your nearest distributor.

PROMPT® Publications

VOM and DVM Multitesters for the Hobbyist and Technician
by Alvis J. Evans

VOM and DVM Multitesters offers concise, clearly illustrated text to explain how digital and analog meters work, as well as their uses on the job, in the workshop, and in the home. Subjects include basic concepts of VOM and DVM meters, multitester measurements, and meter troubleshooting.

$14.95
Paper/144 pp./6 x 9"/Illustrated
ISBN#: 0-7906-1031-0
Pub. Date 9/92

The Multitester Guide
How to Use Your Multitester for
Electrical Testing and Troubleshooting
by Alvis J. Evans

With the instructions provided in *The Multitester Guide*, basic electrical measurements become easily, correctly, and quickly completed. In addition to the functions and uses of multitesters, the easy-to-understand text and clear examples cover such topics as the measurement of basic electric components and their in-circuit performance; the measurement of home lighting, appliance, and related systems; and automotive circuit measurements.

$14.95
Paper/160 pp./6 x 9"/Illustrated
ISBN#: 0-7906-1027-2
Pub. Date 12/92

Call us today for the name of your nearest distributor.   **800-428-7267**

PROMPT® Publications

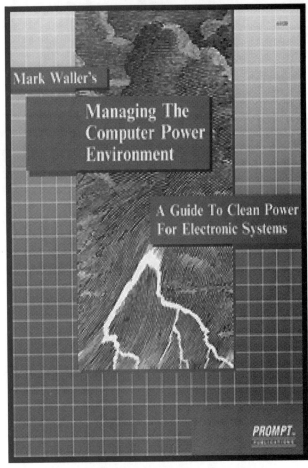

Mark Waller's
Managing the Computer Power Environment
A Guide to Clean Power
for Electronic Systems

This book provides background in electrical technology to help the reader control the quality of the power that drives their computer system. Among the topics covered are utility power, grounding, power distribution units, backup power systems, and conditioners.

$19.95
Paper/192 pp./7-3/8 x 9-1/4"
ISBN#: 0-7906-1020-5
Pub. Date 5/92

Mark Waller's
Surges, Sags, and Spikes
Protecting Your Personal Computer from Electrical Power Problems

In nontechnical language, Mark Waller has written a book for PC users concerned with protecting their computer systems against a hostile electrical environment. Using illustrations with helpful diagrams and photographs, the author takes a comprehensive look at solutions to computer power problems.

$19.95
Paper/240 pp./7-3/8 x 9-1/4"
ISBN#: 0-7906-1019-1
Pub. Date 5/92

800-428-7267
Call us today for the name of your nearest distributor.

PROMPT® Publications

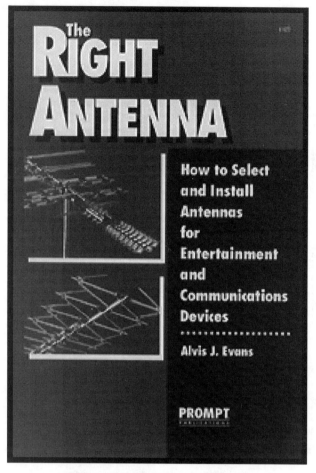

The Phone Book
Money-Saving Guide to Installing or
Replacing Telephone Equipment in
Your Home or Business
by Gerald Luecke and James B. Allen

Tired of the hassle and wait involved in telephone installation? *The Phone Book* presents clear instructions showing you how to install your own phones. Whether you need a single phone or a complex system, this book will guide you with easy-to-understand text and illustrations. Subjects include converting from old to new-style systems, installing business telephone systems, and more.

$16.95
Paper/176 pp./7-3/8 x 9-1/4"/Illustrated
ISBN#: 0-7906-1028-0
Pub. Date 9/92

The Right Antenna
How to Select and Install Antennas for
Entertainment and Communications Devices
by Alvis J. Evans

Television, FM, CB, cellular phone, satellite, and shortwave signals are available in the air to anyone, but it takes a properly selected and installed antenna to make use of them. *The Right Antenna* will give you the confidence to select and install the antenna that meets your needs. Included in the book are sections covering the selection and installation of antennas for specific devices, how antennas work, and how to identify and eliminate television interference.

$10.95
Paper/112 pp./6 x 9"/Illustrated
ISBN#: 0-7906-1022-1
Pub. Date 10/92

Call us today for the name of your nearest distributor. **800-428-7267**

PROMPT® Publications

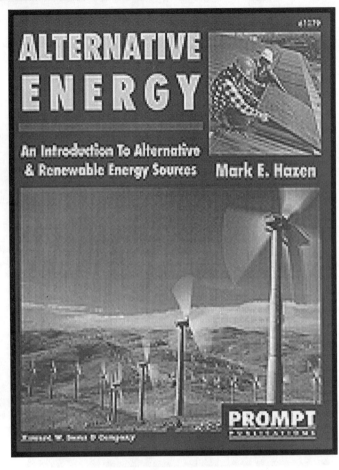

Is This Thing On?
Gordon McComb

*T*akes readers through each step of selecting components, installing, adjusting, and maintaining a sound system for small meeting rooms, churches, lecture halls, public-address systems for schools or offices, or any other large room. In easy-to-understand terms, drawings and illustrations, *Is This Thing On?* explains the exact procedures behind connections and troubleshooting diagnostics. With the help of this book, hobbyists and technicians can avoid problems that often occur while setting up sound systems for events and lectures.

$14.95
Paper/136 pp./6 x 9"/Illustrated
ISBN: 0-7906-1081-7
Pub. Date 4/96

Alternative Energy
Mark E. Hazen

This book is designed to introduce readers to the many different forms of energy mankind has learned to put to use. Generally, energy sources are harnessed for the purpose of producing electricity. This process relies on transducers to transform energy from one form into another. *Alternative Energy* not only addresses transducers and the five most common sources of energy that can be converted to electricity, it also explores solar energy, the harnessing of the wind for energy, geothermal energy, and nuclear energy.

$18.95
PAper/320 pp./7-3/8 x 9-1/4"/Illustrated
ISBN: 0-7906-1079-5
Pub. Date 10/96

800-428-7267 Call us today for the name of your nearest distributor.

PROMPT® Publications

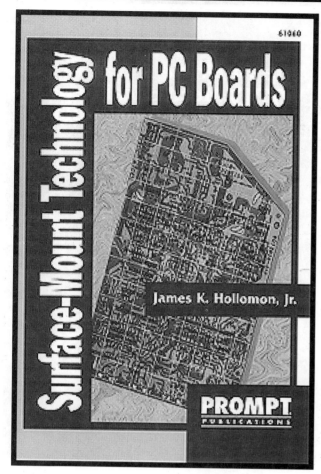

Surface-Mount Technology for PC Boards
by James K. Hollomon, Jr.

Manufacturers, managers, engineers, and others who work with printed-circuit boards will find a wealth of information about surface-mount technology (SMT) and fine-pitch technology (FPT) in this book. Practical data and clear illustrations plainly present the details of design-for-manufacturability, environmental compliance, design-for-test, and quality/reliability for today's miniaturized electronics packaging.

$26.95
Paper/309 pp./7 x 10"/Illustrated
ISBN#: 0-7906-1060-4
Pub. Date 7/95

Semiconductor Essentials
by Stephen Kamichik

Gain hands-on knowledge of semiconductor diodes and transistors with help from the information in this book. *Semiconductor Essentials* is a first course in electronics at the technical and engineering levels. Each chapter is a lesson in electronics, with problems presented at the end of the chapter to test your understanding of the material presented. This generously illustrated manual is a useful instructional tool for the student and hobbyist, as well as a practical review for professional technicians and engineers.

$16.95
Paper/112 pp./6 x 9"/Illustrated
ISBN#: 0-7906-1071-X
Pub. Date 9/95

Call for the complete list of new releases from PROMPT®.   **800-428-7267**

PROMPT® Publications

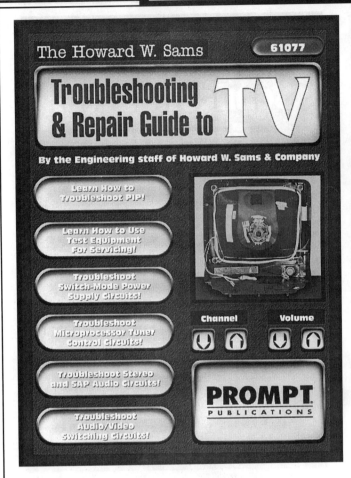

The Howard W. Sams Troubleshooting & Repair Guide to TV
Howard W. Sams & Company

This is the most complete and up-to-date television repair book available! Includes complete repair information for all makes of TVs, timesaving features that even the pros don't know, comprehensive basic electronics information, and extensive coverage of common TV symptoms. Illustrated with useful photos, schematics, and graphs, it covers audio, video, technician safety, test equipment, power supplies, picture-in-picture, and much more.

$29.95
Paper/384 pp./8-1/2 x 11"/Illustrated
ISBN#: 0-7906-1077-9
Pub. Date 6/96

The In-Home VCR Mechanical Repair & Cleaning Guide
Curt Reeder

Like any machine used in the home or office, a VCR requires minimal service to keep it functioning well and for a long time. A technical or electrical engineering degree is not required to begin regular maintenance on a VCR. This book shows readers the tricks and secrets of VCR maintenance using just a few small hand tools, such as tweezers and a power screwdriver. Compiled from the most frequent VCR malfunctions Curt Reeder has encountered in the six years he has operated his VCR repair and cleaning service.

$19.95
Paper/222 pp./8-3/8 x 10-7/8"/Illustrated
ISBN: 0-7906-1076-0
Pub. Date 4/96

800-428-7267 Call us today for the name of your nearest distributor.

PROMPT® Publications

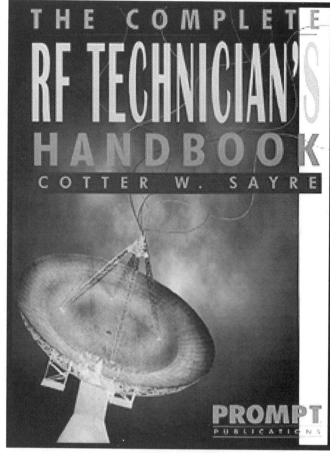

Complete RF Technician's Handbook
Cotter W. Sayre

This book will furnish the working technician or student with a solid grounding in the latest methods and circuits employed in today's RF communications gear. It will also give readers the ability to test and troubleshoot transmitters, transceivers, and receivers with absolute confidence. Topics covered include reactance, phase angle, logarithms, diodes, passive filters, and more. Various multiplexing methods and data, satellite, spread spectrum, cellular, and microwave communication technologies are discussed.

$24.95
Paper/281 pp./8-1/2 x 11"/Illustrated
ISBN: 0-7906-1085-X
Pub. Date 7/96

The Microcontroller Beginner's Handbook
Lawrence A. Duarte

This book will bring information to the reader on how to understand, repair, or design a device incorporating a microcontroller. Examines many important elements of microcontroller use, including such industrial considerations as price vs. performance and firmware. This book not only teaches readers with a basic knowledge of electronics how to design microcontroller projects, it greatly enhances the reader's ability to repair such devices.

$18.95
Paper/240 pp./7-3/8 x 9-1/4"/Illustrated
ISBN: 0-7906-1083-3
Pub. Date 7/96

Call us today for the name of your nearest distributor. **800-428-7267**